World Agriculture and the Environment

About Island Press

Island Press is the only nonprofit organization in the United States whose principal purpose is the publication of books on environmental issues and natural resource management. We provide solutions-oriented information to professionals, public officials, business and community leaders, and concerned citizens who are shaping responses to environmental problems.

In 2004, Island Press celebrates its twentieth anniversary as the leading provider of timely and practical books that take a multidisciplinary approach to critical environmental concerns. Our growing list of titles reflects our commitment to bringing the best of an expanding body of literature to the environmental community throughout North America and the world.

Support for Island Press is provided by the Agua Fund, Brainerd Foundation, Geraldine R. Dodge Foundation, Doris Duke Charitable Foundation, Educational Foundation of America, The Ford Foundation, The George Gund Foundation, The William and Flora Hewlett Foundation, Henry Luce Foundation, The John D. and Catherine T. MacArthur Foundation, The Andrew W. Mellon Foundation, The Curtis and Edith Munson Foundation, National Environmental Trust, National Fish and Wildlife Foundation, The New-Land Foundation, Oak Foundation, The Overbrook Foundation, The David and Lucile Packard Foundation, The Pew Charitable Trusts, The Rockefeller Foundation, The Winslow Foundation, and other generous donors.

The opinions expressed in this book are those of the author(s) and do not necessarily reflect the views of these foundations.

World Agriculture and the Environment

A COMMODITY-BY-COMMODITY
GUIDE TO IMPACTS AND PRACTICES

Jason Clay

Island Press / Washington • Covelo • London

ISLAND PRESS is a trademark of The Center for Resource Economics.

Maps adapted from *Goode's World Atlas*, copyright by RMC, R.L. 02-S-10 www.randmcnally.com.

Library of Congress Cataloging-in-Publication data.

Clay, Jason W.
   World agriculture and the environment : a commodity-by-commodity guide to impacts and practices / Jason Clay.
       p. cm.
   Includes bibliographical references (p.   ).
   ISBN 1-55963-367-0 (cloth : alk. paper) — ISBN 1-55963-370-0 (pbk. : alk. paper)
   1. Agriculture—Environmental aspects. I. Title.
   S589.75 .C53 2003
   363.7—dc22

                                                          2003016786

British Cataloguing-in-Publication data available.

Design by Kathleen Szawiola

Printed on recycled, acid-free paper

Manufactured in the United States of America
10 9 8 7 6 5 4 3 2 1

CONTENTS

PREFACE

Live like you'll die tomorrow; farm like you'll live forever.

There are two basic truths that will shape the future of farming—there is a steady increase in the consumption of food and fiber produced by agriculture, while at the same time there is a steady decline in the quality and productivity of soil around the world. The two trends are on a collision course. This collision will not be avoided by a single solution.

It is this trajectory that explains why a former farmer, anthropologist, and human rights activist now works in an environmental organization focusing on agriculture. It is a question of survival. Most biodiversity lives in the soil rather than on top of it, and most is found in areas of human use rather than parks or protected areas. However, as a result of the increased demand for agricultural products and the use of unsustainable agricultural practices, farmers convert natural habitat into new agricultural lands after they exhaust and abandon the lands that they previously farmed. As a consequence, farming is the single largest threat to biodiversity and ecosystem functions of any single human activity on the planet.

This book shows how this pattern can be broken and identifies activities that producers, policy makers, researchers, market-chain players, and environmentalists can play in the creation of more sustainable agricultural practices within the evolving context of global trade.

The people who know the most about making farming more sustainable are farmers themselves. The most innovative among them are often simply trying to survive economically in an increasingly competitive world. They use resources more efficiently, and they are constantly experimenting with new crops, combinations of crops and practices, and technology. Innovation comes from experimentation. Many producers are actually farming with nature rather than against it. This does not mean that they are returning to the practices of their ancestors, but rather they are experimenting with a mix of old and new approaches that give them better returns with fewer inputs and fewer impacts. Many producers have learned a very important truth—that to save money is to make money. For example, some have found that they make more money increasing soil vitality and fertility and reducing inputs than they do by focusing on increased yields alone.

Many producers have also come to realize, some the hard way, that in a global economy with increased transparency and information what some producers do can affect the reputation of all. As a consequence, producers are organizing themselves to protect their interests and reputations. The most progressive are beginning to organize their industries to share experiences and lessons learned as well as to negotiate as larger blocks with regulators, buyers, investors, and nongovernmental organizations (NGOs).

Most environmental NGOs have no interest in becoming agricultural development agencies. But most have also come to realize that they cannot achieve their missions without ensuring that farming practices become more sustainable. Such groups are now beginning to develop agricultural programs through which they intend to engage producers, but most have not yet developed (much less implemented) detailed approaches through which they will engage agriculturalists. Furthermore, most NGOs are far more comfortable working with governments to develop regulatory approaches to address the negative impacts of agriculture—in short, to tell farmers what they cannot do. Such "stick" approaches are not likely to work by themselves. Furthermore, they provide no incentive for producers to do better than what is required by law.

This book identifies a number of approaches and market-based incentives that would encourage producers to achieve entirely new levels of performance, and as a result raise the expectations of government, and others, about what is even possible. The goal of NGOs should not be to put farmers out of business, but rather to make sure that they or their descendents can still farm the same piece of land in twenty or fifty years without the use of unsustainable inputs. This book identifies areas where agriculture can be made more sustainable globally while at the same time reducing pressure on natural habitats and increasing biodiversity and ecosystem services within areas that are farmed.

Government officials around the world have fewer resources with which to reduce the impacts of agriculture on the one hand or to make it more sustainable on the other. Increasingly, they are asked to do more with less. This book demonstrates how government land use and zoning programs can be based on the productive potential (or the unproductive potential) of areas as well as the value of natural resources and habitat for other purposes. It shows officials how to think about the medium- to long-term costs of allowing, much less giving incentives to, the establishment or continuation of unsustainable agricultural production systems.

Some governments, too, are experimenting with very innovative approaches to support or encourage sustainable agriculture. For instance, they are exploring how to link regulatory structures, licenses, and permits to performance and to better management practices (BMPs) in order to encourage the adoption and use of more sustainable practices as well as the standards by which performance is measured. While not exhaustive, this book offers a number of examples from different types of countries as well as from different types of crops and producers that provide government officials with considerable information about how to think about adapting or incorporating similar approaches in their own countries.

The manufacture and sale of agricultural products throughout the market chain from the producer to the consumer are increasingly centralized and vertically organized. Most of the players are monitoring the increased public concern regarding product quality in general and chemical residues in particular. Where there are consumer concerns there are potential liabilities. This book suggests how food manufacturers and retailers can begin to think about greening their supply chains through the adoption of BMP-based screens that reduce not only their overall liability, but also the environmental impact of their producers, and increase their profitability at the same time.

Finally, there are dozens of research topics and areas that are suggested for each of the crops discussed in this book as well as for hundreds of crops not discussed in this book. Such research could be pivotal in helping to put agriculture on a more sustainable footing. To best accomplish this, however, researchers may need to distance themselves from the money and interests of input suppliers as well as the latest theories of the day that preoccupy academia. There is a tremendous amount of research whose results and findings could be applied immediately and could help producers reduce their environmental impacts as well as increase their profitability. Most would consider such research timely and in the public interest.

While my editor would probably kill me for saying this, this book is not intended to be comprehensive. Libraries have been filled with books written about each of these crops. Rather, the goal of this book is to identify and analyze several concrete examples of ways that a wide range of commodities are being produced around the world that reduce their environmental impacts. The goal is to show that there are new ways of thinking and acting that reduce agricultural impacts. These ways of thinking are relevant to most crops produced on the planet, but they cannot be adopted whole cloth—they will have to be adapted to different crops and circumstances. The most important thing to take away from the book, then, is not what to think in any specific circumstance but rather how to think. And of course, while most of these actions make sense for farmers in their own right, there are also important roles for governments, buyers, environmentalists, investors, researchers, and consumers.

This book will stimulate dialogue and discussion among producers and between producers and others genuinely interested in these issues. Such discussion, based on new facts, will amplify, redirect, and focus the debate on sustainable agriculture. As such, the book should encourage the identification of BMPs from around the world and stimulate their analysis so that the lessons can be more widely disseminated to reduce the impacts of global agriculture and increase its sustainability and profitability at the same time. Our future, and the future of every other living thing on the planet, will depend on it.

Jason W. Clay
24 February 2003

ACKNOWLEDGMENTS

Most of the things I have done since a very early age in my life have influenced this book. By the age of two I was creating farms in my sandbox (which was made of an old tractor tire) complete with ponds, drainage systems, and fields with terraces and contour plowing. I first "drove" a tractor, a Farmal H, that same year. Like most children around the world, I learned about farming from my parents, relatives, and neighbors. The most important of these include W. E. and Doris Clay, my parents, Gordon Howitt, my uncle, and Rex Jameson, a neighbor. As with my six siblings, my first lessons were in the garden, then the barnyards and chicken houses, and finally the fields. In the spring of 1966 the same tractor that I had driven when not yet three turned over and killed my father, and I learned first hand about the darker side of farming and, equally important, the responsibility of making a living.

It seems that the more I tried to escape from farming, the more I was drawn back to it both as an observer and a student. I learned about shotgun houses, cotton, and tenants in southern Arkansas while working on an archeological dig amidst soybeans; corn, beans, and coffee from Tzotzil-speaking farmers in Chiapas, Mexico; coffee, cassava, beef, and horticulture from farmers in the breadbasket of the northeast of Brazil; bananas, sorghum, tea, and cattle from Ugandan refugees; and teff, honey, sorghum, and millet from Ethiopian refugees.

Davydd Greenwood, Milt Barnett, Bill DeWalt, Norm Uphoff, Walt Coward, John Whiting, Sir Peter Bauer, and Ford Runge helped me understand farming and farmers within wider cultural, political, and economic contexts. Amory Lovins, Jim LaFleur, Jose Zaglul, Konrad von Moltke, John Forgash, Allan Nations, Greg Simmonds, Peter Kenmore, Gordon and Anita Roddick, Ben Cohen, Dave Cole, Florence Sender, Josh Mailman, Jeff Dlott, Sue Hall, Diane Osgood, Robins McIntosh, Werner Jost, Joaquin Orrantia, Jose Vicente Mogollon, Rachel Stringfellow, and literally thousands of farmers from six continents, each in their own way, helped me to understand not what farming is or was, but what farming as a business is becoming and might yet be.

However, I would never have tried to pull all this information together in one place without the support of Diane Wood, former Vice President of the Center for Conservation Innovation of World Wildlife Fund (WWF), who provided the greatest part of the funding for writing this book. Others within WWF (presently or in the past) who provided support, information, or encouragement include Barbara Hoskinson, Amy Salzman, David Schorr, Polly Hoppin, Sarah Lynch, Sheila O'Connor, Andrea Ries, Richard Perkins, Elizabeth Guttenstein, Jean Paul Paddack, Garo Batmanian, Bella Rocher, Andrew Ng, Lucy Aquino, Judy Oglethorpe, Sara Christiansen, Doreen Robinson, and Anthony Anderson.

A number of individuals have helped with the research for this book and have my heartfelt thanks. Katherine Bostick certainly stands out for her willingness to see this project through to the end and for catching many inconsistencies. Other researchers from WWF include, in the order of their involvement, Gautham Rao, Miranda Mockrin, Tim Green, Govindarajan Dhanasekaran, and Taryn Fransen. In addition, Martha Alt, Andrea and Jeff Vallina, Jim and Andre LaFleur, and Alexa Clay also assisted with the research, fact finding, and preparation of the book. Panfilo Tabora and students Cleomar Bizonhin, Juan Francisco Chiriboga, Frankys Maikel de la Osa, Guido Durán Maridueña, Karina Garcés Herrera, Diego Garcia Velasco, Xiomara Gonzalez Hernandez, Jose Rafaél Gonzalez, Rebecca Gutierrez Bermudez, Jered Hayes, Edmar Hodgson Sobalvarro, Tania Johanning Villegas, Maria Enith Melendez, Yerling Miranda Jimenez, Kelly Rohlfing, Anna Sommer, Jose Maria Tijerino Picado, Luis Antonio Velex, and Paul Whitsell from EARTH University in Costa Rica assisted with research for tropical crops.

In addition to the support from WWF to write this book, work undertaken for or supported by the Ford, MacArthur, AVINA, and Packard Foundations; the Pew Charitable Trusts; the Interchurch Organization for Development Cooperation (ICCO); the World Bank; the Food and Agriculture Organization of the United Nations; and the Multilateral Investment Fund of the Inter-American Development Bank all contributed to the analysis presented here.

Todd Baldwin of Island Press had the courage to edit and sharpen the text as well as to fight for pages to make the book more comprehensive. Thanks for both. And thanks to Cecilia González and Chace Caven for seeing the editing process through to the end.

Thanks also go to my children, Alexa, Zale, and Hawkins, for their patience and understanding during the research and writing of this work, which required my time and attention during formative periods of their lives. Finally, my deepest thanks for insights and encouragement goes to my wife, Mary Ann Mills. Without her support, the travel, research, and writing for this book would not have been possible.

In 1837 John Deere patented the steel plow that cut through native prairie grasses and converted them to farms, first in the United States and then in distant lands. This plow, more than any other invention, symbolized the human ability literally to turn nature on its head. The steel plow became the foundation for modern agriculture. As the plows and the machines that pulled them got bigger, more and more land could be farmed by fewer people. Every increase in scale and intensity, however, increased environmental impacts as well. Over time it has become apparent that agricultural practices, more than any other single factor, have determined the state of the global environment.

But that picture may be changing. Though the John Deere name is synonymous with mechanized agriculture, the company has started venturing into the realm of sustainable agriculture. John Deere recently entered into a joint venture with a Brazilian agricultural company. This company held land that was producing primarily through cultivation techniques that eliminate tillage. The strategy of this partner is to buy degraded pasture and rebuild the soil to fully productive land for the cultivation of such crops as soybeans. Seeds are planted without turning the soil and organic matter is left on the surface. Such practices reduce soil degradation, erosion, and the use of fertilizers and pesticides, while increasing the soil's retention of water and other agricultural inputs added during production. The system is based on crop sequencing (growing two or three crops in the same year) as well as three-year crop rotations. The ground is planted with grass to build organic matter. Other improvements include keeping marginal lands out of production; areas that are not appropriate for farming are terraced and planted to trees. Perhaps even more interesting, the workers on the farm have an equity position in the company based on their length of employment and productivity. This is the new face of agriculture, and a few corporations like John Deere and its partners are beginning to invest in it. But it will take a long time to become the global norm.

One of the great contradictions of our time is that we know more about and are better able to save spaces and species than ever before. But we are losing both species and their habitats faster than ever, and more often than not the cause is agriculture.

Parks and protected areas, comprising about 5 percent of the land on the planet, have long been recognized as cornerstones of effective efforts to save biodiversity and ecosystems. Yet most species on the planet today live where people are trying to make a living. In general, most parks and protected areas were created to protect geologic formations and areas of striking beauty or cultural value, not biodiversity.

In any event, it is difficult to imagine how the size of parks and protected areas can be effectively maintained—much less increased in area—in the current and foreseeable political climate. Many protected areas are systematically attacked from all sides. Around the world, legal and illegal invasions of such areas are undertaken by oil and gas companies, miners, loggers, and others looking for resources that can be exploited. In addition, about half of the world's current protected areas are surrounded by agriculturalists, many of whom see protected areas as their next fields. The net effect of such actions is to reduce the value of such areas for biodiversity conservation. Even if the areas currently under protection can be maintained, recent research suggests that some 30 to 50 percent of species within them will disappear because their populations are too small to survive over the long run on the protected land available.

By contrast to the land under formal protection, about half of the habitable land on Earth is used for agriculture and livestock production. Because of this enormous scale, agriculture represents both a significant threat and an opportunity to protect biodiversity. One strategy for saving biodiversity is to help producers become more sustainable and productive so they can stay where they are, instead of expanding into pristine areas, while at the same time accommodating more biodiversity on their lands. Agriculturalists are the managers of global lands. They shape the face of the Earth (Tilman et al. 2002).

The environmental costs of agricultural practices (referred to as environmental externalities, ecological footprints, subsidies from nature, or passing environmental costs on to future generations) are usually not measured. When producers are not required to cover the true costs, they pass them on to society. For example, most current agricultural practices reduce the ability of ecosystems to provide goods and services. Clearing natural habitat and soil erosion both reduce carbon sequestered in the environment. This loss of organic matter in the soil reduces the ability of soil to absorb and retain water. Such practices not only increase overall environmental degradation downstream, they also increase the amount of external inputs (especially fertilizer) required to maintain productivity. Not only can more sustainable agricultural practices reduce these impacts, they can also make agriculture a central part of the environmental restoration process.

The goal for sustainable agriculture must be to insure that society benefits not only from the production of food and fiber but also from the maintenance or restoration of ecosystem services such as watershed protection, healthy soil and the biodiversity that depends on both. Globally, land cleared for agriculture is rarely allowed to return to a "natural" state. There are some exceptions in the eastern United States and Europe, but in general, once converted, land is used in one form or another by humans often until virtually nothing will grow there. At that point it may be abandoned, but it will never regain the biological diversity that it once had. The extent of environmental degradation caused by agriculture can still be seen near archeological sites in Central America and Southeast Asia that are a thousand years old. To restore such degraded land to productivity and reestablish other ecosystem services might be possible, but only at great expense.

Sustainable agriculture requires that ranchers and farmers alike be rewarded for producing food, fiber, *and* ecosystem services (Tilman et al. 2002). Globally, the main obstacle to this approach is that current subsidies support unsustainable production systems in one part of the world and make them necessary for survival in the rest. However, if a portion of these subsidies were used to pay for the production or maintenance of ecosystem services, they could increase overall agricultural production, profitability, and viability in the short, medium, and long term.

CURRENT AGRICULTURAL PRODUCTION

Agricultural production has modified the natural landscape more than any other human activity. The land dedicated to agricultural production continues to grow (see Table I.1). Globally, agricultural land use has increased at a rate of approximately 13 million hectares per year for the past thirty years. Much of this expansion has come at the expense of forests (except in North America and Europe). Producers are whittling away at natural habitat on the margins of agricultural areas. Because roads, infrastructure, and urban expansion often come at the expense of agricultural land, agricultural expansion into new areas is even more rapid than suggested by these figures, which reflect only net growth.

There are several factors that determine the overall damage from agricultural production as well as the strategies to address it. Most agricultural commodity production is for basic foodstuffs, and most products are consumed within the country that produced them. Some 90 percent of all arable land is planted to annual crops, which cause more damage than perennials. Because annual crop production methods tend to exhaust the soil in which the crops are grown, producers must continually convert natural habitat to agricultural uses. As soil loses its fertility, land is used for a succession of different crops with fewer and fewer nutritional requirements. This can be visualized as farming down the nutrient chain.

TABLE I.1. Global Land Area by Use
(in billions of hectares)

	1961	1970	1980	1990	2000
Total agricultural area	4.41	4.50	4.72	4.91	4.97
Total arable land	1.27	1.30	1.33	1.38	1.36
Total permanent pasture	3.14	3.21	3.29	3.41	3.48
Total forest and woodlands[1]	4.37	4.33	4.30	4.32	4.17
Non-arable and non-permanent crops	11.70	11.65	11.60	11.54	11.56

Source: FAO 2002.
[1] Data for 1996.

Many think of capital-intensive, high-input production systems when they think of industrial agricultural commodities, and they think that these systems are somehow a distinct category of farm from those that put food on our table. In truth most high-input, intensive production systems are used to produce food crops that are destined primarily for the food industry and that feed most people on the planet.

Of course, agricultural crops are also used to manufacture nonfood products. These include fiber crops such as cotton, hemp, sisal, jute, flax, and wood pulp. Tobacco is also a major nonfood crop. In addition, plantation-grown natural rubber is indispensable in the manufacture of a number of key industrial products. The area of production devoted to these agricultural crops, however, is only a small fraction of that devoted to food crops.

IN SEARCH OF THE "IDEAL" AGRICULTURE

Many environmentalists do not believe that conventional farming can be improved sufficiently to reduce its damaging effects to acceptable levels. They would rather see agricultural producers revert to less intensive forms of low- or no-input agriculture that were common a century ago. These systems of production relied on a mix of crops, trees, livestock, and ground cover and on crop rotations rather than more intensive monocrop production. Through their diversity, such systems offered more protection against pests and the weather. Nutrients were recycled within the system and through livestock. Production was more labor-intensive. Such systems of production, it is said, produced less environmental degradation and were more sustainable than today's intensive, highly specialized agricultural production systems.

Such agricultural production systems have ancient roots. China's Yellow River (Huang He) Valley and Iraq's Tigris-Euphrates floodplains have been farmed more or less continuously for more than 7,000 years. Similar farming systems have been deployed or developed independently in Asia, Europe, and the Americas for 2,000 to 3,000 years. It is clear that some agricultural production systems can be operated over centuries or longer.

Nonetheless, there are several flaws in this "idealization" of less intensive farming, at least as it is often portrayed. Historically (and even in many areas where such farming is still being practiced today), the evidence is not conclusive that it was or is less hard on the land than many current practices. Some of these less intensive farming systems have failed, and often population densities have pushed cultivation levels beyond what is sustainable. There is ample evidence that parts of the Andes, Mesoamerica, North Africa, the Middle East, Europe, South and Southeast Asia, New England and even the Great Plains (to name but a few) were overfarmed to the point of degradation or collapse using "traditional" forms of agricultural production. Even today some of the most "traditional" production systems are found in rural areas with the most severe malnutrition and famine, as well as some of the most severe environmental degradation.

Most importantly, the Earth is currently home to over 6 billion people. Supporting them all by low-intensity cropping—depending solely on recycling organic matter and using crop rotation with legumes—would require doubling or tripling the area currently cultivated. This land would have to come from somewhere—and would most likely mean the elimination of most if not all tropical rainforests and the conversion of a large part of tropical and subtropical grasslands too. Lower-intensity agricultural practices are very labor-intensive, so such reversion would also require the return of a substantial share of the labor force to farming (Smil 2000). These are hardly acceptable alternatives.

During the last forty years global population nearly doubled. Contrary to many predictions, as the population has increased, global food production increased to feed most people. In fact, global per capita agricultural production increased 25 percent, while the amount of land needed to produce this additional food increased by only 10 percent. Table I.2 shows that, with the exception of wheat, the increases in consumption of major food crops are significantly larger than the increase in lands devoted to producing these crops. Where there have been famines, they have been caused by politics and human policies. Moreover, they need never have happened. While that may give little consolation to those who starve, it should give guidance to those who want to prevent famine in the future. Life expectancy has risen dramatically; China, for example, now has a mean life expectancy of sixty-nine years, up from thirty-five years in the 1950s (Chen and Ge 1995). The world prices of nearly every staple foodstuff are, in inflation-adjusted terms, lower than a generation ago. Most are at their lowest level for any time for which there are records.

This apparent bounty is due in large part to the "green revolution" in agriculture. Since 1900 the world's cultivated area increased by about one-third, but because of a more than fourfold increase in productivity, total production has increased almost sixfold. A major portion of this gain can be attributed to selective breeding programs and to an eightyfold increase in external energy inputs, mostly in the form of fossil fuels (Smil 2000). This energy is used for machinery, fuel, and fertilizer and pesticide production. Energy, machinery, and agricultural chemicals have been

TABLE I.2. Feeding a Hungry World

Crop	Contribution to Total Food Demand (%)	Change in Demand 1974–1994 (%)	Change in Area of Production 1974–1994 (%)
Rice	30	71	52
Wheat	18	97	96
Corn (maize)	13	115	72
Cassava	12	40	17
Potatoes	5	115	25
Sorghum	4	54	18
Bananas	4	47	14
Sweet potatoes	4	37	20
Food legumes	3	32	13
Barley	3	79	22
Plantains	2	29	−13
Millet	2	28	12

Source: FAO and CGIAR as cited in *The Washington Post*, 1995.

substituted for labor. Other gains have come from reduced storage losses and increased food distribution to a wider range of consumers over more of the year.

On the other hand, it is clear that the Earth's current population cannot be supported in the American lifestyle, in which an estimated 40 percent of food is thrown away. The issue of feeding the world is not one of overpopulation, but rather a fundamentally different one of overconsumption. Such waste has an undeniable impact on the biosphere through the use of natural, material resources that are required to produce what is wasted.

The current answer to feeding the world is large-scale, high-input, monoculture (monocropping) agricultural production systems, which have existed for only 50 to 100 years. The environmental problems caused by such production systems perpetuate and intensify earlier agricultural impacts. The most damage is caused by habitat conversion (and the corresponding loss of biodiversity and ecosystem functions), soil erosion and degradation, and pollution (from fertilizer and pesticides). These impacts are not new. They result from the expansion of agriculture into natural habitats, shortened or eliminated fallow cycles, adoption of double and even triple cropping schemes, introduction of faster maturing and higher yielding varieties, and use of heavy machinery that causes soil compaction. In addition, the consolidation of smaller farms into huge operations, salinization of soil resulting from improper irrigation practices, use of agrochemicals, inefficient use of larger quantities of water, and consequent creation of more effluents from farming systems also contribute to increasing levels of environmental degradation.

These negative impacts raise serious questions about the long-term sustainability of high-input, intensive agriculture. The increasing dependence on globally limited supplies of fossil fuels is not sustainable. Continuous intensive monocropping may be productive and profitable in the short term, but as it is possible only through the application of increasing amounts of synthetic chemical fertilizers and pesticides, it is not sustainable over time. It is in no small part responsible for the modern form of "shifting cultivation" that results in the moving agricultural frontiers that are found around the world.

And of course this list of threats posed by intensive agriculture does not account for the growing tendency of farmers to turn to genetically modified organisms (GMOs) and the latest round of biotech inputs to increase productivity on ever less fertile land. These too may pose severe threats to biodiversity and to agriculture itself through the creation of noxious pests and weeds or, more importantly, the mutation or loss of beneficial soil microorganisms.

Around the world today, agriculture is practiced by a wide range of producers. Whether farmers sell 100 percent of their product to markets or are primarily subsistence-oriented (producing food for their families and selling surplus into local markets) they all have the potential to cause environmental damage. As producers become more dependent on markets to meet their own wants and needs, they produce what their circumstances will allow them to to obtain the highest returns with the fewest risks. Initially this means selling surplus subsistence production. Over time, however, it means planting cash crops within less intensive production systems.

Eventually even this focus can shift as producers move to intensive monocropping systems, with subsistence crops marginalized into gardens. In many areas of the world there is still considerable local market demand for subsistence crops, but even in these markets what is valued can shift over time. For example, in Africa, production is shifting from such traditional crops as sorghum, millet, and cassava to rice, corn (maize), and wheat.

Despite all the problems with intensive industrial agriculture, it is equally clear that low-input cultivation systems, as they were practiced in the past, cannot meet the current food and industrial needs that people around the world have come to expect from agriculture. Somewhere between these two extremes are systems of production that are more sustainable and productive and that make better use of fewer resources than either the less input-intensive or more intensive systems that currently dominate global agricultural production.

Any use of natural resources has impacts. The problem at this time is that producers have no incentives to reduce their negative impacts. If anything, because there are no disincentives to reduce environmental impacts, producers have every reason to ignore them. The question for societies is which impacts are acceptable, and how to discourage the practices that lead to unacceptable impacts.

THE ORGANIZATION OF THIS BOOK

This book identifies and explores the main threats that key agricultural commodities pose to the environment as well as the overall global trends that shape those threats. It then identifies new practices as well as tried-and-true ones that can increase production while minimizing environmental costs. Many who analyze the environmental impacts of agriculture focus on trade policies that affect specific agricultural commodities traded internationally. There are two problems with this approach. First, most agricultural products are consumed in the producing country and not traded across borders, even in a processed form. Second, the main environmental impacts are on the ground; for example, they relate to production practices, not trade. Trade and trade policies are one way to approach the problem, but only if they can be focused in such a way as to reduce the production impacts of commodities that are not by and large traded internationally.

This book takes the position that working with farmers directly to identify or co-develop better management practices (BMPs) may be far more effective in the short term and may provide better information to inform subsequent trade and policy strategies. While some BMPs may be encouraged by government or even international trading partners, most probably will not. In the end, the protection of endangered species and habitats with high conservation value is often essentially a local or regional issue that involves subsistence farmers or producers connected to local markets rather than international ones.

Another issue that receives considerable attention among those interested in agriculture, poverty, and the environment is who causes the most environmental damage. A common assumption is that large-scale, capital-intensive, high-input commercial farms have more negative impacts than small farmers who are trying to scrape together a living by producing food for their families and selling surplus locally. In fact, both are to blame. An increasing body of evidence suggests that smaller, more marginal producers may actually cause the bulk of environmental damage in both developing and developed countries. This damage can result from farming marginal land, not having efficient equipment (or the money to buy it), or not having good information about better practices.

This book does not attempt to answer the question of whether large-scale, high-input; low-input; or subsistence agriculture causes most environmental damage. Rather, the focus is to identify which practices are more environmentally destructive and whether better practices exist to reduce or avoid those impacts altogether for any of these systems of production. The focus is on primary production directly rather than on the processing of the primary products, except where processing occurs largely on the farm. Likewise, the focus is not on value-added processing through intensive feedlot systems such as those for cattle, chicken, or pigs. Such operations are more similar to factories than to farms and should be subject to the same pollution controls as other factories.

The twenty-one crops that are the focus of this volume include: bananas, beef, cashews, cassava, cocoa, coffee, corn (maize), cotton, oil palm, oranges, plantation-grown wood pulp, rice, rubber, salmon and shrimp from aquaculture, sorghum, soybeans, sugarcane, tea, tobacco, and wheat. These crops occupy most of the land used for agriculture in the world (see Table I.3). In addition, they represent a mix of temperate and tropical crops, annual and perennial crops, food and nonfood crops, meat and vegetable crops, and crops that are primarily traded internationally as well as those that are consumed primarily in the country of origin.

A number of significant crops are not discussed in this book. In many cases, the excluded crops are those whose area of production is in decline, or ones that are not deemed as globally significant as another crop that is included. Some of the more obvious tradeoffs were the inclusion of wheat instead of barley, rye, or oats; sorghum instead of millet; cassava instead of sweet potato; soybeans and oil palm instead of peanuts (groundnuts), sunflowers, canola (rapeseed), olives, or coconuts; and sugarcane instead of sugar beets.

Some of the omitted crops are very important locally. This is the case with such crops as potatoes, grapes, apples, horticulture crops, cut flowers, or sugar beets. The assumption, however, is that the issues and lessons that are raised through the discussion of the crops that are included are transferable to most of the others. And, while no blueprints for sustainability are included, the larger purpose of this work is to help the reader understand how to think about agricultural production and the environment.

The discussion for each crop chapter follows the same outline. Each chapter begins with "Fast Facts" that summarize important comparable information for each crop, including maps of production areas. These facts include: the total area in and volume of production, the average and total value of production, the main produc-

TABLE I.3. Global Area Planted to Crops Discussed in this Report, 1961–2000
(in millions of hectares)

	1961	1970	1980	1990	2000	Percentage Change 1961–2000
Bananas	2.03	2.71	2.78	3.38	4.1	90.1
Beef	3,144.74	3,211.45	3,287.88	3,409.64	3,459.8	10.0
Cashews	0.52	0.86	0.88	1.06	2.7	280.8
Cassava	9.63	11.62	13.60	15.20	16.8	74.5
Cocoa	4.10	4.06	4.42	5.38	7.5	82.9
Coffee	9.76	8.88	10.04	11.30	10.6	8.6
Corn	105.58	113.13	125.69	131.32	138.7	31.4
Cotton	31.86	34.16	34.32	32.97	32.7	2.6
Oil palm	3.62	3.27	4.28	6.08	9.7	168.0
Oranges	1.21	1.60	2.22	3.15	3.6	197.5
Rice	115.50	133.10	144.54	146.93	154.1	33.4
Rubber	3.88	4.62	5.41	6.65	7.7	98.5
Salmon	NA	NA	NA	NA	NA	NA
Shrimp	0.05[+]	0.15[+]	0.25[+]	0.9[+]	1.8	3500.0
Sorghum	46.01	49.41	44.09	41.55	42.0	−8.7
Soybeans	23.82	29.52	50.65	57.13	74.1	211.1
Sugarcane	8.91	11.11	13.29	17.08	19.6	120.0
Tea	1.37	1.69	2.37	2.26	2.3	67.9
Tobacco	3.40	3.77	3.90	4.65	4.2	23.5
Wheat	204.21	208.02	237.19	231.28	213.7	4.6
Wood Pulp	#	#	#	#	10.0	#

Source: FAO 2002.
Note: [+] Estimated by author based on FAO production data.
# indicates data not available.

ing and consuming countries, the percent of production exported, the species name(s), and the main environmental impacts as well as the potential to reduce those impacts.

In addition, each chapter presents (to the extent possible) comparable information about each crop. The discussion starts with an introduction to and history of each crop as well as an overview of the main producing and consuming countries. The main systems of production are described for each crop as well as any processing of the crop that occurs within the area of production. A section is included about the current substitutes for each crop and the impact of substitutes on markets. Market-chain analyses are included for each crop to the extent possible, but because this information is rarely in print, it is not complete. Market trends are also identified and analyzed but these, too, should not be considered definitive, as this is the stuff of crystal balls as well as fortunes to be gained from trading and is, as a result, rather incomplete. Finally, the major environmental impacts are discussed and strategies for addressing them are identified.

Much of the production and trade data in this book is based on statistics from the Food and Agriculture Organization of the United Nations (FAO). This data is generally the best available but is considered by many to underestimate both total production and area in production. In addition, a wide range of figures from different sources are used to illustrate different issues raised in different chapters and some of this data is contradictory. Every attempt has been made to reconcile these numbers, but it has not always been possible.

Price data, too, has been difficult to obtain and more difficult to verify and standardize. In general, prices have been indexed to 1990 U.S. dollar values. World producer prices for individual commodities were calculated by transforming FAO producer price data into world averages. Such data is reported as individual commodity prices per country. A group of countries that represent world production was chosen, taking into consideration the available data. The world producer price for each commodity is calculated from a simple arithmetic average of the chosen representative countries' producer prices.

Libraries are full of information about agriculture in general and about these crops in particular. Furthermore, the world is full of farmers who produce them and who can supply valuable information and strong opinions. In short there is no dearth of information or opinions about these commodities and how they are produced. There is also considerable information (a vast quantity of publications, research, data, and analyses) focused specifically on describing or proposing how to reduce the environmental damage from producing each crop. This volume draws on all of these sources, including my own personal experience with large and small producers in both the developing and the developed world.

Though every attempt has been made to make the crop chapters complete, inevitably there are gaps. Some issues are harder to address for most of the commodities in question. For example, little work has been undertaken to assess the cumulative environmental impact of any single crop in a specific place over time, much less the comparative impacts of crops that produce products that are readily substituted for each other. Even less has been done to evaluate the global impacts of a specific crop or to identify the likely environmental impacts of global trends within an industry. This book offers insights of a different scale and focus and suggests how more comprehensive work on future trends could help those interested in the environment and agriculture better understand issues of economic, social, and environmental viability.

REFERENCES

Chen, X.-S. and K.-Y. Ge. 1995. Nutrition Transition in China: The Growth of Affluent Diseases with the Alleviation of Undernutrition. *Asia Pacific Journal of Clinical Nutrition* 4(4):287–293.

FAO (Food and Agriculture Organization of the United Nations). 2002. *FAOSTAT Statistics Database*. Rome: UN Food and Agriculture Organization. Available at http://apps.fao.org.

Smil, V. 2000. *Feeding the World: A Challenge for the Twenty-First Century*. Cambridge, Mass.: MIT Press.

Tilman, D., K. G. Cassman, P. A. Matson, and R. Naylor. 2002. Agricultural Sustainability and Intensive Production Practices. *Nature* 418 (8 August): 671–677. London: Macmillan.

Washington Post. 1995. Feeding a Hungrier World. February 13.

AGRICULTURAL TRENDS AND REALITIES

1

Missouri half a century ago. By 1956 most
re connected by gravel roads to the main
provided posts for fences and cover for
if not most of their own food, fresh in the
ellars during the winter. Most farms had a
er most raised chickens and had vegetable
en beehives were common as well. Virtual-
ff and erosion, provide water for livestock,
n) for food. Many had woodlots that sup-

ving a mix of crops in rotation with corn as
ans, and oats. Clover was often planted to
st and to build the soil. Midwestern farms
ef cows that were relegated to more rugged
aze on the crop fields after harvest. Corn-fed
re cash from the corn crop. Farming was mecha-
orses to pull plows, discs, harrows, planters, cultiva-
the fields. Cultivators were used to till the soil and kill
. In some instances, row crops like corn and soybeans were
ree years in a row. Weed killers were used on corn, but usually
basis and only if weed infestation was particularly heavy; otherwise,
en normally walked the rows to cut weeds. The average row crop was pro-
with six to seven passes of a tractor pulling different equipment over the
se of three to five months.

By the 1960s different U.S. government programs encouraged farmers to increase
their farm and field size as well as the intensity of crop production. Fewer, more
valuable crops were produced. Not only did farm size increase, so did land value,
fixed investments in machinery, and the overall use and cost of inputs such as fertilizer and pesticides. Fencerows, waterways, and the last vestiges of blue stem prairie
were eliminated in the quest for greater efficiency. Many pasture areas that were
considered too poor quality to farm in the past went under the plow. Erosion increased. Wildlife, once common on farms, was virtually eliminated. Ponds and

streams became loaded with sediment, nutrients, and pesticides. For the first time, farm families became dependent on purchased food. Well water was no longer safe to drink on most farms. The average crop was produced with about the same number of passes of a tractor, but the activities were different. While more efficient soil preparation meant fewer passes, additional tractor passes were needed for fertilizer and pesticide applications.

As farming changed, communities changed. Farmers have always depended on inputs and services from others. Prior to World War II, small crossroad commercial centers existed about every 5 to 8 kilometers (3 to 5 miles) around the countryside. They usually consisted of a blacksmith shop, feed store, general store, and a church. By the 1950s communities flourished at intervals of about 17 kilometers (10 miles). These communities consisted of blacksmith shops, feed stores, schools and churches, grocery stores, and clothing and hardware stores. By the 1970s most commerce was shifting to larger towns spaced about every 50 to 80 kilometers (30 to 50 miles). And farmers became as aware of weather patterns in Europe, Argentina, and Brazil as they were of those in neighboring states.

Similar trends have occurred in other parts of the world. More efficient production has led to lower prices. As prices dropped, market-oriented producers have attempted to increase their income by increasing their holding size as well as the intensity of production. Many smaller producers who found themselves unable to compete with the volume of large-scale producers have identified new crops, found ways to add value to traditional crops, or simply become marginal subsistence farmers. In 2000, for the first time, the number of small farmers in the world declined, implying that many small producers could no longer support their families by producing their own food, or perhaps that life elsewhere was preferable to the marginal, isolated existence of farming.

While the specifics and the speed of the changes have varied around the world, agriculture and its relationship to societies has changed everywhere. Governments have become much more involved not only in agricultural production but also in seed and agrochemical development, product development and promotion, and currency and trade issues. Globally, increased urbanization, the expansion of markets, and increased trade in raw materials as well as manufactured products have stimulated technological changes and increased overall scales of production. At the same time, increasing awareness of global food production systems has made consumers more concerned about the quality of food they eat as well as how it is produced. These same factors have made the food industry ripe for both vertical integration (where a company controls ownership of a product for all or most stages, from production to the consumer) and consolidation.

Mechanization, new inputs such as fertilizers, pesticides and technology, improved crop varieties, and government support and protection have tended to cushion producers around the world from many market realities. Globalization is changing that. In the past producers competed with their neighbors for local markets. As transportation improved, producers competed at regional and even national levels while government protected them from foreign imports. Today most agricultural production is still consumed in the country of origin, but globalization promises to change that, too.

As technology has come to dominate producers' decisions about how to solve problems, responses have tended to focus on a single technology (e.g., seed, fertilizer, pesticides, tillage, or water) or rather simple combination packages of the individual technologies. Subsidies accentuate this response. One consequence is that in the past century, more producers are planting single crops, with fewer rotations. This has resulted in the loss of an estimated 75 percent of global agricultural biodiversity. It is simply too complicated to find ways to improve the production of each of the wide range of plants and animals that have developed in local niches around the globe over millennia.

The overarching goal of agricultural research has been to identify and focus only on those species or varieties promising the most potential for economic gains. India, for example, is rapidly replacing 30,000 varieties of rice with a single variety. By the year 2000, 75 percent of the world's food came from seven crops—wheat, rice, corn, potatoes, barley, cassava, and sorghum. Some 60 percent of the world's food calories came from the first three alone. If soybeans, sweet potatoes, sugarcane and beets, and bananas are added, these crops account for 80 percent of total crop tonnage (Kimbrell 2002). This simplification is shortsighted at best, and fails to take into account the current reality of agricultural production and its future consequences.

THE CURRENT REALITY

For more than 99 percent of human history, people obtained their food by hunting, fishing, and gathering. Over the past 7,000 years that has changed remarkably. Today only 2 percent of all human food energy and only 7 percent of all protein is captured from the wild, and most of this is from water. The rest is produced by agriculture and aquaculture on land.

As a result, agriculture is the largest industry on the planet. It employs an estimated 1.3 billion people and each year produces some $1.3 trillion worth of goods at the farm gate. In the developed world, food prices (in real terms) have fallen by 40 percent over the same period. For example, because of overall increases in per capita income and relatively cheap food, Americans spend only 14 percent of their income on food. Europeans, on the other hand, spend some 44 percent more on food than the rest of the developed world. In developing countries, however, the poor can spend as much as 75 percent of their total income on food.

Not only has the percentage of income spent on food tended to decline in the United States, but the percentage of those dollars kept by farmers has declined as well. In 1900 an American farmer received some 70 percent of every dollar spent on food. By 1990 U.S. farmers received an estimated 3 to 4 percent of the money spent on food. Globally, agribusiness produced $420 billion in 1950, and farmers received a third of it. Researchers estimate that by 2028, the total global market for agricultural production will be $10 trillion and farmers will receive 10 percent of it.

Part of the reason that less and less money goes to the farmer is that more "value" is added to agricultural products than ever before. In the past farmers sold products in open markets and received a large portion of the consumer price. Today, the consumer's cost of food includes manufacturing, quality control, preservation and packaging, labeling, distribution and handling, storage, advertising, compliance with laws and regulations, professional management, and even the cost of air-conditioned supermarkets. While food prices have steadily declined, the cost to manufacture, hold, distribute, and sell food has increased, further squeezing farmers. American farms represent only 0.9 percent of the country's gross domestic product (GDP), but the food market chain—those who sell to and buy from farmers—is about fourteen times as large. The price of a cup of coffee has more to do with the convenience and ambience of where you buy it than the cost of the beans. Similarly, the Coca-Cola company spends more on each can than on what is in it.

As a consequence of increases in productivity and economies of scale, the number of farmers around the world is declining in absolute terms. In the United States farmers represent less than 1 percent of the population, and they not only feed the rest of the population but also produce enough for this country to be the largest exporter in the world. Only 18 percent of U.S. farms produce 87 percent of the food. Farming populations in France and Germany have fallen by half since 1978. In countries belonging to the Organization for Economic Cooperation and Development (OECD), the number of farms is declining by 1.5 percent per year, and farmers and their families now represent only 8 percent of the population. In short, throughout the world there are fewer, more highly productive farms every year.

Past success in increasing food production and lowering costs does not imply that there will be sufficient food in the future. There are several worrying trends. First of all, as discussed later in this chapter, hunger issues are as much about distribution and income as production. Second, production will not keep increasing forever. Global food demand is likely to double over the next fifty years. While total cereal production has doubled in the past forty years, the increase in yield growth rates have declined from 1987 to 2001, indicating that productivity is nearing its genetic and resource limits. The world's population is expected to increase another 50 percent by 2050. Increased affluence (projected as a 2.4-fold increase in per capita real income around the globe by 2050) is leading to increased consumption of meat and animal products, which requires additional agricultural production (Tilman et al. 2002). In the United States it takes 0.42 ha (45,000 square feet) of agricultural land to feed a single person eating a high-animal-protein diet. This model will not work in developing countries where there is only about 0.08 ha (9,000 square feet) of agricultural land per person available for cultivation. Furthermore, per capita land availability is decreasing worldwide.

PROBLEMS WITH LARGE-SCALE PRODUCTION

Most agricultural systems around the world are evolving into larger, more specialized units of production owned by fewer and fewer people. In Brazil, for example, 80 percent of the land is owned by 10 percent of the population. In the United States

163,000 large farms now account for 61 percent of sales, while only 50,000 farms produce 75 percent of all food.

Efficiency is today's key agricultural issue—production per hectare, production per unit of fixed and/or working capital investment, cost per unit of production, cost per unit of key production input, etc. In general, smaller farms (those less than 11 hectares) are more efficient producers than bigger ones in terms of production per area of land. Studies from around the world show that smaller farms almost always produce more product per unit area than larger ones. The cost of production per unit produced increases with farm size. This is true at least in part because smaller farms are usually run by families, and the cost of family labor is not included in their calculations of costs. Midsized and larger small farms, on the other hand, are more economically efficient when labor and technology are included in the calculation.

A 1992 U.S. agricultural census found that smaller farms are two to ten times more productive than larger ones and ten times more productive per acre than farms of 6,000 acres or more. The smallest farms (1.6 hectares or less) were 100 times more productive per acre than farms of 2,400 hectares or more. The problem with such small farms is that most of the farmers, unless they have very valuable cash crops, cannot make a living from farming alone and must subsidize their income with off-farm employment.

In addition, market factors often outweigh local economic or environmental efficiencies in the marketplace. Simply put, it is easier to purchase larger amounts from a smaller number of suppliers. Nowhere is this clearer than with livestock. During the past forty years, global per capita meat consumption has increased by 60 percent. To meet this increased demand, livestock production is increasingly industrialized, with several thousand cattle or pigs or 100,000 chickens often raised in a single facility. Over the past fourteen years, the average size of animal operations in the United States has increased 1.6-fold for cattle, 2.3-fold for pigs, 2.8-fold for eggs, and 2.5-fold for chickens. In Canada pig operations have increased 2.6 times in size in ten years (Tilman et al. 2002).

Such operations come with costs, often in the form of diseases. In 1997 a chicken virus in Hong Kong killed six people and resulted in the slaughter of 1.2 million birds. Outbreaks of foot-and-mouth disease in the United Kingdom resulted in 440,000 animals being put to death in 1967 and 1.2 million in 2001. Bovine spongiform encephalopathy (BSE, more commonly known as mad cow disease) resulted in the slaughter of 11 million animals in 1996 (Tilman et al. 2002).

In North Dakota most "farms" now are greater than 8,000 hectares, but they are not single properties. Most have been pieced together through years of acquisition and often consist of many small farms 50 to 100 miles apart. While spreading out the holdings may reduce localized climatic risks, such farms are less efficient to operate. More importantly, owners cannot afford more environmentally sensitive management practices and cropping patterns. When the land being farmed is spread over such a wide area and the time window for management and cropping is so narrow, it is impossible to monitor the conditions on each plot and move machinery back and forth to deal with small-scale problems. It is simply easier to farm single crops with uniform management interventions. It is clear that the most efficient interventions are made as a result of monitoring and tailoring the response to the observed problem. On such large farms it is difficult to monitor crop conditions and pests for areas that are less than 1 square kilometer. This scale is simply too large for the most effective and efficient management.

These patchwork farms were created not in response to normal market incentives but rather because of government policies. U.S. commodity programs encourage wheat (and corn) producers to acquire more base acres (from which subsidies are calculated) in order to receive higher government payments. As a consequence such farmers may be producing wheat, but what they are really growing is government subsidies.

There are other troublesome issues regarding farm size. As farm size increases, poverty in local communities and absentee ownership increase as well. In addition, as farm size increases in rural areas, crime tends to increase while the number of local businesses decreases (Kimbrell 2002).

Scale issues are not limited to conventional high-input farming. In the United States, at least, organic production is even more concentrated than conventional agriculture. In California, five farms control half of the state's $400 million organic produce market. Horizon Organic in Colorado controls more than 70 percent of the nation's organic milk market. Until recently it produced more than 30 percent of its milk on only two dairy farms (Baker 2002; Pollan 2001). Similarly, in Brazil a tiny fraction of the total number of farms accounts for almost all of the millions of hectares of no-till agriculture.

Productivity does not depend on size alone. Well-managed farms are always more productive than poorly managed farms of the same size. They use fewer chemicals, fertilizers, and antibiotics per unit of production; they also have lower production costs, fewer and less severe environmental impacts, and fewer health problems than less well-managed farms. Because well-managed farms have equal or higher yields, they are more profitable and environmentally preferable.

AGRICULTURE AND SOCIETY

Agricultural production reflects the inequities of societies. There are more than 1.2 billion people on the planet who live in absolute poverty, earning less than $1 a day. Twice that many people survive on less than $2 a day. The Food and Agriculture Organization of the United Nations (FAO) estimates that 830 million people in the world are underfed. Almost 80 percent of the world's hungry live in rural areas and depend on agriculture to make their living. While at one time, wild-harvested food fed many of these people, today half a billion rural poor are landless or lack sufficient land to produce what they eat or the income to buy it. From 1970 to 1990 the number of hungry people in every country except China increased by an average of 11 percent (Kimbrell 2002).

Hunger issues are as much about distribution and income as production. There is enough food for everyone on the planet to have 3,500 calories a day. In fact, there is sufficient food to provide everyone on the planet nearly 2 kilograms (4.3 pounds) of food every day, including 1.14 kilograms (2.5 pounds) of grains, beans, and nuts; 0.45 kilograms (1 pound) of fruit and vegetables; and nearly another 0.45 kilograms (1 pound) of meat, milk, cheese, and eggs (Kimbrell 2002). Over the past thirty-five years, per capita food production has grown 16 percent faster than population. Still, people are hungry. There is growing recognition that agriculture has a major part to play in improving this situation (DFID 2002). For example, in Africa agriculture employs about two-thirds of the labor force, accounts for 37 percent of the GNP and is responsible for half of exports. Still, the sector is doing little to generate wealth among the poor. In South Asia agriculture generates 27 percent of the GNP but also has little impact on reducing inequality.

For developing countries as a whole, per capita agricultural production increased by 40 percent between 1980 and 2001, but growth was uneven. China, for example, quadrupled the value of its agricultural output and overtook the United States as the world's largest agricultural producer. Likewise, India tripled its agricultural output. In sub-Saharan Africa, however, agricultural production fell by about 5 percent over the same period. Africa is the only continent where the number of hungry people has increased in absolute terms between 1980 and 2000 and is projected to increase even further.

Globally, the total production of foodstuffs surpasses total consumption. In 2000 the amount of grain in storage constituted nearly 1.2 years' worth of global consumption. However, for the past three years the world has produced less grain than it eats. According to the U.S. Department of Agriculture, in 2000 the shortfall was 35 million metric tons, in 2001 it was 31 million metric tons, and in 2002 it was an estimated 83 million metric tons. As a result grain stocks have dropped to the lowest levels in thirty years. In 2002 world wheat stocks were estimated at only 23 percent of annual consumption, while rice stocks were 28 percent. Corn was lowest of all at less than 15 percent; these are the lowest stocks for corn since record keeping began forty years ago. Production shortfalls are caused by low prices for producers at planting time (which cause planters to reduce the total area planted), high temperatures (which stress plants and so reduce yields), low temperatures (which delay planting or shorten growing seasons), and reduced or erratic rainfall or falling water tables (which stress or kill plants and so reduce production). With wheat and corn prices increasing by 30 percent or more in 2002, at least the first factor should be lessened until the next harvest.

Increased producer prices will eventually affect prices of processed and manufactured goods as well as animal products. If the poor had a hard time buying food when prices were lower, they will have an even tougher time now. World grain prices have generally fallen since the mid-1990s with the exception of the recent upturn in wheat and corn prices. This should have put more food within the reach of the poor. But the problem of the poor and hungry is their lack of income, rather than the supply of food or its price. Markets that foster the delivery of regular food supplies at lower and more stable prices help create food security and potentially help reduce hunger. Yet the production of this food often actually reduces the income of the rural poor, who are being displaced precisely because they can no longer compete with cheaper food coming into their area. This as much as anything accounts for the large percentage of their income they must spend on food. In India stagnation in agriculture drives poor people to towns and accounts for as much as 30 percent of urban growth.

It would not take a lot to change this picture of entrenched rural poverty. A recent study covering fifty-eight developing countries concluded that a 1 percent increase in agricultural productivity locally would reduce the proportion of people living on $1 a day by 0.6 to 1.2 percent (Thirtle et al. 2002, as cited in DFID 2002). In India a recent study concluded that the average real income of small farmers rose by 90 percent and that of the landless by 125 percent due to increases in local agricultural productivity (Dev 1998, as cited in DFID 2002). Increases in income were mainly attributable to labor productivity gains linked to new technology.

Urbanization also increases global hunger. At the end of World War II only 18 percent of the population in developing countries lived in cities. By 2000 that figure had reached 40 percent, and it is expected to climb to 56 percent by 2030. About 50 percent of the urbanization is due to migration, both from abandoning agriculture and from the lure of potential jobs in cities. Few migrants to cities are able to produce any, much less most, of their food. They are at the mercy of the markets. If they cannot afford to buy, they go hungry. The truly poor on the planet can spend 75 percent of their income on food and still go hungry. Some 1.2 billion people have on average only 150 kilograms of food per person per year, or less than a pound a day.

Shipping the highly subsidized surpluses of developed countries to less-developed countries appears to be a generous way to improve the plight of the hungry. But the solution is not so simple: such shipments lower the value of local production and therefore the income of local producers while they reduce the demand for rural labor or at least the price paid for it. The subsidized agriculture of developed countries is not sustainable in its own right. More importantly, it does not contribute to sustainable food production systems in the developing world. Such food assistance rarely reaches those who need it most, plus it often causes the structural position of those who need it most—the rural poor—to deteriorate even more. Food assistance can undermine the ability of poor farmers to produce and sell food competitively in local markets. Instead of importing surplus produced elsewhere, the food needs to be produced where it will be consumed. If productivity in less-developed countries were boosted, there would be surpluses to sell and markets in which to sell them. This in turn could boost incomes at the local level and enable more people to afford food. One of the most cost-effective ways to do this would be to reduce subsidies and market barriers in developed countries.

Consolidation of farms into ever-larger agricultural production units contributes to poverty by displacing more people every year. Through mechanization, a given unit of land employs fewer people as well. Those who are employed in rural areas tend to own little or no land. Their production on small plots often subsidizes their work for larger landowners. This situation tends to occur until such workers (or their children) migrate to cities. For decades now, most children born on farms have not ended up farming.

There are a number of other social issues involved with food production. Migrant and temporary workers often account for a large percentage of production. In the United States, migrant workers produce half of all food. Such workers are fifteen times more likely to exhibit symptoms of pesticide exposure; 300,000 farm workers in the United States suffer acute pesticide poisoning each year. In addition, the average occupational fatality rate in the United States for all industries is 4.3 per 100,000 workers. However, for agriculture, forestry, and fisheries the rate is more than five times higher at 24 per 100,000.

Finally, food quality and safety are also important social issues. While it is hard to compare current levels of food contamination and overall quality with those from the past, in all likelihood most food is healthier, safer, and fresher throughout the world than at any time in history. Historically, the biggest issue has always been the quantity of food, not its quality. While that is still an issue for a significant portion of the global population, it is no longer the paramount issue for most. More affluent consumers are probably more preoccupied with food quality than ever before. Recent food problems, including pesticide and antibiotic residues, mad cow disease, bacteria such as *E. coli* and *Salmonella*, hoof-and-mouth disease, and contaminated animal feeds have heightened consumers' concerns about their food. These concerns have generally been greater, or at least expressed more vocally in Europe than in the United States.

GOVERNMENTS AND AGRICULTURE

The paramount goal for governments when it comes to agriculture is a simple one: ensure secure and inexpensive food supplies. Countries have chosen to meet this goal in a number of different ways. Governments have sponsored public works programs to increase the amount of arable land, developed infrastructure to allow products to be moved more efficiently to markets, and supported the development of technology to increase food production. They have also created subsidies based on production, or sometimes the lack of it. They have subsidized the purchase of inputs and capital and pursued a wide range of policies to encourage increased, but stable, production (Clay and von Moltke 2002).

Increasing urbanization, particularly in the developing world, complicates the issue of secure and inexpensive food supplies. Urbanization increases the demand for surplus food production from the countryside and, consequently, the pressure on rural areas to produce more. To avoid high food prices and urban unrest, most governments subsidize food prices. This has been true throughout history. Unfortunately, as discussed earlier, cheap food leads to the impoverishment of rural populations as well as to environmental degradation. Historically, the initial response to agricultural "development" has been a dramatic reduction in rural populations through migrations to cities, a process that poses huge risks of social unrest in countries like

China. While this transition took a century or more in developed countries, it is happening much faster in many developing countries. Recently it has become apparent that market prices for agricultural goods can no longer support rural populations. This problem is accentuated as rural populations see the standard of living of urban populations rising and wish to emulate it.

The initial government response to this process is twofold—to continue to increase agricultural output and to seek new markets to raise the incomes of rural producers. But as other countries pursue this strategy, commodity prices deteriorate, which leads to protectionism. Another strategy is to develop new markets for organic or nontraditional crops. While this approach offers a "first-mover" advantage, other producers quickly follow suit (Clay and von Moltke 2002). The fundamental market structure is not changed. In addition, any benefits come with significant risks since the development of new markets is inherently risky due to the costs of innovation and the risks of markets not developing. For those producers who are not protected by government (increasingly the norm in developing countries, and likely to become more common in developed countries over the next twenty to thirty years), the best option is to become more efficient and sustainable. This could mean more efficient use of all inputs, reduction of waste, value-added production, differentiation of production, selling directly to buyers or consumers, building their main asset—soil fertility—through improved management, and developing income from sources other than the sale of product.

Subsidies

More dramatic than the impoverishment of the rural poor is when an entire country cannot meet its own basic food needs. Historically, this has led to riots and political instability. No politician wants to lose his or her job because of food shortages and high prices. Nonetheless, government policies are at the heart of many food and agriculture problems. Politicians and policies, for instance, cause most famines by disrupting production (either through the confiscation of seed and other inputs or through war), by hoarding production so that it is unevenly distributed throughout the country, or by encouraging the production of nonfood crops on prime agricultural lands.

To avoid famine and economic dislocation, countries use different kinds of policies to provide incentives or disincentives for the production of different crops. Subsidies are used to encourage agricultural production. They come in many forms, but collectively they give producers the ability to sell products at prices that are lower than would otherwise be possible. Almost every developed country has found itself subsidizing agricultural producers. The exceptions—New Zealand, Australia, and to some extent Canada—represent special cases since they do not have large rural populations and their natural advantages in certain crops permit them to produce at lower costs than most other countries. Subsidies ensure agricultural surpluses under most conditions, and they allow producers in a country that subsidizes agriculture to reap benefits as producers in non-subsidizing countries are forced out of business (Clay and von Moltke 2002).

In most developed countries, and increasingly in developing ones, there appears to be no alternative to agricultural subsidies. The global population is increasingly urbanized so government priorities remain unchanged—food availability to urban workers at the lowest possible prices. Subsidies achieve that goal without generating revolts in rural areas. To date, the potential for political unrest is far more powerful than economic calculations comparing the efficiency of subsidies with that of alternative policies.

The United States began to subsidize farmers during the depression in 1929. In general, the U.S. government guarantees market prices for key agricultural products. The 2002 farm bill increased future subsidy payments precisely when there was increasing awareness of the negative impacts of subsidies and discussions about ways to reduce them. The United States also spends some $659 million per year to promote its agricultural products and exports. This is more than just selling corn, wheat, and soybeans. It also entails giving $1.6 million to McDonald's to promote Chicken Mc-Nuggets in Singapore and $11 million to Pillsbury to promote its Doughboy brand internationally (Kimbrell 2002).

According to the OECD, global agricultural subsidies amounted to about U.S.$311 billion annually in 2001 (OECD 2002a). The United Nations estimates that the costs in lost revenues to poor countries amount to some $50 billion per year. That sum effectively offsets the entire $50 billion annually in development assistance from all sources.

Developing countries are unable to subsidize their agriculture in the same manner. However, as developing countries become more urban, their governments—given a choice between higher incomes for agricultural producers and lower food prices for urban dwellers—unhesitatingly pick the latter. Rural populations do not overthrow governments, particularly democratic ones. Urban populations do (Clay and von Moltke 2002).

Agricultural subsidies in rich countries reduce production costs or artificially raise the prices their producers receive. These subsidies are often the difference between making and losing money. Throughout the 1970s and 1980s, U.S. subsidies represented up to 30 percent of farmer income. In 2000, subsidies represented on average 100 percent of net profit for farmers in Indiana. Such subsidies lead to overproduction and overexploitation of resources. They also inflate land values.

While there has tended to be a shift away from policies that tie payments directly to production, some 72 percent (down from 82 percent in the mid-1980s) of support to farmers in OECD countries still keeps producer prices above those on world markets (DFID 2002). In 1999 OECD member countries provided $283 billion in domestic agricultural production subsidies. In 2001 that figure had risen to $300 billion. The European Union's Common Agriculture Policy spends some $40 billion per year by paying some 7 million farmers subsidies linked to the amount they produce (Power 2002). Some 80 percent of these subsidies goes to 20 percent of the producers. However, total domestic support to farmers in 2001 amounted to U.S.$93.1 billion in the European Union (OECD 2002b). This compares to $49 billion in total subsidies to farmers in the United States in 2001.

Another form of subsidies is non-recourse loans, in which governments lend money to farmers using future harvests as collateral. The loan rate is based on a set value per unit of production. This calculation assumes certain average yields as well as values for the crop. However, farmers can hold crops and wait for higher prices. Thus the farmer can sell the crop at a higher rate and repay the loan in cash, or default on the loan and forfeit the lower-value crop to the government.

Export supports, or refunds on exports, used to encourage the sale of agricultural surpluses on international markets, are another form of subsidy. These also depress world prices. While this may benefit some consumers, the overall distortion in domestic markets has a negative impact on rural economies as it tends to increase poverty and food insecurity.

Finally, there are several government payments that act like subsidies. The U.S. Army Corps of Engineers spends many times more each year to maintain riverine transportation than the value of the agricultural products that come down the rivers. In addition, local governments often give financial incentives to lure agribusinesses to their areas. For example, Seaboard Corporation located a gigantic confined hog operation in Guymon, Oklahoma, after receiving an estimated $60 million in public incentives (Kimbrell 2002).

The alternative to subsidies is to let markets function without regulation. The risks of such a strategy are numerous and large. Liberalized commodity markets are volatile. The lack of elasticity in supply can cause volatility in demand, and vice versa. This in turn can cause some producers to go bankrupt while others get bought out through consolidation. Effects at the consumer level could be even more troubling—periods of oversupply and low prices alternating with periods of undersupply and high prices. While low prices may be attractive, high prices are politically unacceptable because they create hunger. Additionally, letting existing markets function without regulation means that there will be no resources for rural conservation, since today's markets only supply goods that have market prices and these prices do not include environmental services (Clay and von Moltke 2002).

Protectionism

Contrary to popular opinion, most support to farmers occurs through government manipulation of domestic prices rather than through subsidies. According to the OECD, consumers pay about one-third more for their food than they would without government support for farmers. The question is not whether consumers pay more but who actually gets that extra money. Producers receive some of it, but traders also profit by exporting subsidized goods. Processors also benefit by using subsidized goods to manufacture products that are more competitive on global markets than those produced by companies in other countries where raw materials must be purchased at full market price. The ripple effects of subsidies and market protection throughout the economy can be significant.

Governments support their own farmers through market barriers that include tariffs and taxes on imported goods. In 1998, $456 billion worth of agricultural goods was traded across borders, a threefold increase from 1978. Tariffs on agricultural goods, however, still average about 40 percent of total sales compared with less than 10 percent for manufactured goods. Governments use quotas to regulate the volume as well as the country of origin of imports. Bananas illustrate this point. The European Union (EU) imposes strict quotas and tariffs on cheaper bananas from Latin America while allowing virtually free access to more expensive ones from former colonies in Africa, the Pacific, and the Caribbean. While the World Trade Organization (WTO) ruled against this practice, it still exists—as do similar ones for sugar and other commodities.

Trade barriers and tariffs have enormous impacts. In a report commissioned by the European Union, Nagarajan (1999) estimated that a 50 percent cut in tariffs in both developed and developing countries would generate $150 billion in increased sales of agricultural products from developing countries. At the time, the figure represented three times what those same countries received in aid from all sources. The IMF reports that developing countries lose some $30 billion to agricultural supports each year (UNCTAD/WTO 2002), and the World Bank (2000) estimates that agricultural tariffs and subsidies cause annual losses of $19.8 billion for developing countries (e.g., about 40 percent of the amount they receive in development assistance). Other estimates are even higher. The fact that these figures are so far apart clearly demonstrates the lack of information and transparency regarding tariffs.

Tariffs imposed by developed countries on agricultural products such as meat, sugar, and dairy products from developing countries are almost five times higher than tariffs for manufactured goods (DFID 2002). High and complex tariffs, coupled with increasingly stringent formal and informal product and performance standards, limit developing countries' access to international markets. These same tariffs discourage diversification of domestic production into higher-value items and retard the development of processing facilities in the protected countries.

There are other nontariff trade barriers that affect trade as well. Sanitary requirements to prevent importation of exotic pests and diseases, for example, can be effective trade barriers not because of the standards that are required but rather because of the cost of the testing to prove that the standards have been met. For many producers and exporters in developing countries it is simply not cost-effective to attempt to compete in such markets. In the future, price will probably induce large corporations to work with producers and exporters in developing countries to help them meet these requirements.

Governments are being forced to address subsidies as they affect international trade. WTO negotiations, for example, continue to focus on the reduction or elimination of agricultural subsidies. There is also an assumption by many analysts that more open markets improve farm revenues. Analyses of other commodity markets suggest that it is unlikely that the reduction or elimination of subsidies will improve prices in the long term. In all likelihood advantages will be short-lived, and the prices paid to agricultural producers will continue their downward trend.

International Trade

Some 90 percent of agricultural production is consumed in the country of origin. While a small number of commodities are produced primarily for export, most are produced for domestic consumption. In developed countries agricultural products are often incorporated into manufactured foods or refined raw materials, which are in turn exported. Export of raw materials in the tropics is often assumed to occur primarily with large-scale plantation production of crops like bananas, sugar, coffee, tea, and rubber.

The reality, however, is somewhat different, as much of the international trade in agricultural commodities is counterintuitive. For example, developing countries are net importers for all cereals, and developed countries (with the exception of Japan) are net exporters. Production costs in developed countries tend to be lower for capital-intensive activities such as large-scale grain production. Similarly, but to a lesser extent, developing regions (with the exception of Latin America) are net importers of meat, and developed ones are net exporters (with the exception of Japan, Eastern Europe, and the former USSR).

India illustrates the complexities of international trade. One in four farmers in the world live in India. When the WTO recently forced India to open its markets, imports quadrupled. Cheap (often subsidized) imports from the U.S. but also from Thailand and Malaysia caused prices and rural incomes to plummet. The price of coconuts fell 80 percent, coffee 60 percent, pepper 45 percent; most significantly, most domestic production of edible oil has been wiped out. Imports from the United States (soybeans) and Malaysia (palm oil) now account for 70 percent of India's vegetable oil consumption (Hines 2002).

An increased focus on agricultural exports in India is also threatening the livelihoods of the rural poor. Through small-farm consolidation and increasing mechanization, the number of people living on the land in Andhra Pradesh, for example, is expected to decline from 70 to 40 percent of the total population. In other words, about 20 million people will leave the countryside for urban areas in twenty years (Hines 2002).

A trend that has arisen in the past two decades with regard to international trade is an increase in the production and sale of differentiated products. This shift from bulk to boutique commodity production is an attempt by producers and others in the market chain to create and/or capture niche markets. The notion that any commodity produced anywhere in the world can be exchanged for any other is being challenged. Perhaps the most politicized example of this is the trade in genetically modified organisms (GMOs), discussed below in the section "Technology and Agricultural Production." Many countries and consumers are insisting that GMO products be labeled and kept separate from conventional products.

Markets are further differentiated when products are broken into different sub-products and by-products that have different markets in different parts of the world. For example, in the United States the price of dark chicken meat is less than half that for white meat. There is virtually no market for chicken heads or feet. However, in Asia the market for white and dark meat is reversed. Consequently, poultry companies in the United States sell increasing amounts of dark meat to Asia and import white meat in return. Chicken heads have large and valuable markets in countries such as Thailand, and chicken feet are prized in China.

The issue of product differentiation or segregation is not a new one. Many products have long been valued more or less depending on their variety, age, country of origin, or even the reputation of the producer. Today, such differentiation is a growing part of international markets and trade. Increasingly, producers differentiate their production and target specific markets with highly specialized goods. This has become even more important with the increased use of environmental and health certification as a marketing tool. In addition, consumers are better informed and more demanding than ever before.

Globally, companies are consolidating to achieve greater efficiency and to be more competitive in a global economy. Companies that have significant shares in all the major markets are less affected by producer subsidies and market barriers. At the same time, if the WTO is successful in dismantling those subsidies and barriers, large conglomerates will be well positioned to achieve even more significant economies of scale even as they differentiate their products.

Currency Values and Commodity Production

International currency values are little-understood factors that influence the expansion and contraction of agricultural production in many parts of the world. Currency devaluation stimulates agricultural production for export. For example, the 30 percent devaluation of the Thai baht against the U.S. dollar contributed to a 26 percent increase in Thai exports in 1998 of products such as poultry to Japan and Europe and processed goods to Japan. However, if local industries are dependent on imports, devaluation erodes their competitive advantage.

Local currency devaluations also affect imports and consumption. The August 1998 devaluation of the ruble in Russia led to a rise in the cost of imported meat and contributed to a decline in meat consumption. Over time the devaluation will stimulate meat production and a rise in supply. However, in the short term in 1998 there was a 15 percent decline in meat imports accentuating a 54 percent decline in meat consumption over the past decade. Similarly, the decline in value of the Australian dollar relative to major international currencies has helped that country to export meat and grain in world markets, but imports are now more expensive there.

Given the powerful role of China as the leading producer and importer of agricultural products, the stability of the Chinese yuan is a key issue in global trade and the stability of global agricultural production. If devaluation occurs in China for any reason, all goods will become cheaper on international markets; this will have a tremendous impact on producers of several different commodities in many different parts of the world. For example, China is the world's largest producer of cotton, wheat, rice, and tobacco; the second largest producer of corn, shrimp from aquaculture, and tea; the third largest producer of oranges; and the fourth largest producer of soybeans and sugarcane. If the Chinese yuan declines in value, these goods will be cheaper on international markets. This could encourage exports from China, making producers in other parts of the world less competitive. However, China would also be able to import fewer goods. Currently China is a net importer of corn (even though it is the second largest corn producer in the world), soybeans, rubber, and palm oil. Any decline in imports of these products will have an impact on Chinese consumers and potentially China's political stability in addition to its impact on global markets and producers around the world.

China's admission to the WTO may have impacts on agricultural production that could reverberate around the world. For example, without internal market barriers China's production of rice, wheat, and soybeans would most likely be reduced at least until they could be produced more efficiently and become competitive with producers in other parts of the world. One factor that is helping China at this time, however, is that its currency is undervalued. China also has a trade surplus, and this should provide sufficient force to maintain the yuan at its present value. But this situation can change.

The example of China demonstrates that producers around the globe are more connected than ever before. Changes in currency values in one country, particularly a large producer or consumer like China, the United States, or the European Union, have a ripple effect—even a tidal wave effect—in many other parts of the world. From the point of view of the environment, it is important to emphasize that the main environmental impacts occur during periods of economic expansion *or* economic contraction. Either can be equally damaging.

The Regulatory Context

It is clear that governments must play a key role in reducing the environmental costs of agriculture. Current attempts at this are made in large part through systems of laws and regulations. Such approaches tell producers more about what they cannot do than what they should be doing. This approach is about eliminating worse practices rather than encouraging better ones. Many environmentalists have supported government fixes because there are far fewer governments than farmers, and because governments are often located in the same place as environmentalists—capital cities. Unfortunately, this has caused many environmentalists to focus on getting the stick right rather than looking for carrots or other positive incentives that would promote desired changes.

Most countries have developed command-and-control mechanisms (a set of regulations that govern input use and performance levels of producers) to address many of the environmental problems resulting from agricultural production. For example, one area of regulation includes which pesticides are allowed, under which conditions, and in what form. Other policies support more market-based approaches to changing behavior. These include taxing pesticide use and effluent pollution and eliminating subsidies for fertilizers and fuel. The European Union and the United States both use direct producer payments (subsidies) to encourage the adoption of good practices and to pay farmers to set aside highly erodible or less suitable agricultural land. While these measures were intended to cut production (which did not happen), they did result in a number of environmental benefits.

With the possible exception of water use, few countries have developed programs to encourage the adoption of better practices that reduce environmental impacts. Even water used for irrigation is rarely priced at its true cost and consequently is often wasted. In Chile, Mexico, and California farmers are allowed to trade or sell the rights to water they have but do not need to other buyers, such as other farmers or cities. This system encourages farmers to use their water efficiently in order to sell the surplus and generate additional income. While such systems are a good start and the conservation benefits are real, water is still not part of an open, competitive market.

By focusing on regulating the impacts of individual entities, government command-and-control regulatory systems are not effective in addressing nonpoint source pollution. Thus, while many other industries have reduced their pollution, agriculture has become the largest polluter in many countries. The U.S. Environmental Protection Agency (EPA) has confirmed that this is the case in the United States. The United Kingdom estimates that its environmental costs from farming are as high as $2.5 billion per year. Two-thirds of the cost is from air pollution—nitrous oxide and methane. In fact, nitrous oxide represents almost half of the costs of agriculture's environmental impacts in the United Kingdom (ENDS Report 2000). The European Union has begun to identify similar issues. As agriculture is increasingly identified as one of the largest polluters, the industry will come under more regulation. Given that so much agricultural pollution is nonpoint source pollution, the overall approach to cleaning up the industry will have to focus on improving management at the landscape or ecoregional level, rather than measuring and reducing pollution from specific end-of-pipe sources, as other industries have done.

TECHNOLOGY AND AGRICULTURAL PRODUCTION

Many argue that technology will increase production and feed more people while reducing the variability as well as the environmental impacts of agricultural production. Technology has clearly been important in agriculture. As Table 1.1 indicates, productivity has been boosted considerably; most of these gains have been achieved through selective breeding, domestication, and the appropriate and timely delivery of water and other inputs.

Technology has been particularly important for improving production in non-tree crops such as corn, cotton, rice, soybeans, and wheat. Because trees take longer to develop, they take longer to improve through traditional breeding programs. In addition, most tree crop production cannot be mechanized, so less research has been undertaken on trees. Recent gains from biotechnology suggest that it might be

TABLE 1.1. Global Productivity Increases for Crops Discussed in This Report, 1961–2000
(in kilograms per hectare)

	1961	1970	1980	1990	2000	Percentage Change 1961–2000
Bananas	10,590	11,708	13,307	13,892	16,463	55.5
Beef	11	15	17	19	21	90.9
Cashews	557	599	528	583	593	6.5
Cassava	7,406	8,486	9,129	10,300	10,611	43.3
Cocoa	288	380	377	474	459	59.4
Coffee	464	433	481	537	698	50.4
Corn	1,943	2,351	3,155	3,679	4,274	120.0
Cotton	858	1,038	1,200	1,632	1,670	94.6
Oil Palm	3,771	4,639	6,981	9,982	12,224	187.6
Oranges	13,151	15,683	18,132	15,845	17,330	31.8
Rice	1,867	2,377	2,745	3,539	3,897	108.7
Rubber	546	646	693	784	888	62.6
Salmon	NA	NA	NA	NA	NA	NA
Shrimp (farmed)	#	#	#	#	611	#
Sorghum	890	1,129	1,200	1,369	1,381	55.5
Soybeans	1,129	1,480	1,600	1,898	2,176	92.7
Sugarcane	50,268	54,765	55,302	61,629	64,071	27.5
Tea	720	771	799	1,117	1,302	80.8
Tobacco	1,052	1,237	1,349	1,534	1,610	53.2
Wheat	1,089	1,494	1,855	2,561	2,737	151.6
Wood Pulp	#	#	#	#	#	#

Source: FAO 2002.
Note: # indicates data not available.

possible to improve productivity for trees, especially plantation species. There has also been work on technology to improve cultivation efficiency as well as harvesting and reducing post-harvest losses.

Some crops appear to have reached technological plateaus, as yields have not increased for the last fifteen to twenty years. Wheat yields in the United States and Mexico have leveled off. Rice production in Japan, Korea, and China is declining. Overall productivity of cereals is flat at best. In fact, while there is still potential to increase yields of wheat, rice, and corn, overall yields from breeding have not increased for the past thirty-five years. Most increases have been achieved as farmers have realized the yield potential of the varieties they cultivate (Tilman et al. 2002).

As yields hit ceilings, farmers and agrochemical companies try to find other ways to increase overall production. Pesticides, for example, allow farmers to plant the same crops year after year rather than using fallowing or crop rotation, both of which effectively reduce average production of high-value crops by half to two-thirds.

Plant breeding can improve productivity further. More important, it can help producers address many other issues that directly affect their profits. For example, developing varieties with resistance to troublesome diseases or insects can boost yields. Diseases reduce global production by an estimated 13 percent, while insects destroy another 15 percent and weeds reduce production by 12 percent. In all, some 40 percent of production potential is lost before harvest. After harvest, another 10 percent spoils or is lost to pests (Spector 1998). In sum, half of all production potential is lost.

As technology has become more important for increasing production, farmers have become more dependent on the companies that sell seeds and chemicals. With the advent of hybrid seeds eighty years ago, commercial farmers increasingly purchased their seeds from private companies rather than saving them from the previous harvest. These companies, in turn, capture an increasing share of the value in agricultural markets.

Unfortunately, the success of single interventions in agriculture is often transitory. Improving resistance to pathogens is a good example (Tilman et al. 2002). U.S. corn varieties now have a useful lifetime of four years, half of what it was thirty years ago. Similarly, within a decade or two of introduction, agrochemicals such as herbicides, insecticides, fungicides, and antibiotics lose their effectiveness because of resistance. Insects can develop resistance to agrochemicals within a decade or so. Pathogens can become resistant within one to three years.

According to numerous analysts, the gains in agricultural productivity in the late twentieth century were largely due to improved water use, prompting significant investments in water control (e.g., dams and storage facilities). India has invested more in water control than in any other activity, and China invests more than ten times as much in water control as in agricultural research (Huang et al. 2000).

Another technological trend within agriculture is the increasing reuse of waste (both on and off farm) to replace more expensive inputs. Orange peels, sugar bagasse, blood, bone meal, undigested matter from slaughtered animals, and residues from breweries are all effective sources of low-cost feed and/or soil amendments. Recycling wastes is laudable efficiency at many levels. It may come at a price, however, at least in some instances. Feeding not fully cooked body parts of animals to other animals led to mad cow disease, for example.

Spreading sewage sludge and livestock manure on farms may also expose soil to antibiotics, growth hormones, and other drugs that harm plants (Raloff 2002). The EPA reported that farmers spread 7 million metric tons of sewage sludge (called biosolids) and 3 million metric tons of animal manure on the soil each year (Raloff 2002). Much of the manure comes from feedlots where antibiotics are used more or less routinely. The Union of Concerned Scientists estimates that 2 million pounds of veterinary antibiotics are consumed by animals in feedlots each year and 27.5 million pounds of antibiotics are used as growth-promoting feed additives (Raloff 2002). The entire U.S. population, by contrast, consumes 4.5 million pounds of antibiotics annually. It takes eight days in a holding tank for 50 percent of antibiotics in manure and urine to break down. If forced air (which promotes more thorough aerobic decomposition) is used, then 70 percent can break down over the same period. Unfortunately, manure is not held this long (Raloff 2002). As a consequence, a very large amount of antibiotics is released into the environment, where it is quite likely that they will increase the resistance of bacterial pathogens.

Historically, many of the technological advances that increased agricultural production were supported by government. In developed countries, much of that support now comes from the private sector, but globally, governments still fund about two-thirds of all agricultural research, spending about $33 billion per year, or some 1.04 percent of the value of output in the mid-1990s. Support for such research has increased by 3.6 percent per year in developing countries but only by 0.2 percent per year in developed countries (Huang et al. 2002).

Genetically Modified Organisms

In the 1990s, companies like Aventis, DuPont and Monsanto transformed agriculture with a series of large biotechnology deals. These companies used technologies from the pharmaceutical sector to create new "transgenic" seed varieties (called genetically modified organisms or GMOs) with traits that could not be engineered through conventional breeding programs. Initially, these companies lacked the existing seed lines to use as building blocks, and access to farmers who would buy their products. Therefore, they (among others) spent $8.5 billion buying seed companies and creating joint ventures not only in the United States but also in England, India, South Africa, and Brazil. Seminis, a smaller Mexican company, through a series of acquisitions went from being a small player to a world leader in vegetable seeds.

By 1998, when 33 percent of the U.S. corn crop, 55 percent of the U.S. cotton crop, and 90 percent of Argentina's soybean crop were produced from transgenic seeds, this strategy appeared to be paying off. Even so, the distribution of the technology is still not widespread. Some 96 percent of all genetically modified crops are grown in the United States, Canada, and Argentina. They are also grown increasingly in Brazil, China, and South Africa. Some 5.5 million farmers in developing countries are thought to be using genetic modification technologies (Huang et al. 2002).

Both developed and developing countries have largely been interested in the development of varieties that are insect-resistant or herbicide-tolerant or both (Huang et al. 2002). Only 1 percent of field trials for GMOs in developing countries focused on higher yields. There is some evidence that this strategy is also paying off. Herbicide-resistant soybeans in Argentina have reduced per-hectare costs of production through reduced herbicide use (Qaim and Traxler 2002). Chinese cotton farmers, using a genetically modified strain to produce the biological pesticide *Bacillus thuringiensis* (Bt), have reduced pesticide sprayings for the Asian boll worm from twenty to six times per year. These farmers can thus produce a kilogram of cotton for 28 percent less than the cost to a farmer using non-Bt varieties. Similar cost reductions have been reported in Mexico and South Africa (Huang et al. 2002). U.S. growers using Bt-modified corn saved from $7.00 to $36.25 per hectare ($2.80 to $14.50 per acre) (Carlson 1997). However, if the price of genetically modified products declines due to consumer resistance, then producer savings may be a moot point.

By 1999 consumers in Europe and Japan had made it clear that they did not want GMO crops. By 2002 even American consumers began to express concerns about GMOs, and some food manufacturers (Nestlé, IAMS pet foods, Gerber, Heinz, and Frito-Lay) and retailers (e.g., McDonald's) decided not to use GMOs in their products. Several large European grain millers and traders told wheat industry leaders that they would stop buying wheat from North America if genetically modified wheat were allowed on the market; this move was subsequently echoed by Japanese and American grain millers as well (Cummings 2002). Grocery store chains are beginning to require that their meat suppliers guarantee that they do not use feeds that include genetically modified ingredients. There is, too, an increasing consensus that manufacturers should be required to label their products so consumers can tell which ones contain GMOs.

The interest in (or opposition to) labeling genetically modified food ingredients hinges, in part, on liability issues. Without labeling it is impossible to tell which products contain GMOs, thus minimizing liability exposure starting at the retail level and including manufacturers, refiners/processors, and the major grain trading companies as well. In short, when corporations do not know whether there are genetically modified ingredients in their products, they can claim plausible deniability if ever questioned by consumers. This explains why the efforts to block labeling of genetically modified ingredients are in the United States, where consumers are more litigious. Swiss Re, one of the largest reinsurers, refuses to insure any risks associated with genetically modified food. Some insurance companies have refused to insure any biotech firms against risks associated with genetic modification at any cost (Zepeda 2001).

PRODUCT SUBSTITUTION

Throughout history, food crops have gained or lost popularity based on how easily they can be produced, how expensive they are, or how durable they are. Throughout the past four centuries, seeds and cuttings of food crops from all over the world have been shared with producers. Today, most farmers throughout the world have found food crops that are suited to their growing conditions and that are culturally acceptable. With improved transportation systems and increasing global trade, these food products now compete with each other, often even on local markets. This process is called product substitution. Some products are nearly perfect substitutes (cane sugar and beet sugar). Other products are functional substitutes even though their individual properties may be slightly different (e.g., corn oil for soybean oil or one kind of meat for another).

When different products have similar product characteristics but different prices, there is a much greater chance of product substitution. For example, less expensive canola oil may be purchased instead of the more expensive olive oil. In the end, however, price changes for any single product are limited by the total supply not only of that product but of all substitute products as well.

Product substitution tends to stabilize and even lower prices. As a result, it tends to lead to more single-crop, large-scale production systems, since lower prices are more damaging to small farmers.

Parallel forms of crop substitution occur from a production point of view. Producers, for example, will grow more of one product that has an increasing market and less of another for which the market is declining. In some cases, substitutes are grown in different climates or habitats and therefore have different environmental costs. For example, palm oil substitutes for many vegetable oils, but it is best suited to the moist tropics. When the demand for palm oil increases rapidly, as it has for the past twenty years, this creates incentives for producers to move into tropical forests and convert them to oil palm plantations.

Some crops are substituted for each other as equivalents. Others are substituted for each other over time as tastes and consumer preferences change. Table 1.2 suggests that from 1961 to 2000, the area used to produce fruit and vegetable oil crops has increased more than any other type of crops, implying increases in overall markets for these products. The area utilized for the production of roots and tubers, legumes, and cereals has increased very slightly, while that used for coarse grains (e.g., barley, oats, sorghum and corn/maize) and fiber crops has actually declined. Most of the total gains in production have resulted from increased productivity per hectare.

Table 1.3 shows that the cultivation of different categories of crops changes over time. In the case of declining production, the overall use of certain crops (e.g., oats) declined in absolute terms. In some cases, particular crops (e.g., rye) became less attractive than other substitutes. By contrast, other crops (e.g., canola, cowpeas, olives) have been planted in increasing acreage as price or demand has shifted in their favor.

A few examples from Table 1.3 illustrate these points. The production of oats, for example, declined globally as horses were replaced by machinery. More recently when oats were linked to a healthy diet demand increased, but the upsurge has not offset the loss from the declining market for animal feed. To meet U.S. demand, manufacturers had to look abroad for oat supplies. Rye has lost market share to other cereal grains such as wheat. As subsidies for sugar beets are reduced, beet producers are losing their markets to cheaper sugar from cane. Grapes and potatoes have declined in area planted but have increased phenomenally in productivity, implying that specialized producers have taken over more of the production. In addition, the number of varieties cultivated globally of both species have actually declined, indicating that substitution can also take place within a general category of food such as grapes. The area planted to olives has tripled in forty years as incomes have risen, in-

TABLE 1.2. Global Areas Planted by Type of Crop, 1961–2000
(in millions of hectares)

	1961	1970	1980	1990	2000	Percentage Change 1961–2000
Total cereals	648.23	675.86	717.34	708.16	672.11	3.7
Total coarse grains[+]	328.52	334.74	335.60	329.95	304.26	−7.4
Total fiber crops	38.77	40.95	40.92	37.58	35.03	−9.7
Total fruit	24.40	28.62	32.62	41.06	48.30	98.0
Total oil crops	113.54	131.90	161.98	184.10	222.32	95.8
Total legumes	63.70	63.97	60.71	67.78	67.76	6.4
Total roots & tubers	47.61	48.19	45.95	45.95	52.70	10.7

Source: FAO 2002.
Note: [+] Coarse grains include barley, oats, corn, and sorghum.

TABLE 1.3. Global Area Planted to Comparative Crops of Interest, 1961–2000
(in millions of hectares)

	1961	1970	1980	1990	2000	Percentage Change 1961–2000
Canola	6.28	8.21	10.98	17.59	25.72	309.6
Coconuts	5.23	6.69	8.75	10.04	11.56	121.0
Cowpeas	2.19	5.31	3.15	5.10	9.87	350.7
Grapes	9.33	9.10	9.25	8.02	7.67	–17.8
Oats	38.26	30.68	24.53	20.59	12.85	–66.4
Olives	2.61	3.39	5.13	7.48	8.05	208.4
Peanuts (groundnuts)	16.64	19.49	18.36	19.69	24.29	46.0
Potatoes	22.15	20.77	18.76	17.59	19.94	–10.0
Rye	30.25	19.23	16.11	16.61	9.75	–67.8
Sugar beets	6.93	7.59	8.87	8.66	5.97	–13.9

Source: FAO 2002.

creasing demand for olive oil instead of cheaper vegetable oils. The area planted to cowpeas has nearly quadrupled in the same period because it is a fast-maturing crop, a cheap source of vegetable protein, and it requires little water or other inputs and therefore can be produced in much of the world. In short, cowpeas have performed well and have been planted by farmers as a food crop instead of other crops that had been grown in the past.

FROM FARM TO SUPERMARKET

Globally, consumers are sending clearer signals than ever before about what they want in their food—higher quality as well as healthier, safer, and tastier products. This is sending signals throughout the market chain stretching from the consumer, to the grocery store, to the food manufacturer, to the farm. Most companies in the food industry are currently exploring different ways to insure that they have more control over the production processes for agricultural commodities as well as overall product quality. Certification is one way to do this, as is ownership of an increasing portion of the market chain. In some instances companies are developing their own producer guidelines, which producers who want to sell products to them are required to follow.

Some consumers are turning to organic products to meet their desire for healthier, safer, and tastier products. In 2000 the global market for certified organic products reached $20 billion, about 40 percent of it in the United States. Organic food is the fastest growing food sector, but it still has a long way to grow before it captures significant market share. For example, in Denmark where organic production has the largest overall market share, it still represents only 4 percent of sales.

Even though organic products cost more, consumers seem to be willing to pay for them. In general the unit cost of organic production is higher than for conventional agriculture. In the United States, for example, organic soybeans are twice the price of conventional soybeans. Farmers in the United Kingdom reportedly receive twice as much for organic wheat as conventional wheat, but much of the income is a sub-sidy from the government. One thing is clear: As organic markets grow, the price paid to producers will fall, as with conventional agriculture. Put another way, today's conventional product price will be tomorrow's organic price premium. And the scale of the individual units of production will increase.

Electronic or e-commerce food sales are becoming more common. Rabobank in the Netherlands estimates that by 2003 some 10 percent of the world's $4 trillion in agriculture production will be traded on-line. In 1999 Walnut Acres, one of the largest organic food processors and distributors in the United States, increased its sales directly to consumers from 7 percent through catalogues to 27 percent over the Internet in only nine months (David Cole, personal communication). So far, how-ever, electronic business-to-consumer sales have been far less than expected.

The main on-line sales in the food industry, as in most others, are between busi-nesses. This is because electronic sales are far more efficient than other types of sales, which can involve more human error. The major issue that must be addressed is how to make sales more efficient while still being able to track products back to the producer so that issues of production quality, production practices, and pesticide residues can be addressed sufficiently to meet consumer concerns. In short, while it is increasingly important for each product to have a clear chain of custody, the product must also be efficiently stored and retrieved, handled, and sold. While two-dimensional bar codes can provide a few thousand bits of information, at this time there is no easy way to use a bar code on bulk commodities. In fact, the very thing that makes food products commodities—their ease of movement and substitutability—makes them difficult to trace back to the source. It is likely that be-nign markers may be developed to mix with commodities, but so far that has not hap-pened. Demand, however, will very likely cause effective tracing systems for com-modities to be developed very soon.

Horizontal and Vertical Integration

Since the 1990s, there has been tremendous consolidation in all aspects of the food industry market chain—from input suppliers to producers to processors, manufac-turers, and retailers. Consolidation of these industries has tended to concentrate power in the hands of fewer players.

According to *The Economist* (2000), much of the pressure on agrochemical and seed companies to drop or at least label GMO products has come from a smaller number of more important retailers as a result of the growth and consolidation of the retail industry. For example, five companies control two-thirds of the grocery busi-ness in Germany. Some 70 percent of all Swiss now shop in a single chain every week. The two largest French supermarkets have merged. Nearly half of the largest supermarkets in the United States have been involved in buyouts or mergers since 2000.

The larger the retailer, the more leverage it has over its suppliers, not just on price but also on quality, timing of delivery, and even conditions of production. Because of improved product codes and electronic checkout scanners, retailers now have exact information about consumer purchasing patterns and not just survey data. This information can be used to convince manufacturers to comply with consumer preferences. Depending on the product (e.g., nuts, fresh fruits or vegetables), retailers can sometimes buy directly from producers.

In addition to retail-level consolidation there has also been consolidation in food manufacturing and distribution. In 2002 Unilever bought Bestfoods, Philip Morris/Kraft announced its intentions to buy Nabisco, and General Foods announced its intention to buy Pillsbury. The top ten food companies in the world are listed by total sales in Table 1.4.

An interesting fact about the consolidation of these food companies is that they are all publicly held, which means that they have the additional challenge of pleasing their shareholders as well as their customers. This is a significant shift from the smaller, family-owned businesses that dominated the food industry after World War II. This also affords consumers as well as environmental organizations another significant point of leverage within the industry.

Reduction in Number of Middlemen

Consolidation is also occurring among middlemen, all the different people who handle a product between the producer and the retailer. The number of middlemen—and how significant they are for holding, moving, or transforming the item into subproducts or manufactured products—varies tremendously depending on both the agricultural product in question as well as where it is produced and consumed in the world.

As Table 1.5 demonstrates, the U.S. market for the key agricultural commodities is dominated by only twenty-one firms. Each of the commodities is controlled by only four firms, with the market share held by those four firms ranging from a low of 45 percent for turkeys to 81 percent for beef packing and 80 percent for crushed

TABLE 1.4. International Food Companies Ranked by
Annual Sales, 1999

Company	Annual Sales
Nestlé	$35.1 billion
Kraft/Nabisco*	$34.9
Unilever/Bestfoods*	$32.4
General Mills/Pillsbury	$12.6
PepsiCo	$11.6
Groupe Danone	$9.8
ConAgra/International Home Foods*	$9.6
H.J. Heinz	$9.4
Sara Lee	$8.0
Kellogg	$7.0

Note: * indicates pending mergers.

TABLE 1.5. Concentration of U.S. Agricultural Markets in Four Largest Companies

Market Share	Broilers (50%)	Beef Packers (81%)	Beef Feedlots* (>50%)	Pork Packers (59%)	Sheep Packers* (70%)	Turkeys (45%)	Flour Milling (61%)	Soybean Crushing (80%)	Dry Corn Milling* (57%)	Wet Corn Milling* (74%)
Tyson Foods	x	x		x						
ConAgra	x	x	x	x	x	x	x		x	
Gold Kist	x									
Pilgrim's Pride	x									
Cargill		x	x	x		x	x	x		x
Farmland Beef		x								
Continental			x							
Cactus Feeders			x							
Smithfield				x						
Superior Packing					x					
High Country					x					
Denver Lamb					x					
Hormel						x				
Pilgrim's Pride						x				
ADM+							x	x	x	x
General Mills							x			
Bunge								x	x	
AGP#								x		
IL Cereal Mills									x	
Tate and Lyle										x
CPC										x

Source: Heffernan 1994; Hendrickson and Heffernan 2002.
Note: * indicates 1994 data; all other data for 2000.
+ Archer Daniels Midland
# Ag Processing, Inc.

soybeans. The same three firms (Cargil, Archer Daniels Midland, and Zen Noh) account for 81 percent of all U.S. corn exports and 65 percent of all soybean exports (Hendrickson and Heffernan 2002). Grain-trading giant Cargil recently bought its rival Continental, giving the company 42 percent of all corn and 33 percent of soybean exports from the United States. About 80 percent of all cattle are slaughtered by only four companies. Similarly, four other firms crush 80 percent of the soybeans for oil. Four firms handle 50 percent of all broiler chickens. Concentrations can occur by controlling breeding lines and in other ways as well; 90 percent of all commercial turkeys globally come from only three genetically different breeding flocks (Kimbrell 2002). Firms such as Smithfield, International Beef Processors (IBP), and Archer Daniels Midland (ADM) are not household names but they control much of the initial processing as well as the market chain between producer and retailer.

The firms that dominate the markets tend to do so for a number of different key commodities. For example, ConAgra is one of the four dominant purchasers and resellers of eight different commodities. Cargil and ADM dominate seven and four commodities, respectively. In other parts of the world the concentration of power in just a few firms is even greater, and in a number of instances the firms are the same as those companies listed in Table 1.5. For example, Dean Foods is the largest dairy processor in the United States, more than twice the size of its nearest competitor Kraft, but it was bought by Suiza in 2001 (Hendrickson and Heffernan 2002).

Some companies are attempting to enter all the different aspects of a particular agricultural sector, except for taking on the direct risk of the producer. For example, a grocery store chain might contract with farmers to produce meat or fresh produce while eliminating all the intermediaries that would normally buy and resell the product within the distribution chain. This is vertical integration. Other firms have opted for a strategy of dominating specific value-added manufacturing sectors of the food industry. For example, Table 1.5 indicates that the same companies tend to dominate the meat processing industry—including beef, pork, and poultry—in the United States.

While 1.4 billion people in the world depend on farm-saved seed as their primary source for planting, this is not true in all countries or for all commodities. Until the past decade, the world's $20 billion seed industry was highly fragmented. Much of the fragmentation was based on the fact that different companies were working on different crops in different parts of the world. But within any one country, seed supply may be highly concentrated. In the United States, for example, 40 percent of all vegetable seed is produced by one company, 75 percent of all cereal seed by five companies, 73 percent of all corn by two companies, and 70 percent of all cotton by one company. Only 47 percent of soybean seed is produced by four companies, while farmers save 25 percent of the soybean seed that is planted (Kimbrell 2002).

Many agrochemical suppliers are vertically integrated. ConAgra, the largest distributor of agricultural chemicals in North America, is also one of the largest fertilizer producers. In 1990 the company bought its way into the seed business. It owns more than 100 grain elevators (both local and terminal), 2,000 railroad cars, and 1,100 barges. ConAgra is the largest of the three firms that mill 80 percent of the wheat in North America. It is also the largest turkey producer and the second largest broiler producer in the United States. It owns and operates poultry hatcheries. In addition, it produces its own poultry feed as well as other livestock feed. It markets chicken and turkey under its Country Skillet brand and processed meat under its Banquet and Beatrice Foods brands. It also owns the Swift Butterball, Hunt's, Peter Pan, and Orville Redenbacher brands.

But ConAgra is only the second largest food processor in the United States and the fourth largest in the world. Globally, the largest food processor is Nestlé, which is followed by Philip Morris, the largest food processor (and of course the largest tobacco producer) in the United States. In 1994, Philip Morris owned such brands as General Mills, Kraft Foods, Miller Beer, Louis Rich Turkeys, and Oscar Meyer. Ten cents of every dollar spent on food in the United States went to Philip Morris (Heffernan 1994).

Increasingly, the most significant issue for producers in the food industry is vertical integration. Vertical integration poses two main issues for producers. First, it is often accompanied by a reduction of players in the market, giving producers fewer options. Second, vertical integration allows companies to control the quality of production and institute chain-of-custody monitoring from producer to consumer. This is important not only because of consumer preferences and concerns, but also because of government regulations. Vertical integration allows companies to create or match technological developments with consumer concerns. As markets become more vertically integrated, the number of producer guidelines increases. These are the production practices that producers are required to follow in order to sell into specific markets. The practices are not necessarily environmentally sensitive, however. Producer guidelines do not necessarily encourage the creation, much less the adoption, of better practices. In fact, they often require the prophylactic use of inputs such as pesticides to prevent problems rather than to address them as they arise, and they often require increased specialization as well.

Strict compliance with producer guidelines makes sense for companies that want to ensure a steady supply of uniform food. Such guidelines can guarantee predictable results and reduce any risks of pesticide residue or other potential liabilities. While these systems have been the most common for contract farmers (both for plant and animal crops), they are becoming more widespread. Most chickens are raised according to producer guidelines in vertically integrated operations in developed countries, and pork production is moving in the same direction.

In many parts of the world, farmers are organizing themselves. This is not new. In the past, many farmers' organizations focused on selling products in bulk and buying fertilizers and other agrochemicals in bulk. Most failed because of poor management. Today, however, more cooperatives are being developed as vertically integrated businesses. In this way, farmers can add value to their products and sell them more directly to consumers. To do this effectively, producers need competent business managers, and they need to listen very carefully to consumers' wants and concerns.

A number of the trends outlined in this chapter are pushing agriculture into ever more remote areas on an ever increasing scale. Product substitution and international trade make what would appear to be different commodities compete with each other in the market place (e.g., palm oil and soy oil). This means that the productivity and cost of production of one commodity can affect the price for other commodities as well. Increasingly, producers do not merely have to compete with other producers of the same commodity, but also with producers of any other commodity that can be substituted. In addition, the WTO's interpretations to date regarding international trade indicate that subsidies such as producer payments and market barriers such as tariffs will soon be a thing of the past, which will amplify the effects of substitution. Similarly, technological improvements in resource use, overall farm management, and manipulation of crop material (whether through genetic engineering or traditional breeding programs) will also tend to push the frontiers of where agriculture can be undertaken profitably.

By the early twentieth century, the global agricultural frontier had already expanded over most of the temperate areas with good growing conditions. Many of those areas were relatively quickly abandoned because they were not suited for sustained agriculture with the technologies of the day. This was true of large parts of Europe. It was also true of the more mountainous and forested areas of the eastern United States and the drier areas of the western Great Plains. Not all land is suitable for farming. The environmental costs of farming in the wrong places with the wrong methods are quite high both for farmers and for society as a whole, as when the 1930s dust bowl put an end to cultivation in the western Great Plains. Unfortunately, many of these lessons have still not been learned in many parts of the world.

Throughout the world, most of the land that is best suited for agriculture is already in use. Further expansion is likely to occur on marginal lands that will not support sustained production and so will be quickly degraded and abandoned. The level of degradation makes such land expensive to rehabilitate, so it is usually cheaper to expand into natural habitat.

At this time, the agricultural frontier is expanding into many of the last remaining tropical forests. Much of the Amazon and the Brazilian cerrado (the flat tableland of forests and savannas in the interior of the country bordering the eastern Amazon) is succumbing to peasant agriculture, cattle ranching, soybean production, and several cash crops that are produced until the soil is exhausted and the farmers move on to clear more land. In Southeast Asia the expansion of first rice and rubber, then cocoa and coffee, and now oil palm and pulp plantations has contributed to the loss of both pristine and degraded forests. In West and Central Africa this same cycle of first oil palm and rubber and then coffee and cocoa plantations is also responsible for the conversion of large tracts of natural forests.

There are some encouraging signs that such expansion is not inevitable. These signs come from many different parts of the world and from many different actors. At

the farm level, producers are finding a number of innovative ways not only to continue to farm the same land over a long period of time but also to improve production and income at the same time. While they are still definitely in the minority, the
number of farmers using innovative farming practices are increasing. For example,
some Brazilian farmers have found that they can convert degraded pasture into highly productive soybean, corn, and cotton rotations within five to six years by applying
no-till practices that increase the organic matter in the soil. Their efforts allow them
to increase their assets even more rapidly than they increase their income from soybean production. For example, degraded land is valued at $400 to $500 per hectare.
Land that can produce soybeans is worth $2,000 per hectare. Producers reclaiming
degraded land can increase their assets by up to $300 per hectare per year over the
five to six years that it takes to rehabilitate the land. Improving the value of the degraded land can earn farmers more than the net value of the soybeans or other crops
that they produce on it. Brazilians currently farm about 60 million hectares and are
converting their different forests into agricultural land at a rate of about a million
hectares per year. An additional 80 million hectares of land have been abandoned or
degraded. Much of this land should never have been converted from natural habitat
to agriculture. However, if even 15 percent could be reclaimed for agricultural use,
Brazil's current rate of agricultural expansion could be sustained for twenty years
without needing to clear a single hectare of natural habitat. If productivity is increased on each hectare, then the rate of expansion of cultivated land could be
slowed even more and total production would still increase.

Encouraging signals are not just coming from farmers. While most governments
recognize the need to maintain stable, cheap food supplies, an increasing number of
politicians are beginning to question whether the current system of subsidies and
market barriers is the best way to do that. They are beginning to ask, for instance, if
this is a government payment for a social good, why not pay for improved environmental or social performance rather than to maintain the income level of producers?
The income levels of farmers could be maintained by payments for maintaining watershed quality or carbon sequestration just as easily as for cotton, corn, or soybeans.

Similarly, many within and outside government are beginning to question
whether it would be more effective to use government regulation to encourage innovation in addition to insuring compliance with minimal standards. As it is, the focus is on what could be called the Tiger Woods approach to government: finding a
new stick of a different size to solve every problem. Many are asking whether government can also have carrots in its bag of solutions.

Finally, the private sector itself is beginning to send messages to agricultural producers about what it wants and does not want. As a result of recent food scares in Europe, many of the world's largest food companies are now beginning to work with
their suppliers at all levels to insure product quality and safety. Inevitably this implies
not just product testing but also setting up production systems in which there is reduced risk of product quality being compromised in the first place. Many of these
companies are developing their own producer guidelines. There is a tremendous po-

tential to reduce environmental and social costs through the appropriate development and adoption of such guidelines.

Changes in investment strategies also have the potential to reduce the environmental and social costs of agriculture. It is now apparent to most investors that companies that are better managed, even with the same financial rating, will have better returns both in the short term and over time. This is true of agriculture and aquaculture as much as any other business. For that reason investors and insurers are interested in developing screens that will allow them to evaluate the management capacity of potential borrowers or claimants to reduce risk.

There is more momentum now from all sides—from the producers themselves as well as consumers, buyers, investors, insurers, and governments—to reduce the social and environmental costs of agriculture while making it more financially viable. So if everyone is interested in this, what is holding it up? Simply put, it is the boundaries to existing ways of thinking put up by producers, food companies, banks, and governments. People are trying to address aspects of this problem where they see them, where they have easy access to make some changes, and where they have the most leverage. Unfortunately, not enough people are working across political and commercial boundaries to bring about an effective transformation of agriculture.

Serious environmental issues remain to be addressed, as the next chapter will discuss. There are serious social and political issues as well. As the Oromo people of Ethiopia say, "You can't wake a person who is pretending to sleep." People can no longer ignore the problems conventional agriculture poses to life on Earth or the likely scenarios of where it is headed if left to business as usual.

When I was a child in Missouri, people used to say, "If you don't know where you're going any road will get you there." Producers, consumers, corporations, and regulators may not agree on every goal or technique, but they can surely agree on where they do not want to go. This book is an attempt to create a discussion about where agriculture should go and some of the ways that might be used to get there.

REFERENCES

Baker, L. 2002. The Not-So-Sweet Success of Organic Farming. *Salon.com.* (San Francisco, CA) July 29. Available at http://www.salon.com/tech/feature/2002/07/29/organic/print.html.

Becker, E. 2002. A New Villain in Free Trade: The Farmer on the Dole. *Washington Post* (Washington, D.C.). August 25. Page 10.

Carlson, G. A., M. C. Marra, and B. J. Hubbell. 1997. Transgenic Technology for Crop Protection. *Choices.* Third quarter. Pp. 31–36.

Clay, J. and K. von Moltke. 2002. Changes in World Agriculture and Sustainable Rural Development: An Essay. 30 pp.

Constance, D. H., and W. D. Heffernan. 1991. The Global Food System: Joint Ventures in the USSR, Eastern Europe, and the People's Republic of China. Paper presented at the annual meeting of the Midwest Sociological Society. Des Moines, Iowa. April.

Cummins, R. 2002. The Death of Frankenfoods: Nailing the Coffin Shut. *BioDemocracy News No. 40.* Organic Consumers Association. Available at http://www.organicconsumers.org.

Dev, S. M. 1998. Regional Disparities in Agricultural Labor Productivity and Rural Poverty. *Indian Economic Review* 23(2):167–205.

DFID (Department for International Development). 2002. Better Livelihoods for Poor People: The Role of Agriculture. *DFID Report No. 7946*. London: Department for International Development.

Economist. 2000. Growing Pains: A Survey of Agriculture and Technology. London. 16-page insert. March 25.

———. 2001. Towards Efficient Farm Support. London. June 9. Pp. 12–13.

———. 2001. Patches of Light. London. June 9. Pp. 69–72.

ENDS (Environmental Data Services). 2000. Farming's Environmental Costs Top £1.5 Billion per Year, Says Agency. Report #309, October.

FAO (Food and Agriculture Organization of the United Nations). 2002. *FAOSTAT Statistics Database.* Rome: UN Food and Agriculture Organization. Available at http://apps.fao.org.

Goldberg, G. 2000. Submitted Comments to the U.S.D.A. Advisory Committee on Agricultural Biotechnology. April 27.

Hawken, P., A. Lovins, and L. H. Lovins. 1999. *Natural Capitalism: Creating the Next Industrial Revolution.* Boston: Little, Brown and Company.

Heffernan, W. D. 1994. Agricultural Profits: Who Gets Them Now, And Who Will in the Future? Paper presented at the Fourth Annual Conference on Sustainable Agriculture. Iowa State University, Ames, Iowa. August 4.

Hendrickson, M., and W. Heffernan. 2002. *Concentration of Agricultural Markets.* February. Denver, CO: National Farmers Union. Available at http://www.nfu.org/documents/01_02_Concentration_report.pdf.

Hines, C. 2002. Export Drive Is Sending Poor on Wrong Route. *The Guardian.* 10 June.

Huang, J., C. Pray, and S. Rozelle. 2002. Enhancing the Crops to Feed the Poor. *Nature* 418 (8 August): 671–677. London: Macmillan.

Kimbrell, A., ed. 2002. *Fatal Harvest: The Tragedy of Industrial Agriculture.* Washington, D.C.: Island Press.

Lydersen, K. 2002. On Farms, a No-Till Tactic on Global Warming. *Washington Post* (Washington, D.C.). August 26. Page A-7.

Nagarajan, N. 1999. *The Millennium Round: An Economic Appraisal.* European Commission Economic Paper No. 139. November. Available at http://europa.eu.int/comm/economy_finance/publications/economic_papers/economicpapers139_en.htm.

OECD (Organisation for Economic Co-operation and Development). 2002a. *Agricultural Policies in OECD Countries: Monitoring and Evaluation.* Paris: OECD Publications Service.

———. 2002b. Agricultural Policies in OECD Countries: Producer Support Estimate by Country. Database available at http://www.oecd.org/dataoecd/39/62/2674689.xls.

Pauli, G. 1998. *Upsizing: The Road to Zero Emissions, More Jobs, More Income, and No Pollution.* Sheffield, England: Greenleaf Publishing.

Paden, M. 2002. Ecoagriculture: Blending Parks and Farms. *Human Nature* 7(1). Washington, D.C.: Academy for Educational Development.

Pimentel, D., C. Harvey, P. Resosudarmo, K. Sinclair, D. Kurtz, M. McNair, S. Crist, L. Spritz, L. Fitton, R. Saffouri, and R. Blair. 1995. Environmental and Economic Costs of Soil Erosion and Conservation Benefits. *Science* 267 (24 February): 1117–1123.

Pollan, M. 2001. How Organic Became a Marketing Niche and a Multibillion-Dollar Industry. *The New York Times Magazine.* May 13: 30–65.

Power, C. 2002. Planting New Seeds. *Newsweek.* July 22.

Qaim, M. and G. Traxler. 2002. Roundup Ready Soybeans in Argentina: Farm Level, Environmental, and Welfare Effects. Paper presented at the 6th ICABR Conference on Agricultural Biotechnologies: New Avenues for Production, Consumption and Technology Transfer. Ravello, Italy. July 11–14.

Raloff, J. 2002. Pharm Pollution: Excreted Antibiotics Can Poison Plants. *Science News* 161 (26): 29 June: 406–407.

Smil, V. 2000. *Feeding the World: A Challenge for the Twenty-First Century.* Cambridge, MA: MIT Press.

Thirtle, C., L. Beyers, L. Lin, V. McKenzie Hill, X. Irz, S. Wiggins, and J. Piesse. 2002. *The Impact of Changes in Agricultural Productivity on the Incidence of Poverty in Developing Countries.* DFID Report No. 7946.

Tilman, D., K. G. Cassman, P. A. Matson, and R. Naylor. 2002. Agricultural Sustainability and Intensive Production Practices. *Nature* 418 (8 August): 671–677. London: Macmillan.

UNCTAD (United Nations Conference on Trade and Development). 1999. *World Commodity Survey, 1999–2000.* Geneva, Switzerland: UNCTAD.

UNCTAD (United Nations Conference on Trade and Development)/WTO (World Trade Organization). 2002. The WTO Director General Releases the 2001 Annual Report: Overview of Developments in the International Trading System. *World Tr@de Newsletter.* November. Geneva, Switzerland: UNCTAD/WTO.

World Bank. 2000. *World Development Report 2000.* Washington, D.C.: World Bank.

Zepeda, L. 2001. Don't Ask, Don't Tell: U.S. GM Food Labeling Policy. Paper presented at American Association for the Advancement of Science, San Francisco, CA. 18 February.

AGRICULTURE AND THE ENVIRONMENT

2

In October 1998 Hurricane Mitch struck Honduras, Nicaragua, and Guatemala. The storm killed an estimated 10,000 people and caused damages of $5.5 billion to the local economy. Throughout the region agricultural lands were devastated, but not all farmers suffered to the same degree. Conventional farms using chemical-intensive monoculture practices had 60 to 80 percent more soil erosion, crop damage, and other water-caused losses than those farms that practiced more conservation-oriented forms of agriculture such as polyculture, crop rotation, biological pest control, water conservation, terracing, strip cultivation, and agroforestry. The reason? There are likely several, but among the key factors is undoubtedly the fact that more conservation-oriented farming techniques, though less productive than high-input monocropping systems, are more likely to preserve the integrity and biodiversity of the landscape. This in turn renders it more resilient to otherwise catastrophic events.

Agriculture, like any other use of natural resources, has an environmental impact. Some impacts, however, are more acceptable than others. Many agricultural practices today carry high costs both to society and to producers, and they reduce the long-term viability of agriculture. The decaying health of agricultural lands is only occasionally as dramatic as it was in the aftermath of Hurricane Mitch; more often it manifests as a slow loss of productivity, a gradual shift in the crops that the land can support, or the loss of agricultural lands to other purposes altogether, such as urbanization.

In the United Kingdom, 15 percent of all agricultural land has been lost to urbanization. In the future, the highest losses of land due to urbanization and population growth are likely to happen in China, India, Nigeria, Pakistan, Bangladesh, Brazil, and Indonesia, where half of the increase in world population will occur in the next generation.

Of course, the loss of agricultural land and productivity is not just a contemporary problem. It is ancient: Most civilizations collapsed because they destroyed their natural resources, particularly their soil (Hawken et al. 1999). This is true of the Indus Valley civilization, Babylonian, Egyptian, Mayan, and many others. The importance of good soil was well understood even in Roman times. In 146 B.C. after winning the Third Punic War, the Romans deliberately salted the fields around Carthage so that they would not produce grain again.

The question today is whether countries can avoid salting their own fields with unsustainable practices and thus avoid the fates of those civilizations. Several key environmental factors must be addressed if that is to happen. These include maintaining soil fertility, reducing conversion of natural habitats and the associated loss of biodiversity, reducing pollution (especially from agrochemicals), increasing the efficiency of water use, minimizing climate change by controlling production of heat-trapping ("greenhouse") gases, and finding ways to cut energy use. In addition, it is essential to evaluate the environmental impacts of new technologies such as gene splicing.

SOIL

Sustainable agriculture is really about soil. Soil is more complex than the human brain, and we know even less about it. Soil can reduce or create greenhouse gases, crumble bedrock, and purify all the fresh water on the planet. Healthy soil biota can facilitate some ten times the nutrient uptake and equal or greater biomass production as degraded soils with only a tenth of applied solid nutrients (Kimbrell 2002).

Soil is alive. There is more biodiversity and biomass within soil than there is on top of it. More microbes live in a teaspoon of soil than people on the planet. In the top few centimeters of one square meter of rich healthy soil one can find up to 1,000 ants, spiders, wood lice, beetles, and larvae; 2,000 earthworms, millipedes, and centipedes; 8,000 slugs and snails; 20,000 pot worms; 40,000 springtails; 120,000 mites; and 12 million nematodes (Kimbrell 2002).

A key indicator of soil quality and health is organic matter, which in effect is dead biomass—the decaying remains of plants, animals and animal wastes, and microbes. Organic matter is essential for both the maintenance of soil fertility and structure. (Soil structure is what allows rainwater to soak in and plant roots to penetrate; once soil loses its structure it can become bricklike and impervious to water and most plant roots.) Organic matter in soil declines as a result of the conversion of natural habitat to arable land. Carbon typically declines by 30 to 50 percent in just one or two decades after conversion to field crops. Aeration from cultivation speeds up decomposition rates of existing organic matter in the soil. In addition organic matter is removed in the form of harvested crops rather than being returned to the soil, so it is not replaced at the same rate as in natural ecosystems.

Organic matter is mostly made up of carbon, so soil carbon content is a good indicator of soil fertility. Studies in Michigan indicate that every 1 percent gain in soil carbon content results in a more than 20 percent increase in potential yield. The average carbon content of undisturbed soil is about 2 percent. It falls below 1.5 percent in the first stages of cultivation, and it is less than 0.5 percent in severely degraded soils. Dead biomass (organic matter) makes up 90 percent of all soil carbon. The remainder is made up of living roots, bacteria, fungi, and soil invertebrates.

From the middle of the nineteenth century to the middle of the 1990s—some 150 years—humans converted close to 1 billion hectares of forests, grasslands, and wetlands to farmlands. In virtually every instance, soil erosion rates increased many

them with the gradual release of much-needed fresh water. Large enough forests can even have a distinct impact on local climate.

The wholesale conversion of natural habitat—whether forest, savanna, wetlands, or other—inevitably results in the loss of biodiversity and ecosystem functions. In the United States since the 1930s, depending on the region, from 30 to 80 percent of edge habitat and natural waterways have been removed as farms have gotten bigger (Kimbrell 2002). The disappearance of these woodlots, hedgerows, windbreaks, and grass-covered waterways, plus the draining of wetlands and the channeling of many streams, has eliminated the last vestiges of natural habitat on many farms. These changes, designed to create more uniformity within and between farms, have reduced or eliminated the breeding, foraging, and migration routes of many species. They have also resulted in increased soil erosion. This process has happened in many other parts of the world as well. In addition to habitat conversion, habitat degradation is occurring worldwide from shortened fallow periods, soil erosion, and poor water and nutrient management.

Converting natural habitat and using intensive monocrop farming techniques has turned some species into pests (Kimbrell 2002). This is true of native as well as introduced species—grackles, blackbirds, red-winged blackbirds, sparrows, crows, rabbits, pigs, deer, cane toads, guavas, cheat grass, crabgrass, etc. By providing vast areas of food and by eliminating the diversity that would support competition from a wider array of species, modern agriculture has reduced the number of species that can live in large areas of the landscape while allowing a few to become dominant.

Modern agriculture is reducing the genetic diversity of food crops as well. There are some 7,000 crop species that are available for cultivation, but 90 percent of the world's food comes from only thirty of these. Most domestication and crop-selection programs have focused on higher yields, pest resistance, and fast-growing crop varieties. These varieties now dominate over half of all the land planted to rice, corn, and wheat. The same general trend is true for livestock as well. A sixth of some 3,800 breeds of domestic animals common a century ago no longer exist. This loss in diversity makes all of agriculture more vulnerable to diseases and climatic changes. Agriculture that is vulnerable is likely to have larger environmental costs because as crops fail in one area demand will stimulate production in others.

Agriculture also has substantial impacts on freshwater and marine habitats. For those who are concerned about the global environment, the connections between agriculture and marine sources of pollution and biodiversity loss have been increasingly clear. The dead zone in the Gulf of Mexico (a large and growing area that does not support life) is caused by Midwestern agricultural pesticide runoff. Similar dead zones in the Baltic are also caused by agricultural runoff. Before reaching the sea, this same runoff has already destroyed the habitat of countless freshwater rivers and streams. The main threats to the Great Barrier Reef off the coast of Australia and the reefs off the east coast of Central America are agricultural in origin. In both of these reef areas suspended solids from erosion and pollution from agrochemicals are the main threats to coral reefs and all the species dependent on them. In short, the impact of agriculture is not just on terrestrial habitats or the fields that are cultivated from them but also on the adjacent freshwater and marine habitats that are polluted.

Direct habitat loss and modification are not limited to more traditional forms of crop and livestock production. Today, the fastest-growing form of food production is aquaculture. Shrimp and tilapia aquaculture have caused considerable damage to fragile coastal areas (e.g., wetlands, mangroves, mudflats, salt flats, estuaries, etc.). Large-scale commercial marine aquaculture affects not only the resident biodiversity of the area, but also everything that migrates through the area. Salmon aquaculture, the first of many forms of open ocean net-cage aquaculture, has been shown to have a tremendous impact on benthic communities (the organisms that live in the sediments on the bottom of the ocean), food chains, and particular species through the spread of disease.

EFFLUENTS AND POLLUTION

Once input-intensive agricultural production systems are in place, one of their most damaging impacts comes from the use of agrochemicals. The damage depends on the nature of the crop, its requirements, and the physical and biological environment in which it is grown. In addition, the scale of production and the information available to the farmer also influence the use of agrochemicals. There are, for example, different types of fertilizers and different ways to use them, just as there are different ways to fight pests even among farmers producing the same crop. Some practices are preferable to others because they reduce environmental damage.

In the United States, fertilizer use in 1946–47 was 800,000 metric tons. The next year use had exploded to 17 million metric tons as a result of the perceived opportunity to feed many parts of the world whose agriculture had been disrupted after World War II (Kimbrell 2002). Globally, from 1960–1995 nitrogen fertilizer use increased sevenfold and phosphorous use increased 3.5 times (Tilman et al. 2002). Unless there is a dramatic increase in the efficiency of fertilizer use or unless the costs increase, usage of both will increase another threefold by 2050.

Unfortunately, the effectiveness of fertilizer applications diminishes over time. Synthetic fertilizers can reduce the ability of soil to produce or make nutrients available to plants. The application of concentrated forms of nitrogen by farmers reduces the activity of nitrogen-fixing bacteria in the soil, and it increases the populations of other organisms that feed on nitrogen. As the balance of organisms changes over time, applications become less effective. In the United States in 1980, 1 metric ton of nitrogen produced 15 to 20 metric tons of corn. By 1997 the same amount of fertilizer produced only 5 to 10 metric tons of corn (Kimbrell 2002).

Moreover, there is evidence that only 30 to 50 percent of all nitrogen and 45 percent of phosphorous applications contribute to the growth of the target crop (Tilman et al. 2002). In the early 1990s U.S. farmers applied 56 percent more nitrogen to their crops than was present in the crops harvested (Hawken et al. 1999). Excess nutrients are easily leached from soil by rain, which carries them into streams and lakes. This is especially true for soils low in organic matter, which is the case in much large-scale agriculture. A study by the U.S. Environmental Protection Agency (EPA) shows that about 72 percent of rivers studied and 56 percent of lakes studied are polluted from agriculture. Agriculture is also the main cause of groundwater pollution (see Box 2.1). Nitrates are the main contaminants, followed by pesticides (U.S. EPA 1994, as cited in Soth 1999). A recent study in Europe concluded that agriculture is the main cause of phosphorus pollution in the coastal zones of Mediterranean countries (Ongley 1996).

BOX 2.1. SOME QUICK FACTS ON AGRICULTURAL PESTICIDES

- Amount of all pesticides used each year on cotton: 25 percent.
- Number of pesticides presently on the shelves that were registered before being tested to determine if they caused cancer or birth defects, or were toxic to wildlife: 400.
- Amount of time it takes to ban a pesticide in the United States using present procedures: 10 years.
- Number of active ingredients in pesticides found to cause cancer in animals or humans: 107.
- Of those active ingredients, the number still in use today: 83.
- Number of pesticides that are reproductive toxins according to the California Environmental Protection Agency (EPA): 15.
- Number of pesticides found to cause reproductive problems in animals: 14.
- Most serious cause of groundwater pollution confirmed in California: agricultural chemicals.
- Number of pesticides found in drinking wells of California since 1982: 68.
- Number of California wells affected: 957.
- Number of California farming communities affected: 36.
- Percentage of the total U.S. population supplied with drinking water from groundwater: 50%.
- Number of different pesticides documented by the U.S. EPA to be present in groundwater in 1988: 74.
- Number of states affected by pesticide contamination of groundwater: 32.
- The most acutely toxic pesticide registered by the EPA: aldicarb, used frequently on cotton.
- Percentage of total aldicarb applied in California that was used on cotton: 85 to 95%.
- Number of states in which aldicarb has been found in the groundwater: 16.
- Percentage of U.S. counties containing groundwater susceptible to contamination from agricultural pesticides and fertilizers: 46%.
- Number of people in the United States routinely drinking water contaminated with carcinogenic herbicides: 14 million.
- Percentage of municipal water treatment facilities lacking equipment to remove these chemicals from the drinking water: 90%.
- Estimated total costs for U.S. groundwater monitoring: $900 million to $2.2 billion.
- Estimated costs for U.S. groundwater carbon filtration cleanup: up to $25 million per site.
- Percentage of all food samples tested by the U.S. Food and Drug Administration in 1980 that contained pesticide residues: 38%.

- Of the 496 pesticides identified as likely to leave residues in food, the percentage which FDA tests routinely detect: 40%.
- Average number of serious pesticide-related accidents between World War II and 1980: 1 every 5 years.
- Average number of serious pesticide-related accidents between 1980 and the present: 2 every year.
- Increase in cancer rates between 1950 and 1986: 37%.
- Number of Americans who will learn they have cancer this year: 1 million.
- Number who will die from it: 500,000.
- Cost of cancer to the United States in terms of lost production, income, medical expenses, and research resources: $39 billion per year.
- Highest rate of chemical-related illness of any occupational group in the United States: farm workers.
- Pesticide-related illnesses among farm workers in the United States each year: approximately 300,000.
- Number of people who die each year from cancer related to pesticides: 10,400.

Source: Monsanto 1999.

Some 135 million metric tons of fertilizer are used each year around the world. However, as the impacts of the overuse of fertilizers have become understood, the use of synthetic fertilizer over the past decade has declined per hectare in the developed world. Most fertilizer use is now in developing countries (Kimbrell 2002).

There is good evidence that delaying and reducing the rates of fertilizer application can reduce overall costs and pollution without hurting yields. The efficiency of nitrogen fertilizer use in U.S. agriculture has increased 36 percent since 1980 (Tilman et al. 2002) as a result of public expenditures on research and investments by farmers in soil testing and improved timing of fertilizer applications. In addition, the development and utilization of crops with higher nutrient-use efficiency has improved nitrogen utilization rates. In general, matching a plant's demand with the timing and location of applications has been the overall key to improvements. In some cases, the issue is the relative impact of specific production systems. For example, nitrogen runoff is thirty-five times higher from corn and soybean fields than from pasture. Feeding cattle on grass is one way to reduce overall nitrogen pollution in ecosystems (Hawken et al. 1999).

The use of nitrogen raises other, more fundamental, environmental issues. Artificial nitrogen fertilizers and planted legume crops around the world have doubled the amount of nitrogen in terrestrial ecosystems from a background level of some 110 million metric tons (Tilman et al. 2002; Kimbrell 2002). Similarly, agricultural applications of phosphorous have doubled its availability in the environment. In addition to these applied fertilizers, runoff from intensive livestock operations (cattle, pigs, or chickens) pollutes both aboveground and underground freshwater sources with excess nutrients. The most dramatic impact of these excess nutrients is eutrophication of freshwater ecosystems. Eutrophication involves the explosive growth of algae that can kill fish and other aquatic organisms. Algal blooms are common in the Chesapeake Bay, the Gulf of California, the Gulf of Mexico, and off the coast of

China. Many of these blooms are the result of nitrogen and phosphorous from farms being carried in runoff into streams and rivers.

In addition to fertilizers, there has been a tremendous increase in the use of pesticides. The global use of pesticides is perhaps 5 million metric tons per year, having more than doubled over the past thirty years. In the United States pesticide use nearly tripled from 215 million pounds in 1964 to 588 million pounds in 1997 (Kimbrell 2002). Pesticide use has increased so dramatically because it is required for continuous monocrop production systems. Pesticides allow the reduction or even the elimination of crop rotation, so that farmers can grow large areas of the same valuable crops year after year without rotating the crops with lower-value ones.

Hewitt and Smith (1995) of the Henry Wallace Institute (an Arkansas-based, nonprofit sustainable agriculture research and education organization) cite more than fifty different studies in the United States and Canada alone that document the adverse effects of agricultural pesticides on bird, mammal, and amphibian populations. David Pimentel (1999) has estimated that globally some 672 million birds are affected by pesticides each year and that some 10 percent of these die. From 1977 to 1984 half of all fish kills off the coast of South Carolina were attributed to pesticide poisoning (Kimbrell 2002).

The most toxic "dirty dozen" pesticides (which include DDT and aldrin) are banned or severely restricted in the United States, Europe, and Japan but are still commonly used in many developing countries. Recent treaties on persistent organic pollutants (POPs) will phase out the most harmful pesticides throughout the world. These pollutants, which include many pesticides, not only persist in the environment, they also concentrate within the food chain, causing reproductive, developmental, and immune-system problems in both humans and animals.

Resistance to chemical pesticides is growing among the organisms they are designed to kill. There is evidence that their effectiveness may decline as their use increases. For example, in the United States in 1948 some 50 million pounds of insecticides were used each year and about 7 percent of the preharvest crop was lost to insects. By 2001 nearly 1 billion pounds of insecticides were used, but estimates suggest that insects destroyed as much as 13 percent of the crop (Hawken et al. 1999).

Another often overlooked but important cause of land and water pollution is the improper disposal of by-products and waste generated during production and processing of agricultural crops. Agricultural waste and by-products are often heaped and left to rot where they are created. They can become breeding grounds for pests. Sometimes they are dumped into rivers, where they absorb oxygen as they decompose, which in turn can asphyxiate fish and other aquatic organisms. In other places the age-old practice of burning is used to dispose of agricultural waste, but this leads to air pollution. While agricultural wastes are still a problem in many parts of the world, they are also a potential resource. Reclamation of these wastes for commercial composting operations and creation of other uses and markets for agricultural waste and by-products could reduce pollution, increase or maintain organic matter levels and overall soil fertility, increase farmer income, and reduce spending on fertilizers and other inputs.

Finally, there is the problem of pollution generated by aquaculture. Aquaculture is, by definition, a water-based industry. All inputs, waste, etc. are found in the water. It is difficult to control pollution in land-based pond systems, and impossible to control it in open ocean systems. Depending on the form of aquaculture, the most significant effluents are wasted feed and excrement, dead animals, and molted shells (from growing shrimp). In addition, however, medications are applied directly to aquatic animals, in the feed, or in the water. A number of chemicals and fertilizers are used to condition ponds or to increase food production.

WATER

Due to global increases in population and irrigated agriculture, freshwater withdrawal increased more than sixfold during the last century (from 579 to 3,750 cubic kilometers per year). Demand is expected to increase to 5,100 cubic kilometers per year by 2025 (Shiklomanov 1993, 1998). The increased demand has brought great benefits to agriculture; from the mid-1960s to the mid-1980s, irrigation was responsible for more than half of the increase in world food production (Hawken et al. 1999).

But fresh water supplies are running out. According to the Global Water Project based in Amherst, Massachusetts, water is being pumped out of the ground faster than it can be replenished. The annual depletion of aquifers, due mostly to irrigation, is 163.6 cubic kilometers (Kimbrell 2002). The main cause is irrigated agriculture in America, North Africa, the Arabian Peninsula, China, and India. Most of the irrigation is inefficient and so much of the water is wasted. In addition, agriculture is now in competition with other human water needs, most notably in urban areas.

Water usage can have greater impacts than the mere removal of fresh water from the ecosystem. From 1950 to 1980 more than 35,000 large dams were built. While the largest of these dams were built to generate electricity, the smaller ones were mostly built for irrigation and livestock. A dam can be as damaging to downstream ecosystems as any other impact of agricultural production.

Globally, the area of irrigated agriculture has increased steadily from 47.3 million hectares in 1930 to 254 million hectares in 1995 (Kirda 1999; Shiklomanov 1998, as cited in Soth 1999). While the total area under irrigation is still increasing, the per capita area under irrigation has declined by 5 percent since 1978 (Tilman et al. 2002). Without increased efficiency in water use, previous production gains from irrigated agriculture may well be lost as agriculture competes with other water users in a world with less per capita water available. In most of the world, the impacts of previous and current agricultural practices (e.g., the elimination of watersheds, the loss of organic matter in the soil to retain and release water slowly, and the use of finite water sources in aquifers) are responsible for per capita and in some cases absolute declines in water available for irrigation.

About 40 percent of the world's food is produced on the 16 percent of agricultural land that is irrigated (Tilman et al. 2002). Three crops account for 58 percent of all irrigated land: rice (34 percent), wheat (17 percent), and cotton (7 percent).

Globally, the agricultural sector is responsible for about 69 percent of all freshwater withdrawal, more than twice the amount of industrial, municipal, and all other users combined. Some 75 percent of all irrigated land is in developed countries. In these countries 73 percent of all fresh water is used for agricultural irrigation, but in some countries the share is as much as 98 percent. In Asia agriculture accounts for 86 percent of freshwater use and in Africa 88 percent. Much of this irrigation is inefficient. At least 60 percent of water used for irrigation is wasted. As Table 2.1 indicates, 1 kilogram of rice can be produced using as little as 1,900 liters of water, but in some regions as many as 5,000 liters are used to produce the same quantity.

The efficiency of water use varies from region to region and from crop to crop (Gleick 2000). In the United States, for example, per-unit food production of the same crop requires twice as much water as in Asia because U. S. production is less efficient (Kimbrell 2002). In addition, animals require far more water than plants. For example, one kilogram of corn requires 800 to 2000 liters of water and 1 kilogram of beef up to eighty times that amount or as much as 70,800 liters of water. Daily water use by livestock in developed countries can be quite high—milk cows 154 liters per day, steers 51 liters per day, pigs 9 liters per day, sheep 3 liters per day, and chickens 0.3 liters day (Gleick 2000).

Aquatic animals can require even greater amounts of water than land-based animals. It can take as much as 300,000 liters of brackish water to produce 1 kilogram of shrimp from aquaculture. With brackish water the issue isn't so much the water use per se but all the associated costs of inefficient use—pumping, aeration, the "down" time of conditioning new water until it can be stocked with animals, and treating the nearly constant flow of effluents. Some shrimp farmers can reduce their use of brackish water to less than 500 liters per kilogram of shrimp produced, but the trade-off is that they must compensate by using more energy for aeration.

One of the main reasons that water is used so inefficiently for agriculture is that the real cost of water is much higher than most farmers pay. Other water users tend

TABLE 2.1. Freshwater Requirements for Different Agricultural Products

	Water Requirement by Area (liters/m^2)	Water Requirement per kg of Product (liters/kg)
Potatoes	350–625	500–1,500
Wheat	450–650	900–2,000
Rice	500–950	1,900–5,000
Sorghum		1,100–1,800
Soybeans	450–825	1,100–2,000
Sugarcane	1,000–1,500	1,500–3,000
Chicken		3,500–5,700
Cotton	550–950	7,000–29,000
Beef		15,000–70,000
Shrimp aquaculture	1,000–100,000*	1,000–300,000*

Sources: Soth 1999; Pimental et al. 1999; Tuong and Bhuiyan 1994; FAO 2002; Gleick 2000.
Note: * indicates brackish water.

to pay much higher prices than agricultural users for water from the same sources. An open market for water through auctions or transparent, competitive bidding would increase the price of water to farmers and increase the efficiency with which they use it.

There are known technical solutions to reduce water use through conservation. Some new hybrids require less water than previous ones. For example, some rice varieties can reduce water needs by 20 to 30 percent. If combined with improved irrigation and management systems, water use for rice and other agricultural products can be reduced by as much as 50 percent (Gleick 2000). Sprinkler nozzles are 95 percent efficient. Drip irrigation and improved water management (e.g., demand-driven water supplies) are two effective ways to reduce water use by 30 to 70 percent over conventional flooding systems, and they can increase yields by as much as 20 to 90 percent (Kimbrell 2002). Israel has developed drip irrigation systems that take advantage of both brackish water (focusing on halophytes that prefer brackish water) and wastewater that has been kept separate from sewage. However, only 0.7 percent of worldwide irrigated areas use drip irrigation because given the current value of water, it is cheaper to use it inefficiently than it is to buy all the necessary equipment to deliver and use it more efficiently. The main issue will be to find incentives that encourage other producers to adopt the same techniques before they waste too much more fresh water.

Very little is done to capture and use rainfall in agriculture. Great gains could be made by learning to use rainfall more efficiently by directing its on-land flow so that more can be absorbed into the soil or stored. For example, more water runs off bare soil; soil that is protected by mulch or crop residues slows the rate of runoff. Many small-scale producers use cisterns that catch water draining off roofs plus other forms of ponds and embankments to catch and store rainwater for use another time. Similarly, water-absorbing substances are available that can be mixed into the soil to hold water and release it gradually. While agriculture, even truck gardening, makes little use of these at present, they can reduce watering needs by 50 percent or more. Although these are expensive inputs for the extensive production of low-valued crops, they would be viable for most horticulture and even perennial crops, particularly where water is more scarce or more expensive.

CLIMATE CHANGE

Agriculture is an important contributor to climate change through the release of carbon dioxide and other so-called greenhouse gases into the atmosphere. According to some sources, agriculture contributes about a quarter of the total risk of altering the Earth's climate (Hawken et al. 1999). One estimate suggests that soil microbes produce 85 percent of atmospheric greenhouse gases (Kimbrell 2002). The production of such gases increases when soils are disturbed and when nutrients are added (in short, when modern agriculture is practiced). Soil conservation measures can reduce the production of greenhouse gases by 16 to 42 percent (Kimbrell 2002).

Temperate farmland can have as much as twenty to thirty times the biomass below the surface as above it. Soil carbon can exceed 110 metric tons per hectare. Bad agricultural practices tend to mobilize this carbon by converting it to carbon dioxide, which escapes into the atmosphere. At this time some 2 billion hectares of agricultural lands are low in carbon as a result of current agricultural practices. Globally, the net loss of soil carbon from agriculture from these lands is estimated to have contributed 7 percent of all atmospheric carbon (Hawken et al. 1999). Carbon emissions are also caused by production of agricultural chemicals such as fertilizers and pesticides. Other greenhouse gases also have their origin in agriculture. For example, most nitrous oxide (N_2O)—a greenhouse gas hundreds of times more potent than carbon dioxide—is produced from the interaction of synthetic fertilizers and soil bacteria (Hawken et al. 1999).

Much of the planet's methane (CH_4) emissions comes from the production of livestock and continuously flooded rice paddies (Wassman et al. 2000). One estimate places total methane emissions from rice at some 10 to 15 percent of total global methane emissions (Wang et al. 2000). Other sources suggest that the total methane emissions from rice represent 5 to 30 percent of global emissions. An estimated 1.3 billion cattle produce some 72 percent of all livestock-generated methane (Crutzen et al. 1986, as cited in Hawken et al. 1999).

Ozone depletion is another factor in global climate change. One pesticide alone, methyl bromide, is generally considered responsible for 5 to 10 percent of the Earth's total ozone depletion (Kimbrell 2002). In California a third of the 15 million pounds of methyl bromide used each year is for strawberry production. In other states it is used primarily on tomatoes or potatoes.

Besides contributing to global climate change, agriculture will also be affected by it. Weather extremes and climate variability caused by climate change will tend to limit production. In addition, land use and land use change affect climate change, and it will affect them as well.

Current crop production is taking place within an atmosphere of increasing concentration of carbon dioxide. If all other production variables remained constant, increasing carbon dioxide levels would increase crop yields. Unfortunately, all other variables will not remain the same. Increasing carbon dioxide levels cause global warming. In addition, increasing the concentration of carbon dioxide causes partial closure of plant stomata (the small openings in plant leaves that control the flow of air), which in turn decreases evaporative cooling and can cause leaf temperatures to exceed air temperature (Shafer 2002).

Soybeans provide a good example of how complicated it is to predict the agricultural impact of global warming. Doubling of carbon dioxide in the atmosphere at a constant temperature could increase plant mass by 50 percent and seed yield by 30 percent. However, this increased level of carbon dioxide brings global warming. Soybean seed yields decrease about 10 percent for each Celsius degree above 30 degrees. Declining yields result from fewer and smaller seeds, reduced stores of carbohydrates within seeds, different ratios of fatty acids, and increased vitamin E (Shafer 2002).

Increases in atmospheric carbon dioxide will cause all plants—crops and weeds alike—to increase pollen production (Shafer 2002). For example, experimental carbon dioxide enrichment from 280 parts per million to 600 parts per million increases ragweed reproduction nearly 4.5 times. It is likely that some weeds and other pests may be able to outperform crops under increased carbon dioxide conditions. This could have significant implications for sustainable agriculture with reference to the use of herbicides and insecticides.

Another major impact of climate change will be uncertainty (Shafer 2002). The distribution of crop pathogens and pests depends on climate, as outbreaks are influenced by weather. Variability will most likely increase due to climate change. In addition, host/parasite interactions are known to depend on the concentrations of sugars and other chemicals in plants, and these, too, will be affected by climate change.

It is clear from the evidence to date that the impact of climate change on agricultural production will not be spread evenly around the world. For example, by 2025 if global carbon dioxide increases to 405–460 parts per million and global mean temperature increases by 0.4–1.1 degrees Celsius, then cereal crop yields will likely increase in many mid- and high-latitude regions but will decrease in most tropical and subtropical regions. By 2050 if carbon dioxide increases to 445–640 parts per million and temperature increases by 0.8–2.6 degrees Celsius, then there will be mixed impacts on cereal yields in mid-latitude regions and more pronounced cereal yield decreases in tropical and subtropical regions. However, if by 2100 carbon dioxide increases to 540–970 parts per million and global mean temperature increases by 1.4–5.8 degrees Celsius, then there will be reduced production throughout the world (Shafer 2002).

Proposed solutions for global climate change could affect agriculture. For example, increasing overall soil carbon could have broad environmental benefits on and off the farm while increasing overall agricultural production. By contrast, calls to produce more biomass on farms to substitute for as much as 10 percent of total petroleum needs will take land from crop and animal production and, unless performance is better than the current norm, degrade the land and require the use of agrochemical inputs to maintain production.

Soil is an excellent source of sequestered carbon, and many see a potential to use improved farming practices as a way to take atmospheric carbon and store it in the soil. Globally, there are about 1,500 gigatons of carbon in soils and about 770 gigatons in the atmosphere. Carbon sequestration programs on agricultural lands could have a triple benefit. First, they would help offset global warming. If properly designed, sequestration programs could also provide additional income for farmers. Finally, by enriching the soil with carbon they would increase the productivity of existing farmland, thereby reducing the pressure to expand into other areas. Most agricultural areas, however, are not net carbon sinks (areas that gain more carbon than they lose each year), so if carbon sequestration programs are to provide income for farmers, improved practices are needed or land will have to be taken out of agricultural production and dedicated to carbon sequestration. This could happen either because the price of sequestered carbon warrants it or because governments decide it is important enough to pay for.

ENERGY

Globally, the food sector uses some 10 to 15 percent of all energy consumed in developed countries, and even more in the United States (Hawken et al. 1999). If that energy is derived from fossil fuels, it also contributes to increased atmospheric carbon dioxide and therefore to global climate change.

Much of the energy use in agriculture is in the processing, manufacturing, and food distribution systems. For example, it can take 40 calories of energy to ship 1 calorie of lettuce from California to the East Coast, or 240 calories of energy to ship 40 calories of strawberries from Chile to the United States.

A recent study in Germany suggests that the production of 0.24 liters (a typical cup) of strawberry yogurt entails 9,093 kilometers (5,650 miles) of transportation. The amount of transportation is more than doubled if one includes distribution. Germans eat 0.7 billion liters (3 billion cups) a year (Hawken et al. 1999). In the United States, the food for a typical meal has traveled nearly 2,092 kilometers (1,300 miles) (Kimbrell 2002), but if that meal contains off-season fruits or vegetables the total distance is many times higher.

Production of agricultural chemicals accounts for 2 percent of all industrial energy use, and synthesis of nitrogen fertilizers alone is thought to require half of all energy used for high-yield crops in developed countries. One estimate suggests that a ton of coal is required to create every 2.27 kilograms (5 pounds) of synthetic nitrogen fertilizer (Kimbrell 2002). Between 1910 and 1983 corn yields in the United States increased 346 percent. During the same period, however, energy consumption for agriculture increased 810 percent (Kimbrell 2002).

Other sources of energy consumption in agriculture typically include tilling, planting, applying fertilizers and pesticides, harvesting, and drying. In the United States, for example, crop drying accounts for 5 percent of total on-farm energy use.

Many opportunities exist for improving energy efficiency in agricultural production. While U.S. farms have doubled their direct and indirect energy efficiency since the 1970s, it still can take up to ten times as much energy to produce food as the energy value of the food itself (Hawken et al. 1999). In the Netherlands it requires 100 times as much energy to produce hothouse tomatoes as the energy contained in those tomatoes. Some 75 percent of the energy is used to heat the greenhouses and another 18 percent to process and can the tomatoes. It would require only a third the amount of energy to produce the tomatoes in Sicily and ship them by air to the Netherlands (Hawken et al. 1999). The margins for producers are so small in some agriculture industries that small changes can have a surprisingly large impact. One study found that North Carolina chicken farmers could increase net farmer income by 25 percent simply by switching from incandescent lighting to compact fluorescent lamps (Hawken et al. 1999).

GENETICALLY MODIFIED ORGANISMS

The commercial planting of genetically modified or transgenic crops began in 1996 with around 2 million hectares. In 1997 the area planted to genetically modified organisms (GMOs) increased to more than 10 million hectares. In 1999 about 40 million hectares of genetically modified crops were grown in a dozen countries. By 2001 more than 52 million hectares were planted to GMOs globally. The prediction is that more than 74 million hectares will be planted to genetically modified crops by 2006 (Freedonia, as cited in *The Economist* 2003). While soybeans, corn, cotton, and canola have dominated the area of crops planted to date, the first genetically modified rice will be planted in China in 2003. Transgenic wheat is just around the corner, possibly as early as 2004.

The production of genetically modified crops has been dominated by the United States (71.9%), Argentina (16.8%), Canada (10%), and China (0.75%), primarily because U.S. corporations dominate the technology. Between 1996 and 2001, sales of transgenic soybeans increased from $11 million to $1,090 million; of corn from $15 million to $544 million; and of cotton from $35 million to $480 million (Verdia 2003). Most of the crops were bred to resist herbicides (e.g., Roundup-ready soybeans) or to produce their own insecticidal proteins (e.g., corn and cotton bred to produce their own *Bacillus thuringiensis*, or Bt). Increasingly, however, varieties are being developed for other purposes, including some that require less fertilizer, fewer pesticides, and less water. This latter factor is particularly important for producers in arid environments.

The implications of GMOs for people and the environment are much debated, as they are neither black nor white. As described in Chapter 1, consumers in several countries have made it clear that they do not want such products, despite any possible benefits. Herbicide application per hectare appears to have fallen on genetically modified crops compared to conventional ones, but overall more herbicide is being used because more marginal land has been brought into production and is being planted to genetically modified crops. In addition, resistance appears to be building to some pesticides commonly associated with genetically modified crops, so GMOs may not entirely fulfill their promise of freeing the agricultural sector from the treadmill of developing new varieties before the old ones lose their effectiveness. While this may be good for seed companies, it will not be good for farmers who will find that an even higher percentage of their income goes to pay for research and development.

The net benefit to farmers generated by GMOs is difficult to calculate. In many areas, GMOs allow the use of no-till cultivation practices, in which existing vegetation is killed with herbicides and a new crop is sown through the dead vegetation and the previous year's crop residues. No-till techniques reduce soil erosion and also allow marginal land to be cultivated that would otherwise be susceptible to high erosion rates. Yields across the board have not increased, but yields for specific crops such as Bt corn appear to have increased on average. Where productivity increases, prices may decline and offset most of the potential gains. What will happen, of course, is that this will be another factor that will affect farmers' overall cost of production. To the extent that this allows some producers to lower their overall costs, higher-cost producers will go out of business without government subsidies.

Research supported by the FAO in China suggests that cotton farmers who adopt and use better practices in general improve their yields and reduce their use of inputs by comparison to conventional producers. Similarly, those planting Bt cotton reduce their use of inputs by comparison to conventional producers. However, the use of pesticides and other inputs is decreased most (and net profits are increased most) when producers use improved practices such as integrated pest management (IPM) in conjunction with planting Bt cotton (Peter Kenmore, personal communication).

Aside from the health concerns that some consumers have about GMOs, there are other concerns as well. The GMOs that allow the use of broad-spectrum herbicides will tend to reduce most biodiversity in the field. However, some pests will develop resistance and, over time, this will pose the same problem of generating "superbugs" or "superweeds" as broad-spectrum antibiotics do for human health. By 1997, for example, eight insect pests in the United States had already developed resistance to Bt (Hawken et al. 1999).

There are also concerns that genetically modified crops might interbreed with wild relatives in the field, neighboring fields or even in nearby unfarmed areas and that the added genetic material might contribute to the formation of new pests. In addition, Bt products actually kill neutral or beneficial insects. Bt corn can stunt or kill Monarch butterfly caterpillars. Perhaps more important, but less symbolic, Bt corn can be lethal to green lacewings—important natural predators of the corn borer that Bt corn was developed to combat.

While the impact of "escaped" plant varieties of GMOs can be debated, the impact of genetically modified animals, particularly aquatic ones, is of far more concern. The recent development of transgenic salmon for aquaculture production has brought this issue into sharp focus. It is known that salmon escape from net-pen aquaculture no matter how many protocols have been put in place. So far, the industry has not been allowed to culture such salmon on a commercial scale because of concerns of how escapes might affect wild populations. But hearings are currently under way in several countries that would allow the use of such animals.

It is clear that some genetically modified crops can be produced in such a way as to reduce the aggregate environmental damage from agriculture, at least in the short term. This means that more food and fiber can be produced with genetically modified crops than with other varieties with less damage to the soil (either in terms of erosion or fertility) or to downstream freshwater and marine environments. It appears, however, that producers will be required to support the ongoing research necessary to ensure the development of new varieties in a timely way. Fortunately, the technology allows for the development of new varieties relatively easily by comparison to traditional plant and animal breeding programs. However, the biggest issue is what the genetic modification technology will do to the soil environment and the web of organisms that depend upon it. It is very difficult to predict such impacts when most soil organisms have not yet been named, much less studied. For this reason alone, caution is suggested. More importantly, though, agricultural producers need to develop the capacity to monitor the health of the soil in order to truly understand the impact of GMOs on the environment.

BOX 2.2.   GMO ISSUES IN EUROPE IN MAY 2000

- A series of recent events in Europe will continue to constrain the ability of farmers to sell GMO products (and consequently their willingness to plant GMOs) throughout the world.
- The United Kingdom, Germany, Sweden, and France saw the accidental introduction of GM canola in seed that was labeled as non-GMO. This introduction by AVANTA, partially owned by AstraZeneca, has affected thousands of farmers.
- Traces of GM pollen have been found in honey produced in the United Kingdom. It is difficult if not impossible to sell tainted honey in the United Kingdom. The issue of who is liable is now being determined.
- The Welsh Assembly voted unanimously to ban the sale of all GMO products, or manufactured products that contain them, throughout Wales.
- A three-year study by the University of Jena in Germany found that genes from GMO crops could spread from plants into other forms of life. This will raise health and environmental questions about the safety of GMO products.
- These events reflect the increasing intensity of the GMO debate in Europe. Even before these issues became widely known, corn exports from the United States to Europe decreased from 2 million metric tons in 1997–98 to 137,000 metric tons in 1998–99. Soybean sales declined from 11 million metric tons in 1997–98 to 6 million metric tons in 1999. Japan has also refused to purchase grain if it cannot be guaranteed to be from non-GMO sources.

TOWARD MORE SUSTAINABLE AGRICULTURE

There is no doubt that agriculture has had a greater environmental impact on Earth than any other single human activity. The question is whether new kinds of agriculture can be developed that will produce the food needed to feed an increasing population and still accommodate all the other life forms on the planet.

It is also clear that business as usual in agriculture is not the solution. Conventional agricultural production technologies will not provide the food and fiber needed by populations in the future. Under most systems of agricultural production at this time, it is not a question of if, but rather when, virtually all of the natural habitat on the planet will become degraded to the point that it is no longer productive and then abandoned for future generations to find ways to rehabilitate and repair. Not only is it unfair to pass these costs on to future generations, it is unnecessary. In fact, it is cheaper to farm sustainably now than it ever will be to correct the problems created through unsustainable farming.

There is no single "right" way to practice more sustainable agriculture. Many farmers have found ways to reduce environmental damage, improve production, and increase profitability. How the farmers do this depends tremendously on where they live, what they produce, and where they sell their product. Broadly speaking, though, farmers are beginning to invent, adapt, and adopt a wide range of approaches that are usefully seen as "better management practices." Such practices involve maintaining and building soils, maintaining the natural ecosystem functions on farms, working with nature and not against it to produce products, reducing total input use and using inputs more efficiently, and reducing waste or creating marketable by-products from materials that were previously considered waste.

The ways in which farmers can improve their lots are not limited to production; they also involve market initiatives. Some farmers are experimenting with organic or other ecological labels to differentiate their products and enable them to charge more for them. Some are adding value to their production (e.g., bagging potatoes rather than selling in bulk, processing fruit into juice, selling meat directly to consumers), others are trying to become more vertically integrated (e.g., by buying equity in processing plants or distribution companies) into the market. Some specialize their production while others diversify to reduce risks. A number of specific examples of these new strategies are given in the different commodity chapters.

Likewise, producers around the world can benefit from appropriate government actions. These can include appropriate zoning and siting of operations in areas that are most suited to agriculture or aquaculture expansion, or it can include linking permits and operating licenses to the adoption of specific practices. The governments that help producers most in the future, however, are likely to be those that require certain performance levels but leave it up to the farmer to find the way that is best to achieve these on the specific property in question.

Similarly, buyers, retailers, investors, insurers, and consumers can all support producers who adopt better practices to reduce their environmental and social costs to more acceptable levels. This is not only good for the producer, it also reduces the risks and increases the profits of everyone else who is part of this system. And, of course, the consumer is the biggest winner of all by getting a reliable source of products that are grown with less damage to the environment and that contain fewer substances that may be harmful to human health.

Farmers have learned a lot about sustainable production. Unfortunately, farmers' business is farming, not teaching others to farm better. This book is an attempt to glean examples of lessons that farmers and others have learned around the world while trying to survive in an increasingly competitive business. The focus here is on commodities, because they are what virtually all farmers sell and would like to sell more of at higher prices. This is not to deny that farmers' strategies are complex and involve systems both of their own and others' making. Nor is the goal to provide blueprints for what should be done and how it should be done. The goal here is to help people understand how to think, not what to think. To paraphrase an old adage—give a person a solution and they will solve a problem, teach them to think and they will be able to solve problems that do not yet exist.

REFERENCES

Crutzen, P. J., I. Aselimann, and W. Seilor. 1986. Methane production by domestic animals, wild ruminants, other herbivorous fauna and humans. *Tellus*. 38B:271–284.

The Economist. 2003. Seeds of Change. The World in 2003. London: The Economist Newspaper Limited.

FAO (Food and Agriculture Organization of the United Nations). 2002. *FAOSTAT Statistics Database*. Rome: UN Food and Agriculture Organization. Available at http://apps.fao.org.

———. 2003. *AQUASTAT*. Rome: UN Food and Agriculture Organization. Available at http://www.fao.org/ag/aGL/AGLW/aquastat/main/index.stm.

Ghassemi, F., A. J. Jakeman, and H. A. Nix. 1995. *Salinisation of Land and Water Resources: Human Causes, Extent, Management and Case Studies*. Sydney, Australia: University of New South Wales Press, Ltd.

Gleick, P. H. 2000. *The World's Water 2000–2001*. Washington, D.C.: Island Press.

Hawken, P., A. Lovins, and L. H. Lovins. 1999. *Natural Capitalism: Creating the Next Industrial Revolution*. Boston: Little, Brown and Company.

Hewitt, T. I. and K. R. Smith. 1995. *Intensive Agriculture and Environmental Quality: Examining the Newest Agricultural Myth*. Washington, D.C.: Henry A. Wallace Institute for Alternative Agriculture.

Huang, J., C. Pray, and S. Rozelle. 2002. Enhancing the Crops to Feed the Poor. *Nature* 418 (8 August):671–677. London: Macmillan Publishers.

Kimbrell, A., ed. 2002. *Fatal Harvest: The Tragedy of Industrial Agriculture*. Washington, D.C.: Island Press.

Kirda, C., P. Moutonnet, and D.R. Nielson (eds). 1999. *Crop Yield Response to Deficit Irrigation*. Dordrecht, Netherlands: Kluwe Academic Publishers.

Lal, R. 1995. Global Soil Erosion by Water and Carbon Dynamics. In Lal, et al., eds. *Soils and Global Change*. Boca Raton, FL: CRC/Lewis Publishers.

Lal, R., J. L. Kimble, R. F. Follet, and C. V. Cole. 1998. *The Potential of U.S. Cropland to Sequester Carbon and Mitigate the Greenhouse Effect*. Chelsea, MI: Sleeping Bear Press.

Monsanto. 1999. Fact Sheet On Pesticide Use. Originally published at http://www.biotechknowledge.com/showlib_biotech.php32. Biotech Knowledge Center, Monsanto. Now available at http://journeytoforever.org/fyi_previous2.html, dated March 17, 2000.

Ongley, E. D. 1996. Control of Water Pollution from Agriculture. *FAO Irrigation and Drainage Paper 55*. Rome: UN Food and Agriculture Organization.

Pimentel, D., C. Harvey, P. Resosudarmo, K. Sinclair, D. Kurtz, M. McNair, S. Crist, L. Spritz, L. Fitton, R. Saffouri, and R. Blair. 1995. Environmental and Economic Costs of Soil Erosion and Conservation Benefits. *Science* 267 (24 February): 1117–1123.

———. 1999. Water Resources: Agriculture, Environment, and Society. *Biosciences* 47(2): 97–106.

Reddy, A. K. N., R. H. Williams, and T. B. Johansson. 1997. *Energy After Rio: Prospects and Challenges*. New York: United Nations Environmental Program.

Shafer, S. R. 2002. The Expected Impacts of Climate Change on Global Agricultural Systems. Powerpoint presentation. National Program Leader, Global Climate, U.S. Department of Agriculture, Agriculture Research Service, Beltsville, Md.

Shiklomanov, I. 1993. World Fresh Water Resources. In P.H. Gleick, ed. *Water in Crisis: A Guide to the World's Fresh Water Resources*. New York: Oxford University Press.

———. 1998. World Water Resources and World Water Use. Data archive on CD-ROM from the State Hydrological Institute, St. Petersburg, Russia.

Soth, J. 1999. *The Impact of Cotton on Freshwater Resources and Ecosystems: A Preliminary Syn-*

thesis. Fact Report (draft). C. Grasser and R. Salemo, eds. Zurich: World Wildlife Fund. 14 May.

Tilman, D., K. G. Cassman, P. A. Matson, and R. Naylor. 2002. Agricultural Sustainability and Intensive Production Practices. *Nature* 418 (8 August): 671–677. London: Macmillan Publishers.

Tuong, T. P. and S. I. Bhuiyan. 1994. Innovations Toward Improving Water Use Efficiency of Rice. Paper presented at World Bank Water Resources Seminar. December 13–15, Lansdowne Conference Center, Leesburg, VA.

US-EPA. 1994. *National Water Quality Inventory*. 1992 Report to Congress. EPA-841-R94-001. Office of Water, Washington, DC.

Verdia, Inc. 2003. *Products and Markets*. Available at http://www.verdiainc.com/products.php. Accessed 2003. Copyright 2003.

Wang, Z. Y., Y. C. Yu, Z. Li, Y. X. Guo, R. Wassman, H. U. Neue, R. S. Lantin, L. V. Buendia, Y. P. Ding and Z. Z. Wang. 2000. A Four-Year Record of Methane Emissions from Irrigated Rice Fields in the Beijing Region of China. *Nutrient Cycling in Agroecosystems*. 58:55–63.

Wassman, R., R. S. Lantin, and H.-U. Neue. 2000. Characterization of Methane Emissions from Rice Fields in Asia. I. Comparison among Field Sites in Five Countries. *Nutrient Cycling in Agroecosystems*. 58:1–12.

COMMODITIES

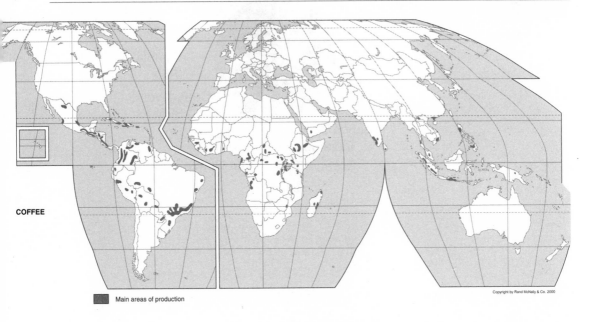

COFFEE

Main areas of production

Copyright by Rand McNally & Co. 2000

COFFEE *Coffea arabica, C. canephora*, and other species

PRODUCTION		INTERNATIONAL TRADE	
Area under Cultivation	10.6 million ha	Share of World Production	76%
Global Production	7.4 million MT	Exports	5.6 million MT
Average Productivity	698 kg/ha	Average Price	$1,510 per MT
Producer Price	$1,130 per MT	Value	$8,441 Million
Producer Production Value	$8,362 million		

PRINCIPAL PRODUCING COUNTRIES/BLOCS (by weight)	Brazil, Vietnam, Colombia, Indonesia, Mexico, Côte d'Ivoire, Guatemala
PRINCIPAL EXPORTING COUNTRIES/BLOCS	Brazil, Vietnam, Colombia, Indonesia, Côte d'Ivoire, Guatemala, Mexico
PRINCIPAL IMPORTING COUNTRIES/BLOCS	United States, Germany, Japan, Italy, France
MAJOR ENVIRONMENTAL IMPACTS	Conversion of primary forest habitat Soil erosion and degradation Agrochemical use and runoff Effluents from processing
POTENTIAL TO IMPROVE	Good Better practices known for both sun and shade-grown coffee Organic, shade-grown, and Fair Trade certifications exist Low prices driving harmful practices

---

Source: FAO 2002. All data for 2000.

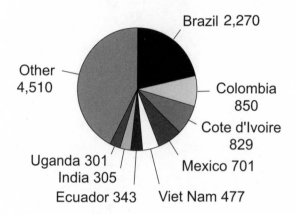

Area in Production (Mha)

Brazil 2,270

Other 4,510

Colombia 850

Cote d'Ivoire 829

Mexico 701

Viet Nam 477

Uganda 301
India 305
Ecuador 343

OVERVIEW

The coffee plant was originally found and cultivated by the Oromo people in the Kafa province of Ethiopia, from which it received its name. Around 1000 A.D., Arab traders took coffee seeds home and started the first coffee plantations. The first known coffee shop was opened in Constantinople in 1475, and the idea quickly spread to other parts of Europe. England's King Charles II raged against coffeehouses as centers of sedition because they were the meeting place of writers and businessmen. Lloyd's insurance company was started in the back room of a coffeehouse in 1689. In fact, coffee shops became centers of political and religious debate throughout the continent, and many were subsequently closed. The owners were often tortured.

Coffee first arrived in Europe from Turkey via overland trade routes. It is not known exactly when coffee first arrived, but it had probably been there some time before coffeehouses became common in the sixteenth and seventeenth centuries. It is possible that coffee was brought in along the same trade routes that were used to transport gold, valuable gums, and ivory from Africa and silk and spices from Asia. In any case, coffeehouses were already established in northern Europe with the sixteenth-century arrival of cocoa, which then spread quickly as another coffeehouse drink.

Over the centuries coffee has gone from a luxury to necessity. Globally, coffee consumption is increasing but not nearly as rapidly as production, so prices are decreasing. In 2002 real coffee prices reached historic lows. Many producers are abandoning coffee plantations; others are destroying them. All of this is happening when markets in developed countries are fixated more than ever on high-quality coffee.

While many consumers are willing to pay more for their coffee, they are actually drinking less of it. Furthermore, increased supply has not been followed by a commensurate decrease in price in most developed countries.

PRODUCING COUNTRIES

Coffee is produced in about eighty tropical or subtropical countries. Some 10.6 million hectares are currently in coffee production. Average annual production is about 7.4 million metric tons of green, or unroasted, coffee. The value-added coffee industry is worth about U.S.$60 billion worldwide, making coffee the second most valuable legally traded commodity in the world after petroleum (McEwan and Allgood 2001). It is a primary export of many developing countries, and as many as 25 million people depend on coffee for their livelihood.

The main coffee-producing countries by area planted, as opposed to total production, are Brazil (2.27 million ha), Colombia (850,000 ha), Côte d'Ivoire (829,000 ha), Mexico (701,326 ha), and Vietnam (477,000 ha). Each of the following countries has between 200,000 and 350,000 hectares planted to coffee: Cameroon, Ecuador, Ethiopia, Guatemala, Honduras, India, Peru, Uganda, and Venezuela. Combined, the top eleven countries account for nearly 74 percent of all land devoted to coffee and 74 percent of global production as well (FAO 2002). Even so, coffee production is less concentrated than many other commodities.

Coffee can still be important from an overall point of land use even if the country is not a major exporter. For example, Côte d'Ivoire and Puerto Rico both have 25 to 49 percent of all their agricultural land planted to coffee. Colombia, Costa Rica, Ecuador, El Salvador, Guatemala, Guinea, Panama, and Papua New Guinea each have 10 to 24 percent of all their agricultural land planted to coffee.

The main coffee producers by volume harvested are Brazil, Vietnam, Colombia, Indonesia, Mexico, Côte d'Ivoire, and Guatemala. These countries are also major coffee exporters. However, coffee is also one of the leading exports (see Table 3.1) in a number of countries that are not the largest producers or exporters. World exports are expected to rise by 11 percent to 81 million bags in 2002, while stockpiled reserves are expected to reach record levels of 27 million bags.

Globally, production averages 698 kilos per hectare. Martinique has the highest per-hectare production with more than four times the global average. Tonga achieves more than three times the global production average while Costa Rica, Zimbabwe, Thailand, and Malawi all produce at more than double the global average.

CONSUMING COUNTRIES

Coffee began as a luxury item, but it has become a basic food item that is now considered a daily necessity for many consumers. Today more than 2 billion people around the world are estimated to drink coffee regularly. Europe's thirst for coffee is the most voracious, as it annually consumes 2 million metric tons or just over 40 percent of all coffee traded globally (*The Financial Times* 2002).

TABLE 3.1. Coffee's Ranking of Total Exports by Value for
Selected Countries, 2001

Leading Export	Second Largest Export	Third Largest Export
Burundi	Angola	Costa Rica
El Salvador	Colombia	Equatorial Guinea
Ethiopia	Kenya	Côte d'Ivoire
Guatemala		Laos
Honduras		Sierra Leone
Madagascar		Yemen
Nicaragua		Congo
Rwanda		
Tanzania		
Uganda		

Source: ITC 2002.

Traditionally, producing countries along with the United States and Europe consume the most coffee by far. The United States consumes 25 percent of internationally traded coffee. However, the amount of coffee bought in the United States has declined in both absolute and per capita terms. Consumption throughout the European Union has also declined, but it is rising in Japan and Russia. The main coffee importers, as shown in the Fast Facts chart, are the United States, Germany, Japan, Italy, and France.

More people throughout the world are drinking coffee. Some 40 percent of the world's population drinks at least one cup of coffee each year. In general consumers first turn to lower-quality robusta varieties, which are used to make instant and mass-market coffee. As markets mature, consumers switch to higher-valued arabica blends, but they do not necessarily drink more coffee.

The industry has its eyes on China as an indicator of future global market trends. The Chinese currently drink about a cup per person per year. If China follows Taiwan, this will increase to thirty-eight cups per year. If it approaches the United States, consumption could reach 463 cups. Sweden has the highest per capita coffee consumption in the world with each person drinking on average 1,100 cups per year. How China progresses will have a tremendous impact on global demand and markets. It is likely, however, that initial impacts will be confined to the lower-grade beans used to make instant coffee.

The trade in coffee is relatively concentrated. In 1989 eight companies controlled more than half of the internationally traded coffee (See Table 3.2). No single company dominates the trade, however. Consolidation of the food industry is likely to affect coffee traders as well.

PRODUCTION SYSTEMS

Coffee is a woody shrub or small tree that can reach 10 meters in height, but under cultivation it is usually pruned to about 2.5 meters to facilitate harvesting. Coffee grows in tropical climates and performs best with good sunshine, moderate rainfall,

TABLE 3.2. The World's Largest Coffee Traders, 1989

Enterprise	Volume (1,000 bags)	Market Share
Rothfos	9,000	12.6%
ED & F. Man Holdings Limited	5,000	7.0%
Volkart	4,000	5.6%
Cargil	4,000	5.6%
Aron	4,000	5.6%
Rayner	4,000	5.6%
Bozzo	3,500	4.9%
Sueden	3,500	4.9%
Total	36,500	51.1%

Source: The Economist Intelligence Unit 1991.

average temperatures from 15 to 21 degrees Celsius (59 to 70 degrees Fahrenheit), no frost, and at altitudes between sea level and just over 1,800 meters (6,000 feet) (Manion et al. 1999).

Coffee matures (begins to flower and fruit) about three years after planting. One main and one secondary flowering season occur per year. Each mature tree produces approximately 2,000 "cherries" per year or 4,000 beans. This is the equivalent of half a kilogram (1 pound) of roasted coffee (Manion et al. 1999).

Coffee is a relatively easy crop to grow, but it is susceptible to a number of diseases and insect pests. At least 350 different diseases attack coffee, while more than 1,000 species of insects may cause the plant problems.

Two of the most significant factors that affect coffee production in any country are the relative costs of land and labor. Because coffee grown in full sun has a productive life of six to eight years and shade-grown coffee eighteen to twenty-four years (even more if plants are cut back and harvested from the new shoots), the relative value of land and labor can shift over time. Historically, most commercial production came from landholdings of 500 hectares or more. Today holdings of less than 5 hectares of planted coffee account for more than half of global production. Small producers are able to substitute unpaid family labor for both paid outside labor and many of the more expensive chemical inputs.

Two species account for the bulk of the coffee produced around the world—arabica (*Coffea arabica*) and robusta (*C. canephora*). Arabica came from the highlands of Ethiopia and was the first type of coffee that was produced for sale. Production of robusta coffee developed after World War II. The two species, and improved varieties developed from them, differ in taste, aroma, caffeine content, disease resistance, and optimum cultivation conditions. Natural variations in soil, sun, moisture, slope, disease, and pest conditions dictate which coffee is most effectively cultivated in which region of the world. The two coffees are compared in Table 3.3. In general, arabica coffee is produced in Latin America while robusta coffee is produced in West Africa and Southeast Asia. However, Brazil is both the world's largest arabica producer and the second largest (after Vietnam) robusta producer.

TABLE 3.3. Comparison of Arabica and Robusta Coffee Varieties

	Arabica	Robusta
Altitude of cultivation	500–2000 m	0–1000 m
Temperature requirements	Moderate	More heat tolerant More sensitive to cold
Humidity requirements	Lower	Higher
Soil requirements	Fertile soil	Poorer soils
Disease resistance	Low	Higher
Flavor profile	Fuller flavor	Weaker flavor
Caffeine content	Lower	Higher
Average price	Higher (up to 30%)	Lower
Labor as percentage of total variable costs	40%	60%
Agrochemical and material inputs as percentage of total variable costs	25%	15%
Overhead as percentage of total variable costs	35%	25%
Proportion of world supply	75%	25%
Main products	High-quality brands and specialty coffees	Instant, flavorings, mass-produced brands

Source: De Graaf 1986, as cited in Manion et al. 1999.
Note: Overhead includes capital, administration, and management.

In the 1990s there was considerable expansion of coffee production into new areas. The new coffee producers, including Vietnam and India, were able to be competitive in spite of low prices because labor was cheap and they could produce robusta coffee on relatively poor soils. Traditional coffee producers, however, such as Colombia, Costa Rica, and Mexico, were able to maintain coffee production in the face of higher land and labor costs by increasing yields from arabica coffee and by focusing on the small but growing markets for higher quality and certified shade-grown and organic coffee.

In the future coffee production will expand in those areas that have low input costs of production (e.g., inexpensive land and labor) with respect to the price that can be obtained for the coffee. Thus, expansion is certain to happen in India and Vietnam and perhaps in Myanmar, Laos, and Cambodia. Future environmental costs of coffee production are likely to be most pronounced in these regions. However, in Costa Rica and Colombia it is unlikely that coffee will hold its own unless a way can be found to certify and market more of it at higher prices so that producers can receive an increasing amount of every dollar paid in consuming countries.

Full-Sun Versus Shade-Grown Coffee

The two main types of coffee production systems are often characterized as "full-sun" and "shade-grown" coffee. Most commodities in the world are produced by genetic varieties that are fairly similar and whose production has very similar methods and environmental costs. This is not the case with coffee. The two main species used for coffee production require different growing conditions.

Full-sun coffee, sometimes referred to as "technified," high-input coffee, tends to be robusta coffee planted in monocrop stands. Robusta originated in West Africa and performs better in hotter and wetter climates. However, few absolute statements can be made about either variety of coffee. In some climates arabica can also be planted in full sun, as it is in parts of Brazil. Shade-grown coffee, by definition, is planted among other, taller trees, often in association with other subsistence or cash crops. Traditionally, shade-grown coffee was part of a small farmer's overall farming strategy. Arabica is most often grown in shade and therefore incorporated into other existing farming and agroforestry systems or polyculture production systems. Increasingly, full-sun coffee is grown both on small farms as well as on large-scale plantations using more chemicals and increased mechanization.

Often presented as two distinct systems of production, in fact, shade-grown and full-sun coffee production systems are different ends of a continuum. Growers employ a range of different techniques depending on economic, microclimate, and farm-specific factors (e.g., finances, farm size, experience with coffee production, history of coffee growing and coffee diseases in the area, etc.).

Since the end of World War II, technological innovations have led to the introduction of high-yielding varieties of coffee that grow best in monocultures with full sun (or close to full sun) and agrochemical inputs. This "sun" coffee is planted in much higher densities. For example, traditional shade-grown coffee is planted in densities of 1,100 to 1,500 plants per hectare, while sun coffee is grown in monocultures of 2,400 to 7,000 plants per hectare (Manion et al. 1999).

Costs of Production of Full-Sun Versus Shade-Grown Coffee

The cost of coffee production varies from one region to another and seems to be more related to local land and labor costs than to the species or varieties produced. Labor costs for coffee account for 40 to 60 percent of the variable costs of production, but shade coffee has higher labor costs per unit of coffee produced. Africa has the highest overall cost of production at U.S.$1.14 per pound. Latin America has an average cost of production of $0.74 and Asia $0.69 per pound (Talbot 1997, as cited in Manion et al. 1999). Typically, costs of production tend to be 50 to 60 percent of the export costs of green coffee (Manion et al. 1999).

Governments play a vital role in determining the profit of growers. Through ad valorem taxes, countries receive an average of 10 to 50 percent of the export value of green coffee (Manion et al. 1999). In most cases these funds are invested back into the coffee industry. However, corrupt states have been known to misappropriate them.

Another important issue that affects the type of coffee production that is undertaken is the value of local currency. This can have two different types of impacts. Many of the chemicals and fertilizers used in coffee production are imported. If local currency values decline, imported inputs become more expensive relative to the value of the raw coffee. Furthermore, increased production reduces market price, potentially leaving producers in a worse position. Over time, this will result in less coffee production, but producers that are dependent on coffee can suffer a severe drop in income in the short term. On the other hand, coffee producers who use low-input, shade-grown methods and who live in countries with low currency values will be inclined to plant coffee much longer than farmers from countries with higher-valued currency because the relative value they can obtain on the global market is higher for them.

Producers respond to coffee prices. If prices are high, they invest in new plantings. In Brazil, if the prices fall, marginal producers will destroy their trees. The present low prices are caused by investments made when prices were higher. Precipitous price falls beginning in 1989 caused growers in Brazil to reduce the number of coffee trees from 4.2 billion to 3.2 billion by mid-1992 (May et al. 1993). As a result, prices bounced back and others began to plant, contributing to the current crisis.

The microeconomics of sun versus shade-grown coffee are not always obvious or consistent. In Nicaragua, for example, sun coffee has significantly higher production costs. Even so, comparing the average profits for Nicaraguan producers of tradition-al, low-input (semi-technified), and high-input (technified) coffee over the past five years underscores why the transition to full-sun coffee is occurring. The yields from the full-sun coffee are more than seven times those of the traditional production systems and nearly three times those of the low-input production systems. Similarly the per-hectare profits of the full-sun, high-input coffee producers, based on a five-year average, are nearly three times those of traditional producers and nearly twice those of traditional low-input producers (Banco Central de Nicaragua and MAGFOR 1997/98, as cited in McEwan and Allgood 2001). In Colombia, however, the data suggests that the reverse is true.

The intensity of production is increasing. In the past, dense plantings of coffee contained 2,400 trees per hectare. Today more than 5,000 trees per hectare are common in many parts of the world, and as many as 15,000 trees per hectare can now be found in parts of Brazil. In the past it was common for coffee trees to be harvested for up to twenty or thirty years, then fifteen to eighteen years became the norm. As noted above, full-sun coffee is now normally grown for only six to eight years. This allows producers to shift crops more quickly in response to changing prices.

Until recently coffee was always picked by hand. However due to the cost of labor and the problems of organizing and managing large labor forces, many commercial coffee farmers are now using machines to pick coffee. Some of the machines beat the bushes in a process that is best described as a "car wash." Other machines in use now in Brazil in the most dense coffee stands actually cut the trees off 7 to 10 cm (3 to 4 inches) above the ground and separate the coffee cherries from the rest of the plant. The coffee plant then regrows for one year and the second year blooms again and then is harvested by cutting it off again. This can be repeated for three or four cycles.

Coffee planted at 5,000 bushes per hectare can produce 3,300 kilograms per hectare each year. Coffee planted at 15,000 bushes per hectare and harvested by cutting almost to the ground can produce 5,400 kilograms per hectare every two years (or 2,700 kg/ha/yr). The advantage of the latter system, however, is that the mechanical picking reduces the need for and the cost of labor. In parts of Brazil, due to specific labor laws, temporary contracted coffee pickers can cost farmers as much as U.S.$9 a day even though the worker only receives 45 percent of that. In addition, producers in Brazil are finding that with intensive plantings, irrigation doubles coffee production and justifies the use of modified center pivot or even the more expensive drip irrigation systems.

Another issue may also be beginning to affect production. It is predicted that the increased numbers of producers who do not use pesticides (either because of their outright cost or to comply with certification guidelines) are actually causing an increase in coffee borer or *broca* and other pests. This has long been a problem in Colombia, but it is now also becoming a problem for El Salvador, Mexico, and other Central American producers. In El Salvador investments in fertilizers and pesticides have declined by as much as 40 percent, and now total production is declining by as much as 15 to 20 percent (I & M Smith Ltd. 2002).

PROCESSING

Coffee processing can have significant environmental impacts. Within twenty-four hours of being picked, coffee should be processed to retain its overall quality. This is the most serious time constraint associated with coffee production. The first task is to remove the seeds from the fleshy fruit of the coffee "cherry." This is done either through wet or dry processing. In the dry procedure, the cherries are dried and then threshed. The amount of water used in dry processing is 1.4 to 14 liters per kilogram of processed coffee depending on the equipment. The main waste is the hulls themselves, which represent 50 percent of harvested weight, and parchment, the thin covering on the seed that represents 12 percent of the harvested weight (May et al. 1993). These materials can be used for fuel, organic matter for soil conditioning, fertilizer, or animal bedding. Since most processing is done at central locations, it is expensive to haul the material back to the farms. Dry processing of the cherries is difficult in many countries of the humid tropics.

In wet processing, machines are used to remove the outer hulls and most of the pulp. The remaining pulp is allowed to ferment for a few hours until it can be easily removed. The beans are then dried either in the sun or in mechanical dryers that are fueled with wood or coffee husks. In Costa Rica, the wet processing system requires 3000 to 4000 liters (3 to 4 cubic meters) of water to process 240 kilograms of coffee.

In El Salvador, where water is scarcer, only one-tenth as much water is used to process coffee. The pulp from wet processing creates a serious waste disposal problem, as discussed later under "Degradation of Water Quality."

Final processing for coffee depends, to some extent, on the market. In Brazil, for example, the domestic market accounts for 40 percent of unroasted beans. This breaks down to 8 to 9 million sacks used by the roasting and grinding industry and 0.8 million sacks for the manufacture of instant coffee. Of the 60 percent that is exported, about 2.4 million sacks are used to make instant coffee while about 15 million sacks are exported as green beans (May et al. 1993).

Tea and hot chocolate are partial substitutes for coffee. Postum is a caffeine-free, cereal-based substitute designed by food manufacturers in the United States to take the place of coffee. This product was developed when coffee prices were high and when consumers were concerned about the levels of caffeine in coffee. A wide range of coffee substitutes (both with and without caffeine) can be found in both the United States and Europe.

In different parts of the world local substitutes have existed for some time as sources of caffeine. Tea in Asia and cocoa in the American tropics were traditional sources of caffeine for large populations prior to the introduction of coffee. Chicory was often used in parts of Europe and Louisiana. Guarana in Brazil and yerba maté in Paraguay, Argentina, Uruguay, and Brazil are popular high-caffeine beverages that partially substitute for coffee. However, from a flavor point of view there are no direct substitutes for coffee.

Although not commonly thought of as such, caffeinated soft drinks such as colas are perhaps the most important substitute beverages for coffee, at least in developed countries as well as those countries that are adopting similar consumption practices (e.g., Mexico). From 1962 to 1989 the percentage of Americans drinking soft drinks almost doubled to 62.1 percent from 32.6 percent; this is precisely when the per capita consumption of coffee was declining.

Similarly, there is also thought to be a correlation between coffee and cigarette consumption. It has been noted that when people cut back on their smoking, they drink more coffee. If this is true, then a market swing in the United States is probably already underway, as the absolute number of smokers is declining. However, similar changes in Europe and China could stimulate considerable increases in demand for coffee.

MARKET CHAIN

The general market chain (the stages between producers and consumers) for coffee includes on-farm growing, harvesting, primary processing and sorting, export, shipping, distribution, roasting, packaging, redistribution to retail stores, purchase by the consumer, brewing, and drinking. The actual number of players can vary considerably within the market chain as one entity can often fill a number of the different functions (see Table 3.4). There are also major differences between coffee market chains for domestic consumption and those for international coffee trade and consumption.

In addition, of course, there can be considerable competition between the different layers of the market chain for a greater share of the value added to the product as it moves from producer to consumer. In some instances this has resulted in bypassing some traditional players altogether. For example, Nestlé established processing facilities in major producing countries such as Brazil and Côte d'Ivoire. Other multinational corporations prefer to undertake processing in the consuming countries.

TABLE 3.4. The Number of Players in the Market Chain for Coffee in
Brazil and the World

Actor	Number in Brazil	Number in the World
Coffee farmers	300,000	5,000,000
Laborers	(producers + laborers = 10% of Brazil's rural population)	14,000,000
Processors	3,000	
Cooperative processors	70	
Roasters	1,200 in the food industry	
Instant coffee processors	11	
Exporters	162	

Source: May et al. 1993.

The International Coffee Agreement was created in 1962 and the International Coffee Organization (ICO) was created in 1963 when the agreement went into effect. The goal of the agreement was to introduce stability in the coffee market and to protect countries (and producers) from coffee dumping and vast price swings. The ICO, which has 55 member countries, put the sixth and most recent version of the International Coffee Agreement into effect on October 1, 2001 (ICO 2003). The agreement will be in effect for six years. During the initial agreement, many consuming countries decided to allow more value-added activities to take place in producing countries. Increasing income in those countries was seen as a way to improve the standard of living of many rural poor and to increase political stability. The fact that higher prices were passed on to consumers was seen to be more than offset by the overall political stability achieved.

During a two-year suspension of the International Coffee Agreement's quota and control provisions in 1989, the system began to change, and the consuming countries assumed greater control of the market. This change had quite significant impacts on where value was captured from coffee. In 1985, $0.38 of every dollar spent for retail roasted coffee in the United States went to the producer countries. Just ten years later in 1995, only $0.23 made it back to the producer countries. This amounted to a 40 percent reduction to producer countries while the retail price of coffee increased by more than 30 percent in real terms.

Even taking into consideration these issues, coffee still brings more money to producers in absolute terms than other commodities such as sugar, tea, bananas, oranges, cotton, or tobacco. The capital also tends to be more broadly distributed to people in producing countries when compared to minerals such as petroleum or bauxite. In 1994 more than U.S.$12 billion worth of coffee (80 percent of world production) was traded between countries. This sum was equal to the entire flow of foreign aid from the United States during the same year.

Depending on the location, the amount of money distributed through the coffee market chain can vary somewhat. According to *The New Internationalist* (1995), for example, growers (including agricultural labor) can receive up to 10 percent of the retail price paid for coffee while shippers and roasters generally receive the bulk of all value from coffee at some 55 percent. In 1997 coffee growers received about 5 percent of every dollar spent on coffee. Farm laborers received about 8 percent, transport and loss accounted for 6 percent, and the value added in the producer country (e.g., processing, grading, bagging) amounted to another 3 percent. By contrast, the value added in consuming countries (e.g., shipping, roasting, grinding, packaging, and transportation) amounted to 67 percent. The retail share of every dollar was about 11 percent (Talbot 1997). In Europe, where there is more competition between roasters, retailers can receive as much as 25 percent of the value of all coffee sales.

By 2002, however, this picture had changed. The combined farmers' and farm laborers' share of the final sales of coffee had slipped from 13 percent to 7 percent. Roasters, retailers, and global buyers, on the other hand, accounted for 29, 22, and 8 percent, respectively of the final price of coffee (*The Financial Times* 2002). In 2002, it was interesting that while the price of coffee had fallen more or less continuously for a few years, the price declines were not passed on to consumers. Instead, players in the chain simply increased their profit margins as a result of the lower prices. Looked at another way, in 2001 coffee exports generated $8 billion for the economies of producer countries, but more than $50 billion for the economies of the consuming countries (McEwan and Allgood 2001).

The coffee trade has become increasingly centralized since World War II. This has culminated with a few giant multinational corporations dominating world trade. By the mid-1990s, for example, two roasting companies, Nestlé (55 percent) and Kraft (25 percent), controlled 80 percent of the market in the United Kingdom. Today, while there is a tremendous rush for better coffee in the United Kingdom, sales of instant coffee still dominate the market (by 87 percent). Globally, the five dominant importers account for more than 40 percent of the global coffee trade (*The Financial Times* 2002). In the rest of Europe similar dominance is common. In France five roasters control 90 percent of the market, while in Italy the top five roasters control 70 percent of the market (*The Financial Times* 2002).

MARKET TRENDS

Between 1960 and 2000 coffee production increased 2.9 million metric tons, or 61 percent. International trade in coffee doubled over the same period. During the same forty-year period, prices declined 57 percent.

Coffee, like cocoa, is a classic commodity that has been studied for years. Both supply and demand respond to changes in prices. There are wide price swings because producers respond to high prices by planting. Because coffee is a tree, once it has been planted producers only need to cover variable costs to continue producing. This means that additional product will cause a long-term price decline until trees go out of production. Of course, the variable costs of large producers are higher than those of smaller ones who provide their own labor, so with declining prices larger producers are more likely to take out coffee sooner than smaller ones.

From the end of World War II through the end of the 1980s, price came to dominate the retail coffee sector, particularly in the United States, and quality suffered accordingly. In the United States in 1962, coffee consumption began a thirty-five-year decline that has only recently come to an end. Beginning in the 1980s and gaining momentum in the 1990s, increasing numbers of consumers began to pay more for specialty coffees, arabica beans, and darker roasts which have now been made available by thousands of independent roasters as well as a few larger retail companies such as Starbucks.

Retail specialty coffee beverage sales in the United States have reached more than $3 billion with another $2 billion in sales of roasted beans. This new "quality-based" coffee industry in the United States represents more than 5 percent of global output. Price increases in 1994 and 1997 did not slow growth in this market, so it is likely to continue to grow for some time. If demand in specialty coffees continues, this will exert pressure to increase coffee production in pristine mountain areas because of the unique flavor profiles those conditions can produce in the coffee grown there. This could lead to considerable habitat conversion and environmental degradation.

Running parallel to the increase in high-quality coffee is a growing specialty market for coffee certified as grown in ways that are environmentally or socially sustainable. A major portion of this developing market is also for organic coffee. Organic coffee is produced without synthetic fertilizers or pesticides, and growers use natural chemicals and predators to keep pests in check. Today, certified organic producers receive an additional $0.15 per pound for their coffee, which can represent a 30 to 50 percent premium depending on local markets if they can sell it as organic. However, as much as two-thirds of certified organic coffee still does not have markets, and producers are forced to sell it at normal market prices through the commercial market. This is a particular problem with smaller-scale coffee producers who have little market clout.

Fair Trade certification is slightly different from, but complementary to, organic certification. Fair Trade importers bypass traditional middlemen and buy directly from producer cooperatives in order to return a larger share of the coffee dollar directly to the producer. Fair Trade coffee focuses more on worker and producer rights and the benefits that they receive from the sale of their product. Today there are more than 500,000 farmers who produce and sell more than 14,545 metric tons (32 million pounds) of Fair Trade coffee. While growing, this production represents only half of one percent of total coffee production. To put this in perspective, the largest single conventional producer in the world, Brazil's Ipanema Agro Industry, has 12.4 million trees planted on 5,000 hectares and produces up to 3,266 metric tons (7.2 million pounds) per year (Manion et al. 1999). Fair Trade programs have developed certification programs to create consumer confidence in product claims. These programs, however, are often subjective and not always verified by a third party.

In addition to these coffee certification programs, a number of other smaller programs have been developed as well. For example there are such general coffee labels as shade-grown coffee and songbird-friendly coffee. The profusion of certification programs and ecological labels, brands, and claims has tended to raise awareness of the many issues related to coffee production but left most consumers rather confused about what each represents, much less which is "best." What would be best for producers and consumers alike is if the different certification programs could get together and agree on one or two standard sets of measurable criteria that were evaluated by third-party certifiers. One way to begin to get to this point is to undertake a side-by-side comparison of the different programs to identify which actually deliver the results that are most important to producing environmentally and socially sustainable coffee.

While only a small part of the coffee market in the past, specialty coffee of all kinds is now estimated at 10 to 15 percent of the global market and expected to grow by some 15 percent per year in the near future (McEwan and Allgood 2001).

While forecasting the coffee market is more of an art than a science, it seems at this time that most increased demand for coffee will come in China, Russia, and Eastern Europe as well as other developing countries. Any sustained expansion in demand will tend to eliminate stocks (probably within a year or two) and have to be supplied from new sources. Traditional markets in the United States and Europe are not expected to contribute to absolute growth, but they are likely to stimulate the production and processing of higher-quality coffee.

ENVIRONMENTAL IMPACTS OF PRODUCTION

The main negative environmental impacts from coffee production include habitat conversion, soil degradation, pesticide use, and degradation of water quality. Each of these impacts is discussed separately.

Habitat Conversion

The most serious impact of coffee cultivation continues to be the conversion of natural forest areas to plant coffee. Increasingly, it is full-sun coffee that is being established in plantations. Natural ecosystems are destroyed as a result of the expansion of sun coffee production. The affected natural systems will never fully recover.

The data suggests that there is a strong correlation between full-sun coffee production and deforestation. Of the fifty countries in the world with the highest deforestation rates from 1990 to 1995, thirty-seven were coffee producers. This is in part linked to the fact that the highest levels of deforestation are in tropical countries where coffee is also grown. Even so, the top twenty-five coffee exporters had a combined average annual forest cover loss of 70,000 square kilometers during the same years (Manion et al. 1999).

The large, monocrop plantations typical of full-sun plantations cause the greatest reductions in biodiversity. Studies in Colombia and Mexico indicate that full-sun coffee plantations support 90 percent fewer bird species than shade-grown coffee.

The severe thinning or clearing of forests for planting shade-grown coffee is also a major concern. Considerable biodiversity is lost both above and below ground. Microorganisms in particular are affected through clearing, soil disturbance, and exposure. Even with shade coffee the number of tree species can be reduced by 80 percent or more. Mammals and reptiles show declines in populations and species diversity relative to natural forests. Bat species are reduced by half or more in agroforestry systems such as shade-grown coffee. Furthermore, species that do better in disturbed ecosystems tend to dominate areas of shade-grown coffee.

Some observers have suggested that because much shade coffee is grown in areas of human habitation that are being deforested, species that can move easily often seek refuge in the shade-grown coffee areas. Migratory bird populations, for example, may be forced to seek shelter in an ever shrinking area, whether they are in transit or at a traditional seasonal resting place. While shade-grown coffee can support high wild species diversity of mobile species in comparison to full-sun coffee or many other agricultural activities, it is no substitute for the preservation of pristine natural areas.

There is no evidence that any area of coffee production, whether shade or full-sun, has ever been allowed to revert back to "natural" forest. Habitat conversion, it seems, is forever. In regions like Paraná in Brazil and Java in Indonesia, shade-grown coffee has given way to full-sun coffee or other agricultural crops altogether. This conversion can mean a reduction in the local labor needs. Those displaced by the conversion of land from coffee production to other crops often migrate to frontier areas (e.g., in the Amazon and Cerrado in the case of soybean expansion in Brazil). Or, overpopulation in agricultural areas can cause the migration of poor farmers or landless people to frontier areas where they plant coffee (e.g., in the outer islands of Indonesia and in central Vietnam). In both instances, the production of coffee contributes to serious declines in both biodiversity and ecosystem functions.

In Vietnam, Papua New Guinea, Laos, Myanmar, and Mexico coffee production is expanding into previously pristine natural areas. Colombia, in turn, has increased production by converting to more sun-grown coffee. It is not clear whether the land used for new producers in China, New Caledonia, Samoa, and Mauritius has come from converting pristine areas, or from conversion of other agricultural lands. There is little data globally to indicate what the previous land use was for new coffee production areas.

Another driving force of habitat conversion is the increasing market for high-grade specialty coffees. These coffees tend to be produced in new, out-of-the-way areas with unique soils and topographies that give the beans unusual flavor profiles. Such coffee is often produced in areas that are too steep or otherwise of too poor quality for the production of other food and cash crops. These are precisely the types of areas that are rich in biodiversity or, at the very least, have become local biodiversity refuges in the face of the expansion of other forms of agricultural production. They are also typically the types of areas that are most prone to erosion. Consequently, the demand for higher-quality arabica coffee may exacerbate environmental degradation. Even the demand for shade-grown and songbird-friendly coffee may not actually reduce the impact of the business if it is produced in previously isolated areas rich in biodiversity. For example, lands that have been set aside for preservation in Mexico, Vietnam, Kenya, Nicaragua, and Indonesia have reportedly been invaded illegally by coffee producers.

Soil Degradation

Historically coffee production in places such as Brazil has been characterized by a frontier, throwaway mentality. Coffee production has tended to migrate across the landscape, as plantations are abandoned and new ones started on fresh soil. Such migration left behind lands that were suitable first for short-term agriculture, then for extensive cattle grazing, and finally were often abandoned once soil degradation and erosion left them unproductive. In some instances, extensive use of fertilizers and other agrochemicals allowed such lands to continue to be used, but with their own particular set of environmental impacts.

One of the most degrading forms of coffee cultivation for soils is the use of herbicides to produce "clean" fields free of other vegetation. The use of herbicides to produce weed-free fields (or rather fields free of any other vegetation except coffee) on the slopes of coffee farms, particularly those at high elevations, is one of the major causes of soil exposure and erosion. Low, creeping cover crops such as the legume *Arachis pintoi* can be used to maintain ground cover and reduce soil erosion and exposure of the soil to sun, wind, and rain.

Pesticide Use

Coffee production in countries like Brazil has involved the extensive use of chemicals to combat pests and diseases. Prior to the 1970s producers used benzene hexachloride 1.5 gamma isomer (BHC) in two sprayings to combat bean borer. Later to combat rust, producers used twenty sprayings of a copper fungicide with BHC and foliar fertilizer. Eventually, BHC powder was replaced by lindane emulsion. BHC and lindane are organochlorines, whose use has since been prohibited due to their persistence in the environment. Many problems owing to chemical poisoning were registered among workers. No one investigated the impact on other species or the residuals in the coffee itself (May et al. 1993).

There has been a dramatic increase in the transformation of production from shade-grown to full-sun coffee. One estimate suggests that half of the coffee produced in northern Latin America had been converted to full sun by 1990. Full-sun coffee is also referred to as "technified," high-input coffee production. This form of coffee production results in lower populations of predaceous insects, increased solar radiation, and reduced nutrient cycling. Technified coffee production also results in a spiraling dependence on agrochemicals such as herbicides, fungicides, nematicides, and fertilizers. In Costa Rica, for example, the government recommends that sun coffee producers apply 30 kilograms of nitrogen per hectare per year compared with shade coffee producers who use little or none. In Colombia, with some 86 percent of coffee production technified, the country applies more than 400,000 metric tons of chemical fertilizers, at least when they can afford them during periods of high international prices.

Degradation of Water Quality

Coffee processing degrades freshwater bodies in many tropical ecosystems. Traditionally, when "cherries" were processed at the plantations, coffee pulp was used as mulch on the crop. Now that processing often occurs farther from the fields, pulp produced from wet pulping operations (which is the preferred and most common processing technique) is increasingly dumped in rivers. In the rivers it is a source of pollution because its decomposition uses much of the available oxygen, and the lower oxygen levels in water lead to fish kills. (This type of pollution is measured as biological oxygen demand, or BOD.)

A study in Central America in 1988 showed that processing 550,000 metric tons of coffee generated 1.1 million metric tons of pulp and polluted 110,000 cubic meters of water per day. This was equated with a city of 4 million dumping raw sewage into the region's waterways. In that period, Costa Rica estimated that coffee processing was responsible for two-thirds of the pollution, as measured by total biological oxygen demand, in its rivers. As freshwater supplies become scarcer and demand for fresh water increases, this issue will become even more important (Manion et al. 1999).

BETTER MANAGEMENT PRACTICES

Historically, there have been many opportunities to learn from experiences on the ground with coffee production. Such experiences provide a context within which key conservation strategies can be developed. A number of better practices have been identified for coffee production. Some examples are briefly described here. These deserve more detailed analysis, so that they can be adapted and used by other producers and encouraged by governments around the world (through linkage to credit, price supports, licenses or permits, etc.). The goal here is to reduce environmental impacts; one of the best ways to do this is to increase the longevity of each planting of coffee so that the owners will not be tempted to move to other areas and convert more habitat for any purpose, whether it be coffee or something else. Equally important is discouraging the conversion of shade-grown coffee to large, monoculture stands of full-sun coffee. Other ways to reduce environmental damage include: diversifying production and sources of income, incorporating fallowing strategies, reducing input use, reducing water use, and reducing soil erosion. Appropriate and detailed conservation strategies will be required, ideally for each key ecoregion or at the very least each country in question.

Halt the Expansion of Coffee Production in Natural Forests

In several areas, but particularly Vietnam and other countries in Southeast Asia, coffee production is expanding into natural forests. Due both to the associated environmental damage and the short-term nature of the investment, this type of expansion of coffee planting should be prohibited. The vast majority of expansion of this type is for the production of robusta coffee because it is more productive in hotter, sunnier climates and on poorer soils. However, given the amount of degraded land or marginal existing agricultural lands that could support robusta coffee trees, there is no reason to clear pristine habitat to plant coffee. With the agrochemicals available today and with improved overall production and management practices, much previously degraded land can be brought back into production.

Another way to halt the expansion of coffee into biodiverse-rich areas around the world is to create and enforce permanent protection status in tropical forest areas that are located on the frontier of expanding coffee-producing areas. These areas can be identified in part due to their biodiversity value, but they can also and increasingly be identified because they are not suitable for long-term, sustained production of coffee. In some areas, zoning may be a useful tool for protecting lands whose slopes or fragile soils make them unsuitable for long-term coffee production. Restricting coffee production, creating protected areas, and implementing zoning regulations are all ways to prevent needless environmental degradation that benefits no one in the end.

Discourage the Conversion of Shade-Grown Coffee to Sun Coffee

For existing coffee-producing areas shade coffee systems are preferable to full-sun production systems. Though shade plantations contain significantly less biodiversity than pristine habitats, they support more species than full-sun plantations. In addition, the shade plantations maintain higher levels of soil moisture, enhance nutrient cycling, and decrease erosion. In short, most ecosystem functions are preserved, even though considerable biodiversity is sacrificed. From a conservation point of view, shade coffee can serve as a useful compromise for continuing coffee production in existing areas or as an intermediary step in habitat restoration, but it is not a natural habitat itself.

The long-term economic implications of the conversion from shade to full-sun coffee are not well understood. For full-sun producers, the increased costs of inputs for producing their coffee are more than offset by the dramatic increase in yields. Hence, production increases. The question is: How low can the price of coffee go before it is too great to be offset even by greater productivity? Full-sun coffee producers are more reliant on expensive inputs and tend to have greater working capital costs if not overall debt. This, too, affects their ability to weather poor prices.

Diversify Production and Sources of Income

With the global drop in coffee prices, there is an increasing awareness that dependence solely on coffee is not a healthy strategy for producers. Instead, diversified agricultural production systems could best protect the incomes and viability of coffee producers, particularly the small producers that are responsible for most of the coffee grown in the world.

If this is the case, then there is a need to focus on integrating high-value crops such as vegetables and fruits that can be interplanted with higher-value arabica coffee. Interspersing coffee with fruit trees, vegetables, and/or ornamentals can diversify sources of income and reduce dependence on a single product. While such diversified production systems do not restore biodiversity to the levels found in native stands, they maintain higher levels of biodiversity than the alternatives, and they tend to yield more financially stable local economies as well.

Another key issue is the development of alternative markets and the ability to supply them. These skills are not common among producers and have been sadly lacking to date in the different, alternative coffee marketing programs. Coffee production and even other agricultural crops may be only one source of income for producers in the future. For example, it is possible that coffee growers could receive payments for carbon sequestration—either in aboveground biomass or by building carbon and organic matter on or in the soil. Studies would need to be undertaken to show the relative value of sun and shade-grown coffee for carbon sequestration. Ecotourism, particularly bird watching, could also be incorporated into coffee-growing areas as another stream of income where shade trees have been left and birds migrate through. As producers in other parts of the world have found, these sources of income could rival or even exceed those from coffee.

Incorporate Fallowing Strategies

Fallowing, in conjunction with enrichment planting of cover crops to build up the soil, is another effective strategy for coffee producers and for conservation. Through planned fallows, soils can be returned to their former vitality in a relatively short time. Fallowing can be seen as an overall investment strategy. Fallowing is a way to generate nutrients at the site that would otherwise need to be purchased. It can be profitable in its own right as legumes build up soil nitrogen levels through nitrogen fixation, and other cover crops recover potassium and phosphorus that had leached to soil depths but can be brought to the surface as both deep roots and mycorrhizae are developed. Through the development of a proper fallow plan, even future shade trees can be planted during the fallow period.

In five to seven years of careful cover cropping it is possible to rejuvenate the same area for intensive use. This is already done with black pepper production in Japanese colonies established in the Amazon, where there is crop rotation every seven years. The black pepper vines are just as healthy and productive as they were some seventy years ago when they were started in the Tome Acu area of Pará state. For small-scale coffee producers, the challenge will be to do this on a rotational basis (perhaps only a few trees or 100 square meters at a time), or to plant cash crops during the fallow to reduce the impact of lost coffee income during the period of rejuvenation.

Reduce Input Use

Some of the best prices for coffee, even in the face of declining overall world prices in 2002, are those for shade-grown highland coffee from Guatemala and Mexico. In these areas, even when the average world price has been declining, the price of fine highland, shade-grown coffee has remained relatively stable. For example, in Nicaragua specialty coffee is currently selling for U.S.$1.20 per pound while the regular price for coffee is $0.55 per pound (McEwan and Allgood 2001).

Unfortunately, the specialty markets that support such prices are not well developed and cannot handle all of the certified coffee that is currently available. Yet if markets can be successfully developed and maintained, biodiversity and habitat improvements can be incorporated into coffee production systems in ways that do not affect overall profitability and that may in fact increase overall producer financial viability through certification. This could happen in several ways—either through a premium paid to the producer for certified product, giving the producer access to more transparent information about actual prices for conventional coffee, or the producer reducing overall costs and/or increasing production through the adoption of better management practices that are required by certification.

In shade coffee systems there is negligible use of pesticides, and both the substances used and levels of use can be dictated by certifiers. Furthermore, there are now management techniques that use microorganisms to manage fungal diseases. Native microorganisms and effective microorganisms (EMs), naturally occurring or applied organisms that speed up the breakdown of organic matter, suppress many fungal problems simply by providing competition to the pathogens.

Reduce Water Use in Processing

Coffee production should minimize water use and prevent water pollution to the greatest extent possible. Both Colombia and Costa Rica are experimenting with low-effluent processing systems that are said to produce coffee of a comparable quality to that of a traditionally washed product. This technology should be encouraged. Processors should screen and recycle the water that they use so that less water is used overall and less organic matter is put into rivers. Saving the pulp to compost or to use as mulch will both increase the organic matter in the soil and help the soil retain more water. These two factors will increase production.

One way to reduce waste is to encourage anaerobic fermentation before washing occurs. This process can decompose mucilage on the seed and makes it easier to wash. An added benefit is that it takes less water to wash the seed as well. This strategy is cost effective for processors but the technology is not well known.

Much of the coffee pulp in Costa Rica is put into windrows for drying and composting even though it can take up to six months for full composting to occur. However, coffee processors have found that by inoculating the waste with microorganisms they can reduce the compost time to less than three months. The compost is then returned to the associate growers.

Effective microorganisms are also being introduced directly into the processing stream so that effluents are largely decomposed by the time the wastewater leaves the plant. The microorganisms digest the waste and speed up the overall decomposition. This reduces the total amount of organic matter with high biological oxygen demand released into local waterways.

Reduce Soil Erosion

It has taken considerable time and a lot of mistakes to identify and analyze better management practices for managing soils in coffee plantations, especially with all of the cultural and geographical variations. In the early years of coffee production in Brazil, for example, people established plantations by planting rows of coffee trees perpendicularly up hillsides. This practice guaranteed severe erosion. It is now clear, for example, that planting on contours around hills and spacing the trees so that they are staggered up hillsides reduces erosion tremendously. The Brazilian government tied coffee-planting loans to such improved practices and noticed an immediate reduction of soil erosion. For example, a comparison of perpendicular and contour planting on steep slopes showed a reduction in soil losses from 4.4 to 3.1 metric tons per hectare in only a few years. Furthermore, contour planting reduced runoff by 25 percent, thus retaining more water for the crop. Contour strips (alternating bands of trees with bands of other vegetation) were also found to provide erosion control, but the most effective practice to reduce erosion was to plant grass between the bushes. This practice was found to reduce soil losses to 0.2 metric tons per hectare and rainfall runoff by 90 percent (May et al. 1993).

OUTLOOK

Coffee is big business and as such attracts big bucks. For example, subsidies have stimulated coffee production throughout the world. Such subsidies will not disappear, so the question is whether they can be used more effectively for poverty alleviation and environmental gain. Since society pays for subsidies, they should accomplish societal goals. One such goal would be a measurable reduction of the negative impacts of production on the natural resource base; another could be an improvement of the overall welfare of coffee producers and those who work for them. In this light, full-sun coffee might be acceptable on degraded land, but clearing forests or even converting shade-grown coffee systems is not acceptable. In any case, if markets exist for full-sun coffee, it should not need to be subsidized.

Subsidies could also be used to encourage farmers to convert to multiple cropping systems that are more ecosystem- and biodiversity-friendly. Financial incentives (either through loans based on better management practices, purchase contracts, or certification) can encourage farmers to make improvements toward this end.

The most successful basis for the development of any strategy to reduce the negative environmental impacts of coffee production would be to develop a better understanding of how the international coffee market chain works for the vast majority of lower-grade coffee that moves through it. A value-chain analysis, from producer to consumer, should be undertaken for the global coffee market in order to identify potential partners and strategic entry points to promote more sustainable coffee production and marketing systems, not just for high-end beans but for mass-market robusta varieties that are sold in cans or processed into instant coffee as well.

Recently, there has been considerable interest on the part of many who support organic, fair-trade, and "ecolabeled" coffees to encourage highly visible companies such as Starbucks to make commitments to purchase certified coffees. Aside from the fact that no one can agree which coffees Starbucks should purchase, this need not be the only or the biggest game in town. BP-Amoco is one of the more progressive companies, and it is clearly positioning itself as a green, socially responsible corporate player. Furthermore, BP-Amoco sells more coffee than Starbucks. Why is it not being targeted?

The coffee industry will shortly launch a campaign to bolster the price of coffee. Any such program should, to the maximum extent possible, reduce the overall environmental impact of coffee production while at the same time insuring that improved prices actually make it all the way to the producers.

REFERENCES

Consumer's Choice Council. 2001. *Conservation Principles for Coffee Production.* May. Available at http://www.consumerscouncil.org/coffee/coffeeprinciples_52501.pdf.

Buzzanell, P. J. 1979. *Coffee Production and Trade in Latin America.* Commodity Program, Foreign Agricultural Service. Washington, D.C.: U.S. Department of Agriculture. May.

de Graaf, J. 1986. The Economics of Coffee: Economics of Crops in Developing Countries No. 1. Wageningen, The Netherlands: Centre for Agricultural Publishing and Documentation (PUDOC).

Durning, A. 1994. The History of a Cup of Coffee. *Worldwatch* September/October: 20–22.

The Economist Intelligence Unit. 1991. *Coffee to 1995: Recovery without Crutches.* Special Report No. 2116. London: The Economist Intelligence Unit. March.

FAO (Food and Agriculture Organization of the United Nations). 2002. *FAOSTAT Statistics Database.* Rome: UN Food and Agriculture Organization. Available at http://apps.fao.org.

The Financial Times. 2002. Growers Left Tasting Dregs of Coffee: Farmers' Share of Income from Sales of More Brands is Falling. May 23. London, England.

ICO (International Coffee Organization). 2003. *History.* Available at http://www.ico.org/frameset/icoset.htm. Accessed 2003.

ITC (International Trade Centre) UNCTAD/WTO. 2002. *International Trade Statistics.* Geneva. Available at http://www.intracen.org/tradstat/sitc3-3d/index.htm. Accessed 2002.

I & M Smith (Pty.) Ltd. 2002. Market Report, June 6. Online report from I & M Smith, Johannesburg: South Africa.

Lingle, T. 1992. *The Coffee Cupper's Handbook.* Available in English and Spanish. Long Beach, CA: Specialty Coffee Association of America (SCAA).

Manion, M., G. Dicum, N. Luttinger, G. Richards, J. J. Hardner, and T. Walker. 1999. The Scale and Trends of Coffee Production Impacts on Global Biodiversity. Paper prepared by Industrial Economics, Inc. for the Center for Applied Biodiversity Science, Conservation International, Washington, D.C. October 15. 60 pages. Draft.

May, P. H., R. Vegro, and J. A. Menezes. 1993. *Coffee and Cocoa Production and Processing in*

Brazil. Geneva: UN Conference on Trade and Development. UNCTAD/COM/17. 27 August.

McEwan, R. B. and B. Allgood. 2001. Nicaraguan Coffee: The Sustainable Crop. Unpublished paper.

The New Internationalist. 1995. Coffee. (Entire issue devoted to coffee.) No. 271. September. Oxford: United Kingdom. Available at http://www.newint.org/.

Olman, S. B., with the assistance of Jenny Reynolds. 1993. *Environmental Impact of Coffee Production and Processing in El Salvador and Costa Rica.* Rome: United Nations Conference on Trade and Development. UNCTAD/COM/20. 27 August.

Pieterse, M. T. A. and H. J. Silvis. 1988. *The World Coffee Market and the International Coffee Agreement.* Wageningen, The Netherlands: Wageningen Agricultural University.

Prayer, C. 1975. Coffee. In *Commodity Trade of the Third World.* New York: Wiley and Sons.

Ridler, N. B. 1980. Coffee and Its Economic Role in Selected Latin American Countries. Desarrollo rural en las Americas. Bogotá, Colombia: Instituto Interamericano de Ciencias Agricolas. May/Aug 12 (2): 157–163.

Seudieu, D. O. 1993. *L'Impact de la Production et de la Transformation du Café, du Cacao et du Riz sur L'Environment de Côte d'Ivoire.* Geneva: United Nations Conference on Trade and Development. UNCTAD/COM/24. 6 October.

Talbot, J. 1997. Where Does Your Coffee Dollar Go?: The Division of Income and Surplus along the Coffee Commodity Chain. *Studies in Comparative International Development* 32: 56–91.

UNCTAD (United Nations Conference on Trade and Development). 1999. World Commodity Survey, 1999–2000. Geneva, Switzerland: UNCTAD.

——. 1993. Experiences Concerning Environmental Effects of Commodity Production and Processing: Synthesis of Case Studies on Cocoa, Coffee and Rice. TD/B/CN.1/15.Geneva, Switzerland: UNCTAD. 22 September.

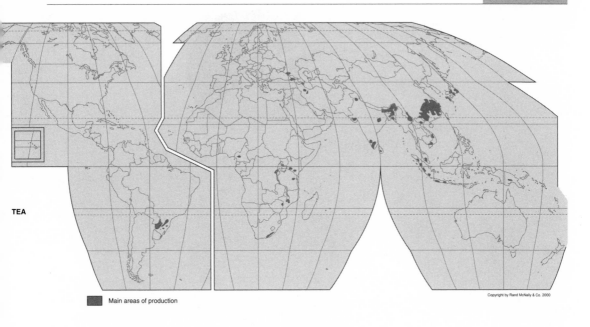

TEA

Main areas of production

Copyright by Rand McNally & Co. 2000

TEA *Camellia sinensis*

PRODUCTION		INTERNATIONAL TRADE	
Area under Cultivation	2.3 million ha	Share of World Production	50%
Global Production	3.0 million MT	Exports	1.5 million MT
Average Productivity	1,302 kg/ha	Average Price	$1,961 per MT
Producer Price	$802 per MT	Value	$2,900 million
Producer Production Value	$2,405 million		

PRINCIPAL PRODUCING COUNTRIES/BLOCS (by weight)	India, China, Sri Lanka, Kenya, Indonesia, Turkey
PRINCIPAL EXPORTING COUNTRIES/BLOCS	Sri Lanka, China, Kenya, India, Indonesia
PRINCIPAL IMPORTING COUNTRIES/BLOCS	Russia, United Kingdom, Pakistan, United States, Egypt, Japan
MAJOR ENVIRONMENTAL IMPACTS	Conversion of forest habitat Soil erosion and degradation Agrochemical inputs
POTENTIAL TO IMPROVE	Fair BMPs exist that reduce overall impacts Some effluents and soil erosion result from need for clean fields Produced in highly biodiverse areas so expansion has large impacts

Source: FAO 2002. All data for 2000.

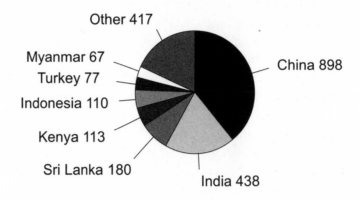

Area in Production (Mha)

Other 417
Myanmar 67
Turkey 77
Indonesia 110
Kenya 113
Sri Lanka 180
China 898
India 438

OVERVIEW

Tea is the leaves of an evergreen shrub that, according to legend, was first discovered in 2737 B.C. by Chinese Emperor Sheng Nong when tea leaves accidentally blew into a pot of water he was boiling. Tea was first cultivated in China more than 2,000 years ago. It was first used as a medicine and subsequently for Buddhist ceremonies. By the fifth or sixth century A.D. it had begun to be drunk for pleasure. Tea consumption spread throughout Chinese culture. In 800 A.D. Lu Yu wrote the first definitive book on tea, the Ch'a Ching (Stash Tea 2002). The three-volume book documented a wide range of tea cultivation, processing, and preparation methods.

By the time of Lu Yu's book, tea was so important that it was taxed. Trade along the Silk Road began to involve such large quantities of tea that the Sung Dynasty (960–1297 A.D.) nationalized the tea trade. That dynasty used tea to barter with nomads for horses. Tea and salt stockpiles were even used to back the Sung Dynasty's paper money.

After seeing the value of tea in China to enhance religious meditation, the Chinese Buddhist priest Yeisei introduced tea cultivation and consumption in Japan. As a result, tea in Japan has always been associated with Zen Buddhism. Tea was embraced almost immediately by the Emperor and spread rapidly from the royal court and monasteries throughout Japanese society.

The Mongols resisted most Chinese foods with one great exception—tea. Even centuries after the Mongols had been driven from China, special tea plantations were cultivated in the north of China for trade with the Mongolians. This tea was pressed into bricks, which were used as Mongol money until the 1920s.

The first mention of tea outside of China and Japan is said to be by the Arabs in 850 A.D., and they are often credited with first trading tea to Europe via Venice in about 1559. However, because of their navigational skills and navy, the Portuguese claim to have made the first European contact with tea in China; they developed a trade route that brought tea to Lisbon as early as 1515. From there it was taken by Dutch ships to France, Holland, and the Baltic countries. When tea was first sold in the Hague, the capital of Holland at that time, it sold for $100 per pound and was distributed through apothecaries. By 1650 the Dutch introduced tea into New Amsterdam (which later became New York). By the time the English took over the colony (the Dutch having traded it for rights to the Banda Islands, the home of nutmeg, which was at that time literally worth its weight in gold), New York consumed more tea than all of England (Stash Tea 2002). The colonists continued to purchase their tea from cheaper Dutch sources rather than from the John Company, the British trade monopoly in Asia. This is the context for the Boston Tea Party where colonists threw tea into Boston Harbor rather than pay a tax imposed by the British on tea purchased through other countries.

After 1650 tea consumption spread quickly in England. Ironically, tea displaced coffee (which had arrived in England earlier) and became the main beverage served at coffeehouses. Such houses were exclusively for men and were dubbed "penny universities" because for a penny any man could buy a pot of tea and a copy of the latest newspaper and exchange ideas with the sharpest men of the times (Stash Tea 2002). These coffee shops tended to become specialized in terms of their clientele (e.g., for lawyers, authors, military men, or businessmen) and eventually evolved into today's gentlemen's clubs. Tea gardens, by contrast, developed to serve both men and women.

Imperial Russia was the scene for another important chapter in the history of expanding tea consumption. China and Russia developed several trade agreements that opened their borders, but trade was limited by the caravan journey of 11,000 miles between the two countries, which took sixteen months. As a result of this long and complicated journey, tea was so expensive that it was only consumed by the wealthy. Gradually tea became more common and spread throughout society. In 1900 the opening of the Trans-Siberian Railroad drastically reduced the price of tea making it, along with vodka, the Russian national drink (Stash Tea 2002). The Russian samovar for serving tea is actually adapted from the Tibetan hot pot and can serve up to forty cups of tea, making tea available all day long in many Russian households.

For tea purists, the United States has had three rather dubious roles in the development of tea. At the 1904 St. Louis World's Fair iced tea was served and became a hit of the fair, and in 1908 New Yorker Thomas Sullivan developed the tea bag (Stash Tea 2002). Nearly a century later in the 1980s tea was first bottled in the United States and sold as a cold drink.

There are three varieties of the tea plant: China, Assam, and Cambodia. However, there are many different strains of these plants, and numerous other factors including soil, climate, altitude, and picking time all affect the flavor and aroma

(China Tea Information 2002). Today there are countless types of tea due to these factors, processing, and the addition of flavors—an estimated 3,000 all told (Tea Glorious Tea 2002)! How tea leaves are processed accounts for enormous differences in flavor. It was originally believed that green and black tea were from different plants. In fact, they are from the same species, *Camellia sinensis* (formerly *Thea sinensis*), and their processing is what differentiates the final product. The rarest tea in the world is white tea from China and Sri Lanka. This tea is picked only at daybreak in very few places. The buds are unopened. White peony tea is made only from tiny buds picked in the early spring. The buds look like small white blossoms, and this is why they are named peony (Tea Glorious Tea 2002).

PRODUCING COUNTRIES

Tea is grown in more than thirty countries. Globally, some 2.3 million hectares are planted to tea, and production from this area was approximately 3.0 million metric tons in 1997. The main tea-producing countries in terms of area planted are China (898,000 ha), India (438,000 ha), and Sri Lanka (180,000 ha). The area in these three countries represents more than 65 percent of all land planted to tea globally and nearly 61 percent of all production. Other significant producers include Kenya (113,000 ha), Indonesia (110,000 ha), Turkey (76,800 ha), Myanmar (66,908 ha), and Vietnam (7,300 ha). These five producing countries hold 19 percent of all land planted to tea and produce 20.4 percent of the world's tea (FAO 2002). The Food and Agriculture Organization of the United Nations (FAO) estimates that by 2010 India, Kenya, and Sri Lanka will produce 70 percent of the world's black tea, and that China will produce 75 percent of all green tea (FAO 2001).

About half of all tea is exported. The five main exporters in 1997 were Sri Lanka (22 percent of global exports), China (18 percent), Kenya (18 percent), India (14 percent), and Indonesia (8 percent). Kenya, Malawi, Tanzania, Zimbabwe, and South Africa account for about 25 percent of global exports, some 250,000 metric tons (Tea Glorious Tea 2002).

Globally, production averages 1,302 kilograms per hectare per year. Bolivia achieves the highest yields with 8,675 kilograms per hectare per year. Zimbabwe (3,667 kilograms per hectare per year) and Nepal (3,632 kilograms per hectare per year) are other countries that also have high productivity.

CONSUMING COUNTRIES

The four largest consumers of tea are India, China, Sri Lanka, and Indonesia. The major tea importers in 2000 were Russia, the United Kingdom, Pakistan, the United States, and Egypt. Motivation for drinking tea varies tremendously. Many, of course, drink it out of habit, often because it is an integral part of their culture. Others drink tea because of its actual or perceived health benefits. Those who market tea to consumers take advantage of these reasons. In particular, health benefits are increasingly touted in tea marketing campaigns.

Some countries have strong preferences for certain teas. For example, although other teas and coffee are growing in popularity, Japan still has a preference for green tea. Japan consumes approximately 20 percent of the world's green tea. In addition, the Japanese prefer a particular type of high-quality green tea. The preference is based not so much upon the region in which the tea is produced as the method of production and processing. The Japanese prefer tea to be processed within an hour of harvest, and they follow a pest control regimen that uses up to fifteen applications of pesticide per year (O'Brien 2002).

In recent years the market share of black tea, although still the largest, has been decreasing while green and oolong tea market shares have been increasing. Oolong consumption has risen in China as well as in Japan, where 18 percent of tea consumed is oolong (Xinhua News Agency 2002).

PRODUCTION SYSTEMS

Tea was originally indigenous to at least China and India. In the wild, tea grows best in warm, humid climates with rainfall of at least 100 centimeters per year. Ideally, the plant prefers deep, light, acidic, and well-drained soil. Given these overall conditions, tea will grow in areas from sea level up to altitudes as high as 2,100 meters.

Cultivated tea is generally a tropical highland crop where it receives some cooler temperatures. Tea is grown in the foothills of the Himalayas as well as in the higher elevations of China, Southeast Asia, and East Africa. The crop also grows quite well in some middle elevations in Bolivia and Guatemala. Since the tea bush totally covers the ground, it is generally a monoculture crop planted in contour rows on highland slopes.

Tea is grown both on large estates and on small farms. Small, privately held farms can be as little as 0.5 hectares or can cover several hectares. In several countries where tea is produced on small farms, cooperatives have usually been formed to build a central processing plant, assist with technology transfer, and to market the product.

A tea estate, by contrast, is a self-contained unit. It is often hundreds, sometimes thousands, of hectares in size. Large tea estates are found in several countries in South and Southeast Asia, East Africa, and Latin America. In addition to the extensive tea plantings, an estate also has its own factory for processing as well as schools, a clinic or hospital, staff houses, gardens, woodlots, reservoirs, places of worship, and guest houses (Tea Glorious Tea 2002).

In Assam, India, one of the best-known tea-producing regions, there are 655 estates that manage 168,000 hectares of tea or 42 percent of all the tea cultivated in India (Tea Glorious Tea 2002). In Darjeeling, tea is produced on 100 estates with some 18,000 hectares of plantings, generally at about 2,100 meters (7,000 feet) elevation. By contrast, in the south of India in Nilgiri, more than 20,000 small farms grow tea on some 37,000 hectares (Tea Glorious Tea 2002).

Tea is grown as a bush that is allowed to grow about 1 meter high. This makes it easier to pick. Bushes are mostly grown from cuttings or clones, which are tended in nurseries until they are ready to be transplanted. Bushes are planted about 1.5 meters apart in rows that are 1 meter apart. On steeper slopes the rows follow the contours to minimize soil erosion (Tea Glorious Tea 2002). On even steeper slopes, terraces are also built to avoid soil erosion.

The tea bushes are trained into a fan shape with a flat top. This is referred to as the "plucking plateau." It is about 1 meter by 1.5 meters in area and takes from three to five years to come to maturity, depending on the altitude. Before the first picking, the bushes are severely pruned. Tea bushes produce for a long time, varying from approximately 40 years for the Assamese variety to over 100 years for the Chinese variety. Bushes will produce even longer than this, but their harvest may not be economically viable (Tillberg 1995).

Bushes are picked, mostly by hand, every seven to fourteen days. Altitude and climatic conditions of the growing area are the two factors determining the length of the regrowth period. A tea bush grown at sea level will replace itself more quickly once plucked than a tea bush growing at a higher altitude where the air is cooler. Only the top two leaves and a single bud are picked from each sprig on the plucking plateau (Tea Glorious Tea 2002). In Japan a mechanical harvester/leaf cutter has been developed that makes the picking of young leaves easier.

The picked leaves are collected in a basket or bag that is carried on the back of the pickers. When this is full it is taken to a collection point where the picked leaf is weighed before being transported to the factory for processing, or "making," as tea blending and manufacturing is known in the tea trade (Tea Glorious Tea 2002). When pickers are on estates and near the factory they will take their leaves to the factory directly to be weighed. On an estate, each picker is credited with the weight of the tea they pick. A skilled picker can pick 30 to 35 kilograms of leaves in a day. This makes about 7.5 to 9 kilograms of processed black tea.

Large estates tend to dominate tea farming in Asia and East Africa and therefore tea is planted as a monoculture crop there. Since tea is a deep-rooted evergreen shrub, soil erosion is rather minimal. However, some erosion still occurs because there are inevitably some exposed areas, such as paths that are used as walkways or areas that are being planted or rehabilitated. Large tea estates can extend as much as 20,000 hectares of monocrop plantations. In such areas, very little biodiversity is visible. From horizon to horizon there is only tea.

While few herbicides are used in tea production, there are several problems with soil-borne fungal diseases and nematodes. As a consequence, a number of applications of fungicides and nematicides on the soil are generally used. When applied over large sloping areas, this can create a big problem of pollution and runoff of the pesticides to the streams and waterways.

PROCESSING

It is in everyone's interest that tea is harvested so as to maintain the high quality of the product and that this quality is maintained right up to the point of use by the consumer. The first stage in product processing and maintenance of leaf quality is with field harvesting and leaf transport. The goal at this stage is to insure that all harvested leaves are acceptable for tea manufacture. This means that foreign matter, including leaves from other plants grown for shade, fuelwood, or windbreaks, should never be present. Pesticides banned either in the country of production or the country of consumption should never be present.

There are three main ways to process tea. These yield green tea, black tea, and oolong tea. All start in more or less the same way. Tea leaves are brittle when fresh and are withered either in sunlight or in warm air. This makes them pliable enough to be handled. The leaves are then rolled, twisted, and slightly broken. Often this is done by machine. The essential oils that give tea much of its flavor come out at this point. If the tea is fired or dried at this point the result is green tea. All Japanese tea and most forms of Chinese tea are green tea. Prior to 1830 Americans drank mostly green tea. That declined, however, so that until recent decades almost no green tea was consumed in the United States. Today green tea is again becoming more common.

Tea leaves turn black when the oils are exposed to air. Oxidization not only darkens the color, it also allows the leaves to develop new flavor compounds. These are commonly known as tannins, though technically they are called polyphenols. This black tea is what most Americans drink today. The British introduced large-scale tea production into many parts of India and Ceylon in the early 1800s. The harvest was processed as black tea.

The third processing method produces oolong tea. The tea is allowed to partially oxidize so that some of the fresh flavors of green tea are still there as well as some of the deeper flavors of black tea. This tea is produced in China and Taiwan, and some consider this middle-of-the-road tea to be the most sophisticated of all.

On arrival at the factory, the leaves are spread on large trays or racks; these are placed in the top of the factory where the leaves wither in temperatures of 25 to 30 degrees Celsius. As the moisture evaporates over the next ten to sixteen hours each leaf becomes limp. Some factories add warm air or fans or both to hasten the process. The leaves are then broken by machine to release the natural juices, or enzymes, that oxidize on contact with the air. "Orthodox" machines roll the leaves and produce large leaf particles or grades. "Unorthodox" machines produce finer cuts or chopped particles. The smaller particles are more suited for quicker brewing products such as tea bags (Tea Glorious Tea 2002).

The broken leaves are then laid out again on trays in a cool, humid atmosphere for three to four hours to ferment, or oxidize, and are gently turned throughout this process. The turning assists an even, golden russet finishing when the fermentation process is complete (Tea Glorious Tea 2002).

After fermentation, the leaf is dried thoroughly by passing it through hot air chambers where all remaining moisture is removed and the leaf turns very dark. The leaf is deposited into chests, where it is stored until it can be graded for size by passing it through a series of gradually larger fine wire meshes. After this, the tea is weighed and packed into chests or sacks for loading onto pallets. Samples of each tea are kept aside, which can be used to market the different lots of tea. Once a "make" has been processed, the factory is thoroughly washed down so that the next lot will not be contaminated by the last one (Tea Glorious Tea 2002).

In the end, most tea is blended with leaves from different pickings and grown on different topography. Such blends can include twenty, thirty, or even more sources of tea, most of them black. Hundreds of flavors can also be added to tea or blended with it. Many of today's common teas actually originated as common grading, flavor descriptors or processing terms (e.g., Pekoe, Orange Pekoe, or Earl Grey) or the names of places (e.g., Yunnan, Assam, Darjeeling). It is the job of the blender, a taster with many years of experience (it takes five years of training to become a novice), to insure brand consistency.

During processing, many teas are flavored. Earl Grey, a blend of black teas, is flavored with oil of bergamot, a variety of citrus fruit whose oils are commonly used to scent perfumes. Jasmine tea is probably the most common scented tea. Traditionally it was made by rubbing jasmine petals into the steamed leaves of green or semifermented tea, sometimes as much as seven times. Manufacturing jasmine tea today tends to involve the rolling of leaves in tumblers with jasmine flowers. Lapsang Souchong is tea smoked with slow-burning pine logs. Mint tea, consumed in North Africa, is actually several varieties of black tea that are mixed with either fresh or dried mint leaves. And chai, a generic name for spiced tea in India, is made with black tea, milk, and sugar as well as some combination of cinnamon, cardamom, cloves, and white or black pepper. Sometimes fresh ginger is added.

Processing tea causes fewer environmental problems than the processing of most commodities. Because tea leaves are rather small and easy to dry, drying is not a very energy-consuming activity. However, it is important that the tea is not over- or under-dried, and it is important that it is not tainted in any way during processing. Solar dryers and passive solar dryers have been developed to dry tea, but these are not terribly widespread. During processing a certain amount of dust, twigs, and organic matter is removed. This organic matter is the only waste that is created. Very little water is used in the processing of tea. Another major processing issue is product quality from a health and safety point of view. Tea must be within acceptable limits of microbiology and free from heavy metals, foreign matter, and any substances that are potentially harmful to consumers.

SUBSTITUTES

There are many substitutes for tea as either a hot or a cold drink. Traditionally, herbal teas were used as medicines, and they still are in many parts of the world. Increasingly, however, herbal teas are being consumed more commonly in their own right even though they are often marketed for their health or well-being benefits. Common herbal teas include chamomile, peppermint, spearmint, rose hips, and lemon verbena. In addition, a wide variety of herbal blends are now sold by distributors of standard teas.

There are several other substitutes for tea that are also sources of caffeine either in the form of hot or cold drinks. Maté is an herbal tea from South America with very high caffeine content. One gram of dried yerba maté leaves contains approximately 15 milligrams of caffeine (Stash Tea 2002). Guarana is a drink in the Amazon that is quite high in caffeine, with 25 milligrams of caffeine per gram of the dried seed (Stash Tea 2002). Perhaps the two most important sources of caffeine are coffee and cocoa. An 8-ounce cup of coffee has 135 milligrams of caffeine on average. While a cup of hot chocolate or cocoa has only 5 milligrams, a 1.5-ounce dark chocolate bar can contain 31 milligrams of caffeine, as much as a cup of green tea. In comparison, a cup of black tea typically contains 50 milligrams of caffeine, but it can vary from 25 to 110 mg depending on how long it is brewed (CSPI 1997).

Perhaps the most important substitutes for tea, however, are sodas. The consumption of sodas has increased more rapidly than any other drink in most countries. This is not only true in Europe and North America but also in developing countries. In Mexico, for example, the population is purported to drink more soda than water. Even with the rapid proliferation of numerous alternatives, tea consumption is holding its own.

MARKET CHAIN

Tea is sold in a variety of ways. From 1706 to 1998 the London tea auction was one of the main auctions for tea and was known for setting the price worldwide. Over time, auctions opened up around the world closer to the areas of production. Today tea is most often auctioned in the country of origin. Auction centers exist in Mombasa, Kenya; Colombo, Sri Lanka; Limbe, Malawi; the north and south of India; Jakarta, Indonesia; and Guangzhou, China. Prices are governed by quality, supply, and demand. Tea brokers act as intermediaries and taste, value, and bid teas on their different clients' behalf. Tea may also be sold by the estate producer at a private sale or while en route to its destination (Tea Glorious Tea 2002). After arriving in the main consumer centers, the tea is taken to the packaging and blending companies.

Tea-packaging companies sell their tea directly to supermarkets and other retailers. They sell through wholesalers as well as teams of salespeople calling on small retail shops and restaurants. Some of the packaging companies have significant market share. For example, Unilever, through such brands as Lipton, has 20 percent of the value of the world market for black tea (Unilever 2003a).

Most tea reaches the supermarket shelf between twenty and thirty weeks after it has been plucked (Tea Glorious Tea 2002). For best flavor tea should be consumed within six months of purchase, or within one year of being picked.

MARKET TRENDS

Global tea production increased from 3.0 million metric tons to 3.1 million metric tons between 2000 and 2001. Increases in tea production, combined with relatively constant consumption and global demand, caused tea prices to continue to decline in 2001. On a global level, the FAO composite price in 2001 was 13 percent lower than the annual average in 2000. In some areas, however, such as Sri Lanka, due to local exchange rates there were price increases in the local currency (CFC 2002).

By 2002, however, there were indications that prices may be rising. Some of the main tea producers—India, Kenya, Bangladesh, and Indonesia—had a decrease in production during the first nine months of 2002 as compared to the previous year. Forecasts for an early winter in India dampened hope that producers would be able to make up for lower overall production in 2002. Sri Lanka was the main exception to declining production as producers there had slight production increases (*The Financial Times* 2002).

Once primarily consumed in Asia and North Africa, green tea is becoming increasingly available around the world. Green tea production is projected to increase at a 2.6 percent average annual growth rate between 2000 and 2010, considerably faster than the 1.2 percent annual growth rate estimated for black tea over the same period. Black tea production estimates for 2010 are 2.4 million metric tons; this is still much larger than the relatively small market share of green tea, which has an estimated production for 2010 of 900,000 metric tons. On the whole, green tea exports are expected to increase along with production increases. Some countries, however, such as Japan, are likely to consume the majority of their domestic production (FAO 2001).

A prime driver of changing consumption patterns is the increased knowledge about and marketing of the health benefits of green tea. Studies show that green tea is high in vitamins and minerals, particularly vitamins B and C, as well as being high in fluoride. Researchers have found that both white and green teas are high in antioxidants; white tea is three times more so than green tea. Both are also used increasingly in anti-aging formulas (Dietz 2002). Among the health claims made of green tea are that it is thought to prevent skin cancer and lower blood pressure. Lotions, perfumes, other beauty products, and energy bars are among the many products on the market today that contain green tea (Stash Tea 2002).

Another potentially important trend in the industry is the increase in organic tea production. Market demand for organic tea in 2001 was more than 3 million kilograms, up from 150,000 kilograms in 1981. The Tea Board in India has recognized this market and is promoting organic production. For example, they plan to convert 100 hectares in each of three regions to organic production as part of a project they are undertaking with the Common Fund for Commodities (Global News Wire 2002).

ENVIRONMENTAL IMPACTS OF PRODUCTION

The main harmful environmental impact of tea production is habitat conversion. This is especially true for tea because much of the habitat used for cultivation is often located in more rugged and remote areas, which tend to be those with the highest biodiversity. Converting rugged natural habitat to tea production has multiple effects. Not only is the number of species reduced, but also, due to the slope of the land, considerable soil is lost before the plantations are fully established to protect the soil. As a consequence, a fair amount of soil degradation can occur. Energy use is another environmental cost. All tea must be dried. Wood is usually the source of energy for this, and as a result drying can lead to localized deforestation. Finally, there is some waste that results from processing tea, but since it is organic matter it can be easily reintegrated as a soil amendment depending on its acidity.

Habitat Conversion

As with any crop that is grown by itself in monocultural production systems on a large scale, habitat conversion and associated biodiversity loss are an issue. In Uganda and Kenya large areas of natural forests were cleared to make way for tea plantations. Tea-growing regions in India were once covered with a variety of grasslands, marshes, and forested areas that hosted a wide range of flora and fauna that included such species as elephants, tigers, and deer. Today the landscape is dominated by vast tea fields (Chaudhuri 2002). Tigers are no longer found in tea-growing regions. In addition, single tea crop cultivation does not support the same ecosystem functions as natural habitat. For example, it has less water retention and increased water runoff and soil erosion than more biodiverse natural habitats.

Agrochemical Use

The chemical inputs applied on tea plantations have had a deadly effect on soil biodiversity while simultaneously polluting river water, killing fish, and harming the animals and people who depend on the rivers for water. Agrochemicals used on tea plantations kill many of the microorganisms that live in soil. Studies in India have shown that as much as 70 percent of soil biota has been lost on tea plantations as compared to nearby natural habitat, especially in areas that workers and machinery pass over (Senapati et al. 2002). The use of chemical fertilizers has resulted in a decline in soil fertility (Fareed 1996).

In India as well as other producing countries, the tea industry has, until relatively recently, used pesticides that had been banned in developed nations. Such chemicals can have effects on human health through runoff as well as direct exposure when in the fields and spraying. Among the pesticides used were synthetic pyrethroids, which, in addition to posing health risks to the immediate environment, can also be quite toxic to fish, downstream organisms, and certain beneficial insects such as bees, and even deplete the ozone layer (Fareed 1996).

Degradation of Soil

Monocrop production and its associated chemical inputs not only reduce soil biodiversity and soil organic matter, but also compact soils (especially in areas that workers and machinery pass over). Compacted soils are low in oxygen. Earthworms can play an important role in oxygenating soil and are commonly used as an indicator of soil health. Researchers have found that tea plantation soil contained between one-third and one-half the number of earthworms per square meter as the nearby natural forest soil. In addition, most earthworms found in tea plantations were not native species to the area (Senapati et al. 2002).

Although well-established tea plants are deep-rooted and provide good ground cover, both of which minimize erosion, areas where tea is being planted or replanted are vulnerable to erosion. A study of soil erosion in Sri Lanka focused on tea, rubber, and coconut plantations. Of these three crops' overall growth phases, tea that was replanted on steep slopes had the highest erosion rates, whereas well-established tea had relatively low erosion rates (UNESCAP 2002). Not only does erosion strip nutrients and topsoil from the agricultural fields, it also causes problems downstream. In Sri Lanka, siltation from erosion is a major problem. Silt fills reservoirs, which reduces hydropower generation and the life of hydroelectric dams (UNESCAP 2002).

In southern India where some tea estates are more than 100 years old, the soil has become impoverished and yields are stable despite the increasing application of fertilizers and pesticides. According to Senapati et al. (2002), soil degradation includes, in addition to those factors mentioned above, reduced cation exchange (a measure of a soil's ability to hold stores of nutrients and release them to plants), reduced water absorption and retention, increased acidity of the soil (pH as low as 3.8, which causes concentrations of aluminum to increase to toxic levels), nutrient leaching, and accumulation of natural toxins from tea leaves, which can begin to alter microorganism soil communities.

The degradation of tea plantation soil is a cycle that feeds upon itself and increases the environmental degradation from tea production. As the soil is degraded, farmers increasingly rely on chemical inputs to maintain productivity. These inputs then contribute to further soil degradation, which leads to decreased productivity, requiring still more inputs to maintain a profitable tea plantation. As the soil degrades, more and more of these inputs are eroded or washed away, entering local water systems and harming the local environment. Some tea plantations are now trying alternative methods to restore soil health and increase productivity from the ground up.

Use of Wood for Drying

Tea processors use various fuels to dry tea leaves. Wood is the most common energy source, though some processors use both gas and wood, or only gas or oil. Large amounts of wood are used to dry tea, and how the wood is harvested has large implications for its environmental impacts. Most of the wood that is used for drying tea comes from harvesting in natural forests. As wood supplies decrease, however, tea plantations are now planting trees to provide their own wood.

In Sri Lanka it takes between 1.5 and 2.5 kilograms of wood to produce 1 kilogram of tea. The tea industry used more fuelwood than any other industry in Sri Lanka in the mid-eighties, some 377,400 metric tons per year and 33 percent of total fuelwood consumption by industry (FAO 1987). By 1992 the tea sector's use of fuelwood had increased to 455,000 metric tons per year, consuming over 43 percent of the fuelwood used by industry. Hotels and restaurants were the second largest wood users, consuming 15 percent of the total and 164,000 metric tons annually (FAO 1999).

Processing Waste

Processing activities in the tea sector do not pose significant environmental problems. In fact, those activities have been categorized as "low polluting" by Sri Lanka's Central Environmental Authority (UNESCAP 2002).

Tea waste is the sole tea processing by-product. It is mixed with lime before being dumped. This alters the soil pH, making it unsuitable for tea cultivation (Fareed 1996), but it is still a valuable soil amendment for any other crop that does not require soil to be as acidic as tea. In addition, tea production machinery is cleaned using a detergent that has a caustic soda base, which is often dumped untreated into local water systems (Fareed 1996).

BETTER MANAGEMENT PRACTICES

Several ways have been identified to reduce significantly the environmental impacts of tea production. Unilever (2003b) has, in fact, developed a set of better practice guidelines to help producers who want to sell to them to reduce their overall impacts. These guidelines are summarized below. In general, the better practices allow producers to encourage and increase biodiversity within their plantings. This is important because much of that biodiversity helps reduce the need for pesticides and other inputs. Other important practices, however, are those that reduce soil erosion and degradation and actually build soil so as to reduce the need for fertilizers and other inputs. In addition, farmers have found more efficient ways to utilize inputs (e.g., fertilizers, pesticides, energy, water, and some of the more toxic chemicals) so that they can get by using less of each.

Conserve Biodiversity

The conservation of biodiversity in the plantation and surrounding areas is important, particularly where plantations are located in areas of high conservation value. The principle should be that the land is being borrowed from nature and that if production of any kind ends on it, the land could be repopulated with a good representation of local biodiversity in a relatively short time. Improving yields on existing plantations can reduce pressure to convert natural habitat to tea plantations. For example, 1.2 percent per year increases in black tea production from 2000 to 2010 are expected to come from improved yields rather than increased planting and habitat conversion.

Producers should reduce pesticide use, abandon illegal pesticides entirely, adopt integrated pest management (IPM) whenever possible, adopt conservation measures for rare or endangered species that are on the farm or that use it as habitat, and work with initiatives that encourage biodiversity. Before new areas are planted, environmental impact assessments should be undertaken and the recommendations followed. While this may be unnecessary for small farms, the principle is that the biodiversity implications must be considered before any new plantings are undertaken.

Riparian areas should be maintained and continue to be dominated by native species. Similarly, areas that are too steep to plant should be left in native habitat. Wherever possible these wildlife habitats should be connected through corridors, not only on the same farm but between farms as well. When planting trees for fuelwood or for windbreaks, native species should be used whenever possible.

Another way to enhance biodiversity within existing tea estates is for producers to abandon tea growing in areas that are unprofitable (e.g., steep slopes, shallow soils, alkaline soils, poorly drained lands, etc.). In many instances, farming these areas takes producers' energy away from more productive parts of farms. Abandoning such areas will often result in higher net producer profits.

Promote Crop Diversity

The number of commercially viable plant species and varieties cultivated for human use is declining each year. An important conservation strategy, particularly with a long-lived plant like tea, would be to save a small patch of 50 to 100 bushes of older varieties of tea in any area that is targeted for replanting. This will save genetic material that may be useful for future tea propagation and production (e.g., to develop varieties that are resistant to diseases that become problems in the future).

Reduce Soil Erosion and Degradation

Soil erosion can be high in tea plantations. They are often planted in areas of considerable slope that receive high levels of rainfall. Erosion is most extensive during periods of planting or replanting, when as much as 75 metric tons of soil per hectare per year can be lost, as compared to 20 metric tons or less for well-managed seedlings or vegetatively propagated tea (UNESCAP 2002). It is important that ground cover be maintained at all times. If an area has to be replanted, the exposed ground should be mulched and replanted with vegetation as quickly as possible.

If the land being planted slopes significantly, then planting should be undertaken on the contour. This is particularly important on slopes that are over 25 degrees (in fact, a rigorous analysis should be undertaken to insure that such slopes are even economically viable for planting in the first place). On particularly steep slopes, single bands of a grass such as napier (Pennisetum purpureum) can be established every five to ten rows of tea to supplement contour planting. The grass can be harvested for mulch or for fodder.

Several other practices can be used to reduce erosion:

- Environmental impact assessments can help identify problematic areas of concern.

- Digging silt pits in newly planted areas can arrest runoff and encourage water retention.
- Mechanical harvesting should be avoided in any areas where soil erosion is likely to be severe.
- Ground cover plantings along field edges should be used to reduce erosion.
- Careful siting and construction activities (e.g., drain design, road and path layout, etc.) can reduce soil erosion significantly.
- Tea prunings can be used to cover all bare soils to prevent soil erosion.
- Soil should not be taken from fields for use in nurseries. This material can just as easily be created from compost mixed with soil in the nursery area.

Senapati et al. (2002) report that several research projects begun in 1991 on six estates in southern India indicate that there are several ways to restore soil fertility and enhance tea production. The application of organic matter to the soil appears to have great potential to increase soil microorganisms and earthworms. In general, more earthworms present in the soil mean higher total green leaf tea yields. As soils are degraded their earthworm populations decline and termite populations increase. As a result, the proportion of termites to earthworms appears to be a good indicator for assessing soil degradation. Yield increases from bio-organic fertilization ranged from 75.9 to 282 percent. This produced profits of U.S.$5,500 per hectare per year compared to conventional cultivation with standard synthetic fertilizers. Such bio-organic fertilizers can be made by composting tea prunings and high-quality organic matter, and mixing the material with earthworms. This was shown to be a more effective way to increase yields than the application of fertilizers alone. Different field trials showed that yields were increased from 79.5 to 276 percent over conventional fertilizer use alone. The vegetative propagation of tea allows both for quicker growing and ground cover establishment.

The adoption of these practices can increase production by 50 percent. However, in addition to increased production, producers also restore their land, improve the quality of their leaf, and conserve soil. The combination of these factors will increase the net value of production both in the short and medium terms.

One of the main measures of soil health and fertility is the content of organic matter, so management should be focused on maintaining or increasing organic matter. In addition, deterioration of soil structure may result from compaction, especially from harvesting mechanization, from changes in pH, from salinity, or from exposure. Mulching with tea prunings, leaf, or other organic matter, planting shade trees, or even planting cover crops for two years prior to replanting of tea plantations can also increase organic matter.

Reduce Fertilizer Use

Financial sustainability of tea production may require the use of fertilizer on some soils, especially over time. In India, for example, higher levels of fertilization are required to make plantations financially viable. However, the principle should be that nutrient inputs should not exceed off-take in the harvested product. Efficiency will depend on the application rate, soil type, soil depth, slope, temperature, and climate. In order to achieve this, nutrient loss through wastes, erosion, effluents, and soil exposure at the time of replanting must be minimized. In addition, because applied nitrogen is so volatile, every effort should be made to increase nitrogen through biological fixation. Application of fertilizer should be avoided within 3 to 4 meters of freshwater systems. Algal blooms in ponds within the farm are an indication of contamination from fertilizer runoff, particularly nitrogen.

In Kenya, about 80 percent efficiency of input use has been achieved by keeping fertilizer applications at an average of 150 kilograms of nitrogen per hectare per year or less. Careful timing of fertilizer applications and monitoring of the crop can also help reduce fertilizer use. For example, in Kenya a yield of green leaf of less than 6,500 kilograms per hectare per year is an indication of inadequate fertilizer; dark green, fleshy and succulent shoots throughout the sorting table are an indication of excess applications of nitrogen.

Ash from eucalyptus (or other fuelwood), from dryers, or from old tea plants is a potential source of potassium, a nutrient essential for plant growth. However, ash is alkaline, and tea plants do not benefit from the added alkalinity (unless soils have become extremely acidic). Instead of using it on the plantations, the ash should be used on the soil of the trees grown for fuel plantations, which will improve fertility in the fuelwood plantings.

The application of organic matter and compost can reduce the requirement for inorganic fertilizer applications. This results in part because the application provides needed nutrients. In addition, however, the organic matter binds with nutrients in the soil as well as those added subsequently, effectively increasing the soil's nutrient-holding capacity.

Another way to reduce fertilizer use is through the choice of appropriate nutrients. For example, ground rock phosphate applied where soil is acidic during land preparation before planting or replanting reduces the subsequent reliance on soluble phosphate fertilizer, which is more easily leached into nearby streams and ponds.

Similarly, if soils are or become extremely acidic (below pH 4), lime should be applied at the time of pruning. Dolomitic rather than burned (quick) lime should be used if available. It will release more slowly in the soil and so tend to require fewer applications over time.

Minimize Pollution from Energy Consumption

Renewable energy resources should be targeted for use since nonrenewable sources such as fossil fuels are not sustainable in the long term. Wherever possible the use of fossil fuels for power generation, vehicles, irrigation engines and factory startup and operation should be minimized to reduce pollution and production of greenhouse gases. Solar, wind, and hydroelectricity should be explored as possible alternatives whenever possible. In some areas, biofuels derived from wastes could supplement fuelwood use. In addition to the source of energy, the boiler and factory energy efficiency should be optimized.

Using wood for drying is a preferred way to reduce nonrenewable energy sources provided the fuelwood is sustainably harvested or derived from managed plantings specifically dedicated for that purpose. In the case of small farms, cooperative fuelwood production schemes could be considered. However, if the only option is to derive wood from protected areas or other fragile forest ecosystems, then fossil fuels should be used as the preferable alternative.

Reduce Water Consumption

Water consumption may be an issue in some areas, primarily as a result of irrigation and especially through extraction of ground water. Both the volume of water used and the ratio of renewable to nonrenewable water used need to be considered. Water can be harvested from building roofs or even from retention ponds built in runoff areas of the property away from streams. It is an important principle to insure that water use is not at the expense of downstream users. In some instances, simply making workers aware of the importance of an issue by measuring it is the first and most important step in reducing overall consumption.

There are a number of ways to reduce use. Drip irrigation uses less water than sprinklers. Water use in the factory can be reduced by condensing and reusing the steam from the drying tea leaves, and by dry-cleaning or brushing the factory lines where product is moved, processed and packaged rather than washing them with water.

Water pollution is also an issue. When caused by inappropriate timing of fertilizer applications or field renovations, it can be reduced by the soil conservation methods described above and especially by increasing organic matter in soil to retain both water and nutrients. Or it can result from high biological oxygen demand, as in the effluent resulting from flushing organic matter during processing. One way to avoid this would be to establish water catchment areas to allow organic matter to decompose and settle out rather than simply allowing it to be flushed down stream.

Reduce Use of Toxic Chemicals

While there will be times when pesticides will be necessary for the production of tea, such chemicals should never be used prophylactically. Integrated pest management (IPM) can be the key to reduced toxicity and more sustainable pest control for tea production. For this to work, however, producers must not only identify the main pests in the different areas of the plantation but also develop management plans for controlling them. This means the development of censuses and analyses of the life cycles and natural enemies. Economic damage thresholds must be established for each pest with appropriate control measures indicated for each. Research on biological control agents (e.g., predators, parasites, biological fungicides, and pheromones) will need to be undertaken on a wide range of tea-growing areas. In addition, positive results and field trials should be incorporated into IPM practices so that improvements can be identified and adopted over time.

Even the most effective IPM strategies will not eliminate the use of pesticides, but they can insure that pesticide applications are kept to a minimum. This in turn will minimize seepage into groundwater or runoff into freshwater systems. In some areas, like East Africa, diseases and insect pests are not a major problem. In such areas, IPM should be used to keep the natural enemies of tea in check and minimize the need for toxic chemicals.

If herbicides are required, safer compounds should be targeted whenever practical. Low volumes and spot applications can be used to reduce overall impacts. Applicators should practice with the equipment to insure they are competent to use it and understand how to protect themselves. For small farms, manual weeding is recommended. This reduces the costs of herbicides, application equipment, and protective clothing.

There are other ways to control pests. In some cases, weeds may become an issue in mature tea plantings as a result of pruning. A longer pruning cycle, or a taller pruning height, results in more complete shading by the tea and thus fewer weeds.

OUTLOOK

Tea consumption is expected to increase moderately over the next few decades. Most of the increases can be accommodated by increasing productivity in existing tea plantations. It is highly likely that many tea consumers, particularly in Asia, will begin to consume increasing amounts of coffee and cold beverages in the future.

While the overall consumption of tea is not likely to increase much, the quality of tea consumed is likely to increase significantly. If the consumption of coffee and other beverages in developed countries is any indication of trends in tea consumption, individual tea drinkers will more likely drink less tea in the future but want a higher-quality product, for which they will be willing to pay significantly more. While this has already happened in developed countries, it is likely that it will begin to happen in China and India as well as those economies grow and consumers have more disposable income. What this means is that production and processing will need to be improved across the board to meet the increasing consumer demand for high-quality teas.

REFERENCES

CFC (Common Fund for Commodities). 2002. *2001 Annual Report.* Available at http://www. common-fund.org/publ/annual/report01.pdf. Amsterdam, The Netherlands: CFC.

Chaudhuri, K. 2002. Tea and No Sympathy. Global News Wire, The Statesman Ltd. *Financial Times Information.* July 28.

China Tea Information. 2002. The Tea Plant. Available at http://www.cnteainfo.com/english/ knowledge/grow/plant.htm. November 14.

CSPI (Center for Science in the Public Interest). 1997. Caffeine Content of Food and Drugs Chart. Available at http://www.cspinet.org/new/cafchart.htm. Press Release, July 25.

Dietz, M. 2002. Tea Change. Nationwide News Party Limited. *Sunday Telegraph* (Sydney). March 3.

FAO (Food and Agriculture Organization of the United Nations). 1987. Technical and Economic Aspects of Using Wood Fuels in Rural Industries. Rome: UN Food and Agriculture Organization. Available at http://www.fao.org/documents

——. 1999. Asia-Pacific Forestry Sector Outlook Study. Working Paper No. APFSOS/WP/43.

——. 2001. Medium Term Outlook for Tea. Committee on Commodity Problems. 14th Session of the Intergovernmental Group on Tea. New Delhi, India, October 10–11. Available at http://www.fao.org/docrep/meeting/003/Y1419e.htm.

——. 2002. *FAOSTAT Statistics Database*. Rome: UN Food and Agriculture Organization. Available at http://apps.fao.org.

Fareed, M. 1996. Tea and Environmental Pollution. *Tea and Coffee Trade Journal*. No. 12, Vol. 168.

The Financial Times. 2002. Tea Bidders Turn Up the Heat. U.S. edition. November 12.

Global News Wire. 2002. Tea Board To Promote Organic Cultivation. *Asia-Africa Intelligence Wire*. November 13.

O'Brien, K. 2002. Green Tea Report. Department of Natural Resources and Environment, Victoria, Australia. Available at: http://www.nre.vic.gov.au. Accessed 2002.

Senapati, B. K., P. Lavelle, P. K. Panigrahi, S. Giri, and G. G. Brown. 2002. Restoring Soil Fertility and Enhancing Productivity in Indian Tea Plantations with Earthworms and Organic Fertilizers. Case study presented at the International Technical Workshop on Biological Management of Soil Ecosystems for Sustainable Agriculture, June, Londrina, Brazil. Workshop organized by FAO and Embrapa Soybean. Available at http://www.cnpso.embrapa.br/workshopfao/cases/Tea%20case%20study.pdf.

Stash Tea. 2002. A World of Tea. Available at http://www.stashtea.com. Accessed 2002.

Tea Glorious Tea. 2002. Tea History. The Tea Council, London. Available at http://www.tea.co.uk/tGloriousT/. Accessed 2002.

Tillberg, M. 2002. The Way of Tea. Available at tea.hypermart.net/teapage.html. Accessed 2002.

UNESCAP (United Nations Economic and Social Commission for Asia and the Pacific). 2002. *Integrating Environmental Considerations into the Economic Decision-Making Process*. Bangkok: UNESCAP. Available at www.unescap.org/drpad/publication/integra/mainpage.htm.

Unilever. 2003a. *TEA: A Popular Beverage*. Unilever Sustainable Agriculture Initiative. Available at http://www.growingforthefuture.com/documents/protocols.htm.

——. 2003b. *Sustainable Tea, Good Agricultural Practice Guidelines for Tea Estates*. Unilever Sustainable Agriculture Initiative. Available at http://www.growingforthefuture.com/documents/protocols.htm.

Xinhua News Agency. 2002. China's Special Teas Promise Rosy Future. *Xinhua Economic News Service*. Lexis-Nexis. September 20.

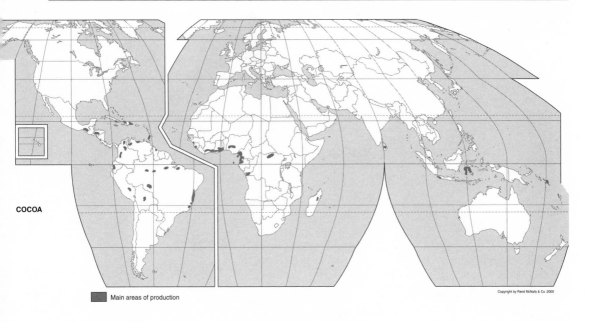

COCOA

Main areas of production

Copyright by Rand McNally & Co. 2000

COCOA *Theobroma cacao*

PRODUCTION

Area under Cultivation	7.5 million ha
Global Production	3.4 million MT
Average Productivity	459 kg/ha
Producer Price	$656 per MT
Producer Production Value	$2,232 million

INTERNATIONAL TRADE

Share of World Production	115%
Exports	4.0 million MT
Average Price	$1,094 per MT
Value	$4,344 million

PRINCIPAL PRODUCING COUNTRIES/BLOCS (by weight)

Côte d'Ivoire, Indonesia, Ghana, Nigeria, Brazil, Cameroon

PRINCIPAL EXPORTING COUNTRIES/BLOCS

Côte d'Ivoire, Ghana, Indonesia, Netherlands, Nigeria

PRINCIPAL IMPORTING COUNTRIES/BLOCS

Netherlands, United States, Germany, France, United Kingdom

MAJOR ENVIRONMENTAL IMPACTS

Conversion of primary forest habitat
Soil erosion
Some use of chemicals

POTENTIAL TO IMPROVE

Poor
Current cocoa prices discourage BMP adoption
Difficult to overcome the constraints of replanting the same area
Full-sun cocoa requires more chemical inputs and produces more
    effluents

Source: FAO 2002. All data for 2000.

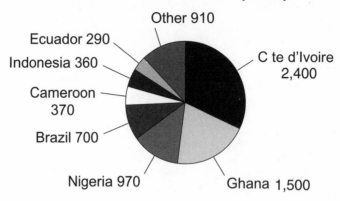

Area in Production (Mha)

Other 910

Ecuador 290

Indonesia 360

Cameroon 370

Brazil 700

Nigeria 970

C te d'Ivoire 2,400

Ghana 1,500

OVERVIEW

Chocolatl—the Aztec word for the drink and the source of our word chocolate—originated along with chocolate itself in the Western Amazon. By 1500 cocoa was the most valuable cash crop in Mesoamerica (Henderson 2001). Cocoa cultivation began in the Americas an estimated 3,000 years ago (Smith et al. 1992). It had already been planted throughout the American tropics by Amerindians at the time of European conquest. The inhabitants, however, consumed the product as a bitter, spicy beverage prepared with hot peppers. Sweetened, solid chocolate was not invented until after cocoa was taken to Europe.

Cocoa was so valuable in ancient Mesoamerica that the beans served as a form of currency, one that literally grew on trees, throughout the markets of the region. The coastal lands, where cocoa grew best, were highly valued by Indians and Spaniards alike. Along the Mosquito Coast of Honduras, cocoa seeds were used as money to buy things in village markets as late as the 1980s. In the state of Bahia and other cocoa producing areas of Brazil, *cacao*, the Portuguese term for cocoa, is still slang for money.

From the beginning of its trade, cocoa was very popular in Europe. Initially, however, it was because of its purported medicinal qualities, for which Amerindian peoples had also used it. It was said to make women conceive, help with childbirth, facilitate digestion, and cure consumption. It was supposed to cure the plague, cough, fluxes, jaundice, inflammation, and kidney stones; it was also supposed to clean the teeth, sweeten breath, provoke urine, expel poison, preserve from all infectious disease, and help emaciated patients gain weight (Henderson 2001). (At least one of those medicinal properties proved to be correct!)

Two varieties of the species, *Theobroma cacao*, are commonly cultivated: criollo and forastero. Cocoa liquor, butter, powder, and cake—the primary ingredients of chocolate—are all derived from the plant's bitter purple seed. Up to a few dozen seeds (about two centimeters long and half that in diameter) are found within each cocoa fruit pod. The leathery pods contain white fleshy pulp in which the seeds are embedded. In some areas, fresh juice from the pulp is consumed locally, but some pulp must be left on each seed in order for it to ferment properly and maintain the value of the seed for making chocolate.

Cocoa was introduced into European markets after the conquest of Mexico and Central America. Almost immediately, it was prepared with sugar and, by 1800, with milk. Due to increased demand, small-scale production spread in the Americas and to the Philippines by 1600. The crop was introduced into present-day Indonesia and India before 1800 (Wood 1991). By 1800 global production was 135,000 metric tons, and Ecuador was the largest producer, followed by Central America and several Caribbean Islands (Hardner et al. 1999).

Cocoa became a plantation crop in the nineteenth century. Advances in cultivation technology and the increasing development of trade with colonies led to the establishment of plantations in Southeast Asia, Oceania, and present-day Sri Lanka. The Portuguese, after losing control of Brazil, spread the crop to West Africa. By the latter half of the nineteenth century production began in earnest as the crop spread to British West and East Africa, Australia, Fiji, and Samoa (Wood 1991).

By the beginning of the twentieth century, today's three primary regions of cocoa production were established—tropical America, West Africa, and to a lesser extent Southeast Asia and Oceania (although Asian production did not become important until much later). At that time, tropical America was responsible for about 80 percent of global production. During the last century, cocoa demand and trade grew dramatically in the United States and Europe, stimulating production throughout the world (Hardner et al. 1999).

The historical spread of cocoa around the world provides an interesting parallel for many agricultural crops. Colonial powers, in conjunction with commercial interests, were in a constant search for new sources of trade and revenue. New crops were introduced throughout the world on the hit-or-miss chance that they would literally take root and provide the basis for local economies. What this meant is that by 1900 cocoa, like many other crops, had been introduced throughout the world and was well known by small farmers. These farmers were simply waiting in the wings until market conditions turned in their favor either to produce cocoa or another crop. As a consequence, any positive market signal encouraged a rapid increase in production.

In the latter part of the twentieth century, the independence of many colonies and the increase of the cocoa market in the post–World War II period resulted in the shift of cocoa production from plantations to small farms of less than 10 hectares.

These farmers already knew how to grow cocoa and could substitute family labor for capital and capital inputs (Hardner et al. 1999). Furthermore, the increase in disease associated with the expansion and intensification of production also tended to favor small-scale production, as small-scale producers could use family labor to reduce the impact of diseases rather than use expensive chemical inputs. Today, most cocoa is still produced on small farms, but production efficiency varies tremendously. For example, small farms in Asia produce five times the yields of their West African counterparts. However, West African producers produce very high quality cocoa.

PRODUCING COUNTRIES

The Food and Agriculture Organization of the UN (FAO 2002) reports that nearly 7.5 million hectares of land were planted to cocoa in 2000. Côte d'Ivoire has the most extensive plantations with some 2.4 million hectares, followed by Ghana with nearly 1.5 million hectares. The other major producers by area of production are Nigeria (966,000 ha), Brazil (697,420 ha), Cameroon (370,000 ha), Indonesia (360,000 ha) and Ecuador (287,300 ha). According to the FAO, these countries account for 87.8 percent of all land planted to cocoa, and 86 percent of the 3.44 million metric tons of cocoa produced annually. Côte d'Ivoire leads all producers with 1.4 million metric tons in 2000; Indonesia follows with 465,700 metric tons, and Ghana is next with 436,600 metric tons. Together these top three produce 67 percent of global production. Seven other countries contribute most of the remaining 33 percent of production.

During the 1990s world cocoa production increased 1 to 2 percent per year, reaching 3.4 million metric tons in 2000. Consumption has increased at double the pace of production, which has largely eliminated the vast cocoa stocks that had accumulated throughout the 1990s. Increased demand has caused price increases as well, as the amount traded in 2000 exceeded total production. In 2002 global market prices were increasing steadily.

The production increases of the 1990s were not evenly distributed around the world. In general, production increased in Africa, stabilized in Asia and Oceania, and declined in the Americas and the Caribbean. There are exceptions, however. Indonesia's production, for example, increased more than tenfold from 1986 to 2000 (33,000 MT to 465,700 MT), boosted by a 300 percent jump in local prices following the winter 1997 currency devaluation (Ruf and Yodding 2001). This encouraged the use of fertilizer and new plantings as well, even though the new plantings would not begin to produce for five to eight years. Yet Malaysian production dropped by more than two-thirds in the same period as many producers found it advantageous to tear out their cocoa plantings in order to establish oil palm plantations (Hardner et al. 1999).

In the same period, Brazilian net exports as a percentage of total production declined due to disease, high production costs, and increasing local consumption. Brazil illustrates how farmers in one country have reacted to declining prices. In

Brazil there are approximately 25,000 producers, each with an average of 25 hectares in production. In addition there are approximately 400,000 permanent rural workers employed by the cocoa industry. Brazil produced about 500,000 metric tons of cocoa from 1986 to 1993, but by 2000 the country produced only 192,949 metric tons. During peak production, the country averaged 700 kilograms per hectare. By 1993 average production had decreased to around 450 kilograms per hectare and was continuing to decline because owners refused to invest money, either in the form of labor or chemical inputs, in a crop that was losing money (May et al. 1993). In Brazil the main limiting factor is neither land nor marketing manipulation, but the cost of labor. The spread of diseases (and the labor implications of combating them) is also an issue.

In many countries, cocoa production has reached its limits. But the factors that limit production can vary tremendously. Some countries have little remaining land that is good for cocoa production. There are few forested areas into which the industry can expand profitably. This is true, for example, in Côte d'Ivoire. Some 2 million hectares in the country (16 percent of its surface area) are already used for cocoa cultivation, and today little forest remains for future expansion of the crop.

Another limiting factor is the productivity of certain cocoa varieties and associated cropping systems. High-yielding, more intensive production strategies can diminish the pressure to convert natural habitat. While most cocoa production averages below 1 metric ton per hectare, the new high-yielding varieties can elevate the yields to as much as 4.5 metric tons per hectare. This strategy can be combined with intercropping or multiple cropping and yields of 2 metric tons per hectare per year can still be expected. Even the lower-yielding of the two strategies more than doubles traditional yields and can decrease habitat conversion (Panfilo Tabora, personal communication).

In Ghana the factors that limit production are different. Ghana has some 2.4 million hectares of land under cocoa cultivation. This is about 10 percent of the country's total area. At this time, Ghana has converted virtually all forests appropriate for cocoa production. There is little room to expand. However, there is another factor that also significantly limits production. In Ghana a government-controlled marketing board still controls producer prices and taxes farmers based on production. Thus, the incentives are all wrong to encourage more sustainable production (Hardner et al. 1999).

CONSUMING COUNTRIES

Worldwide demand for cocoa increased in the 1990s in response to lower prices and increasing incomes in consuming countries. Europe consumes about half of global production; per capita consumption rates there are nearly twice as high as in the Americas (2.4 kilograms per person per year versus 1.3). The United States is responsible for another quarter of all consumption, with annual increases of 3 percent, the highest in the world of major consuming countries (Hardner et al. 1999). Only Eastern Europe has experienced an absolute decline in cocoa consumption since

1988, as governments ceased purchasing cocoa and declining private incomes re-duced purchasing power in those countries. Prior to that time the former USSR was the largest single buyer of cocoa in the world (Hardner et al. 1999). Singapore has increased consumption of cocoa more than 7 percent per year for the past twenty years, but this is most likely accounted for through processing, manufacturing, and re-exporting. The question is whether consumption trends in Singapore will be a harbinger of what is to come in China. If so, this will have significant impacts on global demand and production as well as overall impacts.

In 2000, world imports were dominated by the Netherlands, the United States, Germany, France, and the United Kingdom. These countries account for about half of global imports. However, the first industrialized stage of processing and grinding of the beans is dominated by the Netherlands, the United States, Germany, Brazil, the United Kingdom, and Russia, in that order. Consequently, many imports, at least into the nonproducer countries, also show up as exports (FAO 2002).

PRODUCTION SYSTEMS

The flowers of cocoa are produced on the wood of the tree, either the trunk or main branches. The trees produce large numbers of flowers at certain times of the year de-pending on the variety and local conditions. Only 1 to 5 percent of the flowers are pollinated. The number of seedpods that develop per tree varies from five to twenty-five or more and appears to be directly related to the number of pollinators in the en-vironment. The period between fertilization and pod maturation varies from 150 to 180 days depending on the variety (Laird et al. 1996). Cocoa grows best when rain-fall is between 1,500 and 2,000 millimeters per year, but the range can extend somewhat either way under less than ideal conditions (See Table 5.1). The plant, however, cannot tolerate dry seasons with more than three months of less than 100 millimeters of rainfall per month, and cocoa is also sensitive to waterlogged soils. In short, the pattern of rainfall is more important than the total amount. Temperatures can vary from a maximum of 30 to 32 degrees Celsius to a general minimum of 18 to 21 degrees Celsius. While plants can sometimes tolerate lower temperatures, they will be killed at temperatures below 10 degrees Celsius. Cocoa does not grow well in persistent strong winds; the trees prefer a sheltered location with windbreaks or forest cover to minimize wind (Laird et al. 1996).

TABLE 5.1. Cocoa Cultivation Requirements

Rainfall	1,250–3,000 mm per year
Dry season	No more than three months
Water	Impaired by both waterlogging and extended drought
Temperature	18–32°C, with absolute minimum of 10°C
Wind	Does not tolerate strong or even steady wind (forest cover or windbreaks are essential)
Soil	Deep, fertile, well-aggregated clay loam that is well-supplied with nutrients at surface (not economical on degraded soils)

Source: Laird et al. 1996.

Cocoa can be grown in a wide range of soils. It does best in deep, fertile, clay loam soils. It responds well to surface application of nutrients since the plant has many lateral, surface roots. It is generally assumed that cocoa cannot be produced on previously cleared and cultivated soils. However, improved planting material and cultivation methods now make it possible to cultivate cocoa on such soils (Laird et al. 1996).

Planting material is one of the most important issues affecting cocoa production. Sorting out the genetic material in seeds has been a serious challenge. Fortunately, plant breeders have been relatively successful in tailoring the plant to local growing conditions. Plants can be selected or bred for tolerance to local diseases and pests, seasonality of rainfall, flooding, winds, and acid soils (Laird et al. 1996). The characteristics that plant breeders try to achieve include vigorous growth, early bearing, improved yields, good percentage of bean weight in the pod, and high fat content.

Traditional cocoa plants produce in three years under ideal conditions, but starting with grafted plants rather than seedlings can produce plants that bear earlier. Cocoa trees produce flowers on wood that is two to three years old. Since grafts are made from branches that are two years old, grafted trees produce in the first year. Grafting cocoa seedlings in the field is a new approach to production. This approach uses direct seeding in the field and then uses the seedlings as rootstock for grafting. This approach has shown high productivity within twelve months after grafting and can reach 3 metric tons of beans per hectare in two years. While only recently adapted to cocoa, this technology is now practiced extensively in the Philippines and Malaysia. This system also tends to be undertaken in full-sun, monocrop plantations (raising many of the same issues as full-sun coffee).

There are different ways to produce cocoa in plantations. In Brazil cocoa was originally planted within existing forests. In-forest production is considered the most environmentally positive form of agriculture practiced in Brazil today. But Brazil was one of the first major cocoa producers to create large cocoa plantations. In Cameroon and Nigeria forests were selectively thinned to plant cocoa and other fruit trees. In a relatively short time, a forestlike appearance was regained. This agroforestry system is still the most common form of cocoa production worldwide (May et al. 1993). However, in the 1970s, some growers began to advocate the "clear-cut system" in which all non-cocoa vegetation was removed. This is analogous to full-sun coffee production, and, as might be expected, such producers depend more on agrochemical inputs. But the scale of full-sun cocoa production is much smaller than that of coffee production. At this time, some 70 percent of world cocoa production is still grown by small farmers mostly in agroforestry systems. Some 5 to 6 million of them depend on cocoa for part or nearly all of their cash income.

Cocoa produced for the market is divided into two main categories: bulk or ordinary cocoa from the forastero-type beans and fine or flavor cocoa from the criollo beans. In 1850 fine or flavor cocoa constituted 80 percent of world production; by 1900 it had fallen to 40 to 45 percent, and today it is only about 2 percent of world production (Wood 1987, as cited in Wood 1991).

The economics of cocoa production make it far easier for most large producers to

TABLE 5.2. Labor Requirements for Planting Cocoa in Primary Forest versus Replanting in an Aged Cocoa Farm

Activity in Primary Forest	Labor (person-days/ha)	Activity for Replanting	Labor (person-days/ha)
Clearing primary forest	33	Clearing fallow	30
Sowing cocoa beans	10	Nursery	20
Complementary planting	10	Planting	55
Intercropping	14	Intercropping	20
Initial weeding	3	Initial weeding	16
Replacement of dead seedlings	4	Replacement of dead seedlings	11
Complete weeding	12	Complete weeding	16
Total	86	Total	168

Source: Ruf 1995, as cited in Hardner et al. 1999.

simply push their plantings further into natural forest frontier habitats rather than to replant cocoa in existing plantations and agroforestry plots. This is true for two reasons. As Table 5.2 shows it is cheaper to clear forests than to replant existing plantations. Also, for at least the first few years, newly cleared areas have 15 to 25 percent higher yields than replanted areas (Matlick, personal communication, as cited in Rice and Greenberg 2000).

The cost of labor and/or chemicals to maintain production indefinitely in the same areas or to begin production in previously used or degraded agricultural or pasture areas is deemed too expensive by most producers. Given traditional production levels, world prices were not seen as justifying the expense. The cheapest alternative is simply to clear new forests. However, if the value of forests were to increase significantly, most cocoa production would not be viable without a significant increase in the international price of the commodity.

Once established, cocoa plantations are relatively simple to maintain. Most production activities involve manual labor; these include cutting weeds and clearing undergrowth, thinning trees to open up the canopy (cocoa needs some sunlight), insect control, mulching, fertilizing, harvesting, and on-farm fermentation and drying (May et al. 1993). Unlike many commodities, processing of cocoa begins on-farm. If this level of processing is not undertaken correctly, the value of the cocoa diminishes considerably.

There are many pests that attack cocoa. These include thrips, cocoa mirid species, ants, borers, and other pests as well as witches'-broom, and black pod rot; weeds are also a problem. Increasingly, farmers turn to pesticides to control these, at least when markets are good or credit is available and they can afford to do so. Even the use of pesticides, however, is not always effective. For some pests, like the weird growths known as witches'-broom, the most effective treatment is prompt pruning or the elimination of infected trees. Such labor-intensive pest management measures are expensive for larger planters. In Brazil, for example, some 50,000 trees had witches'-broom in Bahia in January 1991. Because it was considered too expensive to take care of those trees, by April of the same year 250,000 were infected (May et al. 1993). It is for this type of reason that most cocoa is still produced on smaller farms.

While there are some 1,500 insects that feed on cocoa, less than 2 percent of these have become economically significant. A wide range of pesticides is used for these pests. A lengthy list is included in Laird et al. (1996). Cocoa production uses almost all the main categories of chemicals manufactured for pest control—organochlorines, organophosphates, carbamates, and pyrethroids. In some cases, chemicals are used that are banned in the consuming countries. This creates a thriving black market. Of the thirty-two or so most common pesticides used in cocoa, at least nine are included in the Pesticide Action Network's "dirty dozen" (Laird et al. 1996). Pesticide use and misuse are a serious problem in many cocoa-producing areas. In addition to using banned pesticides, the lack of proper training or clothes and inadequate directions on the containers result in exposure and even death for workers. Improper use also causes needless damage to local flora and fauna.

As chemical weed control becomes cheaper and more cocoa is grown under full-sun conditions that encourage weed growth, more herbicides are used. These commonly include paraquat, dalapon, diuron, 2,4,5-T, 2,4-D, picloram, glyphosate (Roundup), and Simazine (Laird et al. 1996). Paraquat and glyphosate are the most common.

Fertilizer stimulates the growth and production of young trees and increases the yield of mature trees, but applications can cause eutrophication if excessive amounts are used. The concentrations of the main fertilizer nutrients—nitrogen, phosphorous, and potassium—are adjusted for specific conditions as well as the amount of shade (Laird et al. 1996). The rate of application also depends on the current value of the crop. As prices increase so do applications of fertilizer. The production of full-sun cocoa requires the application of more fertilizers than shade-grown cocoa.

Because some 50 percent of the cocoa bean is fat, this makes testing for agrochemical residues much easier than for many other crops. Organisms tend to store toxic substances in their fat. Thus, cocoa is particularly vulnerable to pesticide residues (From the organization Toxopeus, personal communication 1994, as cited in Laird et al. 1996).

There are new, highly productive strategies in cocoa growing, which include planting in densities of 4,000 plants per hectare using more compact varieties of grafted plants. These plants are productive for only six to eight years. Another production method is planting at high densities (e.g., about 2,000 plants per hectare) but in single rows intercropped with other crops such as cassava, sugarcane, and sweet potatoes. Intercropped cocoa has more resistance to the major disease problems. However, the plants are also short-lived (about six years of production). In both systems, the cocoa is then pulled out and replanted or production moves to other areas. These strategies, while much more intensive in their use of chemical inputs, can actually be cheaper forms of production because they are more productive and utilize existing cleared areas which require less labor to establish. In addition, the expansion of production is not a major threat to natural habitats. Such strategies are being used on some farms in Malaysia and the Philippines (Panfilo Tabora, personal communication).

In general, large-scale producers pay more for better land, use more paid labor, and have higher fixed and working capital costs compared to small-scale producers (Rice and Greenberg 2000). These factors tend to make larger producers more price-sensitive. If prices decline, they are more likely to destroy cocoa and plant another more profitable crop. It also makes such producers more interested in higher-yielding, shorter-lived varieties.

One advantage that smaller producers have traditionally had is that they have a more intimate knowledge of their plots and even the individual trees. This knowledge is critical for identifying and addressing production problems early, when they are most easily remedied. As a Malaysian researcher has said, "Cocoa is like horticulture, the planter must almost know each tree" (Rice and Greenberg 2000). In Sulawesi small producers achieve yields of up to 2,000 kilograms per hectare per year, more than plantations average but less than half of the yields of many research station field trials. This implies that farmers are not yet close to achieving in the field the known limits of production for cocoa. Even so, pest buildup and declining fertility cause yields on most farms to decline significantly within fifteen to twenty years of planting.

PROCESSING

Cocoa processing begins in the fields. Cocoa is gathered in the fields by workers, mostly women, who are hired specifically for that purpose. Pods are brought together in piles and then broken with machetes to remove the husks. Pulp and beans are initially gathered and later transported by people, animals, or machines in wooden boxes or woven baskets to on-farm fermentation facilities. At these locations the beans are subjected to a five-day fermentation process. The fermentation can be undertaken in baskets, heaps, boxes, or trays. After fermentation the beans are dried. Increasingly, this takes place in artificial dryers heated by fuelwood (May et al. 1993).

Fermentation and drying eliminate astringency and bitterness, imparting the peculiar flavor and brown coloring desired; they also reduce moisture content to 6 to 7 percent (May et al. 1993). A well-controlled fermentation process with inoculation of yeast and other effective microorganisms produces beans with better flavor profiles and storage qualities. Moreover, the beans dry faster because the mucilage that impairs drying is removed by the fermentation process. The reduced moisture content allows the beans to be stored and transported without risk of mold or mildew. No chemical or artificial additives or treatments are employed in processing cocoa beans.

After drying, beans are shipped to commercial centers for direct export or further processing. Many of the main producing countries attempt to add value to their cocoa by further processing the beans. In Brazil, for example, some additional processing is carried out in nine factories. To be competitive on the world market, these factories have a processing capacity of more than 100 metric tons per day. However, because of declining production in the country, Brazil began to import cocoa to keep the factories operating. Even this strategy has not worked; by 1992–93, these factories reported 38 percent idle capacity (May et al. 1993).

During transit, shipping, and storage, cocoa beans are often treated with phosphine to kill pests. This can be done prior to loading as well as during transit. The beans are then fumigated regularly, at least once per year, as long as they are in storage. Methyl bromide is also used for fumigation. This is a severely toxic, cumulative poison. Residues can cause brain damage months after use (Laird et al. 1996).

The next stage of processing after drying is cleaning; the beans are cleaned and all foreign matter is removed. After cleaning, the beans are broken and the resulting fragments, or "nibs," are winnowed. In some cases the beans are processed with alkali to neutralize acidity (this produces what is known as alkalized or "Dutch process" cocoa). The nibs are then roasted and ground and the mass is conditioned at high temperature. At this point processing diverges into two separate product lines, one for cocoa butter and another for fine-pressed cake or chocolate. The former involves filtering, solidification or tempering, degumming, and deodorizing to meet consumer demands in the cosmetics industry. Chocolate is packed in small kibbled cake form, or ground as cocoa powder and marketed directly to end users (May et al. 1993).

For each metric ton of cocoa beans harvested, nearly 10 metric tons of pod husks and pulp are generated. Traditionally this crop residue is discarded, either in small piles in the fields as it is harvested or in larger piles on the margins of the fields where it is left to decompose. Leaving uncomposted pods in the field, however, has been found to spread diseases such as witches'-broom and black pod rot (May et al. 1993).

SUBSTITUTES

While there are no direct substitutes for chocolate (carob has never lived up to its billing), there are substitutes for cocoa butter in the personal care and cosmetics industries. These include coconut oil, palm and palm kernel oils, and babassu oil (from the babassu palm, *Orbignya phalerata*, of Brazil).

An interesting issue raised by cocoa is that its consumption is directly linked to sugar consumption. So when the consumption of chocolate increases, the consumption of sugar increases as well.

MARKET CHAIN

Globally, the cocoa industry employs millions of people in the production sector. There are hundreds of thousands of producers, thousands of buyers at the local level, and hundreds of traders and exporters. A few hundred processors dominate the market. A dozen or so manufacturers of chocolate products dominate the interface with consumers. The profits taken at any point in the system vary considerably.

While the main producing countries control about 80 percent of production, they have not been able to form an effective, cohesive bargaining block.

The control of the final market for cocoa is concentrated in the hands of a very few, very large multinational companies who dominate its processing as well as the manufacture and distribution of chocolate: Nestlé Roundtree, Mars, Jacobs-Suchard, Hershey's, and Cadbury. The Hershey's corporation, for example, imported cocoa equivalent to Brazil's entire exports but obtained ten times the Brazilian suppliers' revenues from its value-added processing and manufacturing. Many of the export houses in cocoa-producing countries are subsidiaries or joint ventures of these same multinationals. These companies are armed with more complete information than any other players in the market chain regarding harvests, purchase terms, and financing. Furthermore, they reduce their risks via hedges in the New York and London commodity exchanges, and they manage their own inventory stocks to prevent, or at the very least buffer, production variations and possible producer price increases.

A 1993 study showed that the value of cocoa exported by the principal producing nations in 1989–1990 was $1.8 billion. By processing the cocoa into chocolate and adding sugar, nuts, and milk, the five dominant corporations obtained gross revenues of $36 billion in 1990. CABI Bioscience (2001) reported that the global trade in confectionery (chocolate has the lion's share of this) is estimated at $80 billion per year.

MARKET TRENDS

Between 1961 and 2000 global cocoa production increased by 183 percent while cocoa traded internationally increased 230 percent. During the same period, prices declined by 68 percent.

By 2000–2001 cocoa prices were at an all time low. This was due in large part to overplanting during the market peak in 1976–77 that stimulated widespread planting. This planting, in turn, resulted in huge surpluses beginning in the early 1990s; these annual surpluses created stockpiles that continued to drive down prices. Cocoa generally takes three to four years to produce after planting and seven years to mature. Production increases for the first twelve years or so and then begins to decline. This means that production from plantings in the 1970s would have peaked in the early 1990s. In 1991 the stockpiled surplus represented 70 percent of one year's production. By 1999 the ratio had dropped to 40 percent, but the surplus stock of cocoa was still more than 1 million metric tons. By 2002 the price of cocoa had increased dramatically in response to increased consumption and declining stocks.

The cocoa market peaked in 1976–77 at a price of about U.S.$3,600 per metric ton and bottomed in 1992–93 at a price of about $800 per metric ton. By 2002 the price was near $1,500. The cycle from planting to retiring trees is about twenty-five to thirty years, but trees can continue to produce at lower levels for some time thereafter. Globally, many producers will probably retire their current, older plantings and replant or convert to another crop within the next decade. Those countries with lower wage rates, particularly in Africa and Asia, will continue to be able to compete even with current low prices or low productivity so long as they have forested areas and are permitted to expand into them. It is not clear whether that will be the case with Brazil and other South American countries or even with Asia, where the price of labor is increasing. If full-sun, highly productive cocoa is planted increasingly, especially in Asia, it will tend to reduce prices considerably and to marginalize shade-grown cocoa in other parts of the world.

European countries have had preferential cocoa tariffs for former colonies. These tariffs discriminated against cocoa products based on the country of origin of production and/or processing. Under the regulations of the World Trade Organization, that will probably no longer be allowed. This, too, will cause shifts in production to favor the lowest-cost producers.

ENVIRONMENTAL IMPACTS OF PRODUCTION

The main environmental problems associated with cocoa production are habitat conversion, forest degradation, soil degradation, and pollution from processing by-products. Each of these is discussed below. In addition, producers use a wide range of pesticides and agrochemicals that have impacts both in the area of use as well as downstream through the impacts of effluent contaminants on freshwater and marine organisms.

Habitat Conversion and Deforestation

The production of cocoa results in deforestation. Best estimates indicate that cocoa production is probably responsible for the loss of some 8 million hectares of tropical forest (Hardner et al. 1999). The climatic and agricultural conditions most suited for traditional cocoa cultivation are precisely those that harbor extraordinary amounts of biodiversity. In fact, most of the land that has been historically cleared for cocoa production is in what would today be called biodiversity hot spots. These include areas in Brazil, Ecuador, Peru, Colombia, Ghana, Côte d'Ivoire, Cameroon, and Indonesia. In the West African countries of Ghana and Côte d'Ivoire, only small patches of original forest cover have been spared in the face of advancing cocoa production.

Average cocoa plantings remain productive for only twenty-five to thirty years, so expansion into new forests is the norm. If nothing is done to prevent it, cocoa cultivation can be expected to cause the deforestation of millions of hectares of tropical forests over the next twenty-five years. Simply maintaining current production levels could well mean the clearing or selective cutting of more than 6 million hectares of tropical forests as the cocoa frontier expands on one side and leaves degraded areas behind on the other. Another question, then, is what will be the next use of those areas currently devoted to cocoa production, and will the environmental impacts be more or less than those of cocoa production?

In Brazil cocoa cultivation is one of the main causes of the conversion of vast tracts (over 700,000 ha in the past century) of Atlantic coastal forests. If one looks at the relationship of deforestation to cocoa production, there is cause for concern. Three periods of deforestation related to cocoa production can be identified in Brazil: 1945–65, 1975–79, and 1982–86. During the first period, deforestation resulted from stagnant cocoa prices. During the second period, deforestation resulted from high prices. And, during the final period, deforestation resulted from declining cocoa prices. In short, deforestation resulted from upward or downward price shifts as well as overall market stagnation (May et al. 1993). On first glance, it appears that cocoa prices have little to do with deforestation. In fact, since cocoa is the only game in town, any change in price can cause deforestation. When prices were flat or were high people planted more to increase their income. When prices were low people increased planted areas or the density of existing plantations in an attempt to maintain their previous income.

Much of cocoa production in Brazil is centered in the state of Bahia. Production peaked there in the 1970s with about 400,000 hectares planted. As cocoa prices fell, agrochemical inputs were no longer financially feasible and marginal cocoa lands fell dormant. Witches'-broom has systematically destroyed cocoa trees throughout the region. Declining prices have left farmers with little money to pay laborers to fight witches'-broom. Debt has become so overwhelming in the cocoa sector that farms have been (and continue to be) sold and/or converted to other uses.

As a consequence, in Brazil today deforestation in cocoa-producing areas is not accelerated by the expansion of cocoa production but rather by its contraction. The low international prices for cocoa are now causing many planters to go in and cut the more valuable shade trees that were left during the initial cocoa planting. Farmers use the funds from these trees to finance the conversion of their farms from cocoa to other agricultural and ranching activities. These alternative cropping systems (generally pasture or annual crops) eliminate virtually all biodiversity, and furthermore have proven to be more short-lived than cocoa-based production systems.

Hardner et al. (1999) predict that at least half of Brazil's cocoa farms will be converted to other uses in the near future. Most conversion will include cutting not only cocoa trees but also the intermixed natural forest remnants within the cocoa farms. Historically, at least, cocoa production slowed deforestation in Bahia, but how much forest will remain in the face of the failure of the cocoa market to rebound remains to be seen. Strategic intervention by conservationists to help save or make viable the Atlantic forest cocoa agroforestry production system could do a great deal for protecting the last pockets of biodiversity within the region. However, unless the land ownership patterns in Bahia revert to smaller units so producers can use their own labor to compete in global markets (which is not likely to happen), cocoa will not be a viable crop in that region.

In Indonesia the rapid expansion of cocoa production opened previously inaccessible tropical forests in such places as Sulawesi and Central Sumatra. New settlements in such areas led to further deforestation even when cocoa went into decline due to low international prices. Small farms expanded from less than 50,000 hectares in 1980 to more than 400,000 by 2000. Cultivation was preceded by the dramatic clearance of forests. In addition to increases in cultivation, the population was increasing in areas with expanding cocoa production. In southern Sulawesi for example, the population doubled in the 1980s and doubled twice in the 1990s. Whether cocoa production ultimately proves profitable or not, most of these immigrants and their children will remain and will put additional pressure on the environment and natural resources.

Forest Degradation

Much cocoa cultivation in the world today is undertaken in agroforestry systems in which some part of the natural forest is left in place. Even so, shade production has considerable impact on the ecosystems where it is established. Biomass and soil fertility declined because of cocoa production in Nigeria (Ekanade 1987). Specific impacts documented include losses of overall foliage cover (reduced by 6.9 percent), reduced height of native trees (a 58.6 percent reduction), reduction in tree girth (a 66.9 percent reduction), tree basal area (88.1 percent reduction), and volume of wood (95 percent reduction). Only tree density and accumulated litter showed a relative increase (by 78 percent and 2.6 percent, respectively) in cocoa plantations relative to natural forests.

Forest mammals, reptiles, and amphibians showed declines both in absolute numbers and species diversity similar to the deterioration of the vegetation matrix. What tends to happen is that some species disappear, and a small subset of species that do well in disturbed areas tend to dominate cocoa production forests.

In Brazil, even in the shade cocoa-planting system where seedlings are planted within native forests, the floral substrata are removed, as are about 90 percent of the original tree species. The impact on sedentary biodiversity can be devastating.

While clearing the understory and much of the forest canopy to plant shade cocoa has significant environmental impacts, experience and research have both demonstrated that sustainable shade cocoa production provides habitat to important forest and migratory bird and mammal species. Sustainable shade cocoa production can play a strategic role in the preservation of forests, forest remnants, and forest corridors—those forested areas that connect larger blocks of intact forest (Knight 1998). Similarly, higher diversity within the cropping system has been found to lead to higher diversity in associated biota, as does lower use of pesticides. Overall, increased biodiversity leads to more effective pest control and pollination. And finally, increased biodiversity leads to more efficient nutrient recycling (Whinney 2001).

Soil Degradation

Cocoa cultivation often exposes soils when forest vegetation is removed prior to planting. Erosion occurs as plantations are established and even during their early years. Once plants mature and tree canopies are reestablished, erosion rates decline. However, studies show that foliage cover is not as complete even in traditional cocoa plantations as it is within natural forests, implying that erosion rates are likely to be higher in cocoa agroforestry plots than in natural forests. Because the leaves of cocoa do not decompose quickly, they can suppress other vegetation. This could make soils more susceptible to erosion.

In addition to erosion, soils in cocoa plantations experience a loss of fertility. Nutrients are exported from plantations in the form of seedpods, but more importantly, the loss of ground cover probably leads to increased leaching. The biotic and soil components of the Nigerian tropical forests where cocoa is being produced have deteriorated considerably (Ekanade 1987). This, in fact, suggests why cocoa plantations must be moved periodically to more fertile, virgin forest areas. However, instead of allowing the forests to regenerate in some form, most abandoned cocoa plantations are cleared and used for conventional agriculture. In this sense, cocoa production is merely the first step in the ultimate deforestation of an area even though the cycle may take twenty-five years or more to complete.

Wastes from Processing

For each metric ton of cocoa beans harvested, nearly 10 metric tons of waste (pods, pulp, etc.) are created. In the past, the waste was often kept in the plantation and used as organic fertilizer or mulch. This practice, however, favors the propagation of witches'-broom and black pod rot unless the materials are properly composted to eliminate diseases. Such waste can also be used as mosquito breeding grounds and can be responsible for the spread of diseases to humans as well.

BETTER MANAGEMENT PRACTICES

Several different but complementary strategies could help reduce the environmental costs of cocoa production. These practices should center on increasing the ability of producers to replant the same areas indefinitely, reducing the use of agrochemical inputs and the creation of wastes, and turning wastes into by-products or substitutes for purchased inputs. Biodiversity can be promoted through interplanting, which can be sold to producers as a means of diversifying their sources of income. Working with producers to adopt better management practices will be most effective when complemented and supported by work with the larger industry, investors, and governments as part of a concerted effort to reduce the negative environmental impacts of the industry.

For cocoa, the identification of better management practices will require that producers and researchers work together to identify, analyze, document, and disseminate information about the most promising practices from around the world. In every instance, the approach should be to identify production techniques that pay for themselves and offset the cost of adopting the more expensive better management practices. It appears that cocoa yields can be improved by more than 40 percent simply by adopting improved practices that allow producers to achieve yields that are within the genetic parameters of the varieties that they cultivate (Ooi et al.

1990). Such practices can be as simple as regular, thorough pruning after harvest to increase yields and reduce pests.

Shape the Expansion and Maintain the Viability of Shade Cocoa

In the near future, unless full-sun cocoa can be produced on existing or degraded agricultural areas, it is likely that cocoa production will continue to expand into existing forests. In these instances, expansion should be encouraged in ways that will reduce its impact on biodiversity and ecosystem functions as well as ensure the financial viability of the industry over time. For example, land use planning and zoning should be undertaken in consideration of what is known about how cocoa can be best produced, over time, with better practices that have already been identified by growers.

Cocoa can also be grown in association with other taller, commercially valuable trees. Cocoa has been grown on vast plantations of coconut, rubber, and oil palm trees. In these systems, the highest price commodity gets the most attention and the others tend to be left to fend for themselves. In many of these systems, since cocoa is often not the principal focus of many large-scale producers, it is often neglected.

In other large-scale agriculture or aquaculture production systems, owners and managers have found that making line workers responsible for specific plots and giving financial rewards to them for increased net profits on their areas can increase profitability as much as fourfold. Such "win-win" incentive programs should be adopted in cocoa production as well. Without such innovations, large-scale producers can never hope to compete with small-scale producers who use unpaid family labor to support their production.

Small-scale growers, however, have a different perspective; cocoa is grown together with other crops with the same care. This strategy would be improved if a more integrated system could be constructed that provides both food and cash crops while utilizing family labor rather than expensive inputs.

The areas of greatest concern and the areas where strategies may be more successful, however, are those where there are still considerable forested areas suitable for the cultivation of cocoa and where the industry may try to expand. For instance, Cameroon has only 0.5 million hectares of forests converted to cocoa production, but an additional 2.5 million hectares of forest land suitable for cocoa. The maintenance of a relatively stable level of production somewhat masks the geographic shift in cultivation from the central and southern regions of the country to the southwest where productivity continues to grow. Unfortunately, the southeast is among the most biodiverse regions in the country. Efforts to stem this shift in production will need to begin immediately and must address the root causes for the shifts in cocoa production, namely, loss of productivity in converted lands in other parts of West Africa.

Increase the Efficiency of Agrochemical Use

While much of the cocoa production in the world at this time is de facto organic, as the price continues to increase a number of producers will find it advantageous to purchase and use increasing quantities of agrochemicals. During periods of low prices, many cocoa producers reduce their applications of expensive fungicides and pesticides. Not only do such practices lead to large crop losses, but also low-level reduced spraying can lead to increased resistance over time. In short, reduced, efficient use of chemicals should not imply their sporadic use, which can be quite damaging.

There are several ways to reduce the use of agrochemicals. One is to certify producers as organic and pay them to use labor instead of chemical inputs to produce their crops. There are formal organic certification procedures, but total organic production globally is still less than 10,000 hectares. However, it would be important to measure the environmental toxicity of several copper and sulfur compounds as well as tobacco extracts that are currently allowed for use by organic producers even though they are highly toxic to other organisms.

There are other ways to reduce chemical inputs as well. Managed spraying systems, using a list of approved chemicals (and excluding ones that are banned in the consuming countries), and ranking the approved chemicals according to their overall toxicity are all ways to reduce the use of the most toxic substances.

Farmers tend to adopt technology packages selectively. Often the highest returns on capital investments are most attractive (Johnson et al. 1999). However, the perceived risks of innovation are often as important as their perceived profitability. The interactive impacts among several variables can also be used to advantage when trying to get producers to adopt better practices. Stepwise adoption of complementary better practices (e.g., the increase of organic matter and the reduction of chemical inputs) can be encouraged as a way to gradually reduce impacts and improve profitability.

There are also a number of biological controls in various stages of development that appear to reduce the need to use agrochemical inputs. For example, nonpathogenic fungi can be applied to cocoa to reduce the levels of infective spores of disease-causing fungi. In Ghana, certain fungus species have been found to inhibit the growth of black pod rot in the laboratory. In Brazil, a commercial formulation of this product has been marketed to control witches'-broom, and producers are very enthusiastic about it (Pesticide Action Network 2001). Another approach involves the introduction of a beneficial fungus into the tissues of the cocoa tree. The fungus does not harm the tree; it helps protect it by attacking pathogens and increasing resistance. CABI Bioscience is investigating several fungi to control witches'-broom in South America (Pesticide Action Network 2001).

Finally, the use of natural enemy species for biological control of insect pests is also being investigated in several countries. In Malaysia, the black ant is being used to control cocoa mirids, a common pest. To date, the main problems with biological controls have been that they kill only a very narrow range of pests, they perform poorly relative to their cost, and the product quality is inconsistent (Pesticide Action Network 2001).

For low-input, small-scale producers, improving shade management can reduce expensive inputs while balancing overall productivity. For example, shade can re-duce weed growth as well as the occurrence of some fungi. Such systems also in-crease the long-term productivity of cocoa and can be used to restore abandoned or degraded land (Rice and Greenberg 2000).

Diversify Sources of Income

Researchers have shown that interplanting low-input cocoa plantings with fruit trees can buffer the impacts of low cocoa prices. The break-even cocoa price for such in-tegrated producers is just over 50 percent of the price needed to break even in cocoa production without fruit trees (Rice and Greenberg 2000).

Similarly, shade trees selectively cut to manage shade can be sold for timber, fuel, or charcoal. The sale of shade trees in coffee plantations has shown that they can compensate for lost yields of 17 percent when prices are high, 33 percent when they are intermediate, and 100 percent when they are low (Rice and Greenberg 2000). Thus, the shade trees offer sources of income at precisely the times when producers need them most. There is no reason to assume similar earnings/loss effects would not apply equally to cocoa.

Managing carbon is another potential income source for cocoa farmers if Kyoto-like mechanisms are ever ratified. Forty-year-old cocoa agroforestry systems in Cameroon fix atmospheric carbon at levels of around 154 metric tons per hectare. Systems that are fifteen to twenty-five years old sequester 111 and 132 metric tons of carbon, respectively. While lower than sequestration rates for primary forests (307 MT/ha), they are far greater than rates for annual crops, even those with associated fallows (Rice and Greenberg 2000). Depending on the price assigned to carbon, se-questration could supply significant income for producers and also an incentive for them to retain shade trees in their areas of production and to reduce chemical in-puts. Provided the carbon can continue to be stored, shorter-term crop rotations tend to sequester more carbon per hectare per year.

Reduce Waste and/or Create By-Products

As described earlier, nearly ten times as much waste from pod husks and pulp are generated for each metric ton of cocoa beans. If properly composted, this material can provide large amounts of organic matter for fields without risking the spread of disease. Alternatively, the pods can be ground and used in cattle feed or the alkaloid theobromine can be extracted from them for sale as a by-product. The pulp that sur-rounds the seeds is increasingly sold for juice, but it is also made into alcohol, vine-gar, wines, and liqueurs (May et al. 1993). Such waste can also be dried and used for fuel. Producers are now exploring the possibility of using this waste for fuel to dry the beans or turning it into charcoal briquettes for sale on the open market.

In recent years, the adoption of very simple and relatively inexpensive crushers, coupled with fermentation of the cocoa hulls by inoculating with effective microorganisms that speed up composting, have shown that there is a potential for returning pod hulls back to the field. The microorganisms used suppress the propagation of other harmful microorganisms and therefore do not contribute to the reestablishment of diseases such as witches'-broom. This could solve the disease problem currently associated with returning pod hulls to the field.

Encourage Full-Sun Cocoa on Degraded Lands

Another way to avoid forest degradation is to change the architecture and planting density through the use of full-sun cocoa. This will only work, however, if trees are planted on existing or degraded agricultural lands rather than newly cleared forests or existing shade cocoa systems. High-density planting is more efficient. In effect, higher yields can be achieved on previously degraded areas without any further impacts on soil fertility or habitat loss.

While cocoa production has been promoted in many countries around the world, the technology being used is the low-density technology that achieves yields of, at best, only 1.5 metric tons per hectare. There are, however, new production technologies that allow those levels to be doubled, or even tripled, to as much as 4.5 metric tons per hectare per year. This strategy is not very expensive. Full-sun cocoa utilizes newer, more compact varieties whose vertical trunks are the primary fruit-bearing areas of the plant, rather than the horizontal branches. These shorter, grafted plants produce more quickly. Pruning allows producers to keep the plants short so that pesticide sprays are more effective. Productivity falls off sharply after about ten years, but if rotated with other crops this system can be used to prevent conversion of natural habitat.

These technologies have been developed and are used by some competitive, private companies. Their strategy is to increase production to the point that labor costs are not as significant a factor in their economic viability. However, labor costs on full-sun cocoa plantations have been reported to be 70 percent higher than on conventional systems (Chok 2001). This means that increased productivity is required to offset such costs.

Larger companies adopting this technology, however, may be swimming upstream. Smaller producers using the same techniques could easily undermine the larger companies because they do not rely on the use of paid labor. Unfortunately, the technology has not filtered down to the smaller producers yet. This is an important bottleneck that could be addressed through the provision of grafted stock and overall production packages to small-scale producers. One place where this is happening currently is in Vietnam, where the government has set up thousands of nurseries to provide grafted cocoa seedlings to coffee producers. Vietnam is the second largest coffee producer in the world, but its production has helped trigger the lowest real producer coffee prices ever. Many coffee producers want to shift production to another crop. It is not clear, however, whether such a dramatic increase in production of cocoa in those areas would not cause a similarly dramatic drop in cocoa prices.

An important leverage point in cocoa production is political. It is virtually impossible to develop effective strategies for working directly with hundreds of thousands of small producers scattered throughout the world. Governments have the ability to encourage producers to adopt better practices through regulatory structures that influence production, or as a condition of concession permits or licenses to use specific areas for cocoa production.

Three countries account for 70 percent of all production. Another seven bring the figure to virtually all traded cocoa. One strategy could be to work with governments to increase the sustainability of existing or planned cocoa expansion. For example, the identification and analysis of better management practices for specific regions could help existing producers reduce environmental problems and increase profits. Such practices could also be the criteria that governments use to zone new areas for cultivation. Such improved practices could serve as the basis for government licenses, permits, or even agricultural credit for the ongoing production of cocoa. Finally, better management practices can be used to make convincing financial arguments for why governments should modify land subsidies or infrastructural support in order to encourage the industry to become more sustainable.

Work with Companies to "Green" Their Supply Chains

A second key leverage point is the marketing and supply system. A careful examination of the cocoa value chain indicates that there are a few areas where most of global production passes through the hands of only a few. These leverage points should be the key targets for affecting the sustainability of production at the local level. For example, cocoa producers traditionally sell their product to one of three buyers—middlemen who aggregate stocks for resale (e.g., Phibro, Sucres at Denrees, Jacobs-Suchard, S.W. Group, Tardivat, Cargil), cocoa butter producers (e.g., Gill and Duffis, Barry, W. R. Grace, Gerkens, Van Houten), or chocolate confectioners (e.g., Jacobs-Suchard, Mars, Nestlé-Roundtree, Cadbury, Hershey's). The last two categories represent very few players. In some areas buyers have near monopolies; in others a few buyers have oligopolistic control of markets. The category of middlemen has the largest number of players, but even there the total numbers are quite small by comparison to the number of producers, especially when dealing with specific regions.

One proposal would be to pressure large multinationals to make conservation investments in pristine forest areas as compensation for the forest destruction that occurs with the production of cocoa. These "forest offsets" would be similar in theory to carbon offsets. Since multinationals are few in number and heavily capitalized, such negotiations would be more simple, cost-effective, and timely than efforts to modify the behavior of hundreds of thousands of small producers.

The problem with this approach is that it assumes that production impacts cannot be mitigated directly, for example, that there are no better ways to produce cocoa. Rather, the goal is to make sure that every company involved protects a forest equal in size to the one that the product they buy will destroy for the establishment of farms. The theory is that there would be no net loss from cocoa production. This is fine when replanting occurs under existing cocoa and shade trees. If, however, cocoa production is indeed a moving frontier, then this option is not feasible. It does not even postpone the inevitable forest destruction for very long unless the set-asides are purchased and put into protected status.

A better approach would be to work with key companies in the supply chain to develop screens for more ecological production that they can use for their purchases. Once the ecologically based screens are created for buyers, they can also be used to reduce the risks of investors and insurers who are also important players with the industry. Such screens will send a signal to producers about what type of products, produced in what way, they want to purchase. While no one knows how to produce cocoa indefinitely on the same piece of land, better practices for the industry are known and those not well known can be made available to those producers who are able to be competitive. Grafting improved varieties, high-density planting, avoiding steep slopes and riparian areas, reducing the exposure of soil during planting, utilizing ground cover, intercropping, reducing input use, reducing waste or converting it to usable by-products, and regular replanting are some of the techniques that can be encouraged. If adopted they would reduce considerably the most common problems from current practices. By working in partnership with producers to encourage the adoption of these practices, buyers can help to maintain their sources of supply well into the future.

OUTLOOK

People are not going to stop eating chocolate. If anything, demand will continue to increase. Unfortunately, traditional production has known environmental problems. Cocoa is not easily or cheaply replanted on the same area, for example. So long as this is the case, traditional cocoa production will continue to expand into natural forests. Every effort must be made to find alternatives. There are two major avenues for this work. The first is to identify and analyze ways to increase and extend production in areas of current use. Research suggests that this may be a promising strategy; because it appears to depend on family labor it may be extremely important for smaller producers.

The second strategy is the development of full-sun cocoa production. This system promises to increase production dramatically per hectare and to reduce the pressure on natural forest conversion from planting cocoa. There are some major drawbacks, however. The system also promises to be far more input-intensive than traditional cocoa production. In addition, while full-sun cocoa can be produced on agricultural land or even degraded lands, it may be far cheaper to undertake in natural forests, degraded forests, or agroforestry areas. In any of these areas, full-sun cocoa would result in a net biodiversity loss. Even so, it is important to examine carefully as an option and to identify the best practices for this type of cultivation that could not only reduce impacts and increase profits, but also improve overall rotation cycles. It is very important to monitor closely what happens in Vietnam with full-sun cocoa as this could very well set the precedent, good or bad, for future cocoa production.

REFERENCES

Akande, S. O. 1993. *The Effects of Producing and Processing Cacao on the Environment: A Case Study of Nigeria.* Rome: United Nations Conference on Trade and Development. UNCTAD/COM/23. 27 August.

CABI Bioscience. 2001. Developing Sustainable Cocoa Production Systems—Briefing Paper. CABI Commodities. Available at http://www.cabi-commodities.org/Acc/ACCrc/ACCrcCOCbp.htm. February.

Chok, D. 2001. *Cocoa Development and Its Environmental Dilemma.* Washington, D.C.: Smithsonian Migratory Bird Center.

CIBLE (Bureau d'Etudes et Recherches Marketing, de Conseils en Gestion Commerciale et Industraille et d'Etudes Socio-Economiques). 1993. *L'Impact de la Culture de Cacao et du Café sur L'Environnement.* Rome: United Nations Conference on Trade and Development. UNCTAD/COM/18. 27 August.

Cocoa Growers' Bulletin. Many issues of this journal have useful articles on the cultivation of cocoa around the world as well as the search for germ plasm in the Americas. Cadbury Ltd. P.O. Box 12, Bourneville Lane, Birmingham B30 2LU, United Kingdom.

Corning, S. L. 1992. *Cacao: Situation and Trends.* LAC TECH Project (LAC/DR/RD). U.S. Agency for International Development. September.

The Economist. 2000. Cadbury Schweppes: Of Sweets and Appetites. May 27. Pages 67–68.

Ekanade, O. 1987. Small-Scale Cocoa Farmers and Environmental Change in the Tropical Rain Forest Regions of South-Western Nigeria. *Journal of Environmental Management* 25: 61–70.

Evans, H. C. 2001. *Disease and Sustainability in the Cocoa Agroecosystem.* Washington, D.C.: Smithsonian Migratory Bird Center.

FAO (UN Food and Agriculture Organization). 2002. *FAOSTAT Statistics Database.* Rome: UN Food and Agriculture Organization. Available at http://apps.fao.org.

Hardner, J. J., G. Richards, and T. Walker. 1999. *The Scale and Trends of Cacao Production Impacts on Global Biodiversity.* Prepared for Center for Applied Biodiversity Science, Conservation International. October 15.

Hebbar, P. K., R. D. Lumsden, U. Krauss, W. Soberanis, S. Lambert, R. Machado, C. Dessimoni, and M. Aitken. 1999. Biocontrol of Cacao Diseases in Latin America: Status of Field Trials. Paper presented at cocoa conference held at CATIE (Centro Agronómico Tropical de Investigación y Enseñanza) Turrialba, Costa Rica. June/July.

Henderson, J. 2001. The Birth of Chocolate, or, The Tree of the Food of the Gods. *Arts & Sciences* 23(1) Fall. Ithaca, N.Y.: Cornell University.

Johnson, E., C. Pemberton, and J. Seepersad. 1999. Adoption Behavior of Private Cocoa Farmers in Trinidad and Tobago. *Journal of Tropical Agriculture* 76(2) April: 136–142. Trinidad.

Knight, C. 1998. Sustainable Cocoa Program. *Plantations, Recherche, Developpement.* Vol. 5: 387–389. November–December. Montpellier, France: Centre de Cooperation Internationale et Recherche Agronomique pour le Developpement (CIRAD)-CP: Service, information, et communcation.

Laird, S., C. Obialor, and E. Skinner. 1996. *An Introductory Handbook to Cocoa Certification. A Feasibility Study and Regional Profile of West Africa.* New York: The Rainforest Alliance.

Matlick, B. K. 2001. *Machete Technology: What Small Cocoa Farmers Need!* Washington, D.C.: Smithsonian Migratory Bird Center.

May, P., with R. Vegro and J. A. Menezes. 1993. *Coffee and Cacao Production and Processing in Brazil.* Rome: United Nations Conference on Trade and Development. UNCTAD/COM/17. 27 August.

N'Goran, K. 2001. *Reflections on a Sustainable Cacao Production System: The Situation in the Ivory Coast, Africa.* Washington, D.C.: Smithsonian Migratory Bird Center.

Ooi, S., Y. Woo, and K. Leng. 1990. Yield Maximization and Conservation in Tree Crop Agriculture with Special Reference to Oil Palm and Cacao. *Proceedings of the International Conference on Tropical Biodiversity.* Kuala Lumpur, Malaysia. 12–16 June.

Padi, B., and G. K. Owusu. 2001. *Towards an Integrated Pest Management for Sustainable Cocoa Production in Ghana.* Washington, D.C.: Smithsonian Migratory Bird Center.

Pesticide Action Network UK. 2001. Compiled by Little, T. Sustainable Cocoa Systems. *Pesticide Management Notes.* London: PMN No. 12, July.

Rice, R. and R. Greenberg. 2000. Cacao Cultivation and the Conservation of Biological Diversity. *Ambio* 29(3) May: 167–173. Royal Swedish Academy of Sciences. Available at www.ambio.kva.se.

Ruf, F. 1995. From Forest Rent to Tree-Capital: Basic 'Laws' of Cocoa Supply in: Ruf, F. and P. S. Siswoputranto (eds.). 1995. *Cocoa Cycles: The Economics of Cocoa Supply.* Cambridge: Woodhead Publishing.

Ruf, F. and Yoddang. 2001. Cocoa Migrants from Boom to Bust in: Ruf, F. and F. Gerard (eds.) 2001. *Agriculture in Crisis: People, Commodities, and Natural Resources in Indonesia, 1996–2000.* Montpellier, France: Centre de Cooperation Internationale et Recherche Agronomique pour le Developpement (CIRAD).

Seudieu, D. O. 1993. *L'Impact de la Production et de la Transformation de Café, du Cacao et du Riz sur L'Environnement en Côte D'Ivoire.* Geneva, Switzerland: United Nations Conference on Trade and Development. UNCTAD/COM/24. 6 October.

Smith, N., J. Williams, D. Plucknett, and J. Talbott. 1992. *Tropical Forests and their Crops.* Ithaca, N.Y.: Cornell University Press.

UNCTAD (United Nations Conference on Trade and Environment. 1999. *World Commodity Survey, 1999–2000.* Geneva, Switzerland: UNCTAD.

———. 1993. *Experiences Concerning Environmental Effects of Commodity Production and Processing: Synthesis of Case Studies on Cacao, Coffee and Rice.* TD/B/CN.1/15. Geneva, Switzerland: UNCTAD. 22 September.

Whinney, J. 2001. *Considerations for the Sustainable Production of Cocoa.* Washington, D.C.: Smithsonian Migratory Bird Center.

Wood, G. 1991. A History of Early Cocoa Introductions. *Cocoa Grower's Bulletin* 44:7–12.

———. 1987. History and Development, Environment, Propagation, Establishment, From Harvest to Store, Quality and Inspection, Production, Consumption and Manufacture in: Wood, F. A. R. and R. A. Lass (eds.) 1987. *Cocoa.* Harlow, Essex: Longman Scientific and Technical.

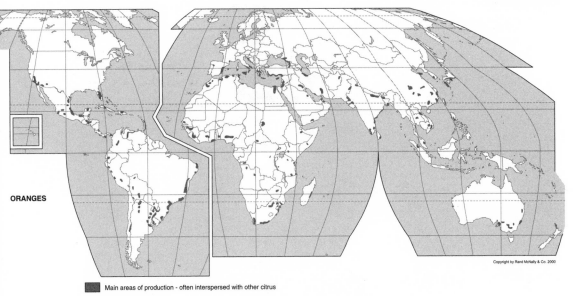

ORANGES

Main areas of production - often interspersed with other citrus

Copyright by Rand McNally & Co. 2000

ORANGES *Citrus sinensis*

PRODUCTION		INTERNATIONAL TRADE	
Area under Cultivation	3.6 million ha	Share of World Production	13%
Global Production	62.4 million MT	Exports	8.1 million MT
Average Productivity	17,330 kg/ha	Average Price	$579 per MT
Producer Price	$219 per MT	Value	$4,691 million
Producer Production Value	$13,662 million		

PRINCIPAL PRODUCING COUNTRIES/BLOCS (by weight)	Brazil, United States, Mexico, India, China, Spain, Iran, Italy, Egypt, Pakistan
PRINCIPAL EXPORTING COUNTRIES/BLOCS	
Oranges	Spain, United States, South Africa, Morocco
Orange Juice (concentrated)	Brazil, United States, Spain, Costa Rica, Belize
Orange Juice (single-strength)	Germany, Belgium, United States, Netherlands
PRINCIPAL IMPORTING COUNTRIES/BLOCS	
Oranges	Germany, France, Netherlands, United Kingdom, Russia, Saudi Arabia
Orange Juice (concentrated)	United States, Canada, France, Korea, Saudi Arabia
Orange Juice (single-strength)	France, Belgium, Netherlands, Germany, United Kingdom
MAJOR ENVIRONMENTAL IMPACTS	Conversion of habitat Use of agrochemicals Soil erosion and degradation Wastes from processing
POTENTIAL TO IMPROVE	High Planting occurs mostly on existing agricultural land BMPs are cost-effective, reduce input use and increase production

Source: FAO 2002. All data for 2000.

Area in Production (Mha)

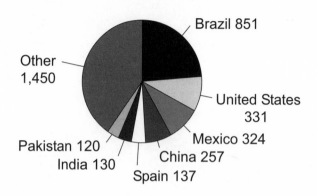

Brazil 851

Other 1,450

United States 331

Mexico 324

China 257

Spain 137

India 130

Pakistan 120

OVERVIEW

For many consumers, orange juice is part of a complete breakfast. Orange juice is perceived as a very nutritious food, one that is particularly important to consume in the winter as a way to avoid colds (though in fact, there are far better and cheaper sources of vitamin C).

The orange is an evergreen tree originally native to Asia. It was brought across the Middle East and North Africa and introduced into Europe by the Moors in Spain. This is where the Seville orange came from. Oranges were first introduced into the New World in the early 1500s but did not become a commercial crop for some 350 years. According to the Food and Agriculture Organization of the UN (FAO 2002), by the year 2000 citrus (which is mostly oranges) was the most-produced fruit in the world with a 22.5 percent share of all fruit produced. The most common oranges cultivated at this time, which provide most of the sweet oranges and juice sold in the world, are several different varieties of *Citrus sinensis*.

There has been a market for orange juice for about 100 years. While fresh oranges are produced throughout the world, it is really Florida growers who created the bulk orange juice market. This was done as a way to sell their product throughout the year in order to avoid glutting the market during the harvest. In the past, Southern California (e.g., Orange County) also produced a lot of oranges, but now real estate development and water shortages have reduced production there. In the 1970s severe frosts in Florida ruined harvests, and buyers began to look further afield to find sources of oranges for juice. At that time, significant production started in the states of São Paulo and Paraná in Brazil's Atlantic coastal forest area. Most of the lands that were ultimately planted with oranges had previously been used to grow coffee.

Overproduction of juice globally now characterizes the fresh and frozen orange juice markets and contributes to declining prices. However, diseases in Florida and Brazil are reducing production and increasing production costs, and production there is expected to decline by 25 percent within the next five to ten years. In all likelihood, this will stimulate a new wave of planting and production in other countries.

PRODUCING COUNTRIES

The countries with the largest areas planted to orange trees include Brazil (850,875 ha), the United States (330,850 ha), Mexico (323,618 ha), China (257,200 ha), Spain (137,000 ha), India (130,000 ha), and Pakistan (120,000 ha). Brazil has about a quarter of all land worldwide planted to oranges and produces more than 28.8 percent of the 62.4 million metric tons of oranges produced globally. The other six large producers have nearly 36 percent of the area planted to oranges globally and produce nearly 36 percent of all oranges annually (FAO 2002).

The data on orange production is somewhat complicated. For example, the primary fresh orange exporters in 2000 by volume were Spain, the United States, South Africa, and Morocco. The main orange juice concentrate exporters were Brazil, the United States, Spain, Costa Rica, and Belize. By contrast, Germany, Belgium, the United States, and the Netherlands are the largest exporters of single-strength, or non-concentrate orange juice.

Globally, Brazil and the United States dominate orange juice production, and combined they accounted for 87.3 percent of total frozen concentrate in 2000 (FAO 2002). Since the 1980s, however, Brazil has been the single largest producer. In 2000 Brazil produced 22.74 million metric tons of oranges, which amounted to 34.4 percent of global production. About three-quarters of all Brazilian oranges are crushed for juice. Brazil produces 47 percent of the world's frozen concentrated orange juice and accounts for more than 80 percent of global exports of frozen orange juice concentrate. Oranges represent 25 percent of total agricultural value in São Paulo state, and 5 percent of all Brazilian agricultural production (Neves et al. 2001).

The United States produced about half as many oranges as Brazil in 2000 (11.9 million metric tons) and accounted for about 18 percent of global production. Mexico, India, China, and Spain add another 6 percent to global production.

However, orange and orange juice production can dominate small countries' exports even if the amount exported is not significant globally. For example, oranges are the second most valuable export in American Samoa, Belize, and Dominica and the third most valuable export in the Cook Islands, Cyprus, Morocco, and Swaziland (UNCTAD 1994).

In addition to frozen orange juice concentrate, most producers also export a number of other by-products. These include frozen orange juice (not concentrated), fresh oranges, pulp pellets (used mainly for animal feed), and essential oils (used as food flavorings and in personal care products and household cleaners). In the case of Brazil, the combined value of all these items is only 7 percent of the value of that country's frozen orange juice concentrate (Neves et al. 2001).

CONSUMING COUNTRIES

The main consumers of orange juice are the United States, Europe, Brazil, Canada, and Japan. The United States first developed the frozen orange juice market and dominated production until the 1980s. It still consumes half of the orange juice produced in the world but now must import an increasing proportion of its juice. Brazil supplies 90 percent of U.S. imports, or about 757 million liters (200 million gallons) per year. The European Union adds about a third of global imports, and Canada and Japan about 5 percent each.

Germans consume an average of 45 liters of orange juice per person per year (Neves et al. 2001) while Americans consume 22 liters and Brazilians only 9 liters. After the United States, the main importers of orange juice are Germany and the rest of the European Union, Japan, Canada, and South Korea. India, Indonesia, Mexico, and a few other countries are also large consumers of orange juice, but because they do not import the juice and their domestic production is not exported they are underrepresented in most statistics.

In addition to being a large producer and consumer of orange juice, the United States is also a large importer of orange juice. In some cases, at least, the juice appears to be repackaged and exported as a U.S. product.

PRODUCTION SYSTEMS

Orange juice is produced from monocrop plantations. In some parts of the world, natural habitat is cleared to plant orange trees. In most countries, such as Brazil, orange juice production comes from plantations that are instead planted on already-cleared lands formerly used for another crop.

The main varieties of oranges grown in Brazil are Pera, Hamlin, Parson Brown, and Natal. In Florida they are Temple, Parson Brown, Pineapple, Hamlin, and Valencia (similar to Pera). The harvest season in Brazil is from July through January; in Florida it is from November through June.

Orange seedlings are started in nurseries and then transplanted into grid planting systems in the field. Organic and/or synthetic fertilizers are applied during the planting. Heavy liming is required for the production of oranges in many areas as well. In some countries, other crops are planted between the rows before the trees grow to maturity. Once the trees begin to cover the area, these other crops are eliminated and grass or other ground cover is grown to prevent soil exposure and erosion. Leguminous ground cover is used in some areas.

Production begins in a few years, but more serious production occurs after five to seven years. Trees will produce for thirty or more years depending on how well they are maintained. Every year several activities must be undertaken to ensure both short- and long-term production. Orange trees are pruned at least once per year to encourage proper spacing of branches and to maximize production. Oranges require direct exposure to sunlight for the fruit to ripen properly. Traditionally, several forms of agrochemicals are used to produce oranges. Fertilizers and lime are applied each year or as needed. In addition to these tasks, ground cover and irrigation systems must be maintained.

During periods of declining prices, producers often cut production costs by forgoing some of these expenses. As prices pick up, producers tend to pay more attention to their trees, but the productivity of untended orchards cannot be turned around overnight. As a consequence, production on untended estates tends to continue to fall even as prices rise. In other instances, producers with a longer-term view appear to be willing to cover their marginal costs in bad years in the hopes of having a few good years. Unfortunately, generally declining prices have tended to squeeze producers' profit margins over time.

On a per-hectare basis the orange industry is one of the biggest users of pesticides. Trees are sprayed several times during the flowering and fruiting season to prevent insect damage. Chemical spraying is also undertaken to prevent mold, which can affect the leaves and the overall heath of the trees. In Brazil orange trees account for 6.5 percent of all pesticide use, but they account for the largest pesticide use per hectare of any agricultural crop (Neves et al. 2001).

Oranges are usually picked in the early morning or late afternoon in order to avoid having the picked fruit sit in the heat of the fields during midday. Oranges are picked by hand and consolidated first into wooden boxes and then into larger tractor-pulled trailers. Small producers tend to transport their production to elevated pickup points along major roads so that the trucks can pass underneath and gravity can be used to load the product. Trucks from a processing plant pass by the pickup point, or by the farm, to collect the fruit. Trap doors are opened in the storage bins so that gravity will allow the oranges to flow down into the larger trucks below.

Because heavier oranges produce more juice, the price paid to producers depends partly on the weight of the fruit. While farmers are paid according to the solid weight of their oranges, the standard unit of measure of oranges is a "box." Each box weighs more or less 40 kilograms (90 pounds) depending on the overall juice content. Farmers receive a premium or are penalized depending on the weight of the box, which gives an indication of the average weight of the oranges contained within.

Citrus groves are huge investments to establish. Costs have been estimated at U.S.$17,500 or more per hectare in developed countries (Barham 1992). It generally takes about five years for the trees to produce substantially and about ten years of operation to pay off the original investment. Discounted and averaged over the cost of the grove, initial capacity costs are approximately 50 percent of the annual grove operating costs (e.g., weeding, trimming, fertilizers, pesticides, and irrigation). Harvesting and transportation costs are generally equal to the annual operating costs (Barham 1992). The high cost of entering or leaving the market tends to discourage large fluctuations in area planted and output, other than those caused by freezes in Florida or drought in Brazil.

Producers are attracted to oranges because of the profits. In Brazil oranges produce 1.7 times the income per hectare as coffee, nearly 3.5 times that of sugar, more
than 5 times that of soybeans, 6.5 times that of corn, and 9.5 times that of wheat
(Neves et al. 2001).

In the 1980s Brazil and the United States accounted for 98 percent of the U.S.
supply of frozen orange juice concentrate. Brazil's share rose rapidly from 10 percent in 1980 to 50 percent of the U.S. market in the late 1980s, which amounted to
a third of the global market. Since then, other suppliers (such as Belize, with 1 percent of U.S. market share) have reduced Brazil's margin to the low 90 percent range
(Barham 1992).

The largest orange juice producers in the world are Brazilian companies; Montecitrus is one example. These companies dominate the global market. Montecitrus
and other large Brazilian companies are already producing certified organic orange
juice and dominate that market as well. They expect to produce a combined 12 million boxes (45,000 metric tons, or more than 100 million pounds) of organic citrus
per year. One trend of note is that by 2002 the cost of orange production had increased relative to price in Brazil and the United States. Both countries are expected
to reduce production in the near future at least relative to international demand.

Citrus is an ideal crop for production on smaller plantations because, unlike
large-scale producers, small producers can use their own labor to cover many of the
expenses associated with establishing plantations. The major bottleneck for small
producers, however, is the cost and volume of production required to run a processing facility. Given that such facilities now exist in many areas, perhaps it is time to
contemplate a new form of either cooperative ownership or employee stock option
plan (ESOP) for processing plants. Given the international interest in chain-of-
custody issues, even juice manufacturers or grocery store conglomerates may decide
that it would be wise to invest in such facilities as a way to ensure supply quantities
and quality.

PROCESSING

The orange processing industry is highly concentrated. Four Brazilian family-owned
companies control more than 50 percent of global supply. Three factors account for
this: the importance of market connections to the viability of processing plants, the
substitutability of products produced in different countries, and the fact that there
are few differences in processing throughout the world.

Oranges should be delivered to the central processing plant and processed within
twenty-four hours of picking. Due to their overall volume, oranges are transported
from the farms to central processing plants by large semitrailer trucks. In the case of
very large estates or plantations, the central processing plant may be owned by the
farm itself.

Oranges are pressed in such a way as to maximize the juice recovery and minimize
the release of acid from the peel. The overall value of oranges is based on three factors: their total juice content (determined by weight per volume), juice color, and

juice sweetness. Fruit juice sweetness is measured in units called "brix," which are an indication of the sugar content in the product. In general, heavier fruit with juice that has a good color and is high in sugar is the most valuable. In fact, in the United States, up to 10 percent of orange juice can come from tangerines to enhance the color.

Once processed, orange juice can be sold fresh, fresh frozen (e.g., not concentrated), pasteurized, or made into a concentrate that can also be sold fresh or frozen. Concentrate is produced by using evaporation to reduce the water content of the juice from 89 percent to 34 percent. (This leaves the soluble solids content of the product at about 66 brix for Brazilian juice; from the United States it tends to be a bit lower, at around 42 brix.) Most juice that is traded internationally is concentrated and frozen, and most of it is blended from the juice of many producers selling into a central processing plant. In many cases, juice is also mixed in other countries as well, for example, a certain percentage of orange juice from Brazil can be mixed with that produced in Florida and still sold as U.S. product.

Because the harvesting seasons in Brazil and the United States complement each other, there is, in effect, fresh fruit throughout the year. This means that frozen orange juice concentrate does not have to be stored for long periods of time. Freezing and transporting liquid are both energy-intensive. Orange juice is one of the lower-value agricultural commodities that is frozen and transported great distances. This is why the concentration method was developed, so that water was not being frozen, shipped, and stored around the world.

SUBSTITUTIONS

While juices from different fruits have very different flavor profiles and nutrients, almost any juice can be a substitute for any other juice. Through advertising and branding programs, even markets for the same juice can be differentiated. Branding and advertising, however, tend to reduce the portion of the juice dollar received by the producer and may reduce the absolute value to producers as well.

What is happening is that the overall consumption of juices, soft drinks, bottled water, and other liquids has skyrocketed. Orange juice consumption has continued to grow, but it is losing market share to many other beverages. In Brazil, which is the largest orange juice producer in the world, orange juice now represents only 4 percent of the total beverage market (Neves et al. 2001).

Overall, demand for orange juice has continued to grow. In fact, blends of orange juice with other fruits are increasingly common on the market. This has not resulted either in a decline in the consumption of orange juice or an increase in returns to producers.

MARKET CHAIN

The orange juice market chain is dominated by large international juice companies. Such companies tend to be heavily invested in costly processing plants in most major areas of production. In Brazil, some of the largest producers also own processing facilities and have been able to compete on international markets. Most of the big juice companies work directly with supermarket chains, retail food services and outlets, and institutional buyers. In some cases, joint ventures have been developed that involve the entire chain from producer to retailer.

There has also been increasing concentration of the orange juice industry at the processing level. The expansion of the Brazilian orange processing industry into Florida has been very calculating. Several of the largest processing plants have been purchased. Initially, independent growers are paid higher prices for their product. Paying higher prices deprives smaller processors of supply, eventually forcing them out of business. Once the competition is eliminated the larger processors drop the prices paid to producers to rates that are lower than they had been prior to the initial purchase of the first processing plant.

Most recently, the juice companies have been subject to buyouts, as many grocery store chains are beginning to buy equity in some of their key suppliers as a way to guarantee supply and quality as well as potentially hamper their competitors. This, coupled with increasing concerns about product quality, may well change the overall market chain, concentrating power in a few key players while making the entire chain more vulnerable at the same time.

MARKET TRENDS

Orange juice consumption is increasing; in fact the consumption of juices in general is increasing. Consumption of frozen orange juice concentrate is expected to expand at only about 1.5 percent per year or less in the United States and Canada. However, growth in the rest of the world is expected to be about 4 percent per annum (J. F. Chaumont 2002).

The only FAO-designated food category to increase in both area and total production faster than vegetable oils since 1960 is fruit production. Between 1961 and 2000 orange production increased by 193 percent, while the volume of oranges traded internationally increased by 198 percent. During the same period, the price of orange juice declined by 61 percent (FAO 2002).

Through the 1990s Brazil came to dominate more and more of the international market as an increasing amount of the orange juice produced in the United States was consumed in that country. There are signs, however, that orange productivity in both the United States and Brazil will decline. Since consumption is not likely to decline proportionately this will most likely be accompanied by a shift of production to other areas. If this new production takes place in natural habitat it will have substantial environmental impacts.

The price of orange juice is rather sensitive to changes in total production. For example a 10 percent increase in production will tend to reduce price by more than 10 percent at the wholesale level. Some projections suggest that aggregate global production will increase faster than demand for the next several years. This is happening because many younger trees are coming into full production.

One of the major trends is for Brazilian orange juice and concentrate to be blended with that produced in the United States and then sold domestically or even exported as an American product. Sources within the industry indicate that as much as 25 percent of the orange juice sold in the United States is actually produced in Brazil.

The United States currently has a tariff in place to protect the domestic orange juice industry from lower-cost Brazilian producers. This tariff currently amounts to about $0.20 per gallon. If the tariff is eliminated, Florida growers will receive that much less for their orange juice. It is doubtful that, given their higher cost of production, they could still compete. However, producers in Belize, Costa Rica, and Mexico benefit from this tariff as well. If it is eliminated, then production in those countries will no longer be viable either.

At this time, the cost in Florida to produce the equivalent of a gallon of juice delivered to a processor is U.S.$0.99 per gallon. With the tariff in place, the cost of delivering a gallon of juice from Brazil is about U.S.$1.06. However, if the tariff were to be eliminated, the Brazilian industry could deliver juice to Florida at a cost of only $0.78. Under current regulations of the World Trade Organization, this tariff will have to be phased out. Such a phase-out would mean a significant shift in where oranges are produced globally. Such a shift in production would have environmental implications, not only due to the expansion of orange cultivation into areas of natural habitat in some areas but also because of contraction in others where producers could be forced to move into much more damaging annual crop production in order to make the same levels of income.

Five families in Brazil control the citrus industry. If the tariff in the United States is removed, they are expected to relatively quickly double their plantings, just as they did in the 1970s and 1980s after freezes destroyed orange production in north and central Florida. However, four of the five key Brazilian citrus-owning families also now own 40 percent of all processed production in Florida. It is not exactly clear where they stand on the tariff issue (Layden 2002).

ENVIRONMENTAL IMPACTS OF PRODUCTION

The production of oranges has two negative environmental impacts: the conversion of natural habitat for the establishment of orange groves and the use of inappropriate production practices which have impacts for as long as production occurs. Significant changes in biodiversity and ecosystem functions result both from the conversion of natural habitat as well as the ongoing degradation of soil and the elimination of biodiversity within producing orange groves. The degradation of soil also makes it more difficult for producers to use inputs (e.g., fertilizers, pesticides, and water) efficiently and to control the levels and content of effluents coming from orange groves.

Habitat Conversion

The citrus industry in Belize illustrates some of the important environmental issues that arise from the orange juice industry. From 1984 to 1994 the industry expanded from an area of less than 12,000 hectares (30,000 acres) planted to oranges to some 23,000 hectares (57,000 acres), nearly doubling the total area in one decade. As the industry has expanded it has grown out of the better-suited river bottoms up the steeper slopes along both sides of river valleys and into forested areas of the watershed. Most of these lands had not been used for agricultural production. And even though oranges are a tree crop, the areas they are planted in are marginal, at best, for orange production. If such lands are not managed very well there is likely to be significant soil erosion and downstream siltation, as well as little long-term orange production. And of course, conversion of natural habitat always brings a reduction in species diversity.

In one area of Belize, some 450 small farmers are involved in orange production. More than half of all the land planted on their holdings (average size of about 6.8 hectares or 17 acres) is planted to oranges. About a quarter of the farmers own their land, another quarter have short-term leases, and half have long-term leases from the government. These farmers clear an average of 2.2 hectares (5.4 acres) per year. It is estimated that as many as 240 hectares (or some 600 acres) of mature forests on hillsides are cleared by these farmers each year. Most producers are focused exclusively on orange juice production. As prices decline, focus is shifting to other products as well.

It is not clear that hillside plantings are sustainable either economically or environmentally in the long term. What is clear, however, is that hillside production requires greater inputs (e.g., fertilizer and labor) than in the lowlands. It is also clear that hillside production causes greater environmental damage because the soils are more erodible. At current prices of U.S.$3 per box, it appears that hillside farms are viable only when farms are small enough (e.g., less than 2.5 hectares or 10 acres) that all the labor can be provided by the family.

Impacts of New Varieties

Another potential development in the orange and orange juice industry is the adoption of new, higher-priced varieties that have more significant environmental impacts. One example is the growth in demand of blood oranges that require cold periods to develop their deep red color. This plant could do very well on the steep highland areas of Latin America, Africa, and Asia and therefore must be viewed as a potential threat to such areas. Currently, some of this land is devoted to high-quality shade-grown coffee. However, with the extremely low price of coffee, these areas might be converted to such a higher-value crop. The potential range of blood oranges, however, goes even further since they actually require colder temperatures to produce the highest-value fruit. Most of this land has not yet been converted for agricultural use. For now, however, blood orange demand is limited primarily to Europe and is supplied mostly by Spain.

Even for conventional oranges, however, there is a general consensus after decades of production that quality citrus with its unique, deep orange colored juice does not develop in tropical areas. Such color is the result of low temperatures that trigger the orange coloration, and these temperatures are not found in typical tropical areas. Therefore, expansion of citrus for both juice and fresh fruit is expected to continue only in areas with a consistent period of cold temperatures as is found in Southern Brazil and for which even parts of Belize are also known. In the future, however, it should be easy to plot out those tropical areas that have the appropriate temperature and rainfall and that might be targeted for expansion, particularly if diseases known to be affecting production in Brazil cannot be brought under control.

Even Honduras, which sits next to Belize, has been unsuccessful in its attempts to produce export-quality orange juice and other citrus. The industry has simply not been able to compete in producing the quality product that is required for premium prices. Honduran citrus projects have ended up supplying local markets. The country's exports to the United States are limited to lower-grade backup juice or periods when there is a freeze in the main producing areas either of the United States or Brazil.

Use of Pesticides and Inorganic Fertilizers

Orange production requires more intensive use of a wide range of pesticides than any other major commodity including bananas. Only horticultural crops have higher input use per hectare than oranges. However, oranges produced for juice require far fewer pesticides than those sold as fresh fruit.

In Belize a number of pesticides are used in the cultivation and production of oranges for juice. In 1994 most commercial orange cultivators in Belize reported the frequent use of the following pesticides: paraquat, fosetyl-al, 2,4-D, glyphosate, aldicarb, diuron, propiconazole, ethoprop, ethion, malathion, phoxim, terbufos, and chlorpirifos. Paraquat, diuron, and glyphosate are broad-spectrum herbicides that kill all types of vegetation; the use of these three chemicals suggests that the technical recommendations to producers have been to maintain clean fields of monoculture crops with the possible exception of species planted for ground cover. Clean production systems have tremendous impacts on biodiversity (both resident and mobile, both in the soil as well as on it) as well as soil erosion and the overall need for chemical inputs.

The application of inorganic fertilizer, coupled with a clean fields approach to cultivation, has had perhaps the single largest impact on the environment. Orange production on clean fields results in ever decreasing levels of soil carbon. As a consequence, any chemicals applied to the soil are more likely to be washed off as effluents before they can be utilized by plants. Even the application of foliar fertilizers, while meeting the needs of the trees, may result in the degradation of soil so that it has diminished ability to utilize and take up the fertilizers. In addition, chemicals subsequently leach into waterways and lagoons.

Processing Waste

Considerable waste is generated by orange juice production. While the weight of the waste is less than that of the juice, the volume is far greater. Such wastes have become very large disposal and environmental issues in many countries. In fact, most processing plants have mountains of peels and pulp that begin to smell. Waste also can become vectors for the breeding of insects that can cause diseases.

BETTER MANAGEMENT PRACTICES

Given that orange production is likely to expand to many parts of the world as production declines in the United States and Brazil, it will be important to work with producers and governments in those areas to minimize the overall environmental impacts of production. Areas of steep slope should not be converted from natural habitat to orange production. In several of the major producing countries (e.g., the United States, Brazil, and Spain) production areas are mostly gently rolling or flat. In these countries, orange groves are rarely planted in areas with more diverse features such as steep slopes. In the past, in fact, orange production has taken the place of other crops. However, as demand expands, some countries may be tempted to encourage citrus production in suboptimal areas that are home to a wider range of biodiversity that should be protected.

In addition to the adoption of better practices during orchard establishment, they should also be employed during ongoing management. For example, ground cover should be maintained at all times and every effort should be made to utilize cover crops that increase both nutrients and organic matter in the soil and that reduce soil erosion. The adoption of many, if not all, better practices is driven by efforts to cut costs in an increasingly competitive industry. Innovation is driven by economics not ethics. There is little or no room for error in many markets.

Nowhere is this more obvious than with water. Given the shortage of water in many orange-producing areas, the reduction of overall water use will also be an issue for many orange producers. In Florida, for example, orange producers are required to install and use drip irrigation. All the water used on one Florida farm is recycled and reused. No surface water is allowed to enter the environment directly. In other farms, the goal of water management is to ensure that the quality of water leaving the farm is better than what is coming in and effluents are currently measured for levels of phosphorus, nitrogen, suspended solids, and key pesticides.

Reduce Use of Agrochemicals

One strategy used by many citrus growers to reduce their dependence on and use of chemical pesticides has been to incorporate new varieties of oranges that are more disease-resistant. They also graft more productive varieties onto hardy rootstock. These strategies are viable options for those who want to control some of the more difficult citrus diseases.

In Japan, one of the most capital-intensive citrus production systems has been developed for satsumas (also called Mandarin oranges). Many satsuma farms are covered entirely by greenhouses. Even the drain water is collected and pumped back onto the crops. Very few pesticides are used. Since the greenhouses are intermixed with citrus grown in open, existing groves, one barely notices them. In short, every effort has been made to close the production system and this has resulted in a very clean, low-input production system that recycles most of its waste and by-products. While such a system could not be imported whole cloth into many orange-producing areas producing bulk juice, it is likely that many specific practices may. For example, catching and recycling water in a water-scarce world will make increasing sense. Likewise, capturing agrochemical inputs in water runoff for reuse makes sense.

In Florida, one orchard has reduced spraying from an average of 22 to 24 times per crop to 10. Orchard managers use IPM, the breeding and release of ladybugs, and the incorporation of hursatelia to reduce chemical interventions. Most of these approaches require that producers anticipate and prevent problems rather than following a strategy of trying to catch up with them later. Consequently, producers must monitor their fields and anticipate issues by paying close attention to warning signals. Increasingly, such farms employ professional scouters to undertake this work. The company reports that professional scouters are more attentive to details and changes than regular employees who are responsible for a number of different activities. Scouters are kept on retainer and their sole purpose is to visit fields and observe which pests are becoming problems in which areas of the groves. One farm manager reports that through better timing and targeting, he has reduced spray costs by 20 to 25 percent with scouting.

A Florida producer suggests that the reduction of pesticide use has resulted from a fundamental change in attitude. Most producers have found that even if they are producing only oranges, they need to be creating polycultures. What this means is that in addition to growing oranges, producers have to create or maintain conditions for beneficial organisms as well as problematic ones. For this reason, whenever producers are forced to use chemicals such as copper and sulphur compounds, they kill the beneficials, too. The goal is to provide conditions where both beneficial and problematic organisms can exist but in balance. Spraying is an indication that the proper balance has not been maintained.

In order to monitor pests and treat them more effectively, producers need information that allows them to make more precise analyses and interventions. One 1,600-hectare (4,000-acre) Florida orange farm, for example, is divided into lots, the smallest of which are 110 acres. These lots allow managers to trace problems and input use to specific areas and even specific workers. Managers now realize that they need to keep records for even smaller areas so that they can make informed production decisions. With better, targeted data they can see that some areas require more inputs and efforts than their yields justify, and they are beginning to cut down trees in such areas.

Producers prune and manipulate trees to increase production or to reduce the overall cost of production. For example, trees are trimmed around the edges and the top to make picking easier. In addition, many producers "skirt" trees, a process that consists of trimming the lower branches so that the lowest branches are half a meter from the ground. This allows air to flow into the tree branches which in turn reduces fungus and mold and the necessity to spray to eradicate it. One negative side effect, however, is that the increased space lets in more sunlight and as a consequence more weeds grow under the trees which are then killed with herbicides. The lesson here is that there are tradeoffs between BMPs and the question becomes which practices are less harmful than others. In this case, the question is which types of sprays are less toxic, fungicides or herbicides.

Reduce Soil Erosion and Degradation

Regardless of the terrain on which orange trees are planted, grass lanes are recommended as a way to minimize erosion and to contain chemical-laden runoff. One recommendation is to keep only a small ring of cleared area under each tree that can serve both for fertilization and for harvesting.

Deliver Better Information and Incentives to Producers

In the global market, Costa Rica is a small producer of orange juice. The country experimented with organic orange juice, but it was not an overwhelming success. In the past, the strategy of the main orange juice company and the farmers producing for it had been to focus on processing and marketing their product rather than to work to improve the production base—the trees. Over time production declined, increasing numbers of trees became diseased and died, and the cost of production increased while overall production declined. Many producers were forced to abandon the businesses altogether.

Tico Fruit, the most successful juice processor in Costa Rica, has taken a slightly different path. The lessons that can be drawn from its experiences could be significant for producers in other parts of the world. Costa Rica has 14,000 hectares of oranges planted on more than 1,000 farms. Tico Fruit accounts for 75 percent of all orange juice produced in the country. While it owns only 25 percent of all land planted to oranges, it produces 50 percent of the country's juice from its own production. It buys oranges from other producers to produce an additional 25 percent of all production in the country.

Tico Fruit has come to two important realizations. First, it cannot produce all the fruit that it can process. And, second, healthy trees require very few chemical inputs. As a result of these realizations, the company has begun outreach programs with the farmers that sell to it. The company knows that the best way to increase its output of juice is to help all of its suppliers increase their production of oranges as well as their net profits from orange juice production.

Toward this end, the company has its buyers provide technical assistance to each farmer in order to help them improve production. This assistance takes three forms—plantation diagnostics, recommendations of specific technical packages or plans of action to address the problems, and support during implementation. The company's assistance is undertaken on a farm-by-farm basis. Tico Fruit buys all the materials required by all farmers in bulk and then extends them in the form of credit to local producers. In this way, the company is assured an increasing supply of fruit, and producers benefit from the lower prices that result from buying in bulk. As a result of the company's efforts, Tico Fruit producers have increased their net income to U.S.$4.20 per tree, as compared with a net income of $1.01 for other producers, as shown in Table 6.1.

In general terms, the health of trees is achieved by making nutrients available to them. For this reason, the company's approach is to feed and maintain the soils in such a way that the nutrients are released for the trees to use. The main inputs are lime and chicken manure. These inputs, when applied correctly, encourage soil microorganisms that are important to healthy soil and trees.

Company employees/extension workers explain to the farmers that the more synthetic fertilizers they use, the more they will have to use over time as they reduce both the nutrients created in the soil as well as their trees' ability to absorb them. Use of highly concentrated, synthetic fertilizers causes imbalances in nutrients, overall nutrient availability, and microbes in the soil. To be corrected quickly, the imbalances are treated with other fertilizers and producers lurch from one application to another. A better long-term strategy is for producers to find a better balance between tree needs and soil dynamics so that the soil and the tree can work together to meet the tree's needs.

Tico Fruit wants the producers to supply the company with increasing amounts of fruit. For this reason the company is cautious about converting to organic production methods. Company officials are not convinced that organic production is sustainable either environmentally or financially. A recent study supported by the FAO comparing organic and conventional orange production in Spain suggests that net returns from organic production are considerably lower for producers because many costs are higher and overall production declines over time from organic orange production systems (Igual and Izquierdo 2001). As it is, the company claims that it has helped its suppliers reduce their overall use of agrochemicals by 75 percent or more, which contributes to the overall increase in their profits. The timely application of limited amounts of agrochemicals appears to be essential to this success.

The Tico Fruit program to reduce the use of agrochemicals and their impacts within the orange juice industry is an excellent example of how a buyer can work

TABLE 6.1. Orange Juice Production and Costs per Tree in Costa Rica
(in U.S. dollars)

	Tico Fruit Producers	Other Producers
Cost per tree to establish plantation	3.00	1.00
Production (boxes/tree)	4	1.3
Average weight per box (kilograms)	2.3	2
Solid weight per tree (kilograms)	9.1	2.7
Price per solid weight	.50	.50
Gross revenues per tree	10.00	3.00
Harvesting cost per tree	.70	.70
Total net income per tree	4.20	1.01

Source: Interview with Tico Fruit staff, 2001.

with its producers to improve their production practices. Such partnerships can result in reduced environmental impacts, and both producers and buyers can increase their profits. Other companies, large or small, should travel to Costa Rica to see this program and adapt the lessons learned to their own circumstances. Such industry-to-industry exchanges are often the most effective way to explain how and why a program works as well as the financial implications.

Reduce Processing Wastes

In the United States and Brazil, methods have been devised to grind the waste from orange juice production so that it can be incorporated into animal feeds. It is, in fact, for this reason alone that Florida has long been one of the states with the most beef cattle finished in feeder lot operations in the United States.

Such waste processing programs are not limited to the United States, however. The Tico Fruit processing plant in Costa Rica is the most advanced in all of Latin America. It produces feed pellets from its orange waste that are sold to the cattle industry. What was a potential waste problem and a disposal cost is instead a stream of income for the company and a useful product for neighboring farmers. The operation, in fact, produces a net revenue stream for the business and has a positive impact on the bottom line. In Belize, by contrast, peels from orange juice production are still a major waste problem.

Aside from converting pulp to animal feed or using it as organic fertilizer amendments, it can also be used as a source for essential oils that have high commercial value as fragrances and flavorings. Pectin, which is important in commercial production of jams and jellies, can also be extracted from the pulp. The orange juice industry also produces seeds that contain antibacterial and fungicidal properties and so can be used to manufacture innovative new pesticides for many crops. The isolation and production of these valuable by-products is now an increasing part of the industry and is integrated into the processing system in such producing countries as the United States and Brazil. Unfortunately, the production of some of these by-products requires significant investments and/or scales to be profitable. This could tend to make larger processing companies more financially viable than small ones. Still, as all these ways suggest, organic waste can be minimized.

OUTLOOK

The production and consumption of orange juice globally is relatively stable. If anything, orange juice is only just holding its own in the market as the number of other juices available in fresh, frozen, and concentrated form has increased dramatically. However, with diseases and lower production in Brazil and the United States it is likely that production may well shift to other parts of the world. In Brazil, orange juice expansion took place at the expense of other crops, such as coffee. This does not pose a serious threat to the environment. In other places it might expand at the expense of natural habitat, and this would be a more serious issue.

It is still too early to determine whether the expansion of the blood orange will pose significant environmental risks. While that fruit is quite valuable, it is not clear that it can be produced easily in areas that are linked to international markets.

Finally, some of the preliminary work in Florida, Brazil, and Costa Rica suggests that orange juice production can be undertaken with far fewer impacts and far greater net profits if relatively simple and inexpensive better management practices are adopted. These can easily be supported by processing plants whose owners also have a self-interest in a long-term, stable supply of product from growers.

REFERENCES

Barham, B. L. 1992. Foreign Direct Investment in a Strategically Competitive Environment: Coca-Cola, Belize, and the International Citrus Industry. *World Development* 20(6) June: 841–857.

FAO (Food and Agriculture Organization of the United Nations). 2002. *FAOSTAT Statistics Database*. Rome: UN Food and Agriculture Organization. Available at http://apps.fao.org.

Hall, M. 1994. Assessment of the Potential Risks from Agriculture on the Coastal Areas of the Stann Creek District. Thesis. Imperial College of Science, Technology and Medicine. Centre for Environmental Technology. University of London. September. 132 pages.

Igual, J. and R. Izquierdo. 2001. *Economic and Financial Comparison of Organic and Conventional Citrus-Growing Systems*. Rome: United Nations Food and Agriculture Organization.

J. F. Chaumont Dried and Frozen Foods. 2002. Orange Juice, Worldwide. Available at http://www.jfchaumont.com. Accessed 2002.

Layden, L. 2002. Local Citrus Growers Say Free-Trade Accord Could Leave Them Dry. *Naples Daily News*. Naples, Florida. October 21.

Neves, M. F., A. M. do Val, and M. K. Marino. 2001. The Orange Network in Brazil. *Journal for the Fruit Processing and Juice Producing European and Overseas Industry (Fruit Processing/Flussiges Obst)*, 11(12) December: 486–490. Schönborn, Germany. Available at http://www.fearp.usp.br/deptos/adm/docentes/fava/homefava/pdf/Citrus%20Network%20Final%20Version.pdf

UNCTAD (United Nations Commission on Trade and Development). 1994. *Handbook of International Trade and Development Statistics, 1993*. Geneva, Switzerland: UNCTAD.

———. 1999. *World Commodity Survey, 1999–2000*. Geneva, Switzerland: UNCTAD.

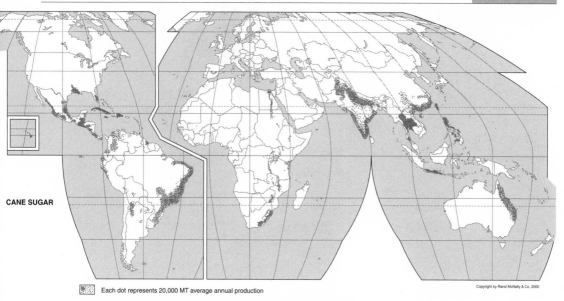

CANE SUGAR

Each dot represents 20,000 MT average annual production

Copyright by Rand McNally & Co. 2000

SUGARCANE *Saccharum officinarum*

PRODUCTION	
Area under Cultivation	19.6 million ha
Global Production	1,255.8 million MT sugarcane
	131.96 million MT sugar (from beet and cane)
Average Productivity	64,071 kg sugarcane/ha
Producer Price	$21 per MT
Producer Production Value	$26,217 million

INTERNATIONAL TRADE	
Share of World Production	16%
Exports	35.0 million MT
Average Price	$229 MT
Value	$8,016 million

PRINCIPAL PRODUCING COUNTRIES/BLOCS
(of sugarcane by weight)

Brazil, India, China, Thailand, Pakistan, Mexico, Australia

PRINCIPAL EXPORTING COUNTRIES/BLOCS
(of sugar from cane)

Brazil, Australia, Thailand, Cuba, South Africa, Guatemala

PRINCIPAL IMPORTING COUNTRIES/BLOCS
(of sugar)

Russia, Indonesia, Japan, Korea, United States, United Kingdom, Malaysia

MAJOR ENVIRONMENTAL IMPACTS

Conversion of primary forest habitat
Soil erosion and degradation
Agrochemical use
Organic matter from processing effluents

POTENTIAL TO IMPROVE

Poor
Price too low to improve industry or genetics
BMPs are known but producers are set in their ways
Subsidies for sugar beets and cane and market barriers in
    developed countries are disincentives for producers to change

Source: FAO 2002. All data for 2000. *Note:* Raw sugar is equal to 17% of sugarcane by weight. Sugar is produced from sugarcane and sugar beets. Sugarcane is a grasslike plant grown in the tropics. Sugar beet is a tuber grown in temperate climates. While this chapter focuses on sugarcane, FAO statistics do not distinguish sugar from cane or from beet.

Area in Production (MMha)

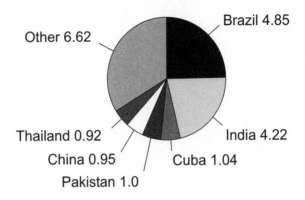

Brazil 4.85

Other 6.62

Thailand 0.92

China 0.95

Pakistan 1.0

Cuba 1.04

India 4.22

OVERVIEW

It is not known precisely when sugar was first made by boiling the stems of the plant *Saccharum officinarum*. However, the plant and the technology are known to have originated in India. The word "sugar" has been traced back to Arabic (*sukkar*) and Sanskrit (*sarkara*). Initially, sugar was used for religious ceremonies and as a medicine to treat ailments ranging from leprosy to gallstones (Swahn 1995).

It appears that in about 500 B.C. residents of present-day India began to make sugar syrup, which was then cooled in large flat bowls to make crystals that were easier to transport and store. In the local language the pieces of crystal were called *khanda*, which is the source for our word candy (Swahn 1995). These pieces were lifted out of the bowls and put in bags where they were squeezed to remove the remaining liquid. These cakes of dried crystals were ideal for transport and trade.

By the fourth century A.D., sugar production had spread throughout India. By the fifth century, the Chinese were growing and making sugar. In the sixth century, sugarcane was cultivated in Persia. The Persians invented sugar loaves, which took their classic cone shape from the conical clay vessels used to crystallize the sugar. These vessels had holes in the bottom to allow liquid to escape. The conical loaves were associated with the sugar trade for more than a thousand years. In fact, Rio de Janeiro's distinctive mountain is known as Sugar Loaf because of its distinctive conical shape and because sugar was an early and commercially important product of the colony (Swahn 1995).

When the Arabs conquered Mesopotamia, they also gained access to sugar production technology that was subsequently spread by the Moors throughout northern

Africa, Sicily, the Middle East, and even into Spain. However, sugar's breakthrough into non-Moslem Europe owed more to the Crusades than to Spain. Crusaders to the Middle East became acquainted with and liked sugar. This subsequently created demand for the product. When the Normans conquered Sicily in 1072, Europeans controlled their first sugar-producing area. As demand emerged, the Italian ports of Genoa and Venice became the main trade ports for Europe (Swahn 1995).

Spain and Portugal gained considerable knowledge of sugar from the occupying Moors. After the Moors were expelled, they used this knowledge to spread sugar use and production throughout the tropics. Within fifteen years of Columbus discovering the Americas, the Spaniards began to plant sugar in Hispaniola (modern Haiti and the Dominican Republic) (Swahn 1995). Sugar, more than any other crop, encouraged the rapid expansion of the slave trade. Because of Spain's control of the Low Country (Holland), all sugar from the New World came into Europe through Rotterdam, where it was processed from brown, conical loaves into white sugar. At that time, sugar was so expensive that it was only consumed by the elite. As a status symbol and during special occasions, sugar crystals were often tied in small bundles and suspended over the tables of those who could not afford to consume it.

Today, although sugar has very little nutritional value, it (or other sweeteners) is found in almost all processed foods. Sugar makes up 20 percent of the calories consumed by Americans, who eat nearly half a kilogram (about a pound), on average, every 2.5 days.

PRODUCING COUNTRIES

Sugar is produced from sugarcane and also from sugar beets. Sugar produced from cane is the focus of this chapter, except where the statistics cannot be separated or when discussing subsidies or product substitutes that have an impact on cane sugar prices and/or production. The Food and Agriculture Organization of the UN (FAO 2002) identifies 103 countries that produce sugarcane. Globally, 19.6 million hectares were devoted to the crop in 2000. Brazil and India have by far the most land devoted to sugarcane with 4.8 million and 4.2 million hectares, respectively. Four other significant producers—Cuba, Pakistan, China, and Thailand—each have around a million hectares. Brazil and India account for 46 percent of all land globally devoted to the production of sugarcane. Cuba, Pakistan, China, and Thailand collectively account for 21 percent of all land planted to sugarcane. The total of these six countries, 67 percent of all production, is precisely the same proportion they represent of land planted to sugarcane. Table 7.1 shows that in several smaller countries sugarcane occupies more than half of all land devoted to agriculture.

In 2000 average yields, globally, for sugarcane were 64,071 kilograms per hectare per year. While no country produces twice the global average, several—Burkina Faso, Chad, Egypt, Ethiopia, Malawi, Peru, Senegal, Swaziland, and Zambia— average production of more than 100,000 kilograms per hectare per year. The overall uniformity of yields globally implies that there are no significant new technologies or innovative production systems available that are sufficiently widespread to

TABLE 7.1. Percentage of Agricultural Land Devoted to
Sugarcane Production, 1994

50 Percent or More	25–49 Percent	10–24 Percent
Antigua	Cuba	Congo
Bahamas	Fiji	Costa Rica
Barbados	Jamaica	Dominican Republic
Belize	Martinique	Haiti
Guadeloupe	Puerto Rico	Liberia
Mauritius	St. Kitts/Nevis	Papua New Guinea
Réunion	St. Vincent	Swaziland
	Trinidad/Tobago	

Source: FAO 1996.

boost yields in a whole country and that most of the existing technology is fairly well distributed throughout the world.

The FAO reported global production to be 1.26 billion metric tons of sugarcane or 132 million metric tons of sugar in 2000. Global sugar production by volume (including sugar beets) is dominated by the European Union, India, Brazil, the United States, China, Thailand, Australia, Mexico, Cuba, South Africa, and Pakistan; these countries combined account for some 70 percent of production globally. World exports are dominated by Brazil, the European Union, Australia, Thailand, and Cuba.

As Table 7.2 demonstrates, sugar is Cuba's leading export, as well as a primary export for a number of countries that are not among the top producers of sugar or sugarcane. Annual sugar exports represent some 30 percent of global production (FAO 2002).

World sugar production has exceeded consumption for the past six consecutive years according to the International Sugar Organization (ISO). This has led to a fourteen-year price low. The last time this happened was in the mid-1980s. It took three years for the surplus to be sold when the markets turned around. The previous surplus happened even when Brazil, the largest producer, had reduced sugar exports in order to produce fuel alcohol from sugarcane because of increasing petroleum prices. The current overproduction is related to Brazil's reentry into the international sugar market through the reduction of fuel alcohol production after successful efforts to find petroleum and lower world prices for oil. There is simply too much productive capacity throughout the world at this time, so chances of sugar prices increasing are slim for the near future.

CONSUMING COUNTRIES

Globally, sugar is considered a staple food by many consumers. Sugar, refined or otherwise, is used in most processed foods currently on the market globally. While there have been considerable efforts to find cheaper vegetable and chemical substitutes as well as more expensive organic or healthy alternatives, the rate of consumer

TABLE 7.2. Sugar's Ranking of Total Exports by Value for Selected Countries, 1990–1991

Leading Export	Second Largest Export	Third Largest Export
Cuba	Dominican Republic	Bahamas
Guyana	Guadeloupe	Jamaica
Belize	Trinidad/Tobago	Panama
St. Kitts/Nevis	El Salvador	Cape Verde
Réunion	Guatemala	Congo
Fiji	Barbados	Malawi
	Antigua/Barbados	Mauritius

Source: UNCTAD 1994.

acceptance is less than expected. Perhaps part of the reason is that the cost of some substitutes for conventional sugar, organic sugar, for example, is five to ten times the price of sugar on the world market. For most consumers, the flavor of substitutes, their availability or price, or the changes of consistency or texture that they impart on finished products are not acceptable. Consequently, hundreds of billions of pounds of sugar are consumed each year.

India is the leading sugar-consuming country, followed closely by the European Union. The United States, Brazil, and China also have high per capita levels of sugar consumption.

The main importers of both raw and refined sugar are Russia, Indonesia, Japan, Korea, the United States, the United Kingdom, and Malaysia. The U.S. quota system for imports limits total supply to U.S. markets. Imports from China, Indonesia, and Russia have not been enough to reduce the overall stocks on world sugar markets.

Price supports in the United States and the European Union keep sugar at just over 44 cents per kilogram (20 cents a pound). The international price, however, is usually half or even less than that. In 1998 raw sugar prices fell to 15.4 cents per kilogram (7 cents per pound), and in 1999 prices fell to less than 9 cents per kilogram (4 cents per pound). These are the lowest sugar prices since the mid-1980s. There are specific reasons. In addition to Brazil's increased sales mentioned in the previous section, the Russians and Indonesians are importing less sugar due to their financial crises.

PRODUCTION SYSTEMS

Sugarcane is grown in tropical lowland climates. It is produced almost exclusively between latitudes of 30 degrees south and 30 degrees north and is most concentrated between 20 degrees. While sugarcane production is often thought of as being produced only on islands or in coastal areas, it is also grown on extensive areas of former tropical forests in countries such as Uganda. Because sugarcane is a grass, most producers feel that it can be grown even on the steepest hillsides. Sugarcane requires intense sunlight and at least 1,650 millimeters of rainfall. Furthermore, the rainfall must be distributed throughout the year. Otherwise sugarcane requires considerable irrigation. The plant performs best in nutrient-rich soils with a high water retention capacity and with pH values that are weakly acidic to neutral. Overall nutrient requirements are quite high. While pest and disease problems have been reduced through breeding programs, biological pest controls are increasingly important at least in part because they lower overall production costs. Sugarcane is mostly grown in large monocrop plantations.

The first planting of cane matures in fourteen to eighteen months. Subsequent harvests occur every twelve to fourteen months. Productivity declines after each harvest, so the useful life of a planting does not usually exceed four to five harvests, but can reach 20. Cane life can be reduced even further through mechanical harvesting, which can pull up as many as 2–10 percent of the plants per harvest.

Throughout the world, most of the activities associated with planting, cultivating, and harvesting sugarcane are done by hand. After digging a shallow trench, cuttings of sugarcane stalks are laid side by side, slightly overlapping, and then covered with soil. The cane soon sends up shoots that can grow to as tall as 6 meters with stalks that are 5 centimeters thick. Rows are planted in a parallel pattern, separated by a meter or less of land. Sugarcane fields are weeded, usually by hand, two to three times during the first year, and then harvested after twelve to eighteen months. One metric ton of cane produces as much as 125 kilograms of refined sugar.

Fields must be weeded between each cutting. Production declines over time, but the overall return on investment makes it cheaper to harvest the declining yields than to replant the crop each year. Nitrogen-based fertilizers are applied to increase production, especially during subsequent years of production. In most parts of the world, sugar plantations are burned to eliminate the dead lower leaves of the plant and to kill or remove snakes before harvesting by hand.

Machines have been developed to open the furrows for planting, to cultivate the crop between harvests, and even to harvest the cane. However, in many areas where sugar is grown, labor is cheaper than machinery and more efficient. Furthermore, the most efficient machinery is too large to negotiate many of the hilly areas or is too heavy and sinks into the soft wet soils that are considered ideal for sugarcane cultivation. Mechanical harvesters are used in parts of the United States and in places such as southern Brazil, but the mechanical harvesters pull up cane, forcing landowners to replant more quickly. In addition, up to 10 percent of the harvest from mechanical harvesters is waste material compared with 1 percent when harvested by hand.

The major technological innovations regarding sugar production have occurred during the past century, particularly with regard to transportation and processing systems. Prior to these innovations, a single animal walking around a small screw press was the way most sugarcane was pressed. The juice was then cooked down and poured into molds to form hard brown cakes ("rapadura") of a uniform size and weight. For local trade, the molds used were often made from wood and were flat and rectangular so that smaller amounts could be sold. These processes were not very efficient. The presses, for example, left a lot of juice in the cane as well as impurities in the extracted juice. Much of the innovation of the past 150 years has focused on improving the process of extracting and refining sugar on or near the plantations where the cane is produced.

Over the past century, more efficient metal roller presses were developed that increased the quality of production. In addition, the technology allowed far more cane to be processed. However, these innovations were much more expensive and required larger quantities of cane in order to be economical in operation. This had the net effect of bringing much larger areas of cane under the influence of single factories that were owned by individuals wealthy enough to afford the up-front investments in the new technologies.

As the areas brought under the control of these sugar factories increased (and, consequently, as the market standards of refined sugar increased as well), the ability of small farmers to continue to compete in sugar processing declined. In the past, sugarcane workers had been given land to grow their own food and animals. As the competition for land for sugar increased, as local populations increased, and as the global markets for sugar both increased and became more competitive, the compensation for workers shifted to wages. Increasingly, sugarcane was cultivated as a monocrop for as far as the eyes could see.

Over decades the production of sugarcane has expanded and contracted depending on the global price for sugar as well as the price for other crops that could be produced on the same land. Cotton was one of the crops often substituted for sugarcane, depending on how favorable the international price was for one vis-à-vis the other.

During this process of the consolidation of the industry, sugar-producing areas became characterized as the "haves" and the "have-nots." As technology improved sugar cultivation and processing, as other regions of the world have been brought into production, and as a number of natural and chemical substitutes have been developed, the price of sugar has fallen. Individual landowners and/or factory owners have been able to maintain their standards of living only by eliminating competition from their own ranks as well by as maintaining their work force in conditions of semislavery. Many analysts have suggested that the production of sugarcane has caused more misery than any other crop on the planet.

International prices are low and workers are paid poorly. In some cases, their wages do not cover the calories that they burn on the job. Working conditions, whether on the production or processing side, are among the most hazardous of any agricultural industry. In Northeast Brazil—the largest and most populous impoverished area in the Western Hemisphere and one of the longest-standing sugar-producing regions of the world—sugarcane workers have the lowest life expectancy of any group and their children have the highest infant mortality rates. Even in the United States, where sugar prices are usually double and sometimes triple international levels, traditional sugar harvesting has been described as "the most perilous work in America" due to the snakes, sharp machetes, dust and ash, and heavy raw materials (Wilkinson 1989).

PROCESSING

The cane is cut in the field and transported to the sugar mills. Timing is essential. The longer the cane sits the more the sugar in the stalk converts to starch, and as a consequence less sugar can be extracted. Farmers are paid for the quality of their cane. At the factory, the cane is crushed between heavy, toothed metal rollers. This yields most of the sugar juice that is subsequently processed into sugar. In addition, however, the ground plant parts are subsequently leached to yield even more sugar.

Sugar represents a mere 17 percent of the biomass of the sugarcane plant; the remaining 83 percent is "bagasse"—the generic term for everything that is left after the sugar has been extracted. Bagasse is often incinerated and therefore can contribute to global warming. It is sometimes sold and used for fuel, animal fodder, or soil amendments. Burning is not necessarily bad; the burning of bagasse as fuel can reduce the need for other fuels that may release more carbon. In addition, if the bagasse is allowed to decompose it may produce methane that could be even worse for global warming.

When sugar mills are flushed—usually once a year—a tremendous amount of organic matter is released. Usually the mills are washed out and the organic matter is dumped straight into streams. The decomposition of this matter reduces the oxygen levels in the water and can result in fish kills. This is a particular problem in tropical rivers that are already low in oxygen. For example, in 1995 the annual cleaning of sugar mills in the Santa Cruz region of Bolivia resulted in the deaths of millions of fish in local rivers.

SUBSTITUTES

There are several other crops that produce sugar and sweeteners. These crops include sugar beets as well as corn or sorghum, which produce nonsugar sweeteners. Sugar from beets, however, is identical for all intents and purposes with sugar produced from cane. It is substitutable in recipes and confections and the two are substituted for each other globally. Beet sugar could not be produced competitively, much less exported without subsidies.

In general these sugar substitutes developed as a way to avoid dependence on imported sugar from the tropics. The environmental impacts of each are somewhat different from those of sugarcane, but because the substitutes can be substituted in the marketplace for many different uses depending on price, it is important that they be discussed in any overall discussion of sugar.

More than fifty countries produce sugar from beets cultivated on 6.8 million hectares. Sugar beets tend to be produced in countries with cooler climates and limited growing seasons (e.g., countries with frosts and/or distinct rainy seasons). The countries that dominate sugar beet production are in temperate climates, for example, the Ukraine, Russia, the United States, Germany, and Turkey. All have 500,000 hectares or more of production. These five producers account for half of all land in sugar beet production. In general, however, sugar beet production appears to be undertaken on a fairly limited basis primarily aimed at supplying domestic markets.

Sugar beets can be grown in any area that supports root crops. In the United States, they tend to be grown in drier areas and are often irrigated. In other parts of the world, they are only produced in rain-fed areas.

Another substitute for cane sugar as a sweetener is corn syrup. The United States is one of the largest corn syrup producers in the world. The market for corn syrup (and sugar beets) in the United States was created as a result of a policy of price supports for sugar. The artificially high sugar price has stimulated the production of corn syrup and sugar beets, and it is doubtful that much sugar from beets or sweetener from corn would be produced if the United States dropped its price support for sugar.

Sorghum can also be used to make molasses and other sugar substitutes. It is not clear how much of the current sorghum planted in the world is being used for this purpose. Sorghum can be grown in rotation with corn, but due to its lesser value tends to be grown on the drier edges of corn producing regions or within less-productive areas.

A number of artificial sweeteners have also been discovered. NutraSweet and saccharin have both been developed as calorie-free substitutes for natural sugar produced from either sugar beets or cane.

Finally, a number of natural sweeteners are being developed. Stevia has been extracted from the plant of the same name in Paraguay and Bolivia. It has 3,000 times the sweetening power of sugar. It is being grown in Canada as a crop substitute for tobacco. Monsanto and other corporations are interested in developing it if they can find a way to patent the process for extracting it. (Their patent on NutraSweet recently ran out.) In addition, Xylitol, another natural sweetener, has recently been introduced into the market. It is extracted from hydrolyzed hemicellulose, the "black liquor" from the waste from pulp and paper mills. Xylitol is 50 percent sweeter than sugar, does not create plaque on teeth, and is low in calories. It is quite likely that within a decade a viable alternative to sugar will be discovered. Whether it is widely accepted or not will depend on how it substitutes for sugar chemically in baking and manufacturing processes. Any significant substitution, however, would generally lead to lower sugar prices and the conversion of some producers, at least, to other crops.

MARKET CHAIN

Much of the sugar in the world is produced on land that is owned by the same companies as the factories that refine the sugar and add value to the production. In Florida, for example, two corporations grow more than 65 percent of the sugarcane produced statewide. However, most sugarcane in the world is produced by growers who sell it to the sugar mills. There are rarely two or more mills close to growers. As a consequence, there is no significant competition between buyers for cane. Factories determine prices and often use outsourced sugar from independent cane growers to improve their overall profitability.

Traditionally, factories sell their sugar to national suppliers, wholesalers, distributors, or traders. The larger factories and traders tend to export whatever quotas or allotments are allowed from the country. In the past, governments tended to dominate sugar markets, both internal and external.

Most sugar goes into confections, whose manufacturers tend to want just-in-time delivery. While many will forward-contract product to ensure delivery and to lock in prices, they do not want their capital tied up in stored product that also requires expensive space. For this reason, traders, wholesalers, and distributors tend to hold most of the product until required by others. Only a small proportion of all sugar is sold directly to the consumer.

MARKET TRENDS

World sugar production has increased 181 percent since 1961. International trade in sugar has grown from 20.6 million metric tons to 35 million metric tons over the same period, an increase of 70 percent. Meanwhile prices have declined in real terms by 46 percent during the same period (FAO 2002).

The world sugar market is suffering from oversupply. This is partly the result of decreased consumption of sugar in developed countries. It also results from the increased production of sugar in developing countries and stable production of sugar in developed countries as a result of agricultural subsidies and market protection. In addition, the increasing presence of artificial sweeteners has dampened overall demand for sugar.

Sugarcane and sugar beets would not be grown in developed countries but for subsidies and market barriers. In the United States, for example, subsidies guarantee domestic prices and import policies keep cheaper, foreign sugar out of the country. As a consequence, U.S. prices are normally twice that of global prices and sometimes as much as three or four times as high. Ultimately, it is the U.S. consumer who pays this price.

Subsidies and market barriers also ensure the production of sugarcane in areas where it should not be planted. This includes the Everglades. There are more than 180,000 hectares of sugar planted in the Everglades Agricultural Area, which blocks the natural flow of water through the Everglades. According to the U.S. General Accounting Office, Americans pay an average of $800 million to $1.9 billion in subsidies and price supports for two main companies (and other smaller ones) to plant sugarcane in the area. In addition, the government pays millions of dollars more to buy back the sugar that these companies cannot sell. The industry also uses hundreds of billions of gallons of South Florida water for irrigation and processing and pays only minimal water taxes (Grunwald 2002).

The contradictions of the sugar economy are clear. The price of U.S. sugar is set at two to four times the international price. Americans are forced to buy sugar that domestic companies could not sell anywhere else. In addition, however, they are also required to pay to clean up the environmental problems of the industry. Finally, they will be asked to buy the lands that are valued at artificially high prices due, at least in part, to sugar subsidies.

Like the United States, Europe also pays farmers to produce sugar beets at an artificially high price. However, Europe goes even further to encourage more production than can be consumed in Europe. It then subsidizes the export of this sugar onto the world market, where it competes directly with unsubsidized production from developing countries. Thus, the European Union not only denies developing countries access to its sugar market, but it competes with them for other markets as well. In both cases, subsidized domestic prices and reduced access to developed-country markets reduce the price paid for sugar in the rest of the world and, as a consequence, tend to increase the environmental impacts of its production.

Sugar from sugar beets cannot compete with sugar from cane in the global marketplace unless sugar beet production is subsidized or cane sugar is subjected to a tariff. This works the same as subsidies but distributes costs differently. Governments, especially those of the United States and the European Union, continue to subsidize sugar beet production. It is not clear how long such subsidies will be tolerated under the regulations of the World Trade Organization. In addition, both the United States and the European Union protect domestic sugar production in developed countries through production subsidies and market barriers. Europe even supports subsidized exports of beet sugar that compete internationally with exports of cane sugar from developing countries. If there is political will to negotiate these issues in overall trade policy, then the production of sugar beets will decline and much of that nearly 7 million hectares will gradually be converted to other uses. More importantly, sugar markets in developed countries will open up to cane sugar from tropical producers.

ENVIRONMENTAL IMPACTS OF PRODUCTION

Sugar has arguably had as great an impact on the environment as any other agricultural commodity. Most of the environmental damage was loss of biodiversity, the result of wholesale conversion of habitat on tropical islands and on coastal areas. While the impact of this conversion can never be documented because it happened hundreds of years ago, in all likelihood considerable endemic flora and fauna unique to the many thousands of islands on which sugar was planted was lost.

The cultivation of sugar has also resulted in considerable soil erosion and degradation as well as the use of chemical inputs to correct the resulting problems. As a consequence, sugar has also had an important impact on other ecosystems. For example, sugar production has changed coastal hydrology. Siltation from soil erosion has clogged coastal ecosystems, especially coral reefs and sea grass beds, which are important to a wide range of species. Nutrient runoff from sugar cultivation has led to nutrient loading and eutrophication of freshwater and marine systems. Finally, sugar mills are cleaned periodically, and the organic matter that is flushed can tie up all oxygen in nearby rivers as it decomposes. This in turn asphyxiates fish and other aquatic organisms.

Habitat Loss

It is quite likely that the production of sugarcane has caused a greater loss of biodiversity on the planet than any other single agricultural crop. First, with nearly 20 million hectares in cultivation, sugarcane has more area devoted to it than most cash crops produced in the tropics. Second, sugarcane production has caused the clearing of some of the most unique and biodiverse regions on the planet. For nearly 500 years, tropical forests, the entire natural habitat of thousands of islands, and millions of hectares of fragile coastal wetlands around the world have been cleared or otherwise converted for planting sugarcane.

In fact, it is quite likely that but for sugarcane, any map of globally significant, biodiverse ecoregions would look quite different. For example, because of sugarcane the Caribbean is not considered significant biologically, nor are any of the islands (except New Guinea) in greater Southeast Asia. Even in areas where sugarcane is grown that have been identified for priority biodiversity salvage work, its cultivation has shaped the strategic prioritization of ecologically significant sites for conservation activities. Priorities in the Everglades, for example, do not include the sugarcane production areas, nor do the priorities in the Atlantic coastal forest of Brazil, the largest sugarcane-producing area in the world.

In short, sugarcane production has altered forever the landscape in many unique parts of the world. A brief glance back at Table 7.1 indicates the overall importance of the crop relative to all other forms of agricultural land use in a number of countries. A dozen countries around the world devote 25 percent or more of all their agricultural land to the production of sugarcane.

Soil Erosion and Degradation

During land preparation, there is a tremendous impact on soils as they are laid bare to be planted with cane. Aside from being stripped of any protective cover, the soils dry out, affecting overall microorganism diversity and mass, both of which are essential to fertility. Exposed topsoil is easily washed off of sloping land, and even on lands with minimal slope nutrients may be leached from the topsoil.

In some areas, such as the Everglades in the United States, the production of sugarcane has contributed to the subsidence of the land. This can result both from the removal of groundwater for irrigation, or the drying out and compaction of land that had previously contained high levels of organic matter.

Sugar processing harms the soil as well. The continual removal of cane from the fields gradually reduces fertility and forces growers to rely increasingly on fertilizers to replace it. The removal of plant matter from the fields makes the production of sugarcane unsustainable as it is currently practiced. In most of the world, sugarcane production is little more than a "mining" operation that strips the resource base. Bagasse, the organic matter left after crushing the liquid from it, is put to work as fuel for the cauldrons or sold as animal feed. If returned to the fields at all it is only in the form of ash, which is of little benefit to soil microorganisms.

Effluents

Silt from eroded soils and nutrients from applied fertilizers often foul local water supplies. Another problem with sugarcane production is nonpoint source pollution of water with pesticides, which is caused either by drift from spraying or by percolation of water through the soil. Effluents are also created from sugarcane processing, as discussed in the next section. Effluent flows into water supplies, and into important ecological areas such as the Everglades, need to be reduced. However, corrective measures may have their own environmental costs. From 1980 to the crop year 2000–01, area planted to sugar in Florida increased from 130,000 to 183,000 hectares (320,700 to 460,000 acres). There has even been a 10 percent increase in the area planted to sugar since 1995. Because of environmental concerns with water quality, large areas previously planted to sugarcane have been removed from production. Consequently, production has intensified in the remaining areas and expanded onto sandy soils, which by 2001 represented 22 percent of all sugarcane cultivation. Production in those areas is high initially, but because such soils are easily leached, production can only be maintained over time with increasing applications of fertilizer (University of Florida 2002).

Processing Waste, Emissions, and Wastewater

In addition to the impacts from production, there are a number of environmental problems at the mill. These fall into three categories—wastewater, emissions, and solid waste. Wastewater includes the water used to wash all incoming cane (10 cubic meters for each metric ton of cane), water from the boiler house used to concentrate the sugar and evaporate the water, and water from cleaning all the equipment. Perhaps the greatest environmental threat from processing occurs when mills are cleaned and thoroughly washed out, which occurs once or twice per year. The resultant impacts are not from toxic chemicals, but rather from the release of massive quantities of plant matter and sludge. As these decompose in freshwater bodies they absorb all the available oxygen, which in turn leads to massive fish kills.

In addition, mills release flue gases from the combustion in the boiler rooms. The flue emissions also include soot, ash, and other solid substances. Ammonia is released during the concentration process.

BETTER MANAGEMENT PRACTICES

Sugarcane growers in a number of different countries have been attempting to reduce the impacts of sugarcane production, both with and without the help of government. On the one hand, cane growers in Australia have developed two separate sets of guidelines to meet and exceed government environmental requirements, the Canegrowers Code of Practice for Sustainable Cane Growing in Queensland and the Canegrowers Fish Habitat Code of Practice (see http://www.qff.org.au/) (Canegrowers 1998). Similarly, growers in Florida and Louisiana have developed their own improved practices to meet increasingly strict environmental regulations. On the other hand, sugarcane producers in countries such as Zambia have been forced to address some of their effluent issues because they threaten the assets of other downstream resource users (e.g., hydroelectric dams, local communities).

Most of the better management practices for sugar production involve the reduction of soil erosion and the building of soil to ensure long-term production with the use of fewer inputs. Building up levels of organic matter in the soil can also reduce the need for other key inputs such as pesticides, fertilizers, and water. One of the key ways to reduce the input use and to build the soil is to increase organic matter by not burning sugar fields prior to harvest. Finally, there are a number of ways to reduce wastes and effluents from processing. Each of these better management practices is discussed below.

Implement Soil Conservation Practices

Sugarcane is currently grown on many steep slopes and hillsides (as in Northeast Brazil and many other regions). Many of these areas should be taken out of production because of the high rates of soil erosion that result from cultivating them. In a number of instances, removing these areas from production and replanting them to trees (e.g., fruit, nuts, wood) would actually encourage increased production in the adjacent, better-suited agricultural lands. Increased attention of producers on their better lands would tend to increase total production more than when producers focus on reducing losses on poorer soils. Put another way, producers will increase overall production when they focus on raising the average production level on the better lands rather than trying to obtain marginal production levels on less-productive lands. In addition, reforesting hillsides would improve overall water retention and hydrology and provide more gradual water release, which could improve yields and reduce the need for supplemental irrigation.

At the very least, implementing standard conservation techniques, such as contour plowing and terracing, in many parts of the world would decrease soil erosion and degradation and actually allow soil to be rebuilt over time. Such practices would also contribute to greater water retention. Soils should be covered at all times to keep topsoil from washing away, so that soil composition and vitality are not degraded. Any areas of slope should be planted before periods of heavy rains and irrigated, if necessary, until the rains arrive. Riparian areas should be left intact so that the plantings are not washed out, the soil eroded, biodiversity lost, and wildlife corridors destroyed.

Additional practices can be incorporated into overall management strategies to improve productivity in the short, medium, and long term. These include crop rotation, (e.g., rice in Florida), green manuring, and enriched fallowing or nutrient banking. These practices should be considered as investments for future savings and increased profits, as they will reduce the need for purchased agrochemicals in the future. Enriched fallowing, for example, uses deep-rooted perennials to draw nutrients up to the surface where they can be utilized more effectively by shallow-rooted commercial crops such as sugarcane.

Several conservation strategies could contribute to greater income for sugar plantations. The planting of fruit trees, for example, would not only provide food for wildlife, it would also give sugar plantations the ability to do value-added processing of jams, jellies, and juices. Cellulose from trees grown on such areas could be fed into paper pulp processing plants along with the bagasse to make the quality and consistency of paper more uniform.

Improve Pest Control and Management

Most sugar producers can improve their overall pest control and management systems. One way to do this is through integrated pest management (IPM) practices that allow producers to reduce the overall impacts of pesticide use. First and foremost, producers should plant pest-resistant cane varieties to reduce the need for pesticides. When pesticides are necessary, producers should identify and use those that are least toxic to control the pests on their crop. The pesticide used should be the most targeted one available rather than a broad-spectrum formulation. This will reduce the potential buildup of pest resistance, particularly of nontarget species. Similarly, there should be no prophylactic use of pesticides. Scouting and periodic monitoring allow producers to apply pesticides only when and where they are most needed, thus reducing overall use. Economic thresholds can be used to determine when pesticide applications are used. In other words, the losses from some pests may not justify the use of pesticides at all. Pesticides should only be applied at or below recommended dosages. They should not be applied when wind will cause drift and should be avoided during the rainy season or just prior to large forecasted rains. Finally, filter strips of vegetation should be planted around fields not only to control erosion, but also to reduce dissolved pesticide flows into surface or ground water (LSU 2002).

Eliminate Burning Prior to Harvest

Burning of cane fields prior to harvest should be abandoned. The practice of burning fields prior to harvest kills much of the wildlife that has managed to survive in sugarcane fields to that point. If the fires are not monitored, they can easily get out of control and burn into neighboring areas. Often what is burned are riparian areas or slopes that are too steep to plant. Both of these areas, however, can be rich in biodiversity that can be destroyed by uncontrolled fires. In some countries it is against the law to burn cane fields and violators are fined severely. But this is not the case in most developing countries.

More important from the point of view of producers, not burning fields prior to harvest improves profits. When growers abandon burning practices, they can harvest some 5 percent more sugar that had previously been lost as a result of burning. This more than compensates for the marginal labor increases involved in harvesting.

Finally, when fields are not burned, organic matter builds up, as much as 20 metric tons of organic matter per hectare from the leaves that are left in the field. Spraying the cane debris with microorganisms that hasten decomposition can break up the vegetable matter into manageable fractions that are more quickly reintegrated into the soil. This partially decomposed organic matter can act as a mulch for the crop. Mulch offers the advantage of holding in moisture, bonding with fertilizers and pesticides, reducing weed growth, and increasing productivity and net profits by reducing overall input use.

Reduce Nutrient Loading and Water Pollution

In some areas, progress has been made in reducing the water pollution from sugarcane production. As a result of a lawsuit, the Everglades Forever Act was created to require the state of Florida to build the world's largest system of artificial marshes to act as biological filters (biofilters) to remove nutrients in runoff entering the Everglades. In addition, the sugar industry was required to reduce the phosphorus content of its effluent by 25 percent. Over the past six years, the industry has actually reduced the phosphorus in its effluent by more than 56 percent (Grunwald 2002). Companies were able to achieve these results by reducing their overall use of fertilizers, using retention basins to hold water longer on the properties, and cleaning their ditches and canals more often.

In 2001, phosphorous levels in farm effluent were 64 parts per billion. This level was reduced to 30 parts per billion after the water left the constructed biofilters. While this is a good start, and well below the concentrations of 400 parts per billion in Miami tap water, most scientists agree that levels in the Everglades need to be reduced to 10 parts per billion or less if the ecosystem is to recover (Grunwald 2002).

While it is clear that the sugar companies are working to reduce their impacts, many still question whether it is enough. The industry, for example, is paying only one third of the cost of creating the artificial wetland biofilters. Instead of funding the cleanup, the industry spent $30 million to fight a proposal of taxing sugar $0.01 per pound to pay for the cleanup. There are now plans for the government to buy and retire 24,000 hectares of sugarcane land (Grunwald 2002).

Reduce Wastes and Effluents from Processing

Sugar processing wastes can also be treated so that they have far fewer harmful impacts. For example, before it is released to streams or waterways sludge can be treated with microorganisms ("activated") so that it decomposes more quickly. Microorganisms already exist that can be used to accelerate decomposition. A redesign of holding lagoons would allow them to be activated more easily for early and rapid decomposition. The treated effluent could then be returned (pumped) back to the soil both as a fertilizer and a source of energy for soil microorganisms.

Bagasse is another 20 metric tons of organic matter that is produced per hectare. Fiber represents almost 50 percent of the biomass of bagasse. This fiber could be used to make paper (as is already done in India), or alternatively for cement board additives. Sugarcane produces a fiber harvest once a year. But sugarcane plantations in any given area tend to be harvested over much of the year. If sugarcane came to be used as fiber, it is not clear whether the sugar or the fiber would have the highest value.

OUTLOOK

Sugar consumption is increasing globally. Economic growth and increases in disposable income in developing countries will increase sugar consumption because sugar is an ingredient that is used increasingly in fast food, prepared foods, and drinks of all kinds, the types of food that are consumed more as income increases. However, production has increased even faster than consumption, and this is likely to continue as many producers in many parts of the world have made significant investments in the cultivation of sugar.

The one outstanding factor that could affect sugar production globally would be a change in policies in the United States and the European Union. These would include the elimination of production subsidies, price supports, market barriers, and export subsidies. If these policies were altered in such a way that eliminated market protection, then sugar in developed countries (from beets or from cane) would have to compete with cane sugar produced in the tropics. While it is clear that production in the tropics would expand, it is not clear that a significant portion of that expansion would be at the expense of natural habitat. However, the price of sugar would probably increase to the point that production would be intensified in many developing countries. This would have harmful environmental impacts, but at least there would be more money available to address those impacts. Furthermore, markets in developed countries might play an instrumental role in providing incentives to clean up the industry.

REFERENCES

Canegrowers. 1998. Code of Practice for Sustainable Cane Growing in Queensland. Brisbane, Australia: Canegrowers. Available at http://www.canegrowers.com.au/environment/codeofpractice.pdf. Accessed 2002.

Clay, J. 1995. Good Sugar, Inc. A Proposal to Multilateral Investment Fund—Inter-American Development Bank. Arlington, VA.

———. 1994. A Proposal for the Good Sugar Program of Rights & Resources. Arlington, VA: Rights & Resources.

FAO (Food and Agriculture Organization of the United Nations). 1996. *Production Yearbook 1995*. Vol. 49. FAO Statistical Series No. 130. Rome: UN Food and Agriculture Organization.

———. 2002. *FAOSTAT Statistics Database*. Rome: UN Food and Agriculture Organization. Available at http://apps.fao.org.

Grunwald, M. 2002. When in Doubt, Blame Big Sugar. *Washington Post*. June 25.

Louisiana State University (LSU). 2002. *Sugarcane Production Best Management Practices (BMPs)*. Baton Rouge: LSU Agricultural Center.

Polopolus, L. C. 1993. Dispelling the World Price and Sugar Subsidy Myths. Paper presented at the 1993 International Sweetener Symposium, Squaw Valley, Calif. August 17.

Swahn, J. O. 1995. *The Lore of Spices—Their History, Nature, and Uses around the World*. New York: Crescent Books.

UNCTAD (United Nations Commission on Trade and Development). 1994. *Handbook of International Trade and Development Statistics, 1993*. Geneva, Switzerland: UNCTAD.

University of Florida. 2002. *Profitable and Sustainable Sugarcane Production in Florida*. Gainesville: University of Florida Cooperative Extension. FL 129. Available at http://extensionsmp.ifas.ufl.edu/fl129.htm.

Wilkinson, A. 1989. *Big Sugar: Seasons in the Cane Fields of Florida*. New York: Alfred A. Knopf.

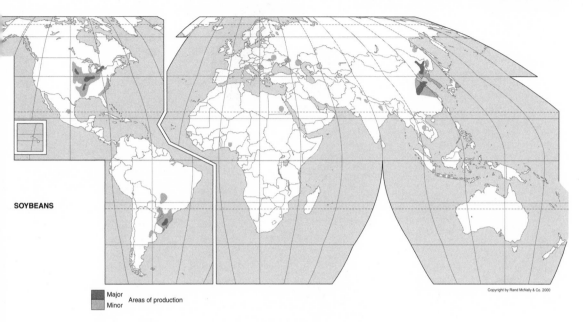

SOYBEANS

Major / Minor Areas of production

Copyright by Rand McNally & Co. 2000

SOYBEANS *Glycine max*

PRODUCTION		**INTERNATIONAL TRADE**	
Area under Cultivation	74.1 Million ha	Share of World Production	57%
Global Production	161.2 Million MT	Exports	91.6 million MT
Average Productivity	2,176 kg/ha	Average Price	$204 per MT
Producer Price	$195 per MT	Value	$18,728 million
Producer Production Value	$31,477 million		

PRINCIPAL PRODUCING COUNTRIES/BLOCS (by weight)
United States, Brazil, Argentina, China, India, Paraguay, Canada, Bolivia, Indonesia

PRINCIPAL EXPORTING COUNTRIES/BLOCS
United States, Brazil, Argentina, Paraguay, Netherlands, Canada, Bolivia

PRINCIPAL IMPORTING COUNTRIES/BLOCS
China, European Union, Japan, Mexico, South Korea, Thailand, Indonesia

MAJOR ENVIRONMENTAL IMPACTS
Conversion of natural habitat
Soil erosion and degradation
Agrochemical use
Genetically modified seeds

POTENTIAL TO IMPROVE
Good
BMPs are being identified and current prices can cover cost of adoption
Conservation tillage and zoning can reduce main impacts
Possible to produce more on less land with fewer inputs

Source: FAO 2002. All data for 2000.

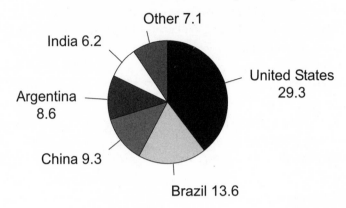

Area in Production (MMha)

Other 7.1

India 6.2

United States 29.3

Argentina 8.6

China 9.3

Brazil 13.6

OVERVIEW

Soybeans were first cultivated in China perhaps as many as 6,000 years ago, making soybeans one of the first domesticated food crops. During the Zhou Dynasty (eleventh century to 256 B.C.) and the Qin Dynasty (221 to 206 B.C.) soybeans became one of the main food crops in the Yellow River Valley. By the Ming Dynasty (1368 to 1644 A.D.), soybeans were grown throughout the country.

The soybean is the "king of beans." Dry, it contains 38 percent protein—twice as much as pork, three times more than egg, and twelve times more than milk. Furthermore, the protein in soybean has a more complete range of essential amino acids than most other foods. In addition, the dry seed contains 18.4 percent unsaturated fat. Many soybean products (e.g., miso, soy sauce, tempeh, and bean curd) originated in China. For example, bean curd (tofu) was invented in the Han dynasty (206 B.C. to 220 A.D.). The technology then spread to Japan around the year 700 A.D. (Tengnas and Nilsson 2002).

Soybeans, along with silk, tea, and porcelain, were one of the earliest exports from China. China displayed soybeans at the Vienna Fair in 1873, and the product then became better known to the outside world. At the beginning of the twentieth century, China began to export soybeans. Production in China reached 11.3 million metric tons in 1936 and accounted for 80 to 90 percent of the world market (Tengnas and Nilsson 2002).

Initially, soybeans were used primarily for human consumption and still are in parts of Asia. In the United States, though, soybeans have not been grown for direct human consumption until recently. Rather, they are grown primarily to provide cheap edible oil and high-protein animal feed. Soybean meal was first a by-product from the crushing of soybeans for oil. Soybean oil is now the most consumed oil in the world. Soybean meal was developed as a substitute for fish meal that left no fishy taste. Confined poultry and pork operations developed as a result of the availability of a relatively cheap, nutritious feed source for animals. After World War II soybean production was introduced in many parts of the United States. By the 1960s the cultivation of soybeans had expanded rapidly, displacing many other crops and becoming the main rotational crop in association with corn. Soybean production was encouraged by agricultural extension agents because the crop is a legume and fixes nitrogen in the soil for uptake by subsequent crops of corn. Through the development of different seed varieties, improved nutrient input packages, and mechanized planting and cultivation, a monocrop soybean production system was developed in the United States. This monocrop technology has been adapted to local conditions and spread throughout the United States and the world, including some of the world's most biodiverse ecoregions. Of particular concern from an environmental point of view is the rapid expansion of soybean cultivation into the natural habitat of the Brazilian cerrado (a relatively flat, mixed woodland and savanna area of central Brazil). Soybean production has spread even more rapidly throughout Brazil than it did in the United States.

PRODUCING COUNTRIES

The Food and Agriculture Organization of the United Nations (FAO) lists 82 countries that produce soybeans. In 2000 there were 74.1 million hectares of soybean production globally. The United States led all producers with more than 29.3 million hectares. Brazil was second with nearly 13.6 million hectares in production. China was third with more than 9.3 million hectares, Argentina was fourth with 8.6 million hectares, and India was fifth with nearly 6.2 million hectares planted to soybeans. These five countries accounted for 90 percent of land planted to soybeans globally and 92.1 percent of total production (FAO 2002).

The main producing countries by weight are the United States (75.1 million metric tons), Brazil (32.7 million metric tons), Argentina (20.2 million metric tons) and China (15.4 million metric tons). In 2000 these four countries accounted for 81 percent of global production. In addition India, Paraguay, Canada, Bolivia, and Indonesia are significant producers. In the past, much of Paraguay's reported production was probably trucked across the border from Brazil to avoid taxes. However, now soybean production in eastern Paraguay is increasing dramatically at the expense of the Atlantic coastal forest in the region.

While Paraguay's overall production is considerably less than that of the largest producers, soybeans are planted on more than 25 percent of all agricultural land in the country. The following countries devote from 10 to 25 percent of all cultivated land to soybeans: Argentina, Brazil, Indonesia, North Korea, and the United States.

Of the largest soybean producers, the United States has the highest average yields, 2,561 kilograms per hectare, though some of the smaller producers achieve even better results. Brazil produces on average 2,400 kilograms per hectare and Argentina 2,340 kilograms per hectare. China's yields are only 1,656 kilograms per hectare. India is far behind with average yields of only 986 kilograms per hectare.

Switzerland reports the highest yields in the world with 4,000 kilograms per hectare. Italy reports the second highest yields, producing more than 3,576 kilograms per hectare. Ethiopia reports yields of 3,571 kilograms per hectare. Though total production is low, the average yields reported in these three countries exceed those of all other countries by a wide margin. If the statistics are accurate, further investigation might indicate why their systems are so much more productive than in other countries (FAO 2002).

Yields have increased considerably through genetic improvements and the use of pesticides and fertilizers. For example, in Brazil production in 1940–41 was 651 kilograms per hectare per year. By 2000–01 average yields had increased to 2,720 kilograms per hectare (Tengnas and Nilsson 2002) or some four times higher as reported by the FAO in 2000 (FAO 2002). Two new varieties in Brazil are expected to increase yields to 3 to 4 metric tons per hectare per year (Tengnas and Nilsson 2002).

Soybean exports can be significant sources of income for both large and small countries. Oil seed cake from soybeans is Argentina's leading export; Paraguay's leading export is oil seed cake and its second largest export is soy-based animal feed. Soybeans are the second largest export for Cambodia, and soybean cake is the third largest export for Brazil (ITC 2002; UNCTAD 1994).

The area of land in soybean production has grown at a rate of 3.2 percent per year in recent years even though prices have been declining. From 1995 to 2000 Paraguay led all countries in percent growth of area planted in soybeans with 62 percent, followed by Brazil (47 percent) and the United States (33 percent). Soybeans now occupy the largest area of any agricultural crop in Brazil with 21 percent of total cultivated area. The area planted to soybeans in Brazil has increased by 2.3 million hectares since 1995 for an average increase of 320,000 hectares per year. To put this another way, since 1961 the area under soybean production in Brazil has increased 57 times while production volume has increased 138 times (World Bank 2002).

CONSUMING COUNTRIES

Most soybeans, whether for human or animal consumption, are processed before use. Even with some direct human consumption in Asia and elsewhere, only a small percentage of soybean solids are consumed directly by people. In the United States less than 1 percent of soybean solids produced are consumed by humans (Schnittker 1997). On the other hand, most soybean oil is used to make a wide variety of food products that are consumed directly by humans. About 2.5 percent of oil production is used in paint, soap, plasters, and other inedible products.

Much consumption of soybeans occurs in the countries of production. However, exports in 2000 accounted for 57 percent of total production, and exports have grown by 45 percent since 1995. The European Union and China account for 59 percent of the international trade in soybeans. However, while the European Union imports have remained stable for the past six years, the imports in China have in-creased 342 percent during that period (World Bank 2002).

PRODUCTION SYSTEMS

In most countries, soybeans are planted (or drilled) in rows. In the United States the rows are spaced about a half meter (1.75 feet) apart, and beans are planted about every 5 centimeters (2 inches) in the row. In Brazil the rows are planted closer to-gether. Like most other annual crops, virtually everything about soybean cultivation is mechanized, from the preparation of the soil to planting, cultivation, application of chemicals, and harvesting. Cultivation practices such as no-till, conservation tillage or direct planting all refer to ways to reduce or eliminate soil preparation. As a result they leave far more organic matter on the surface, reduce total machinery used, and reduce overall costs of production.

Efforts have been made to "drill" soybeans by planting them in rows only a few inches apart, a technique similar to that used for wheat. It was thought that this planting technique would reduce soil erosion, but that did not prove to be the case. In addition, this planting technique requires considerably more seed while reducing production per plant. It does not make sense financially. Likewise, there have been attempts to plant two crops per year in some areas, but even in the tropics the com-bined yields of the two quicker-maturing crops do not generally equal that from a longer-growing single crop.

Varieties differ somewhat regarding the time required for maturation. In the Unit-ed States, where most production areas have a shorter growing season, most soy-beans mature in about 95 to 120 days. Plants grow about 1 meter tall and slightly less across. In Brazil, the varieties planted mature in 115 to 145 days, and the longer growing season produces larger plants. The beans come from dozens of pods at-tached to the main stem or significant branches of the plant. Each pod contains three to four soybeans. Production in Brazil's Cerrado region, with its longer grow-ing season, is higher (by some 20 percent) than in the United States because the larger plants produce more significant branches that flower and set seeds.

Agrochemical use in soybean production is relatively low compared to other an-nual crops. Cotton production, for example, uses eight times the applications of chemicals per acre. In the United States soybeans were grown on 20 percent of the agricultural land but accounted for only 12 percent of total pesticide use in 1995, down from a peak of 24 percent of total pesticide use in 1982. Herbicides make up a large proportion of the chemicals used. A detailed report on herbicide use in the sev-en main soybean-producing states in the United States indicates that nearly all soy-beans planted in the United States are treated with herbicides at some point. Most of these herbicides are applied directly to surface soil rather than by airplanes, as sprays, or in irrigation water (Schnittker 1997).

In Brazil, pesticides, including the herbicide glyphosate (Roundup), are used on soybeans as well. Most of the pesticides used in Brazil are similar to those in the United States. In the southern part of the country the genetically modified variety bred to withstand herbicides (commonly called the "Roundup-ready" soybean) is estimated to be grown on 60 percent of all land in soybean production—even though it is illegal there (Leibold et al. 2001a). In the northern part of the country, glyphosate and similar herbicides are used in no-till operations, and producers spend from U.S.$40 to U.S.$50 per hectare for herbicides (Leibold et al. 2001a, 2001b).

Herbicides are not the only pesticides used. Some insects and diseases are particularly problematic in Brazil because of the year-round warm climate. Planting in narrow rows, as is common in Brazil, accentuates these problems. About 90 percent of soybean seed is treated with fungicide to prevent mildew and mold. Other diseases include stem canker and sudden death syndrome, which is associated with no-till practices and high rainfall. Soybean cyst nematodes were discovered in Brazil in 1992 and some 2 million hectares are now infected. Insects also cause problems. Some producers are using biological controls for some insects; others are spraying several times during the growing season, which adds substantially to the cost of production (Leibold et al. 2001a; 2001b).

Harvesting occurs once per year. Soybeans are harvested and threshed mostly by combines that move through the fields, cutting the stalks and separating the grain from the pod and the rest of the plant. After being combined the beans are transferred to waiting trucks and transported to on-farm bins or directly to grain storage depots. The depots are usually connected by rail or barge lines to more distant warehouses, processors, and/or markets.

Soils and weather during the growing season and any postharvest storage or handling problems on farm after harvest largely determine the grade, which in turn determines the price that a producer gets for the beans. When soybeans are sold, the farmer receives a receipt for the weight of the beans. This receipt also records the average moisture content (to more accurately calculate dry-weight volume) and the quality of the beans (e.g., presence of weed seed and other vegetable matter, dirt, and stones to determine if the price should be reduced due to contamination by foreign matter). At the commercial grain elevator, soybeans are sorted according to quality until they can be dried or cleaned to the same standards.

In the United States some 380,000 farmers produce soybeans, and this crop accounts for about 20 percent of all cropland. The farm-gate value of soybeans in the United States alone is $15 to $16 billion, or about 16 percent of farm receipts from all crops. As the soybean is processed and moves through the system its worth increases to several times that value. In the United States, the average amount of land planted to soybeans on a farm is about 100 hectares, while in the Brazilian cerrado the average size is 1,000 hectares (Stringfellow 2000). On average, land in the cerrado can be purchased and put into production for about U.S.$625 per hectare (McVey et al. 2000a.). The purchase price in Iowa is several times this. In addition, Brazil has about 94 million hectares (234 million acres) of land in the cerrado alone that would be suitable for production of crops like soybeans, while the United States has almost all of its suitable land in production at this time (McVey et al. 2000a).

Brazilian soybean yields in the areas of expansion are superior to those in the

United States, 3.03 metric tons per hectare (45 bushels per acre) compared to 2.56 metric tons per hectare (38 bushels per acre) in the United States. Production in Brazil could be increased even more by reducing harvest losses, retiring old equipment, training harvesters, and changing harvest conditions. In addition to longer growing seasons that produce larger plants, rainfall in Brazil is more predictable and continues well into the growing season. Inputs are cheaper so producers use more. Another major factor has been the development of cultivars that have greater resistance to diseases and insects, and more cultivars are in development at this time. The government is developing training programs for equipment operators. All of these factors will help Brazil to achieve the overall goal of increasing production by 680 kilograms per hectare (10 bushels per acre) within ten years (Leibold et al. 2001c).

The total nonland cost of production per hectare is higher in the state of Mato Grosso, Brazil (U.S.$1,040) than in Iowa (U.S.$870). However, the relevant comparison is cost per metric ton or bushel. In this case, Brazil has an advantage because productivity is higher in Mato Grosso. Brazil's greatest competitive advantage, however, is its cost of land. The land cost for soybeans in Iowa is $87 per metric ton ($2.38 per bushel) higher than for Mato Grosso. Total soybean production costs including land are $78 per metric ton ($2.13 per bushel or 61 percent) higher in Iowa than in Brazil (Baumel et al. 2000). Given that the current average price for exported soybeans is $204 per metric ton, this difference in cost is very significant. As the price of soybeans declines, this discrepancy will challenge Iowa producers' ability to remain competitive and financially viable.

In the Brazilian cerrado crop rotation is common. Rice or pasture is usually planted first to condition the land and build up levels of organic matter in the soil for subsequent soybean production. After soybean rotations have been started, cotton and corn are the preferred rotation crops during the three-year cycle. As a part of this production cycle, area in cotton has increased 600 percent in the cerrado (Leibold et al. 2001c). Even longer-term rotations are also becoming common. For example, many cattle ranchers lease their land (payment is based on shares of production so that both risk and gain are shared) to soybean producers for a period of up to five years to rebuild the soil and then return the land to pasture for up to seven years.

Transportation costs, too, vary between the United States and Brazil. Producers in the United States, for example, have a clear advantage to ship to Asian markets and a smaller advantage to ship to Europe. To date, transportation infrastructure in the United States has benefited from larger public investments and subsidies than those in Brazil.

Soybeans are valuable. Even so, in the United States subsidy payments to guarantee minimum prices for soybeans are an important reason that so many producers continue to plant them on such a large scale even in the face of declining prices. The expansion of soybean production in other parts of the world is also linked to government support. Subsidies for credit, inputs, technology, and infrastructure have been important for the expansion of soybeans in such countries as Brazil. In 1996 there were only 1,800 hectares of soybeans in Rondônia in the western

Amazon, but the area planted increased to 14,000 in 1999. In the eastern Amazon in the state of Maranhão the area planted to soybeans increased from 89,100 to 140,000 between 1996 and 1999 (Fearnside 2000).

Tremendous subsidies have been invested in and are planned, too, for Brazilian infrastructure that is intended primarily for soybean production. This will stimulate habitat conversion. Infrastructure projects that are planned or underway include the Madeira Waterway, the Itacoatiara soybean terminal, part of the North-South railway, and the BR-333 highway linking southern Maranhão and Minas Gerais. Projects that have been developed on paper include four canals/waterways, three railways, and two major roads. Other projects are being considered.

Genetically Modified Soybeans

In recent years, genetically modified soybeans have been developed and widely adopted by producers in the United States and, to a lesser extent, in Argentina, Canada, China, and even Brazil. Because transgenic soybeans raise important issues both from the point of view of production as well as markets, it is important to highlight some considerations at this point. Herbicide-tolerant ("Roundup-ready") soybeans, developed and sold exclusively by Monsanto, became available only in the mid-1990s. Today they account for more than half of all soybeans planted in the United States. Brazil is engaging in a debate regarding whether transgenic crops (including soybeans) will be legal even though some reports suggest that more than half of the soybeans planted in the south of Brazil are already genetically modified varieties (Leibold et al. 2001a).

Genetically modified soybeans offer considerable appeal for producers. With transgenic varieties of soybean, producers report that they use fewer chemical inputs, especially herbicides, because they can time the use better. They also save money because they make fewer passes over the field and have less wear and tear on their machines. These factors appear to lower on-farm production costs (although there is only data at this time for one to four years, depending on how long producers have been using the technology). There are reports of a 10 percent overall increase in soybean production, but these reports are anecdotal.

From an environmental point of view, use of genetically modified soybeans also has some positive impacts. Transgenic soybeans allow producers to use no-till cultivation practices for the first time in areas of continuous cultivation. Because the soybeans have been bred to tolerate a broad-spectrum herbicide, weeds are no longer managed by plowing them under to kill them at the beginning of the growing season or by cultivating the soil during the growing season when weeds normally grow. Now, a one-time spraying of herbicides can kill weeds after the soybeans are growing. This means that organic residue is left on the surface to decompose, building up levels of soil organic matter from year to year. The organic matter on the surface also acts as mulch. It holds water like a sponge, protects the soil from the sun, inhibits weed growth, and protects the soil structure. Inputs are more effective when they become attached to organic matter and are released more slowly. In this way they are not washed away with the first rain. Since "Roundup-ready" soybeans permit the use of no-till production practices, they can reduce soil erosion, in some reports by as much as half.

On the other side of the issue, there are several concerns about use of genetically modified organisms (GMOs). First, there is a general concern that transgenic crops may cross with other plant species. This could create pesticide resistance and, as a result, super weeds or pests. In addition, the application of the same herbicide (in the case of soybeans, glyphosate or Roundup) over long periods of time will most likely create resistant weeds. This problem could very easily force producers onto an herbicide treadmill. Another concern with crops such as transgenic soybeans is that the applications of pesticides can also kill life in the soil that is essential to the maintenance of good soil structure.

It is not exactly clear what the net gain or loss is from herbicide-tolerant soybeans. Even so, there appears to be enough promise in the general approach to proceed cautiously with it. The major issue, however, is that neither Monsanto nor the producers have proceeded cautiously. There are serious concerns that insufficient information exists or is being collected about the potential long-term impacts of GMOs in general (and transgenic soybeans in particular) before they are released on a wide scale. Insufficient field trials, insufficient monitoring, and perhaps most important, insufficient transparency (access to the corporate research protocols and results that document the benign impact of the technology) were all part of the release of this GMO as well as others.

Perhaps the most important issue raised for soybeans by GMO research is how the same technology used in producing herbicide-tolerant varieties can be used to improve existing soybean breeding programs as well as the production of conventional soybeans. GMO technology could allow traits to be selected virtually overnight by comparison to the hit-or-miss techniques of traditional breeding programs.

Conventional plant breeding created a fivefold increase in new soybean varieties certified by the U.S. Department of Agriculture (USDA) between 1961–64 and 1991–94. The development of soybean varieties that tolerated low levels of soil phosphorus and high levels of aluminum was critical to the expansion of soybean production throughout the greater Amazon. Whether traditional or gene-spliced, soybean-breeding programs have implications for where and how soybeans can be planted, how productive they will be, how long it takes them to produce, and how resistant they will be to pests and climate shifts.

PROCESSING

Very little soybean processing is done on the farm or in the immediate vicinity. Most local processing includes drying with natural gas to reduce the moisture content and blowing or screening to clean the beans of stems, dirt, rocks, or other foreign matter introduced during the harvest.

Soybeans are processed by dehulling and flaking the beans and then separating the soybean meal or cake from the oil. This entire process is called crushing. A given weight of soybeans will produce 18 to 19 percent (by weight) oil and 73 to 74 percent meal, or 35 percent protein by gross weight. This varies somewhat depending on the quality and the variety of the bean. The oil is used primarily for human consumption; the meal is used primarily for animal feed.

The United States exports about 35 percent of its raw soybeans before processing. By contrast, Argentina and Brazil add value to most of their crop; they process about 80 to 85 percent of their soybeans and export most of the products.

Argentina, Brazil, and the United States dominate world soybean markets. In 1999 they accounted for 80 percent of soybean production and 70 percent of soybean oil. Soybean oil faces intense competition from palm oil producers in Malaysia and Indonesia, who can produce oil at about half the cost of soybean and other seed oil producers (Stringfellow 2000). The only way soybean crushers can compete with oil palm mills is by exploiting economies of scale. Consequently, the capacity of soybean crushing mills is much larger than those for oil palm. Soybean crushers can process 500 metric tons per hour. The heavy financial investment needed for crushing plants has led to growing concentration in that part of the market chain. Three multinational corporations now own more than half of the capacity in the three largest soybean-producing countries (Stringfellow 2000). Even so, it would be hard for soybean-crushing operations to compete on the price of oil by itself.

SUBSTITUTES

Two main products are made from soybeans—soybean oil and soybean meal or cake. Substitutes exist for each of these, but the various oils have distinctive tastes, and the substitutes for soybean meal have protein characteristics that make each better or worse for certain uses. Different products can be substituted broadly, but not exactly one-for-one. Ultimately, the balance of use will change due to genetic improvements or to changes in taste, price, or consumer preference. Since it is possible to change or eliminate taste or create other characteristics, price is increasingly coming to dominate the picture.

Productivity and mechanization are the main competitive advantages for soybean producers over palm oil producers. Over the past twenty-five to thirty years average soybean yields have doubled in Brazil and increased by 50 percent in the United States. In Malaysia average palm oil yields have actually declined over the past fifteen years. This has more to do with labor requirements than with genetics, however. In Brazil one worker can farm 250 hectares of soybeans, while on a mechanized palm oil estate in Malaysia one worker is required for every 12 hectares. In Brazil it takes 0.07 days of labor to harvest what will make 1 metric ton of soybean oil. In Malaysia it takes two days' labor to harvest 1 metric ton of palm oil. If yield gains and mechanization continue to favor soybean production, then the ratio of soybean oil costs to palm oil costs will fall from 2:1 to as little as 1.35:1 over the next decade (Stringfellow 2000).

Not all oils are substitutes. Lauric (oils rich in lauric acid, primarily coconut and palm kernel oil) and marine (e.g., fish oil) oil prices, for example, behave quite differently from those of the leading vegetable oils. Among the main traded vegetable oils, major deviations in Rotterdam prices (a major market where the prices of various oils tend to be quoted and compared) persist for only a short period (e.g., palm oil's large discount in 1993–94) (Stringfellow 2000). Table 8.1 shows some of the substitutions that are possible.

Soybeans provide about 28 percent of the world's vegetable oil. The market substitutes for soybean oil are canola oil, sunflower oil, corn oil, cottonseed oil, palm and palm kernel oil, coconut oil, olive oil, and various animal fats. Oil palm has by far the highest productivity of any of the major vegetable oils in terms of yields per hectare. Production is five to ten times as much as soy, canola (rapeseed), or sunflower oil. Other oilseeds must make up for the lack of oil production through the income from the meal by-products. This favors soybeans in particular (Stringfellow 2000).

World production of the ten leading oilseeds (not net oil weight) was estimated at 288.2 million metric tons in 1998–99. Soybeans represent 53.7 percent of all oilseed production. Cotton represents 11.3 percent, sunflower represents 9.2 percent, canola (rapeseed) represents 12.8 percent, and peanuts (groundnuts) represent 12 percent. Table 8.2 gives totals (in million metric tons) for the top ten sources of vegetable oil.

TABLE 8.1. Applications and Substitutions of Palm, Seed, and Fish Oils and Animal Fats

Application	Competing Fat or Oil
Spreads, margarines, and vanaspati[a]	Partially hydrogenated seed oils (e.g., soybean oil, etc.) Lauric oils[b] Partially hydrogenated fish oils
Shortenings	Partially hydrogenated seed oils Partially hydrogenated fish oils
Confectionery fats	Partially hydrogenated seed oils Exotic fats[c]
Frying fats	Animal fats
Ice cream fats	Lauric oils Partially hydrogenated seed oils
Applications and Substitutions of Liquid Fractions of Palm Oil	
Frying oil	Seed oils (e.g., soybean oil)
Salad oil	Olive oil Seed oils Partially hydrogenated and fractionated oils[d]

Source: Stringfellow 2000.
[a] A hydrogenated vegetable fat used as a butter substitute in India.
[b] Lauric oils are mid-range oils in the C12 to C14 range (many natural oils are composed of a wide variety of oils from C2 to C22). They are more versatile, higher-priced oils and are often made from palm oil, palm kernel oil, and coconut oil.
[c] Exotic fats are mostly produced from tropical oil seeds and are high (50–60 percent) in saturated fatty acids. Exotic fats include butters made from such seeds as illipe, shea, sal, mango, and kokum. These fats tend to be cheaper than similar, better known fats (e.g., cocoa) and are substituted for them by food, cosmetic, and healthcare product manufacturers.
[d] Fractionated oils are oils that have been separated to select for specific fatty acids. Most natural oils have a wide range of oils which make them less than ideal for many uses. It is easier if they are more uniform. For example, fractionated oils can be created with either low or high melting points.

TABLE 8.2. Global Vegetable Oil Production, 1998–99
(in millions of metric tons)

Type of Oil	Amount Produced
Soybean oil	24.3
Palm oil	17.2
Palm kernel oil	2.2
Cottonseed oil	3.8
Sunflower oil	9.4
Canola (rapeseed) oil	13.0
Peanut (groundnut) oil	4.4
Linseed oil	0.7
Olive oil	2.5
Coconut oil	2.8
Sub-total	80.4
Total from 17 species*	103.7

Source: UNCTAD 1999.
* The other seven oils include corn, castor, tung, sesame, and hempseed. These five are clear (e.g., reported by UNCTAD and FAO); walnut and grapeseed are most likely the remaining two.

Many product ingredient lists indicate that a product is made from x, y, or z vegetable oil. This means that the manufacturer is free to substitute any of the listed oils, depending on the price, without changing the packaging. Partly as a result of such substitutability, canola oil has made quick strides in the market and gained considerable market share. In 1999, 27 million hectares were planted to canola (rapeseed), with the combined acreage planted in China, India, and Canada amounting to more than 70 percent of the total acreage. India and China produce 60 percent of the world's peanuts (groundnuts). The United States and Nigeria account for another 12 percent. These are used both for their oil and as a source of vegetable protein.

Cottonseed oil is a valuable by-product of cotton production, but cotton producers are not paid separately for their seed. Only in the Aral Sea area of the former USSR was cotton produced primarily for seed. In that instance the vegetable oil was a strategic raw material used as a lubricant by the military. Consumer preference, at least in most developed countries in recent years, has tended towards substituting vegetable oils for most of the animal fats used in the more obvious forms of human consumption. However, animal fats along, with coconut, palm, and palm kernel oil are used in the manufacture of numerous personal care products (e.g., soaps, shampoos).

Soybean oil can also be substituted for fish oil in the diets of terrestrial livestock. For the most part, however, soybean and other vegetable oils cannot replace fish oil in the diets of carnivorous aquaculture species (e.g., salmon and shrimp), although many are attempting to find ways to do that. One way would be through genetically modifying the oil-producing plants so that their seeds incorporate the feed characteristics required by marine animals.

Production of soybean meal worldwide amounted to 105.8 million metric tons in 1998–99. The top four producers are the United States, Brazil, Argentina, and the European Union. (Europe produces some soybeans but also imports the beans raw and makes its own meal.) The other main producers are China, India, Japan, and Taiwan. Globally, annual production of all oilseed meal totals about 187.7 million metric tons. Cottonseed meal accounts for 8 percent, sunflower for 6 percent, and canola (rapeseed) for 10.7 percent. With soybean meal, these three constitute more than 80 percent of global oilseed meal production (FAO 2002). Most oilseed meal is used for animal feed.

Soybean meal or cake is used primarily in the diet of livestock raised for meat, particularly chickens and to a lesser extent pigs. The main substitutes are ground corn and sorghum meal and fish meal. Cornmeal is used in many livestock operations, especially feedlot production of cattle, pigs, and chickens. The percentage of meal from different crops will vary as a result of both price and palatability. Soybean meal is considered too rich for many animals and must be "cut" by adding different meals such as corn or sorghum. Corn is a cheaper substitute, but it does not have the same nutritional profile as soybean meal. Sorghum is also gaining considerable ground as an animal ration. In particular it is used for chickens and to some extent pigs. Cassava is increasingly imported to the European Union countries for use as animal feed where it is used to cut the more protein- and oil-rich soybean meal. The largest international cassava supplier is Thailand.

Fish meal, bone meal, ground hair and feathers, and several other meat-packing by-products are increasingly substituted, at least in part, for soybean meal in livestock diets. There is increasing consumer resistance to these forms of substitution, however, as some meat-packing by-products have been linked directly to mad cow disease in England and France. Soybean meal is increasingly substituted for fish meal in aquaculture feeds, but no way has been found yet to eliminate the use of all fish meal. Up to this time, most competition has been in the poultry industry, where the two meals are readily substitutable. For this reason, the prices of soybean meal and fish meal are relatively linked in the marketplace. As the aquaculture industry grows, they are likely to become less closely linked.

MARKET CHAIN

The market chain for soybeans is relatively simple. The producer delivers the product to a warehouse either at harvest or after storing to dry the beans and/or to see if the price will improve. Soybeans are stored in a grain elevator until they are transported via rail, truck, barge, or boat closer to a point of processing for storage until needed. The soybeans are then transported to the processing plants or to ports for export and processing in another country (Schnittker 1997). Table 8.3 shows the relative numbers of primary players in the soybean market chain in the United States, as well as the change in value of the product as it moves along this chain.

There are two main processing streams, one for soybean meal and another for the oil, but many other products flow from these two original products. Processing is done by a small number of companies. In the United States, three companies (Archer Daniels Midland, Cargill, and Bunge) and one cooperative (Central Soya) control more than 80 percent of United States' soybean-crushing/oil-processing

TABLE 8.3. Players in the U.S. Soybean Market Chain

Function	Number of Players	Value of Product (per metric ton)
Producers	380,000	$257
Grain elevators	Thousands	$272
Central elevators/ports	85	$294
Oil processors	144 (4 do most)	$294
European Union importers	<1 dozen	$312
U.S.-based animal feed and food manufacturers	Thousands	$331

Source: Schnittker 1997.

capacity and a similar share of worldwide capacity (Schnittker 1997). These companies expanded to control a larger market share in the past twenty years when expensive technologies were beyond the financial capacity of many smaller firms. The dominant processors are also the leading exporters of soybeans and soybean products.

MARKET TRENDS

Between 1961 and 2000 soybean production increased by 499 percent and international soybean trade increased by 1,492 percent. During the same period the price of soybeans decreased by 53 percent. Soybeans' share of the world oilseed market has been growing steadily and now accounts for 55 percent of the total, up from 45 percent at the start of the decade (FAO 2002).

The International Food Policy Research Institute (IFPRI) estimates that world demand for meat will rise by 63 percent from 1993 to 2020, and that demand for soybeans used in part for animal feed will increase by 66 percent by 2020. This is greater growth than for other cereals, which combined are expected to increase by 40 percent over that period (Schnittker 1997).

Soybean meal has comprised about two-thirds of the value of soybeans in recent years, with oil about one-third. This situation developed over the past thirty to forty years as the demand for protein for animal feed increased rapidly and as the production of other oil-rich seeds such as palm oil, canola, and sunflower weakened the demand for soybean oil (Schnittker 1997).

In a number of fast-growing developing economies (e.g., India, China, Pakistan), growth in production of vegetable oils lags behind growth in consumption. As a result, imports of such oils are growing fast. However, even in developed countries demand for vegetable oils is growing strongly as lifestyle changes (e.g., more fast food, ready-to-eat snacks, and processed foods) also increase consumption of vegetable oil (Stringfellow 2000). Demand growth for vegetable oils in developing and developed countries will drive production and trade by 2010 to about 150 percent of 1999 levels.

The 1997 Asian financial crisis stopped the soybean market increases for both oil and meal in their tracks. The second factor that affected the global market was the liberalization of farm policies in the United States. This, combined with high prices

in 1997, created the conditions for a record harvest in all the major soybean producing countries of the Americas.

Subsidies, price supports, market barriers, and other policies aimed at protecting
soybean producers or food processors in developed countries are ultimately regres-
sive taxes. That is, they are wealth transfers from the entire population of those countries to specific producers and processing industries. It is not clear if this practice will
be tolerated once the full impact of such policies is understood by the taxpayers of
developed countries.

During the past ten to twenty years China has gone from being a soybean exporter to being the world's largest importer of whole soybeans and oil and a large importer of soybean meal. It is this increased demand that is stimulating the production
in the Amazon and other remote areas. This trend was due, at least in part, to the direct investment of the United States Soybean Producers' Association, which spent
considerable time in China demonstrating the value of soybean meal in animal and
aquaculture rations. The association's goal was to eliminate China as a possible competitor producing soybeans for the global market by increasing domestic demand so
much that the country would become a net importer. While China has become one
of the largest consumers of soybeans in the world, it is also producing increasing
quantities of soybeans and has encouraged producers to switch to soybeans from other crops such as tobacco.

The other factor that is now affecting the soybean market is consumer concern
with GMOs in general and transgenic soybeans in particular. Responding to consumer concerns, the European Union has required that products containing GMOs
be identified for consumers. In addition, the European Union is attempting to require that GM soybeans be separated from other varieties so that those purchasing
them for feed or as ingredients can make a choice about the products they buy and
that ultimately make their way to the consumer. As a consequence, the overall trade
statistics are clear—markets for soybeans from the United States have declined due
to the inability or unwillingness to segregate GM from traditional soybeans. Consequently, where GM soybeans had a slight market preference in the late 1990s (even
up to 30 cents per 27-kilogram bushel), by 2000 non-GM soybeans were already beginning to fetch a higher price in the marketplace (Stringfellow 2000). Producers in
the United States are responding and some of the land currently planted to GM soybeans will be replaced with traditional varieties. This in fact has created market
scarcities for seed for traditional varieties and created the need for storage facilities
that can keep the different soybean varieties separate.

At this point, countries that can guarantee that they are not producing or exporting GM soybeans can capture a premium price in the marketplace because of consumer concerns in Europe and Japan about transgenic organisms. China, by contrast, has said it is not concerned about transgenic soybeans.

Aggregate statistics represent thousands of individual purchase decisions. For example, the French grocery store chain Carrefour recently announced a sizeable
contract to purchase Brazilian soybeans to provide feed for their meat suppliers. In

this case a retail chain, responding to consumer demand, is working directly with its suppliers to produce a product that it will be able to sell to its customers. They have insisted that the product be GMO-free.

At the same time, the European Union recently approved the ban of animal feed made from animal bone and meat powders (Tengnas and Nilsson 2002). This decision is likely to lead to an increase in the consumption of soybean meal for animal feed. However, the fear of GMOs may be no less than that of mad cow disease, which spawned the ban on animal products in feed in the first place. It is likely that this will ultimately dampen demand for soybean meal.

Since 1994 soybean oil and palm oil have been priced almost exactly the same. Since they are substituted for each other, their prices are linked. A scarcity of one triggers purchases of the other. The price of soybeans at this time is declining. This appears to be related to two factors—the increased production of soybeans in Latin America and China and the declining price of palm oil based on increased planting in Indonesia, Malaysia and Papua New Guinea.

Another factor has affected overall soybean market trends. Each soybean product tends to have an impact on the pricing of the whole set of products. If the price of oil is high, then the price of meal can be lower to gain market share. If, however, the price of oil is low, then the price of meal must be high or soybeans decline in value. Unlike most oilseeds, soybeans contain only about 18 percent oil. It is not profitable to produce the oil if there is not also a market for the meal. The meal is, in contrast to most oilseeds, quite high in protein—from 44 to 48 percent.

ENVIRONMENTAL IMPACTS OF PRODUCTION

In the United States and the European Union, the production of soybeans poses a few rather distinct environmental problems. These have to do with the use of agrochemicals in production and the degradation of soil through the use of chemicals, erosion, or compaction. Runoff resulting from soybean production can include high levels of agrochemicals, suspended soil, and organic matter. This can be a major source of freshwater and groundwater contamination. Elsewhere in the world the major environmental problems associated with soybean production include the conversion of natural habitat, soil erosion, and the ever increasing use of pesticides.

Conversion of Natural Habitat

In the United States and Europe the decision to plant soybeans does not usually entail a decision to clear natural habitat; soybeans are produced, by and large, on areas previously used for agriculture. Producers have merely made a choice to produce soybeans rather than another crop that had been grown previously. This is not the case in many tropical countries, however, where the cultivation of soybeans often is part of the process of converting extensive areas of natural habitat to agriculture for the first time. This is true of Brazil, Argentina, Bolivia, Paraguay, and Cambodia. In these instances, producing soybeans destroys natural habitat and nearly all the flora and fauna found there. In Latin America soybean cultivation has taken place at the expense of natural savannas and tropical forests.

In addition to direct habitat conversion, soybean production in pristine areas also requires the construction of massive transportation and other infrastructure projects. The infrastructure developments unleash a number of indirect consequences associated with opening up large, previously isolated environments to population migration and to other land uses. This infrastructure contributes directly and indirectly to habitat conversion. In Brazil, for example, plans are underway to build eight industrial waterways, three railways, and an extensive network of highways. Such infrastructure is not used just for soybeans. Estimates suggest that collateral impacts may be as much as six times those resulting directly from soybean production, particularly in areas like the Amazon where isolation had previously been the limiting factor for development (Fearnside 2000).

Some soybean producers clear forests themselves. Others buy the land from small producers, often colonists, who have already cleared it. These same small producers then move further into the frontier and clear more land. In Brazil, soybean cultivation displaces eleven agricultural workers for every one finding employment in the sector. In the 1970s, 2.5 million people were displaced by soybean production in Paraná state and 0.3 million in Rio Grande do Sul. Many of these people moved to the Amazon where they cleared pristine forests (Fearnside 2000). More recently, the expansion in the cerrado involves the displacement of very few people because the area has not been widely inhabited.

In Brazil the savannas and cerrados are the most at risk. These areas have biodiversity that rivals equivalent areas of Amazonian forests, but only 1.5 percent of such lands are in federal reserves. Unfortunately, they can be easily converted into vast expanses of soybean fields. Even during the first year, however, agrochemicals must be provided for the crop to be financially viable. The soils are often so poor that within two years, virtually all nutrients are provided through applied lime and fertilizers. The soil is stripped of virtually all fertility and only serves to hold up the plants.

Soil Erosion

Globally, some progress is being made on the issue of soil erosion. One 1996 study in the United States showed that soil erosion associated primarily with soybean and corn production in the Midwest fell from 37.5 metric tons per hectare in 1930 to 19.5 metric tons per hectare in 1982 and to 15.75 metric tons per hectare in 1992 (Schnittker 1997). This rate is easily still a few times greater than is sustainable (defined as a creation of soil greater than or equal to that lost through erosion).

Despite the progress, there is reason for concern as lands classified as "highly erodible" are now being used for soybean production. In the United States, the Conservation Reserve Program actually paid producers to take highly erodible land out of production. It now appears that the development of herbicide-tolerant ("Roundup-ready") soybeans has encouraged many producers to plant at least some of those lands again. The soybean varieties genetically modified to tolerate herbicides allow producers to employ no-till and conservation tillage production systems to minimize erosion, even on the most erosion-prone areas. However, the net environmental impact of this change in cropping has yet to be determined. The chief fear with the highly erodible lands is that, despite improved techniques, soil erosion will once again become a problem.

The Brazilian National Development Bank has warned that "without well-defined technical criteria" the soil in many areas of the Amazon could be rendered unusable by soybean cultivation. Soybean production also causes soil compaction. In Bolivia, where soybean cultivation has been increasing since the 1970s as a result of investments in crop substitutes for coca production, degradation is already severe. Initially, soybeans could be cultivated without fertilizer or lime applications. By the late 1990s, however, more than 100,000 hectares of former soybean lands were abandoned to cattle pasture because the soil was exhausted. The three Mennonite settlements that had farmed soybeans had moved further to the north to clear more forests to, once again, plant soybeans.

Groundwater Contamination from Fertilizers and Pesticides

In the United States the Environmental Protection Agency (EPA), has recently acknowledged that agriculture is the major source of surface water quality problems in 72 percent of impaired rivers, 56 percent of lakes, and 43 percent of estuaries (US-EPA 1994, as cited in Soth 1999; Faeth 1996). In the United States the main herbicide used with GM soybean production is glyphosate (trade name Roundup). While glyphosate has been touted by the manufacturers as benign, and they can back up their claims with research they have supported, other studies suggest otherwise. For example, these studies allege that the chemical has been linked to reproductive disorders, genetic damage, liver tumors, disrupted embryo development, and developmental delays in mammals (e.g., Cox 1998).

While some producers claim that one application of herbicide is all that is needed for an entire growing season with herbicide-tolerant soybeans, studies show that both the total amount of herbicide used and the number of applications have increased. Chemical usage summaries from the National Agricultural Statistics Service of the United States Department of Agriculture (USDA 1991) show that in the United States total herbicide use on soybeans increased from 56.4 million pounds in 1995 to 75.2 million pounds in 2000. The use of glyphosate (Roundup) increased from 6.3 million pounds to 41.8 million pounds in the same period. In 1995 glyphosate was used on 20 percent of the soybean crop, but by 2000, just four years after the 1996 release of Roundup-ready soybeans, it was used on 62 percent of the crop (Tengnas and Nilsson 2002). Furthermore, the number of applications increased from one application per crop to 1.3 applications.

There is concern (but not yet evidence) that agrochemicals such as the herbicides trifluralin (Treflan), lactofen (Cobra), fomesafen (Reflex), bentazon (Basagran), imazethapyr (Pursuit), sethoxydim (Poast), and clethodim (Select) will contaminate lakes and lagoons in the Brazilian Amazon River floodplains. During the dry season, the waterways dry up and any contamination within the separate water bodies would become more concentrated (Fearnside 2000 and Leibold 2001a).

In the Brazilian Amazon high humidity and heavy rains have already caused the spread of fungus and blights. This, in turn, is resulting in the increased use of fungicides. Similarly, as production continues in the same area over a number of years, pest populations will increase, which will be followed by an increase in the use of chemical controls. Because Brazil has no frost, the different pests will adjust more rapidly to whatever chemicals are used to prevent them than they do even in the United States, where pests have rapidly developed resistance to the chemicals used to control them.

The FAO and others estimate that 25 percent of all pesticides used in Brazil are used in soybean cultivation, and that in 2002 an estimated 50,000 metric tons of pesticides were used by Brazilians on soybeans (World Bank 2002). Because of the rapid expansion of area planted to soybeans, pesticide use is increasing at a rate of 21.7 percent per year. However, the growth in pesticide use is increasing even faster than the growth in either area cultivated or overall soybean production. While part of this can be explained by the lack of frost and pests developing resistance to pesticides due to increased use, there are other factors involved. Production is expanding into areas with insufficient labor and pesticides are used to reduce labor costs. Areas planted to soybeans are becoming larger and therefore mechanization makes the application of pesticides more cost-effective. More work needs to be done on analyzing the full range of agrochemicals used in the cultivation of both traditional and GM soybeans as well as their long-term movements in the environment, their impacts, and the development of resistance to them.

Extraction of Limestone

In Brazil, the lime requirements of growing soybeans in the Amazon alone could lead to considerable destruction of natural resources. Lime (a source of calcium) is applied to soils to counteract acidity, because neutralizing soil acidity makes existing nutrients more available to plants such as soybeans. The mining of limestone requires the removal of considerable overburden (natural cover, soil, etc.) to gain access to limestone deposits. In addition, large amounts of energy are used to cook the limestone and make it into agricultural lime. In the Brazilian savanna areas, 4 to 6 metric tons per hectare of lime are required to produce soybeans. In cleared forest areas, only 2 metric tons per hectare of lime are required, initially at least (Fearnside 2000). This raises two issues. The first is the production and transportation of the lime itself. The second issue is the incentive to shift production into cleared forest areas that do not require the initial application of so much lime.

BETTER MANAGEMENT PRACTICES

There are a number of conservation strategies that can reduce the impact of soybean production. These include creating protected areas in areas of soybean expansion and using zoning to restrict expansion to degraded or abandoned agricultural areas.

The identification and adoption of no-till practices can reduce the soil erosion caused by soybean production, as can linking the adoption of such practices to government subsidy programs. A related policy initiative to reduce the harmful impacts of the industry is to remove subsidies that encourage soybean expansion for artificial markets. Clearly, one conservation strategy should be to identify and analyze the implications of soybean expansion for natural habitat. Finally, strengthening command-and-control regulatory systems can reduce the environmental problems associated with soybean cultivation. Each of these strategies is discussed separately. However, their cumulative impacts are greater than their individual ones.

The strategies that are most appropriate to reduce the impacts of soybean cultivation will vary considerably from one country to another. China has produced soybeans longer than any other country. Today, the country's production is seriously affected by pests. Some 8,800,000 hectares of land have been affected by losses estimated at some 32,900 metric tons. In one of the areas most used for soybean production, the following measures are taken to prevent diseases: crop rotation, deep plowing, late planting, manuring, and the application of pesticides as needed (Tengnas and Nilsson 2002). These strategies, at the other end of the spectrum from input-intensive no-till cultivation with or without transgenic varieties, suggest that techniques will vary widely from country to country and will depend on specific conditions and pests.

In the United States, the better management practices (BMPs) that have evolved are quite different. In this country precision agriculture and the targeting of agrochemicals to address specific needs have resulted in a reduction of the overall average use of chemical inputs per hectare. Likewise, soil erosion has been reduced by standard conservation techniques including terracing, strip cropping, planting cover crops, maintaining waterways, and improving road construction and machine access. Organic matter content in the soil has been improved through mulching and conservation tillage programs such as reduced tillage and spring tilling. No-till production leaves virtually all crop residues on the surface while reduced tillage (or low-till) leaves some 15 to 35 percent of crop residues on the surface (Schnittker 1997). Conservation tillage provides more ground cover, more available waste grain for food, and less disturbance of nesting sites than conventional tillage. All of these practices have a net positive impact on biodiversity and ecosystem functions. Most of these practices also increase yields and profits.

There are a number of impediments to the adoption of BMPs, however. At a recent meeting in Iowa hosted by Natural Resources Defense Council (NRDC 2001), a number of issues were identified that affect overall pesticide use. Perhaps the most important is that the adoption of BMPs requires a greater time commitment to management, yet farm size is growing in most soybean-producing areas so there is less time available to devote to each hectare of production.

The creation of protected areas on expanding soybean frontiers could be an essential source of protection for biodiversity and fragile ecosystems in areas that are often not suited for long-term soybean production. In Asia and Latin America where soybean production is expanding into tropical savannas and forests, one strategy would be to identify key biodiversity sites and work to set them aside as protected areas as a condition of the further expansion of soybean production. For example, in the cerrado of Brazil less than 2 percent of the land is under any form of protection. In the Atlantic coastal forest of Paraguay there is a similar lack of protection.

It is also possible that individual producers would choose to set aside areas of their farms that are less productive because they are simply not profitable to farm. Focusing on more productive areas may allow such producers to increase net profits while reducing overall impacts.

Another way to accomplish many of the same objectives of permanent protection would be through conservation easement programs. In this strategy, producers would be paid (by a government or some other interested party) some portion of the value of their land if they agreed to adopt a specific BMP approach. Such a system might specify preferred practices as well as those to avoid. Or the approach might focus on results and leave it to the producer to figure out how to achieve them. Practices that could be encouraged include getting producers to leave hedgerows intact, not to destroy waterways, and to reseed or otherwise repair waterways where soil erosion has occurred. Other requirements could include prohibiting the use of certain classes of pesticides or insistence on the use of conservation tillage or no-till production.

An example of a conservation easement is the Conservation Reserve Program (CRP) in the United States, which has allowed for a considerable reestablishment of biodiversity on areas of former cropland. In this case, the government purchases a "no-use" right for a period of ten years on some 14.5 million hectares, at least half of which are highly erodible. In some cases, trees have grown back in these areas, and it is doubtful that they will be used for crops even after the end of the ten-year contract. In part this happened because the payments were higher than necessary. The first CRP contract payments often exceeded the local rental value of the land by 30 to 50 percent. About half of the land in the initial program was highly productive and was included more for political reasons than environmental ones (Runge and Stuart 1998). Ways to improve the program include targeting leases only for highly erodible areas, identifying new areas that would form the basis of an overall conservation strategy, encouraging farmers to plant trees or to allow trees to grow and collect carbon sequestration payments as well, and paying less for conservation easements on lower-quality lands.

In many parts of the world soybean crops are grown amongst other crops, not surrounded by natural habitat. In these situations, permanent protection may be irrelevant. Instead, zoning regulations could promote soybean production on lands already in agricultural production or on degraded or abandoned land, rather than on environmentally sensitive land.

Use Zoning to Restrict Agricultural Expansion

In Brazil, where soybean production is expanding rapidly and on a large scale, there is little reason for expansion into natural habitat. First of all, many of the natural habitat areas are not suited for sustained soybean production. If the price of soybeans continues to decline, the area suitable for competitive production will decline even more. Such areas should be zoned so that they will not be used for soybean production. Appropriate financial analyses showing the mid- to long-term financial costs of farming such areas could help generate the political will (both with producers and, more broadly, with civil society) to zone such areas from soybean production.

In addition, Brazil should identify areas that should be zoned for production. Brazil has less than 14 million hectares in soybean production. However, more than 5.5 times that amount of land is abandoned or degraded agricultural land or pasture. Not all of this land is suited for soybean production. Some is hilly or badly eroded, and some was not cleared well. These lands would be more expensive to bring back into production. However, much of the land could be made productive again with up-front investments that are no larger than those required for clearing land. Producers in the cerrado have shown that degraded pasture and agricultural land can become productive soybean lands within two to five years. Furthermore, such land is cheap to buy and within as little as three years can be worth three times the purchase price. The financial case for rehabilitating such lands, however, has not yet been made. A few case studies could be very useful to convince producers, government officials and even lenders that such strategies are not only viable but profitable and creditworthy.

Adopt No-Till or Conservation Tillage Practices

No-till production is increasingly common in both Brazil and the United States. In Brazil it has increased in the cerrado from 180,000 hectares in 1992 to 6,000,000 hectares in 2002. Producers have found that no-till techniques within certain planting sequences each year as well as longer-term crop rotations allow producers to increase production by 10 percent. However, they also allow producers to reduce use of lime, pesticides, and fungicides by 50 percent or more, and the use of other chemicals by 10 percent. In short, the net return per hectare is almost 50 percent higher than that of producers using conventional methods.

Less machinery is required for no-till planting than for conventional tillage. Even so, for farmers who have already invested large amounts in machinery for conventional cultivation, this could be a burden. In addition, while no-till cultivation requires less machinery, it requires some specialized pieces that would have to be purchased. However, the new machinery could be phased in over time or custom planters could be hired to plant the crops. In general, there do not appear to be any significant financial barriers to the adoption of no-till technology. If anything, the main barriers are cultural—producers are not comfortable with the new technology because it runs counter to how they have farmed in the past.

In addition to the financial returns from no-till, there are also a number of conservation gains. In Brazil conventional tillage typically causes soil losses of some 23.6 metric tons per hectare per year. With no-till, soil erosion can be reduced to as little as 5.6 metric tons of soil per hectare per year. The rainfall runoff on fields under conventional tillage is typically on the order of 137.6 millimeters per month. With no-till practices the runoff can be reduced to about 42.4 millimeters. The reduced runoff is the result of crop residues on the soil surface slowing the movement of water, allowing more time for the water to be absorbed by the soil and stored for later plant use or released more slowly over time.

There are also benefits at the landscape or ecoregional level. These have been estimated for the cerrado of Brazil, where the practice is most common, and are shown in Table 8.4. The study summarizes the estimated benefits of adopting no-till agriculture techniques in Brazil on 35 percent and 80 percent of a total cultivated area of 15.4 million hectares. While such studies are always somewhat theoretical, the authors did not include increased yields as a likely positive impact. Even so, the numbers are interesting and give insights into the possible benefits of this practice if adopted on a wide scale. For example, many of the benefits are mutually reinforcing—e.g., more organic matter means better utilization of other inputs (fertilizers, pesticides, water, machinery, energy, and irrigation systems) and thus fewer expenses for them or impacts from them. The findings also suggest that government

TABLE 8.4. Annual Economic Benefits of No-Till Adoption in Brazil
(in millions of U.S. dollars)

Categories of Impacts	35% No-Till Adoption	80% No-Till Adoption
On-farm benefits	356.1	791.4
Incremental net benefits of no-till versus conventional tillage	332.9	739.7
Irrigation pumping economy	23.2	51.7
Off-farm reductions in public expenditures	62.1	138.0
Maintenance of rural roads	48.4	107.6
Municipal water treatment	0.5	1.1
Incremental reservoir life	9.2	20.4
Reduced dredging costs in ports and rivers	4.0	8.9
Off-farm environmental impacts	184.1	409.1
Greater aquifer recharge	114.4	254.1
Carbon credits for diesel economy	0.6	1.4
Irrigation water economy	6.6	14.8
Carbon sequestration in soil	59.5	132.2
Carbon sequestration in surface residues	3.0	6.6
Benefits to integrated no-till and livestock systems	784.0	1,742.2
Total benefits	1,386.3	3,080.7

Source: Landers et al. 2001.

support for the conversion to no-till practices would be more than offset by societal benefits. For example, fewer roads would be washed out from runoff, and there would be less siltation of rivers and lakes and fewer impacts on local sources of drinking water. All of these benefits of no-till would result in fewer government expenditures to fix the impacts of conventional tillage.

The current rate of soil erosion in the United States (which averaged 15.75 metric tons per hectare per year in 1992) could be halved with the adoption of no-till cultivation and other basic conservation practices (Schnittker 1997). The reductions described earlier in the discussion of soil erosion were not accomplished by growing fewer row crops. In fact row crop cultivation has intensified in the United States. Rather, investments in a variety of conservation measures such as constructing terraces, strip cropping, contour tillage, and rotations led to the reduction in soil erosion rates. By 1994 nearly 40 percent of crop acreage in the United States was under some form of conservation tillage compared with only 3 percent in 1984. No-till cultivation was in use on 12 percent of row crops in the United States, and other forms of reduced tillage on 26 percent of planted crops. In addition, about 18 percent of the most highly erodible acres were entered in the Conservation Reserve Program (CRP), where producers were paid not to cultivate those areas (Schnittker 1997).

In the future it may be possible to provide payments to encourage particular land-use practices that link conservation tillage to carbon offsets. Conservation tillage offers the possibility not only of reducing carbon loss from the soil as a result of cultivation, but also of increasing soil carbon in the form of organic matter, with positive impacts on both soil productivity and greenhouse gas reductions. For example, Reicofsky (as cited in Tengnas and Nilsson 2002) reports tests in which carbon loss following conventional plowing was 13.8 times as much as soil that was not plowed. Carbon loss from four different conservation tillage methods averaged 4.3 times the loss from unplowed soil (Tengnas and Nilsson 2002). Another way to structure such a program would be to pay producers with subsidies for building up specific amounts of carbon in their soil.

Encourage Fallowing and Crop Rotation

Enriched fallowing and fallowing with crop rotation are BMPs that can help rejuvenate soils that have been degraded. Fallowing (planting a cover crop and then leaving the land out of production for a period of time) allows the land to recover. It can be an economically profitable investment rather than a period of financial loss without production. Fallowing builds up organic matter in soil, creates surface litter that acts as mulch, and builds up populations of beneficial soil microorganisms. Furthermore, the deep roots of some cover crops can bring to the surface nutrients such as potassium and phosphorus that are trapped in deeper recesses of the soil.

Periods of fallow and crop rotations can include nitrogen-fixing plants such as legumes, or pasturing livestock on the land, to build up significant soil reserves of nitrogen. Fallowing and crop rotation can generate annual savings that are equivalent to the profits of annual production. Fallowing also increases habitat, albeit temporarily, for many different species. However, if fallowing is undertaken on a larger scale, then habitat can be maintained within a landscape that will benefit many different species. Finally, when an area is returned to cropping after fallowing, yields increase and pesticide and fertilizer costs are reduced.

Minimize Fertilizer and Pesticide Use

Practices to reduce fertilizer, insecticide, and herbicide use are certainly possible. As described above, planting legumes as part of a fallow or crop rotation system can reduce the need for applied nitrogen fertilizers. Another way to minimize the use of agrochemical inputs is to adopt precision fertilization and pesticide application systems. These systems avoid excessive applications by targeting the timing and the location of applications on an as-needed basis. In some cases, pesticides can be applied through irrigation systems. To minimize the use of fungicides, microbial inoculates that diminish the impact of pathogenic fungi can be sprayed. While no-till has been shown to reduce the use of some agrochemicals, even some of the most harmful ones, it does seem to lead to an increased reliance on glyphosate.

Another factor that affects agrochemical use in the United States is land tenure. About 50 percent of farmland in major soybean-producing states like Iowa is rented (NRDC 2001). Landowners often prefer to rent to those who maintain "nice, clean fields," for example, ones treated with herbicides so they are free of weeds.

Increased management and time commitments are major issues for producers. Producers are reluctant to adopt precision chemical applications if they require more time. One farmer reported herbicide reduction techniques took about 1 hour per acre for soybeans (1.3 hours per acre for corn) and saved $12.50 to $17.50 per hectare ($5 to $7 per acre). Using ridge-till cultivation (a form of reduced tillage) reportedly saves producers $12.50 to $17.50 per hectare ($5 to $7 per acre) as well. Producers were not willing to maintain these practices because the labor required was at the busiest time for the producers and their families, and it was too expensive to hire others to do the work (NRDC 2001). From their point of view increased labor costs exceeded other savings. Reducing the overall time required for such practices (or increasing the costs of not adopting such BMPs through taxes or pollution fines) would encourage farmers to make the investment.

Incentives for producers to reduce pesticide applications could include cost-share funds where the government agrees to cover part of the cost, one-on-one technical assistance, insurance premiums that underwrite the risk of crop damage or yield reduction, "green" product labeling, and the development of retail markets for low-pesticide products. In both Brazil and the United States it has been suggested that pesticide applicators, whether contractors or landowners, should be trained. There have also been formal calls for licensing applicators. (This is already required for restricted-use pesticides in the United States, but not for those chemicals classified for general use.)

Reward Custom Applicators for Using BMPs

The rise in use of custom applicators is partly in response to increased pesticide regulations. By law, use of restricted pesticides is limited to licensed applicators. If producers don't want to go through the bother of becoming licensed, then they have no choice but to hire custom applicators. Such applicators need to be given strong incentives to adopt BMPs and to reduce the overall impact of both their practices and the chemicals they apply.

The issue, however, is not entirely related to regulations. Labor costs are increasing, which makes management more expensive. As a consequence, in the United States, at least, there is a rise in the use of custom applicators. For example, custom applicators now apply 50 to 60 percent of all pesticides in Iowa. In general, applicators are not committed to BMPs (NRDC 2001). Rather they are often interested in getting through the job as quickly as possible because they are paid by the area sprayed.

Although careful crop rotation can reduce herbicide and fertilizer use, to work effectively it might take three to four crops. The economics of subsidies and the lack of markets for small grains in the United States discourage crop rotation, however (NRDC 2001). Instead, producers are increasingly relying on a package of services provided by custom applicators to reduce the pest problems that result from continuous soybean cultivation or more lucrative soybean and corn rotations. The package includes a rate of pesticide application per hectare as determined by the applicator but also based on manufacturer recommendations. These recommendations are notoriously high.

In Iowa, the state's extension service has shown producers that insecticide rates could be reduced by 50 to 75 percent through integrated pest management (IPM) practices (NRDC 2001). Producers and custom applicators will reduce application rates based on their own or a trusted source's experience. However, both will be concerned about potential liability for a damaged or unprotected crop. Producers will only adopt BMPs such as IPM if they believe that they will result in a good crop without undue pest problems and an average or better yield.

Custom applicators provide services on a wide range of issues from financing to delivering the crop to market. Because they are increasingly part of the application of various inputs, they must be involved in the development of viable reduction strategies. Custom applicators are trusted by producers. It is assumed that they have the latest information and research results, are well-trained, and have the latest equipment that is dedicated to the task. In general, custom applicators will incorporate BMPs into their programs if they perform well for the producers and thus do not threaten the reputation of the applicator (NRDC 2001).

Link Adoption of BMPs to Government Subsidy Programs

Once better practices have been identified, they should be encouraged. One way to do this is to make the adoption of BMPs one criterion on which governments base producer payments or subsidies. Making such payments contingent on compliance with the adoption and implementation of BMPs would insure that the payments achieve concrete results that are better for society as a whole as well as producers. In the United States, this could be done through the existing Conservation Reserve Program; in the European Union it would fall under the Common Agriculture Policy (CAP). Brazil could link credit for rehabilitating degraded or abandoned land to the guarantee that BMPs would be adopted and used by producers.

Many countries already make some subsidies and producer payments conditional on specific activities required of producers. Pressure could be applied to help insure that there is political will to link improved practices with such payments. It is also possible that BMPs could serve as the basis for a labeling system, spurring consumer preference and perhaps resulting in slightly higher payments to producers. For example, certified non-GM soybeans already command a premium on U.S. markets and are virtually required for market entry in Europe and Japan. Organic soybeans command an even higher premium.

Government-sponsored conservation programs could be more effective if they linked government support to proven conservation results that also make production systems more sustainable over time. If subsidies are going to be a socially meaningful part of agricultural policy rather than just a transfer payment or an overall price support, then they should at least guarantee a variety of public goods. Broadly, positive environmental impacts would fall into this category. This approach is not without precedence in the United States and other countries. In the United States, for example, producers have been required to comply with a range of different conservation programs in order to receive benefits. Such programs have been responsible for the following:

- the increase of conservation tillage since 1989,
- the continuation of the carrot-and-stick approach that producers have known since the 1930s,
- the development of criteria used to determine subsidy payment eligibility,
- the rationale for continued federal incentive payments to producers after 2002, and
- the tightening up of the administration of the program each year.

With the exception of the first item, however, the actual on-the-ground results have been far less than might have been hoped.

Eliminate Soybean Subsidies and Market Barriers

Subsidies aimed at encouraging soybean production and export irrespective of the environmental cost or current market demand for the product should be eliminated. Price supports for soybeans in the United States are precisely the kind of subsidy that flies in the face of unregulated markets and that ultimately will cost most of society a lot yet ultimately will fail to protect American producers from cheaper soybeans from other parts of the world, especially Brazil.

Virtually no country is immune from harmful subsidies. Take Brazil, for example, a country with a number of advantages for the production of soybeans. In Brazil several subsidies are being considered. The governor of Roraima, the northernmost state in Brazil, has proposed a twenty-year tax exemption for all soybean producers (Fearnside 2000). The idea is to increase production from close to zero in 1999 to 200,000 hectares by 2005 with an overall investment of some U.S.$300 million. Production would be exported by road through Guyana. If the road through Guyana is completed and paved, then considerable logging would probably take place in that country as an indirect consequence of soybean production. It will be essential to oppose these subsidies if the harmful environmental impacts of soybean production are to be checked.

Likewise, market barriers that tend to protect domestic producers or processors such as those in Europe should also be eliminated. Such programs encourage production in countries that are less well suited and overall are less productive. They also cause producers from more productive countries to attempt to cut costs even more (often at the expense of the environment) so that they can compete with domestic producers (e.g., Europe or China) that are protected with import duties or tariffs.

OUTLOOK

Brazil and Paraguay are the two countries where soybean production is expanding the quickest and, at least in the case of Brazil, where large amounts of land suitable for production are still available for expansion. Many strategies are being pursued to encourage production, but little is known about what environmental impacts might result if such strategies are successful. In addition, there has been little thought given to the overall market impacts of increased soybean production. For example, what would happen to the global price for soybean meal and oil? What would happen to lands that had been cleared of native habitat to plant soybeans if the markets crashed?

In the rush to make money from soybean production, many states in Brazil are trying to encourage investments to support the industry. For example, plans for soybean production in the state of Acre, in western Brazil, are used to justify the construction of the Road to the Pacific (Fearnside 2000). No study, to date, has shown what the environmental impact of such a road might be, much less whether trucking soybeans over the Andes could even compete economically with American soybeans in the Asian market.

Three factors appear to be shaping soybean production and expansion—animal protein consumption, subsidies, and the global economy. Soybean meal is the feed of choice for animals in a world where demand for animal sources of protein is growing very rapidly. So long as that demand increases, the production of soybeans will increase. Where and how soybeans are grown has more to do with agricultural subsidies and market barriers than other factors. While such policies will change, it is not likely to be in the near term, and it is not likely that the new policies will undermine current production, at least in the United States. Finally, both consumption and subsidies are directly linked to the global economy. If the global economy undergoes a significant downturn, then the ability of consumers to consume animal protein and the ability of governments to continue to support subsidies are both likely to be eroded. This would affect overall demand for soybeans as well as production.

REFERENCES

Baumel, P., B. Wisner and M. Duffy. 2000. Brazilian Soybeans—Can Iowa Farmers Compete? *Ag Decision Maker: A Business Newsletter for Agriculture.* December. Iowa State University Extension. Available at http://www.extension.iastate.edu/agdm.

Cox, C. 1998. Herbicide Factsheet: Glyphosate (Roundup). *Journal of Pesticide Reform.* Volume 18 (3). Updated April 2003. Available at http://www.pesticide.org/gly.pdf.

Faeth, P. 1995. *Growing Green: Enhancing the Economic and Environmental Performance of U.S. Agriculture.* Washington, D.C.: The World Resources Institute.

———. 1996. *Make It or Break It: Sustainability and the U.S. Agricultural Sector.* Washington, D.C.: The World Resources Institute.

FAO (Food and Agricultural Organization of the United Nations). 2002. *FAOSTAT Statistics Database.* Rome: UN Food and Agriculture Organization. Available at http://apps.fao.org.

Fearnside, P. M. 2000. *Soybean Cultivation as a Threat to the Environment in Brazil.* A Report to Conservation International. 5 February. 72 pages.

ITC (International Trade Centre) UNCTAD/WTO. 2002. International Trade Statistics. Available at http://www.intracen.org/tradstat/sitc3-3d/index.htm. Accessed 2002.

Landers, J., G. S. de C. Barros, M. T. Rocha, W. A. Manfrinato, and J. Weiss. 2001. Environmental Impacts of Zero Tillage in Brazil—A First Approximation. Paper presented at the First World Congress on Conservation Agriculture. Madrid, Spain. 1–5 October.

Leibold, K., P. Baumel, and B. Wisner. 2001a. Brazil's Soybean Production—Production Inputs. *Ag Decision Maker: A Business Newsletter for Agriculture.* October. Iowa State University Extension. Available at http://www.extension.iastate.edu/agdm.

———. 2001b. Brazil and Iowa Soybean Production—A Cost Comparison. *Ag Decision Maker: A Business Newsletter for Agriculture.* December. Iowa State University Extension. Available at http://www.extension.iastate.edu/agdm.

———. 2001c. Brazil's Soybean Production. *Ag Decision Maker: A Business Newsletter for Agriculture.* September. Iowa State University Extension. Available at http://www.extension.iastate.edu/agdm.

McVey, M., P. Baumel, and B. Wisner. 2000a. Brazilian Soybeans—What is the Potential? *Ag Decision Maker: A Business Newsletter for Agriculture.* October. Iowa State University Extension. Available at http://www.extension.iastate.edu/agdm.

———2000b. Brazilian Soybeans—Transportation Problems. *Ag Decision Maker: A Business Newsletter for Agriculture.* November. Iowa State University Extension. Available at http://www.extension.iastate.edu/agdm.

NRDC (Natural Resources Defense Council). 2001. *Summary Report.* Herbicide Use Reduction Meeting Notes. August 22. Washington, D.C.: Natural Resources Defense Council.

Runge, C. F. 1994. *The Grains Sector and the Environment: Basic Issues and Implications for Trade.* Rome: Food and Agriculture Organization of the United Nations.

Runge, C. F. and K. Stuart. 1998. *The History, Trade and Environmental Consequences of Corn (Maize) Production in the United States.* Report prepared for the World Wildlife Fund. 1 March.

Schnittker, J. 1997. *The History, Trade, and Environmental Consequences of Soybean Production in the United States.* Report to the World Wildlife Fund. 110 pages.

Smil, V. 2000. *Feeding the World: A Challenge for the Twenty-First Century.* Cambridge, MA: MIT Press.

Soth, J. 1999. *The Impact of Cotton on Freshwater Resources and Ecosystems—A Preliminary Synthesis.* Fact Report (draft). C. Grasser and R. Salemo, eds. Zurich: World Wildlife Fund. 14 May.

Stringfellow, R. 2000. The Global Oils and Fats Industry. Presentation at the World Wildlife Federation Workshop on Palm and Soy Oil, Zurich. October.

Tengnas, B. and B. R. Nilsson. 2002. *Soya Bean—Where Does It Come From and What Is Its Use?* A report prepared for the World Wildlife Fund of Sweden. Draft.

UNCTAD (United Nations Commission on Trade and Development). 1994. *Handbook of International Trade and Development Statistics, 1993.* Geneva, Switzerland: UNCTAD.

———. 1999. *World Commodity Survey 1999–2000.* Geneva, Switzerland: UNCTAD.

NASS (USDA National Agricultural Statistics Service). 2001. *2001 Agricultural Statistics.* Available at http://www.usda.gov/nass/pubs/agstats.htm.

US-EPA. 1994. *National Water Quality Inventory.* 1992 Report to Congress. EPA-841-R94-001. Office of Water, Washington, DC.

World Bank. 2002. Soybeans: World Bank 2, Chart 4 "Brazilian Crop Area Change." Unpublished Powerpoint presentation from unknown conference.

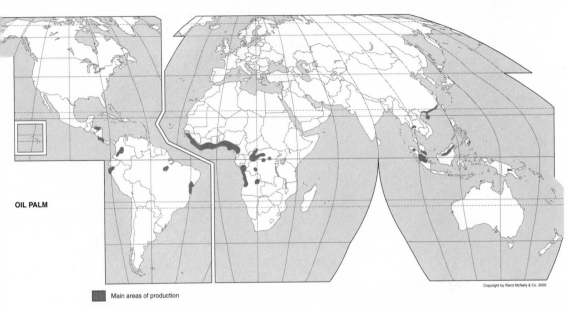

OIL PALM

Main areas of production

Copyright by Rand McNally & Co. 2000

PALM OIL *Elaeis guineensis and E. oleifera*

PRODUCTION		INTERNATIONAL TRADE	
Area under Cultivation	9.7 million ha	Share of World Production	49%
Global Production	118.5 million MT (Fruit)	Exports	8.7 million MT
	17.7 million MT (Oil)	Average Price	$456 per MT
Average Productivity	12,224 kg/ha (Fruit)	Value	$3,969 million
	1,844 kg/ha (Oil)		
Producer Price	$81 per MT		
Producer Production Value	$9,560 million		

PRINCIPAL PRODUCING COUNTRIES/BLOCS (by weight)	Malaysia, Indonesia, Nigeria, Thailand, Colombia, Côte d'Ivoire, Ecuador
PRINCIPAL EXPORTING COUNTRIES/BLOCS	Indonesia, Malaysia
PRINCIPAL IMPORTING COUNTRIES/BLOCS	European Union, India, China, Pakistan, Japan, Singapore, Egypt, Bangladesh
MAJOR ENVIRONMENTAL IMPACTS	Habitat conversion, particularly tropical forests Soil erosion and degradation Threats to key species Effluents from processing
POTENTIAL TO IMPROVE	High Land use planning and zoning can reduce many impacts BMPs are known Investors and buyers are interested in reducing the impact of production BMP-based certification programs are in development

Source: FAO 2002. All data for 2000.

Area in Production (MMha)

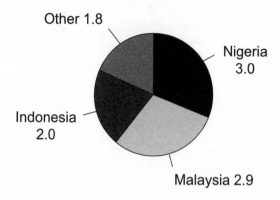

Other 1.8

Nigeria
3.0

Indonesia
2.0

Malaysia 2.9

OVERVIEW

Cultivation of vegetable oil crops has increased faster than any other major type of food or industrial agricultural crop in the past forty years. Likewise, per capita human consumption of vegetable oils has increased more rapidly during the past thirty years than any other food. Economic growth is certainly one reason that more consumers, particularly in China and India, can afford to purchase more vegetable oils. However, this trend also suggests that more people prefer to have a higher percentage of their food prepared with vegetable oils.

While trailing soybeans, canola (rapeseed), and sunflower in area cultivated, by 2000 palm oil was the vegetable oil most produced and traded internationally by volume (FAO 2002). Palm oil (both crude palm oil and palm kernel oil) accounted for 40 percent of all vegetable oils traded, against 21 percent for soybean oil in 1999.

Palm oil can be separated into a wide range of distinct oils with different properties that can be used in a variety of products which, in the past, contained animal or other vegetable oils. Palm oil is used as a cooking oil; is the main ingredient for most margarine; is the base for most liquid detergents, soaps, and shampoos; and, in its most dense form, serves as the base for lipstick, waxes, and polishes. It is even used to reduce friction during the manufacture of steel.

Oil palm cultivation is expanding more rapidly than almost any other agricultural commodity. Cultivation originated in West Africa, where oil palm trees were originally interplanted in traditional agricultural production systems along with other annual and perennial crops. Production was for subsistence or trade within the region. By 1961 trade in palm oil had increased substantially, and Nigeria had 74 per-

cent of the world's plantations (FAO 2002). By the early 1970s monocrop plantations of oil palm had increased dramatically in Malaysia and Indonesia. By 2000,
Malaysia and Indonesia accounted for just over half of the world's total plantation
area, and Nigeria accounted for just over 30 percent. Production is expanding into
Southeast Asia, Oceania, and South and Central America, with dramatic consequences for biodiversity.

PRODUCING COUNTRIES

The main oil palm producing countries in 2000 were Nigeria (3.0 million hectares
under cultivation), Malaysia (2.9 million hectares), and Indonesia (2.0 million
hectares). It should be noted that the estimates of land planted to oil palm, especially in Indonesia, vary considerably. The figures used here are those reported by the
government to the Food and Agriculture Organization of the United Nations and
used in their FAOSTAT statistics database. Other sources, used later in this chapter
to provide specific details or insights, put the figure for planted land in Indonesia
much higher. In any case, these three countries account for more than 81 percent of
all land planted to oil palm and an equal percentage of total production. However,
Malaysia and Indonesia account for 80 percent of all palm oil traded internationally.
In 2001 Malaysia alone produced 50 percent of the world's palm oil and captured
about 61 percent of total trade. In 1999 only 8.3 percent of the palm oil produced in
Malaysia was used domestically, as compared to 47 percent in Indonesia and 100
percent in Nigeria (FAO 2002).

While oil palm cultivation has spread to several places in Latin America, a majority of the world's production has shifted to Asia and the Pacific. Generally higher
per-hectare yields, government initiatives and support, intensive farming practices,
selective breeding, and lower labor costs in the Asia/Pacific region have pushed production to that area. In fact, oil palm production in Asia and the Pacific is undertak-
en almost to the exclusion of any other oilseed. Low production costs resulting from
increasing productivity and increases in area planted to the crop are an additional
spur to production in the region. Indonesia and Malaysia are the most cost-efficient
countries in the world for establishing and running palm oil plantations. This is due
to the high yields, year-round harvesting, low labor costs, favorable climate, and
good soils (Casson 2000).

Crop breeding since the 1960s has resulted in a trebling of average yields globally. Costa Rica has the highest average yield in the world with more than three times
the global average. Nicaragua and Colombia produce just over and under, respectively, twice the global averages (FAO 2002). These yields are due to the development of hybrids by crossing American oil palm (*Elaeis oleifera*) with African oil palm
(*E. guineensis*). Such hybrids are more compact and produce more bunches per tree
per year. As a result, they are able to produce more than double the average yields of
other varieties.

Box 9.1 shows the main oil palm producers in Indonesia. These companies are
large, diversified, and each has corporate ties to a number of other companies and

commodities. In addition to those companies actually producing palm oil, at least in Indonesia, it appears that the financial sector (particularly Dutch and, to a lesser extent, American banks) plays an important role in the expansion of the industry. In 1997 the ten largest Indonesian oil palm conglomerates owned 2.9 million hectares of land, with 723,000 hectares already planted. The area planted represented 45 percent of Indonesian palm oil plantations in 1997 (Casson 2000).

BOX 9.1.   INDONESIAN OIL PALM PRODUCERS

1. Astra Agro Lestari. 27 oil palm plantations in Kalimantan, Sumatra, and Sulawesi; 176,000 hectares as of 1996; planned expansion to 260,000 hectares by 2000; small farms produce 40 percent of Astra's output.
2. Salim Group. Indonesia's largest conglomerate (the group consists of 450 companies); 240,000 hectares in 1995; planned expansion to 400,000 in Sumatra, Kalimantan, Sulawesi, and Riau Islands.
3. Sinar Mas Group/PT SMART. 64,000 hectares in Sumatra and Kalimantan; 40,000 hectares in production; whole or partial interest in 25 plantations.
4. Raja Garuda Mas Group. From 1995–97 it invested U.S.$491 million to increase expansion of its plantation area of 363,000 hectares.
5. Barito Pacific. In joint venture with Astra Group (see above) owns several palm oil plantations.
6. Bakrie Brothers. 88,000 hectares in Sumatra; plans to expand another 88,000 hectares in Kalimantan.
7. PT Perkebunan Nusantara. Largest state-owned plantation company in Indonesia with some 30 percent of country's oil palm production; 14 subsidiaries and 97 plantations located throughout the country with 130,000 hectares of oil palm.
8. SOCFIN (Société Financiere des Caoutchoucs). Franco-Belgian holding company with oil palm plantations in Malaysia and Sumatra. Partially owned by the Bollore Group; also has oil palm interests in Cameroon; developed first oil palm plantations in Indonesia and has nearly 40,000 hectares.
9. PT PP London Sumatra (LonSum). 54,000 hectares of oil palm and rubber in north Sumatra; developing 75,000 hectares of oil palm in East Kalimantan and 10,000 hectares of oil palm in South Sumatra. Also developing small-scale oil palm estates.
10. SIPEF Brussels-based multinational with investments in oil palm as well as other agricultural plantation crops. Major supplier of palm oil to the United Kingdom, Germany, and Belgium. 25,030 hectares of oil palm; second largest foreign-owned group in Indonesia.
11. Cargill Indonesia. Investing $45 million in first oil palm plantation and mill in South Sumatra under the name of Hindoli. Planted 1 million oil palm trees and helped small farmers plant an additional 2.4 million.

Source: Wakker 1998.

CONSUMING COUNTRIES

Indonesia is the largest palm oil consumer in the world and the fifth largest consumer of all vegetable oils. The world's largest palm oil importers are the European

Union (26.4 percent of global imports), India (23.5 percent), China (9.8 percent), Pakistan (7.1 percent), and Japan (3 percent). In India, China, and Pakistan palm oil is used primarily in food preparation and for cooking. In the European Union, Japan, and the United States, palm oil is used primarily for nonfood purposes. Within Europe the main importers are the Netherlands (7 percent of global imports), the United Kingdom (3 percent), and Germany (3 percent) (FAO 2002).

The chart in Figure 9.1 indicates the difference between the levels of palm oil production and consumption in the main palm oil producing and consuming countries. Nigeria, Thailand, and Brazil produce most of their own palm oil and do not export significant quantities. Malaysia, on the other hand, exports more than 90 percent of its production.

The main traded products from oil palm plantations are crude palm oil (CPO), which is extracted from the tissues surrounding the kernel (the mesocarp), and palm kernel oil (PKO) extracted from the kernel. PKO is more highly saturated than CPO, but both can be separated during the refining process into different types of oils with different properties. (Like petroleum, the different nature of the various fractions of palm oil is due to different lengths of the chains of carbon atoms that make up each fraction.) CPO and PKO are refined into bleached deodorized palm oil and olein—the raw materials for cooking oil, margarine, vegetable shortening, ice cream, and other foodstuffs. A third refinery product, stearin, is used for soap production. In addition to food products, large quantities of CPO are used by the steel industry to lubricate and prevent surface corrosion during the cold rolling of steel

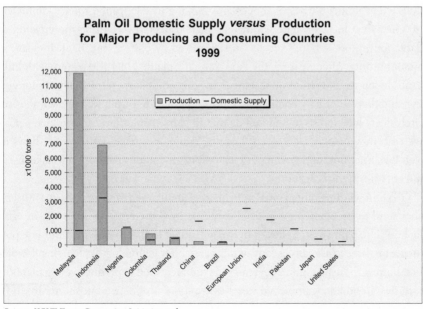

Source: WWF Forest Conversion Initiative, n.d.

Figure 9.1

plate. Cosmetics (including lipsticks, hand creams, liquid soaps, and shampoos) and candles are also produced from CPO and PKO.

In Africa and Brazil the unrefined CPO with a slightly yellowish color is preferred for cooking over more highly refined palm oil or other varieties of vegetable oil. This oil has a distinct flavor that is very much associated with different styles of cooking. The undeodorized and less refined oil also has a much higher content of vitamins A and E and is therefore healthier for the consumer.

PRODUCTION SYSTEMS

Palm oil was first produced in West Africa, where it is still an important source of oil in the diet. It dominates traditional cooking in West Africa as well as in Brazil and other countries where large numbers of slaves originating from West Africa were transported. Traditionally, oil palm was planted in household slash-and-burn gardens that were allowed to revert back to forest cover after one or two years of crop production. Planting oil palm in these areas created agroforestry systems that were harvested over several decades. The trees' fruit also attracted game that was hunted as a source of protein. Since machinery was not available in the past, the oily seeds were pounded and then boiled in water. The oil was skimmed from the surface of the water and stored for later use. Since it was not deodorized, this oil had a strong flavor and for that reason was not considered a viable alternative food product or ingredient in most export markets. More recently, ways have been developed to deodorize palm oil, and this has opened up its markets considerably.

Outside of its native habitat in West Africa, oil palm is mostly grown in plantations in Asia and Latin America. In Southeast Asia these plantations are usually established as monocultures in concessions ranging in size from 4 square kilometers (400 hectares) to 729 square kilometers (72,900 hectares). During the creation of plantations, most standing vegetation is removed by cutting, mechanical clearing, and/or burning. After clearing, the land is usually planted in a grid pattern with little regard to topography. In some instances, logging companies receive permits for cutting, and then subsidiary companies apply for permits to be allowed to clear the "severely degraded" forests to plant oil palm. Oil palm plantations can produce a positive cash flow in only eight years. Logging companies have positive cash flows in even less time, but high grading or clearcutting operations cannot undertake a second cut for fifty years or more.

Oil palm seedlings are planted in fields after about one year of growth in a nursery. They tend to be planted on grids of 8 meters by 8 meters with 143 trees in each hectare. The palms start to flower in two and a half to three years and continue producing for three to as many as forty or fifty years. New technology is changing this picture, however. Tissue culture decreases the time required to produce young plants in nurseries. In addition, improved varieties are more productive but shorter-lived. In the past, trees were most productive until twenty-five to thirty years. The newer varieties tend to be most productive for only fifteen to twenty years. The new, shorter-lived varieties do not grow as tall and, consequently, are easier and cheaper to harvest.

The land needs to be carefully prepared for planting and maintained afterwards. It is prepared by plowing the soil and weeding, either mechanically or with herbicides. Fertilizer is very important to obtain the current high yields of palm oil production. The cost of fertilizer constitutes about 40 to 60 percent of the total maintenance cost, or 15 to 20 percent of the total production cost of the palm oil seeds (Syamsulbahri 1996).

Increasingly, producers plant leguminous cover crops to supply nitrogen for the trees, reducing the need for purchased fertilizers. This can occur until the branches of the palms extend to form a canopy, which creates too much shade for the legumes to grow well. Cash crops can also be planted under or between the trees until the palm trees begin to form a closed canopy. An area 2.5 meters in diameter immediately around the trunk of the tree is kept clear of vegetation. This is where the fertilizers are applied in order to be most efficiently taken up by the trees.

When created in isolation or when the first oil palm plantation in a region is being established, a large area is involved. The investment for building the factory to make palm oil is sufficiently large and the capacity of the factories is high enough to warrant large plantings. Once an oil processing factory is established in an area, plantings can be much smaller for those producers who sell their seeds to the factory rather than processing them on their farms. In the past, there was no small-scale crusher/expeller that allowed small farmers to add value to their produce or that allowed for smaller and more environmentally benign plantings. However, in the past fifteen years smaller crushers/expellers have been developed that can efficiently process the kernel production of some 400 to 1,000 hectares, a typical output for a group of small farmers. Such expellers allow producers to convert their seeds into crude unrefined and unbleached CPO and PKO. These small mills are in place in Costa Rica and Ecuador, where the oil palm plantations are smaller and scattered more in the countryside and do not (at least in Costa Rica) constitute large plantations of thousands of hectares (Panfilo Tabora, personal communication).

Most oil palm plantations in Malaysia are in peninsular areas. They have been established through either direct forest habitat conversion or the conversion of former rubber estates. At this time, oil palm plantations are expanding rapidly in East Malaysia (Sabah and Sarawak). This is likely to be where any expansion of oil palm plantings in Malaysia will take place in the future, but even here land suitable for plantations is getting scarce.

In the 1990s, due to rising land and labor prices in their country, Malaysian palm oil companies began to invest abroad, particularly in nearby Indonesia. In all, some fifty Malaysian companies have established joint ventures with Indonesian plantation companies as well as with companies in Cambodia, Thailand, Papua New Guinea, and the Solomon Islands. Table 9.1 compares the 1997 costs of producing CPO in Colombia, Côte d'Ivoire, Indonesia, Malaysia, and Nigeria with global averages. From this table it is clear that Indonesia, and to a lesser extent Malaysia, are the low-cost producers. This is true because of the productivity of labor relative to its cost. It is also interesting that the costs of establishing plantations in Nigeria are three times as much as anywhere else in the world. It is not clear if this is due to the

TABLE 9.1. A Comparison of Production Costs for Crude Palm Oil, 1997

(in U.S. dollars per metric ton)

	Colombia	Côte d'Ivoire	Indonesia	Malaysia	Nigeria	World Average
Establishment	71.2	69.5	64.3	60.7	224.5	72.1
Cultivation	91.2	136.1	72.5	75.7	113.7	79.3
Harvest/transportation	78.9	33.8	40.2	45.1	90.7	47.3
Milling costs	106.1	105.3	82.6	98.3	130.7	96.6
Kernel milling costs	6.9	7.7	7.2	7.6	8.2	7.5
Kernel oil and meal credits*	−58.2	−54.0	−60.0	−61.9	−65.6	−61.5
Total	296.1	298.4	206.2	225.5	502.2	241.6

Source: PT Purimas Sasmita 1998, as cited in Casson 2000 and LaFranchi 2000.
*Kernel oil and meal are products from milling and have been included in this calculation as a credit (income) back to the system.

TABLE 9.2. Costs of Palm Oil Production (Excluding Milling) in Indonesia, 1997

Operation	Cost (U.S.$/MT)	Percentage of Total Costs
Harvesting	24.29	13
Maintenance	18.78	10
Fertilizer	38.74	21
Transportation	12.00	7
Processing	19.35	11
Overhead	22.00	12
Depreciation	47.00	26
Total	182.16	100

Source: Voituriez 2001.

cost of land, the cost of doing business in the country, the cost of labor, or a combination of the above.

Table 9.2 breaks down the costs of production in Indonesia and calculates the share of each toward the overall cost of production. The two main costs are depreciation and fertilizer (47 percent of total costs) followed by harvesting and overhead (25 percent of total costs). What the table does not show, however, is the financial impact on a processing plant of buying product from independent growers, or what price would be paid. Since this is a common practice, it is likely that independent growers are used to lower a plant's fixed costs per unit of production. In effect, this subsidizes on-farm production while providing independent growers with markets. The price paid to independent growers is probably improved when there are more buyers in an area to increase competition.

In Indonesia, from 1967 to 1997 oil palm was one of the fastest-growing subsectors of the economy. During this time the area planted increased twentyfold, and there was a 12 percent average annual increase in crude palm oil production. Through 1988 the Indonesian government estates were the largest palm oil producers. In 1989 private estates surpassed production on government estates, and in 1991 small-farm production surpassed that from government estates as well.

By 1997 government estates accounted for 449,000 of the 2.2 million hectares of oil palm in Indonesia, small farms made up some 813,000 hectares, and private companies held the rest (Casson 2000). Many of the private company holdings were still immature, which means that their production will increase and they will dominate total production and trade figures even more in the future. In 1997 the ten largest Indonesian palm oil conglomerates held over 2,900,000 hectares of oil palm concessions, of which some 723,000 hectares were planted (Casson 2000).

Oil palm has by far the largest area in estates for all crops in Indonesia. The area of oil palm plantations increased to 2.4 million hectares at the beginning of 1998, and plans were underway to establish another 1.5 million hectares by the end of that year. Many of the forest fires in late 1997 were started by plantation companies that wanted to speed up the "conversion" of tropical forests to plant oil palm. For example, of the 176 companies accused of forest burning, 133 were oil palm plantation companies (Lebbin 2000).

More conservative estimates suggest that the area planted to oil palm in Indonesia may double by 2005 (Lebbin 2000). With concessions of 5.5 million hectares already granted by 1997, the projection of doubling current planting levels is quite realistic. In addition, another 40 million hectares of forested land in Indonesia's outer islands have been slated by the government as available for conversion to cash crop production. In 1998 there were fifty foreign investment projects (80 percent of which were in collaboration with Malaysian companies) with plans to establish 900,000 hectares of oil palm through an overall investment valued at U.S.$3 billion. By mid-1998 only 600,000 hectares had been established.

In Indonesia oil palm production increased more than tenfold from 1975 to 1994—from 397,000 metric tons to 4 million metric tons. This rate of increase is likely to continue as more trees reach maturity. By the end of 1997 only two-thirds of the planted area had matured and were productive; one-third was still too immature to produce oil. It takes three to four years for oil palm to begin to produce. This means that about 800,000 hectares had been planted after the 1993–94 planting period. If demand is relatively constant, the more recently planted areas will increase production substantially as they mature and reduce prices accordingly.

Oil palm can be planted on degraded land, but plantations are normally established directly on newly cleared forest lands. In some cases, logging forests appears to provide the capital for establishing oil palm plantations. In other instances, oil palm concessions are obtained by logging companies that never intend to plant oil palm. Instead they want to log the areas without having to pay the fees to operate a logging concession. Oil palm plantations are often established subsequently on these degraded lands as well as in other forests that may have been selectively cut by illegal logging operations. These factors appear to be related in many individual cases, but a clear connection between logging for timber or pulp and the establishment of oil palm plantations has not been established at this time.

PROCESSING

Palm seeds spoil within forty-eight hours of harvest. Consequently, they must be processed quickly. Given the condition of roads in most producing areas, they cannot be transported great distances. One of the reasons that so many of the oil palm plantings are in large plantations is the size of the fixed investment formerly required for processing plants, particularly those in use in Asia. In the past, processing plants were on a scale that required as much as 30,000 to 40,000 hectares of trees to support them. Today the newer mills being constructed are competitive with only 10,000 to 15,000 hectares of trees, and in many instances most of these plantings are managed by small producers and not the milling company itself. Good roads, of course, would extend the radius from which processing plants could draw their product.

In parts of the Americas, smaller crushers/expellers have been adapted for processing oil palm, and these require as little as 400 hectares of trees to operate competitively. These processing plants have a payback period of approximately nine years. The crushers/expellers can produce crude oil, which is stable and can be transported much longer distances for refining (hydrogenation) and bleaching. Smaller mills of this type are currently in operation in Costa Rica and Ecuador.

Large-scale producers who purchase considerable quantities of labor and machinery make very little money if they only grow oil palm seeds for sale. Small producers, by contrast, who do not need machinery and who rely primarily on family labor, can generate profits selling their oil palm seeds to processors. Consequently, a system has evolved in Asia in which larger companies encourage small producers to plant oil palm and produce seeds because the companies can then rely on them for part of the volume they need to make their processing plants viable economically. These large mills, however, have traditionally had a monopoly on the areas where they buy the oil seeds.

Similarly, the limiting factors in oil palm plantation development in Latin American countries such as Ecuador, Colombia, and even Brazil have traditionally been the high cost of the fixed investment in processing plant technology and the high cost of labor. Smaller mills have been developed, but many companies still operate the older, larger units and cannot replace them without losing money. In addition, the cost of labor in most parts of Latin America is much higher than in Asia. This remains an issue for most producers who want to expand production beyond sales into local markets.

Palm fruit bunches yield a number of different products. Yields of various products from 1 hectare of oil palm trees are detailed below in Table 9.3. Of the marketable products, crude palm oil and palm kernel oil represent about 25.3 percent of the weight of palm bunches brought to the processing plant. Palm kernel meal, the substance that remains when the oil has been removed from the kernel, comprises some 3.7 percent of total weight, and can also be sold for animal feed. The remaining by-products are waste. These include the empty palm bunches (21.1 percent of total weight) and water (about 50 percent of total weight). More efficient companies produce less waste (e.g., they are able to extract a higher proportion of oil and meal

TABLE 9.3. Yields from One Hectare of Oil Palm Trees

Average yield from fruit	19.0 MT of fruit
Crude palm oil from fruit pulp	4.2 MT CPO (22.1% by weight)
Oil from palm kernel	0.6 MT PKO (3.2% by weight)
Palm kernel meal	0.7 MT meal (3.7% by weight)
Empty palm bunches	4.0 MT palm bunches (21.1% by weight)
Water lost during processing	±9.5 MT water (50% by weight)

Source: Blix and Mattson 1998, Cederberg 1998, and Tengnas and Sveden 2002.

TABLE 9.4. Germany's Oil Palm Industry

Oil refiners	Meister Markenwerke, Noblee & Thorl, Walter Rauh Nuesser Ol und Fett, and Deutsche Cargill
Chemical products	AKZO Nobel, Henkel Material Wirtschaft
Foodstuffs	Dr. Oetker, Maylip Nahrungsmittel
Refining food and nonfood products	Unilever
Food	Nestlé
Cosmetics and detergents	Colgate-Palmolive, Procter & Gamble, L'Oreal, Johnson & Johnson, Beiersdorf, Unilever/Lever GmbH, and Avon

Source: Wakker 1998.

than less efficient companies) and turn much of the waste into marketable by-products. They also take the empty palm bunches back to the field where they are used as mulch and reduce the use of some inputs. In general, however, the issue of how to dispose of waste (especially the water) in ways that cause the fewest impacts has not been adequately addressed. Incidentally, palm oil companies are not the only ones that haven't found good ways to solve this problem. The issue of waste water is also a major issue in olive oil and wine grape production.

Even after the CPO and PKO are extracted from the palm fruits considerable processing is still required to produce consumer products. Table 9.4 lists some of the companies that are involved in turning palm oil into various types of consumer products in Germany. A striking thing about the table is the number of easily recognized transnational corporations that are involved in the different products.

SUBSTITUTES

Palm oil is a ready substitute for other vegetable oils for cooking (e.g., soybean, corn, canola, and cottonseed oils) depending on price and availability. It is also a substitute for cocoa butter and coconut oil for most personal care and cosmetic products. Hydrogenated palm oil can be substituted for animal fats in many foods to reduce dependence on animal fat and increase product stability and shelf life. Animal fat is expensive to produce, and the volume of production has remained stable while overall demand has increased. Palm oil, by contrast, is increasing in volume produced and decreasing dramatically in price. According to palm oil processors, however, they could continue to make good profits even if the international price of palm oil were to decline by half.

One substitute for palm oil that deserves special mention is soybean oil. As described in the chapter on soybeans, this oil is a by-product of soybean meal, which is increasing in demand as a source of protein in animal feeds and as a substitute for fish meal in both aquaculture and animal feeds. The demand for soybean meal is driving an increase in the production of soybean oil as a by-product, which in turn adversely affects the price for both soybean and palm oils.

MARKET CHAIN

There is some evidence that the palm oil market is becoming more vertically integrated, as traders and processors are extending their influence into production itself. This will probably continue to happen so long as there are relatively few risks associated with producing the oilseed. When this vertical integration happens, the larger companies will concentrate more on milling and refining the oil and dominating exports, imports, and distribution of uniform oil products that perform the same in different formulations and that can as a result be readily substituted for one another.

There is also some evidence that the recent consolidation of the manufacturing and retail of food and personal care products may extend vertically into the distribution and refining areas. Unilever's operations, for example, tend to run the full gamut up to but not including retail, although they recently sold their oil palm plantations so they are not now directly involved in production. Cargill's operations tend to run the gamut from production to refining the oil for use in manufacturing finished products without getting involved in the manufacturing sphere. The Nestlé corporation refines the oil and uses it to manufacture finished retail products. For example, Nestlé's canned condensed milk manufactured in its Malaysian operations contains some 25 percent palm oil as do nondairy creamer and other products. Other markets in the United States and Europe are less forgiving both about nondairy items being sold as dairy and about overall fat content in general.

Globalization and consumer scrutiny have pushed companies to provide, and consequently to demand, more transparency with regard to their products as well as the ingredients they use. European grocery store chains want to be able to understand and defend the social and environmental impacts of the ingredients in the products they sell. This trend towards accountability for both product quality and production practices has enabled larger companies to influence the entire market chain, including segments of it in which they do not invest directly. A palm oil roundtable in Europe, initiated by WWF, currently includes food processors such as Unilever and grocery store chains such as Migros and COOP in Switzerland.

MARKET TRENDS

Though the average annual world prices for CPO and PKO have fluctuated considerably, overall the price of palm oil has fallen from U.S.$1,102 per metric ton in

be reduced considerably. Similarly, if the economies of India and Pakistan are disrupted for any reason, their demand will fall as well.

The Southeast Asian financial crisis led to a 59 percent increase in palm oil exports in 1998 as producers and governments alike attempted to generate hard currency earnings at the expense of local consumption. This caused local prices to soar and large tracts of forest to be rapidly converted for oil palm plantations.

To counter the increased cost of palm oil for local consumers, the Indonesian government increased the export tax from 40 percent to 60 percent of export value. This slowed the growth of the industry. In 1999 the government estimated that only 177,000 hectares of oil palm would be planted. However, with the 2000 announcement that the export tax would be reduced, the government gave the industry a green light for renewed expansion.

The oil palm industry, for its part, is poised to expand, at least in Indonesia. The share value of companies has stabilized since the financial crisis. Planting targets are increasing. Most importantly, the government has lowered interest rates and changed regulations to encourage the expansion of the industry. Furthermore, joint ventures and marketing cooperation between Indonesian and Malaysian producers will allow them to maintain the price of palm oil and increase their dominant share of the world vegetable oil market at the same time. In Indonesia the government is encouraging oil palm development in the eastern islands (e.g., Kalimantan and Irian Jaya), but the industry is more interested in developing palm oil plantations in Sumatra. The bottom line, however, is that all these factors spell the end of any forests where expansion takes place.

At some point between 2012 and 2015 Indonesia is projected to overtake Malaysia as the leading palm oil producer in the world. Indonesia has been planting oil palm for some time, with the highest planting rates in the 1990s. In the early 1990s Indonesia planted up to 200,000 hectares of oil palm per year. Immediately after the economic crisis in 1997 and 1998 the planting rate slowed to 40,000 to 60,000 hectares per year, but it then began to pick up and by 2000 had increased beyond the rates for the first half of the 1990s. If the market continues to increase and if acceptable profits can be made even at lower prices, the planting rate is certain to increase even more.

ENVIRONMENTAL IMPACTS OF PRODUCTION

The main environmental problems from oil palm production are habitat conversion, threats to critical habitat for endangered species, use of poisons to control rats, and pollution from processing wastes.

Most of the world's oil palm trees are grown on a few islands in Malaysia and Indonesia. These islands have the most biodiverse tropical forests found on Earth. What is particularly striking about oil palm is that it is so much more productive, per hectare, than any other vegetable oil. Given the extraordinary productivity of oil palm trees, it should have been very easy to maintain representative areas of biodiversity within areas of production and to ensure that corridors were maintained that would allow larger animals (especially elephants, rhinoceroses, and tigers) to move between larger parks and protected areas. This did not happen. Now that prices have declined so much, it should be possible to retire the unproductive, unprofitable plantation areas and return them to a more natural state to maintain biodiversity (or at least connect protected areas) and restore ecosystem functions. There has been great reluctance to do this, to date, even though it makes economic sense.

Habitat Conversion

In Africa oil palm has been a subsistence crop for generations. As such it tends to be an agroforestry crop that is interplanted with other cash and subsistence crops. In most cases, this type of production does not have a large impact on biodiversity. More recently the establishment of vast monocrop oil palm plantations in Asia and Latin America, as well as in West Africa itself, threatens vast tracts of tropical forests with high conservation value.

Nowhere is this problem of forest conversion more acute than in Indonesia. In Indonesia, even though there are 20 million hectares of abandoned agricultural land appropriate for the establishment of oil palm plantations, this land is not being planted. Instead, in the 1990s concessions for plantations were granted mostly in forests. Planters feel that it is more expensive to plant in grasslands or in degraded areas because they will have to add so much more chemical fertilizer. The cost of clearing forests is subsidized from the sale of timber from concession areas. Some oil palm production plantations were converted from other uses such as former rubber plantations whose production is now less valuable than in the past. However, it is mostly primary forests that are being converted. Being relatively easy to clear, peat forest areas are one of the main areas of conversion. They were the sites of many of the forest fires in Indonesia in the late 1990s. Peat forests are even less suitable for conversion to plantations than other tropical forests. Peat forests have very high water tables so often palm oil plantings have to be made on elevated pedestals that prevent the roots from being in standing water. As a consequence, many trees fall over for lack of support. If an entire peat forest area is cleared, then the area can dry out so that palm oil production is more successful.

The Malaysian government has successfully lobbied for rubber plantations to be classified as "forest" by the FAO. Such areas are classified as part of the "permanent forest estate," which obfuscates the amount of natural, biodiverse forest that is actually left in the country. There is a chance that the same case might be made for oil palm plantations in the future. Once land is classed as "forest," developers can continue to convert more biodiverse natural forests to monocrop plantations without it ever showing up in any statistical sources. This would have a major impact on biodiversity wherever such crops as oil palm trees are planted. In addition, however, the people who would be lured to the forests as oil palm workers and harvesters would tend to have an additional impact on biodiversity through killing or harvesting other species.

There is a direct relationship between the growth of oil palm estates and deforestation in Malaysia and Indonesia. In the Kinabatangan watershed area of Sabah, Malaysia, large areas of previously logged forests have been converted into oil palm estates. In Indonesia oil palm plantations have also been created illegally within a number of different protected areas (see Table 9.5). Habitat conversion from natural forests to oil palm plantations has been shown to have a devastating impact not only on the tropical forests with the most species of trees per hectare but on other plant and animal species as well. For example, there are nearly eighty mammal species found in Malaysia's primary forests, just over thirty in disturbed forests, and only eleven or twelve in oil palm plantations (Wakker 1998). Similar species reductions occur for insects, birds, reptiles, and most important of all for soil microorganisms.

Burning and Air Pollution

The establishment of oil palm plantations in Indonesia and to some extent Malaysia has been cited as the major cause of the air pollution that affected many non-producing areas of Southeast Asia including Singapore and other cities in 1997. The smoke was so bad that airports were closed for days at a time. Once started, many of the fires in peat forests burned uncontrolled both underground and above ground for months. A recent letter published in *Nature* presented research that suggested that the 1997 fires were one of the main sources of CO_2 emissions globally in a year that had more emissions than any other on record since record keeping started in 1957 (Page et al. 2002). The authors estimate that the Indonesian fires released 0.81 to 2.57 gigatons of carbon into the atmosphere. This represents 13 to 40 percent of the mean annual global carbon emissions from fossil fuels. While this practice of burning has since been outlawed in Malaysia and Indonesia, it is still common in other parts of the world where plantation establishment is now occurring.

Threats to Critical Habitat for Endangered Species

Of all the agricultural commodities described in this book, oil palm poses the most significant threats to the widest range of endangered megafauna. These include the Asian elephant, the Sumatran rhinoceros, and the tiger. It is rare that these three very different species are found in one place, yet they coexist in peninsular Malaysia and Sumatra in precisely those areas where oil palm plantations are expanding. In addition the orangutan, tapir, sun bear, and other primate and bird species are affected by the expansion of oil palm plantations in tropical forests. With all these species, the primary issues are the incompatible conversion and use of the habitat and the elimination of wildlife corridors between areas of genetic diversity. Rhinoceroses and tigers will not be found in the types of disturbed areas that are created in oil palm plantations.

Elephants are a slightly different story. While elephants are affected by forest clearing, they are willing to inhabit disturbed areas and even oil palm plantations. As a consequence, they are considered a nuisance by plantation managers. Elephants like to eat the tender new shoots on oil palms as well as the oil-rich palm seeds if they

TABLE 9.5. A Sample of Indonesian Parks and Reserves Affected by Oil Palm Plantations

Location	Status	Area	Year	Observation
Gunung Leuser	National Park	200 ha	1994	
Siberut	Buffer Zone of World Heritage Biosphere Reserve	70,000 ha	1997	
Bukit Tigapuluh	National Park	12,000 ha	1998	347 families forced off their land inside the park by oil palm company
Kalimantan	Intended for conservation		1998	2 oil palm companies
			1998	21 oil palm estate companies operating in forests intended for conversion
Inhu	Protection Forests	1,500 ha	1998	Encroachment by oil palm companies reported
Bukit Barisan Selatan	National Park			
Kerinci Seblat	National Park			Encroachment reported on two sides

Source: Lebbin 2000.

can get to them. Thus, they not only "eat the profits" but also damage the trees doing it. In some areas elephants have destroyed 20 percent or more of plantations as large as 5,000 hectares. As a consequence, in areas of known elephant populations deep trenches are dug surrounding entire plantations to prevent elephants from entering the farms and destroying the crops. To be effective, these have to be maintained regularly. In other cases elephants are fenced out with electric fences and barbed wire. Still, elephants often find ways into the plantations. In many instances they walk up unprotected rivers and streams. The conflicts are not always benign, either for the elephants or the workers. In at least one instance, an elephant killed a plantation manager. It is not known how many elephants have been killed over conflicts in oil palm plantations.

Soil Erosion

Traditional practices used to establish oil palm plantations can lead to considerable soil erosion. Erosion occurs during forest clearing and plantation establishment when the soil is left uncovered. However, erosion has been accentuated by planting trees in rows up and down hillsides rather than on contours around them, by not properly siting or constructing infrastructure such as roads, and by establishing plantations and infrastructure on slopes of more than 15 degrees.

Erosion can also be encouraged when clearing is not undertaken properly in the establishment of plantations. As late as 2001 in the Riau Province of Sumatra in Indonesia, fallen trees were bulldozed into piles that went straight up and down the hillsides (as opposed to contour rows). Such practices tend to funnel the water into channels and thereby increase soil erosion.

It is expensive for plantations and local governments to correct problems caused by erosion. Eroded areas require more fertilizer and other inputs including repair of roads and other infrastructure. Municipalities have additional expenses in terms of road maintenance but also from increased flooding and the removal of silt deposits as well as the dredging of rivers and ports. In addition, there is some indication that the impacts of soil sediments on local fisheries cause municipalities to lose tax revenues.

Use of Pesticides

Rats are the most common mammals found within oil palm plantations. Rats are attracted to the plantations because they feed on the oil palm seeds. They flourish there because all of their natural predators are removed during the initial forest clearing. Traditionally, snakes and other potential predators are systematically eliminated if they make any attempt to recolonize the oil palm plantations.

Once established, rats are very difficult to remove from plantations. In the past oil palm plantation managers used poisons indiscriminately to eliminate them. This indiscriminate use also poisoned other animals that were attempting to recolonize the plantations. Today, more enlightened companies raise and release owls and other predators to control rats on the plantations. They also instruct workers not to kill pythons and other snakes that eat rats.

The use of other pesticides on plantations is rather minimal, with a few notable exceptions. For example, the *Oryctes rhinoceros* beetle, Ganoderma, stem rot, other beetles and even bagworms can require treatment. Some herbicides are still used, however, particularly when plantations are being established. Once the trees grow and produce a canopy that shades the ground, the use of herbicides is greatly reduced.

Use of Fertilizers

Palm oil production requires less fertilizer per unit of output than other oilseed crops. However, it could require even less. The constant removal of nutrients from the plantations in the form of fruit bunches requires fertilizer inputs so that production does not decline over time. For that reason the standard nutrients nitrogen, phosphorus, and potassium, plus other trace elements, are applied regularly to oil palm trees. A number of factors affect the amount of any type of fertilizer that is applied to oil palm plantations. The key variables include the amount or type of ground cover, the slope of the land cleared for the plantation, and whether the empty fruit bunches or other organic matter are used to mulch the area where fertilizers are applied. If these factors are not addressed then more agrochemical inputs will be required because those used will tend to leach out of the plantation and into freshwater systems.

BETTER MANAGEMENT PRACTICES

There are several effective strategies to reduce the problems caused by oil palm plantations. Given the rate of expansion of the industry and its impact on key ecosystems and species, it is important to make a concerted effort to engage the industry and independent researchers to identify and adopt cost-effective strategies to reduce the overall impact of the industry. In all likelihood successful strategies will need to address many different issues and the concerns of a wide range of stakeholders.

The decision-makers regarding the expansion of oil palm plantations include both company and government personnel. Increasingly, regional governments control permitting and concession agreements. The viability of proposed oil palm operations is, of course, also influenced by other players such as buyers, lenders, investors, and even local communities. Successful strategies to influence the establishment of oil palm plantations should understand the constraints of each of these players. Most damage from oil palm plantations results from where they are located. If expansion can be limited to appropriate sites, many of the problems common to the industry could be eliminated.

An important conservation strategy for oil palm will be to assist with the development of industry investment screens that would encourage investments in more sustainable palm oil businesses (e.g., those that incorporate some of the better manage-

ment practices or BMPs described below). This is important because expanding and
running palm oil businesses requires considerable capital. For example, all major
Dutch banks have oil palm investments in Indonesia. Plantations that are well-sited
and have adopted practices that increase their efficiency, reduce their input use, reduce their waste, and create valuable by-products from waste are less risky investments. Approaching investors about ways to reduce their risks could be a useful
point of departure. The same risk-reducing rationale would be of interest to those
who insure palm oil companies. Poorly managed operations are more likely to have
conflicts with neighbors, or increased liability as a result of flooding or fires.

Another effective conservation strategy will be to work with large-scale, sympa-
thetic vegetable oil purchasers to employ the same type of BMP-based screens as
have been developed for investors and insurers as conditions for their purchases.
These screens could become conditions of letter-of-credit purchase orders. This approach would signal to producers that buyers are interested in purchasing palm oil,
but only that which is produced in a more sustainable way. Unilever, Migros, and
Ecover are companies that have already expressed interest in this approach and are
trying to do it by themselves. It would be less expensive and more credible if a wider
range of stakeholders developed appropriate BMP-based screens for general use.

Some of the specific management practices that need to be identified, analyzed,
and discussed with producers and others as part of the development of an overall
strategy for reducing the impact of oil palm plantations and increasing their economic viability are discussed below. The list, however, is not intended to be exhaustive.

Use Land Use Planning to Protect Critical Areas

Areas identified as critical habitat for endangered species and areas of exceptional
biodiversity need some form of protection. Setting aside preserves is not sufficient,
especially since palm oil production is already encroaching onto some of these sup-
posedly protected areas. More effective zoning, land use planning, and enforcement
on the ground will be the cornerstone of successful strategies to reduce environmental damage from oil palm cultivation. However, the devil will be in the details of
plans refined and implemented on specific sites. Plans should include setting aside
areas of high biodiversity as well as those that are important for the maintenance of
ecosystem functions (particularly along rivers and on steep slopes). Forest restoration
could also be considered in subsequent replanting cycles depending on prior poor
productivity, declining prices, or soil erosion.

Any conservation strategy to address the expansion of oil palm plantations will require an initial mapping of existing plantations showing ecological and species priority areas as overlays, as well as concessions that have already been granted but not
yet developed. This information is essential for any type of land use planning and
zoning; it is also a useful starting point for engaging the industry as well as government officials. At the conclusion of this exercise, it would be possible to identify the
most important areas that should be protected either due to their importance as

biodiversity preserves or as corridors for the movement of key species. This would force those interested in conservation to identify areas that are not key for protection and that could be used for oil palm or other appropriate development.

Zoning is, of course, not only an exercise to be done at the larger landscape or ecosystem level. It can also be usefully undertaken on a single plantation. In this case, it is important to identify minimal-size, viable forest fragments of biological significance within oil palm concessions. This information should then be reviewed with owners and managers to identify the most appropriate options for developing forest corridors within their plantations or connecting them to other intact forest areas in order to mitigate the development impacts of oil palm plantations.

Enlightened palm oil producers might even be convinced to allow Sumatran rhinoceroses to be introduced to unplanted areas of a plantation, with appropriate measures to prevent worker injuries and poaching. Stray rhinoceroses that have wandered out of protected areas could be used to stock such areas. Prior to this, however, basic research would be needed on local vegetation characteristics and food availability. If successful, rhinoceroses bred on such plantations could be used to restock reserves or other areas. Without advocating captive breeding, the vastness of palm oil plantations plus the intermittent availability of "stray" rhinoceroses and the clear failure of other projects already holding rhinoceroses in captivity make such an experiment worth considering. If undertaken appropriately, this could generate another stream of income for palm oil plantations.

Not all zoning needs to be done by government. Estate owners or even associations of small holders can zone their own lands to reduce their impacts and improve their profitability. A few oil palm producers are beginning to understand that fighting rivers and steep slopes actually lowers their overall production because they spend most of their time focusing on the least productive areas instead of the most fertile ones. The traditional strategy, in effect, was to bring the production levels of the problem areas up to the average. The new strategy is quite different. By leaving (or zoning) such areas (e.g., riparian areas or steep slopes) for wildlife corridors and watershed and stream protection, producers actually increase their net profits because they focus their attention not on the problems but on raising the average production on most of their plantings. But this shift in thinking, and the record keeping that would support it, has not happened on most farms.

Make the Economic Case that Supports Conservation

Conservationists need to make the case that supporting wildlife and protecting the environment also make economic sense, and they need to address producer concerns in any conservation effort. The elimination of wildlife corridors has a number of direct costs, only some of which are environmental. The environmental costs in terms of biodiversity loss, loss of ecosystem functions, and degradation of downstream environments are quite high but hard to quantify. What is clear, however, is that plantation managers are increasing their own costs by not taking such factors into account. There are a number of examples, including the following:

Oil palm trees regularly lose branches as they grow, which creates waste on the ground beneath the trees. In addition, empty fruit bunches from oil palm mills must be disposed of; they are generally burned or converted into mulch. All in all, the weight and volume of waste far exceeds the commercially viable products produced from palm oil seeds. In addition, overly mature trees (twenty-five years or older) are felled and either left lying on the ground or gathered and burned or chipped. The trunks of trees can also be used for lumber or made into fiberboard. Palm kernel waste is a major raw material, and is processed into cake for animal feed such as food for pigs. Many plantation owners now put most of the processing waste back onto the farm as mulch. There is still considerable potential for preparations of other by-products such as fertilizers.

Use Integrated Pest Management and Biological Controls

Better management practices should include detailed methods for addressing the main pests in each area. Integrated pest management (IPM) should be adopted to ensure that the least harmful method of pest control is used and pesticide application is kept to a minimum. Only pesticides that are approved in the country of production and the country of consumption should be used. In general, the least toxic and least persistent pesticide should be used to address each problem. One way to achieve this is to develop an overall point system in which producers are given a "total pesticide toxicity allowance" to be used for all needs. In general, however, chemicals should be used only as the last resort. The equipment for applying these chemicals should use as little as possible with effective targeting while minimizing drift.

One of the best ways to develop an appropriate IPM system is to undertake a census of the main pests. This should include an understanding of both the pest's life cycle and its natural enemies. The next issue is to understand what levels of infestation cause economic losses. These would be the action thresholds, and no pest control would be required until infestations reach these action thresholds.

Owls are effective predators of rats, the main mammal pest in oil palm plantations. Owl boxes can be established and monitored for occupancy. Snakes can also be introduced or encouraged. If poisons are used, it is important to choose chemicals that are not toxic to predators that may inadvertently consume poisoned rats. Maintaining adequate populations of predators will reduce the need for poisons.

Integrated pest management is already being used by some producers to reduce the use of pesticides. Workers on the Golden Hope Plantations Berhad in Malaysia have been using IPM measures since the early 1980s. They have found that the following IPM measures reduce pests significantly:

- close monitoring of disease and pest infestations allows them to be more easily controlled with or without few chemical inputs,
- planting species that support or attract natural enemies of oil palm pests helps minimize pest problems,

- proper shredding and rapid decomposition of old trees suppress the pest *Oryctes rhinoceros* (Rhinoceros beetle) from breeding,
- use of a biological control, a native baculovirus, to attack *Oryctes rhinoceros* has been proven 80 to 95 percent effective,
- growing thick legume cover crops helps suppress pests from breeding in the debris, and
- encouraging barn owls and snakes helps to reduce rat populations.

It is clear that some of the IPM practices have to be adjusted as other management practices shift. For example, with the zero-burning policies discussed in the following section, additional control measures are needed to keep pests such as beetles and bagworms in check. These and other IPM efforts should be further documented and as appropriate shared with other producers and government officials.

In some instances, pesticides will be necessary to insure profitable yields. Producers should be encouraged to evaluate the types of pesticides they use in order to increase the efficiency and reduce the environmental impacts of use. This would allow for the identification of specific chemicals and application practices that should be discouraged or even banned.

Eliminate Burning

Better clearing practices do not involve burning. One way to insure this is to enforce burning regulations. It is equally important, however, to identify and disseminate information about the best ways to clear without burning. The goal should be to help identify what companies should do, not just what they should not do. Since 1989 Good Hope Plantations (1997a) has found that eliminating burning is practical for replanting or new plantings. With this method, useful parts of trees are harvested and the remainder are left on the ground where they can be spread out to provide protective ground cover, or piled into rows to prevent runoff and erosion. More than 30,000 hectares have been planted with this technique. The main issue of concern with zero burning is that it might lead to the infestation of beetle pests and stem rot disease. Plowing, pulverizing debris, or planting legumes minimizes this risk.

The main benefit derived from zero burning in Malaysia is that nutrients tend to be released more slowly during decomposition so that they can be utilized by the new trees. This reduces per-hectare inorganic fertilizers needed at the time of planting (e.g., nitrogen by 738 kilograms, phosphorus by 205 kilograms, potassium by 848 kilograms, and magnesium by 487 kilograms). The organic matter also improves the soil. When organic matter is used properly it helps with terracing and the reduction of runoff.

One study found that in 1993 the zero burning technique reduced costs for establishing plantations from 1,070 to 1,415 ringgits (the Malaysian unit of currency) when compared with plantations where burning was used. This is primarily because zero burning reduces the fallow time needed by eliminating the need to dry the cleared forest material for burning. Thus, producers get a portion of a crop that much faster. This method also exposes soil far less than other methods, and it lets replanting occur gradually throughout the year whenever there is sufficient rainfall for the seedlings (Golden Hope Plantations Berhad 1997a).

Encourage the Use of Renewables

When solid wastes from processing palm oil cannot be recycled back to the fields, they can be burned as a source of energy to reduce the amount of nonrenewable energy sources imported to plantations and mills. In addition, solar power or biomass should be explored in lieu of importing fuels from off of the plantations. Whether such alternatives are acceptable often depends on the cost of fuel. Biomass and effluent ponds produce methane, a very potent greenhouse gas, which should be captured and used as a fuel rather than released into the atmosphere. While methane often is not cost-effective as a substitute for diesel within the plantation operation itself, it can substitute for natural gas, especially when bought in small quantities by workers and used for cooking.

Reduce Water Use and Nutrient Loading of Freshwater Bodies

Plantations should irrigate efficiently. In nurseries for palm oil seedlings, drip or perforated tube irrigation systems are preferable to sprinklers. The use of water in processing mills should also be minimized, and every effort should be made to recycle an ever increasing amount of the water that is used.

Operations should divert rainwater from the factory effluent stream to minimize water treatment requirements and the dilution of nutrient-laden wastewater that requires treatment. Ponds and water catchment areas can be constructed to collect rainwater and reduce the amount of water taken from natural bodies.

Mill effluents should be concentrated and reused. For every metric ton of oil produced 2.5 metric tons of effluent are generated, which have an average biochemical oxygen demand (BOD) of 25,000 parts per million. In Malaysia, the BOD level must be below 100 parts per million before effluent can be legally discharged into streams. However, Golden Hope has found that returning the palm oil mill effluent to the plantations actually saves them in costs for fertilizers (nitrogen, phosphorous, and potassium) while avoiding pollution and pollution taxes. The sludge cake produced when the effluent is allowed to settle can also be used to make biogas or livestock feed. An oil mill processing 60 metric tons of fresh fruit bunches per hour is capable of producing about 20,000 cubic meters of biogas per day, enough to generate about 1,000 kilowatts of electricity. Utilization of biogas, however, is limited because of the low cost of natural gas in Malaysia at this time (Rahman bin Ramli 1996).

The changes in Malaysia's effluent standards for palm oil mill effluents from 1978 to 1984 are shown in Table 9.6. As this history of effluent standards suggests, they can be strengthened over time. However, this is only likely to happen if specific,

TABLE 9.6. Palm Oil Mill Effluent Standards for Watercourse Discharge in Malaysia

Parameter	Standard A 7/1/78–6/30/79	Standard B 7/1/79–6/30/80	Standard C 7/1/80–6/30/81	Standard D 7/1/81–6/30/82	Standard E 7/1/82–1/31/83	Standard F 1/1/84 onward
BOD(ppm)	5,000	2,000	1,000	500	250	100(50)+
COD(ppm)	10,000	4,000	2,000	1,000	–	–
Total solids (mg/l)	4,000	2,500	2,000	1,500	–	–
Suspended solids (mg/l)	1,200	800	600	400	400	400
Oil and grease (mg/l)	150	100	75	50	50	50
Ammoniacal nitrogen (mg/l)	25	15	15	10	150Y	100Y
Total nitrogen	200	100	75	50	300Y	200Y
pH	5.0–9.0	5.0–9.0	5.0–9.0	5.0–9.0	5.0–9.0	5.0–9.0
Temperature (°C)	45	45	45	45	45	45

Source: IPAS Training Center, no date.

+ This additional limit is the arithmetic mean value determined on the basis of a minimum of four samples taken at least once a week for four consecutive weeks.

Y = Value on filtered sample.

measurable targets are given and then effective monitoring takes place to ensure that targets are met. What can also be more effective is not to be prescriptive, i.e., to set the standards and let the companies find the best way to achieve them.

OUTLOOK

The consumption of vegetable oils is increasing faster than that of any other major agricultural commodity with the exception of fruit. Production of a whole range of vegetable oils, including palm oil, has expanded even more rapidly as producers have attempted to increase their income from these lucrative markets. As a consequence, the price of vegetable oils has declined rapidly. Palm oil is both very productive and relatively cheap to produce. Palm oil processors insist that they could still make money if the price were to be cut in half again as it has been in the past decade. While palm oil processors may continue to make money, other oilseed producers will not. Sunflower production in Argentina, for example, has declined precipitously in favor of soybeans where at least the value of the soybean meal cushions producers from the glut of palm oil on global markets.

The question is not whether palm oil will remain competitive on global markets. It will. The question is whether palm oil can be produced in ways that are more harmonious with preserving biodiversity and ecosystem functions on the one hand, and on the other whether production can be undertaken on scales that will benefit the majority of the people who live in rural areas. There is some indication that investors, manufacturers, and retailers, particularly in Europe, are keen to use their influence to see that this happens. If so, what happens with palm oil could be a precursor of what could happen with other agricultural commodities.

REFERENCES

Alwy, Mustafa. 1998. *Palm-Oil Plantation and the World Bank/IMF: Its Relation and Impacts in Indonesia.* Paper prepared for the Asia Pacific Peoples' Assembly. Malaysia. November 10–15. Artikelatin. 7 pages.

Anonymous. 1999. The Oil Palm Industry in Malaysia. Available at http://www.kpu.gov. my:1025/commodities/txpalm.html.

Benbrook, C. M., J. Wyman, W. Stevenson, S. Lynch, J. Wallendal, S. Diercks, R. van Haren. No date. Monitoring Progress in Reducing Reliance on High-Risk Pesticides in Wisconsin Potato Production. *Journal of Potato Research.* Madison, WI: World Wildlife Fund/Wisconsin Potato and Vegetable Growers Association/University of Wisconsin Collaboration. Available at http://ipcm.wisc.edu/bioipm/reports/toxpaperfinalnoln.pdf.

Blix, L. and B. Mattson. 1998. *Miljoeffekter av jordbrukets markanvandning.* (*Environmental Impact of Land Use in Agriculture: Case Studies of Rape Seed, Soybean and Oil Palm.*) SIK-Rapport No. 650 1998. Goteborg, Sweden: Institutet for Livsmedel och Bioteknik.

Casson, A. 2000. *The Hesitant Boom: Indonesia's Oil Palm Sub-sector in an Era of Economic Crisis and Political Change.* CIFOR Program on the Underlying Causes of Deforestation. Bogor, Indonesia: Center for International Forestry Research (CIFOR). Available at http://www.cifor.org/publications/pdf_files/occpapers/op-029.pdf.

Cederberg, C. 1998. *Life Cycle Assessment of Milk Production. A Comparison of Conventional and Organic Production.* SIK-Rapport No. 643 1998. Goteborg, Sweden: Institutet for Livsmedel och Bioteknik.

Chandran, M. R. 2001. Global Trends in Agriculture for the Twenty-First Century. In E. Pushparajah, ed., *Strategic Directions for the Sustainability of the Oil Palm Industry.* Kuala Lumpur, Malaysia: The Incorporated Society of Planters.

Cheah, S. C. 2000. Biological Technologies for Improving Plantation Tree Crops: The Oil Palm—A Case Study. In E. Pusparajah, ed., *Plantation Crops in the New Millennium: The Way Ahead.* Kuala Lumpur, Malaysia: The Incorporated Society of Planters.

Douvelis, G. and R. Hanson. 1998. *Limited Palm Oil Availabilities Benefiting U.S. Soybean Oil Exports.* Available at http://www.fas.usda.gov/oilseeds/circular/1998/98-03/special.htm.

FAO (Food and Agriculture Organization of the United Nations). 2002. *FAOSTAT Statistics Database.* Rome: UN Food and Agriculture Organization. Available at http://apps.fao.org.

Golden Hope Plantations Berhad. 1997a. *The Zero Burning Technique for Oil Palm Cultivation.* Kuala Lumpur, Malaysia: Golden Hope Plantations Berhad.

——. 1997b. *Integrated Pest Management in Oil Palms.* Kuala Lumpur, Malaysia: Golden Hope Plantations Berhad.

Ipas Training Center. No date. *Land Application of Palm Oil Mill Effluent.* Ipas Training Center.

LaFranchi, C. 2000. Oil Palm in Indonesia: Dynamics of an Industry That Threatens Biodiversity. Draft report. Washington, D.C.: Center for Applied Biodiversity Science, Conservation International.

Lebbin, D. J. 2000. Conservation Crisis in Indonesia. A Graduation with Distinction Paper. Environmental Science and Policy. Duke University.

——. 1999. *Conservation and the Palm Oil Industry in Indonesia.* Asian Rhinoceros and Elephant Action Strategy (AREAS) Discussion Paper. World Wildlife Fund–U.S. August 12.

Levine, J. S., T. Bobbe, N. Ray, R. G. Witt, and A. Singh. 1999. *Wildland Fires and the Environment: A Global Synthesis.* Environmental Information and Assessment Technical Report 1. UNEP/DEIA/&EW/TR.99-1. Geneva, Switzerland. 46 pages.

Oil World. 1999. *Oil World 2020: Supply, Demand and Prices from 1976 through 2020.* Hamburg, Germany: Oil World.

Page, S. E., F. Siegert, J. O. Rieley, H.-D. V. Boehm, A. Jaya, and S. Limin. 2002. The Amount of Carbon Released from Peat and Forest Fires in Indonesia during 1997. *Nature* 420 (November): 61–65. London: MacMillan Publishers.

Potter, L. and J. Lee. 1998. *Oil Palm in Indonesia: Its Role in Forest Conversion and the Fires of 1997/98.* A Report for World Wildlife Fund–Indonesia Programme. Department of Geographical and Environmental Studies, University of Adelaide, South Australia. October.

——. 1998b. *Tree Planting in Indonesia: Trends, Impacts, and Directions.* Occasional Paper No. 18. Bogor, Indonesia: Center for International Forestry Research (CIFOR). Available at www.cgiar.org/cifor.

Rahman bin Ramli, A. 1996. *Towards Zero Emissions: Maximizing the Utilization of the Biomass of the Oil Palm Industry in Malaysia.* Kuala Lumpur, Malaysia: Golden Hope Plantations Berhad.

Sunderlin, W. D. 1999. *The Effects of Economic Crisis and Political Change on Indonesia's Forest Sector, 1997–99.* CIFOR. Available at http://www.cgiar.org/cifor.

Syamsulbahri. 1996. *Bercocok Tanam Tanaman Perkebunan Tahunan.* Yogyakarta, Indonesia: Gadjah Mada University Press. (Agronomy of some perennial species).

Tengnas, B. and E. Sveden. 2002. *Palm Oil—Where Does It Come From and What Is the End Use?* A report prepared for World Wildlife Fund–Sweden.

Teoh, C. H., A. Ng, M. Abraham, and M. R. Chandran. 2001. Trade-Related Environmental Challenges for the Palm Oil Industry. Paper presented to the Malaysian Palm Oil Association annual meeting.

UNCTAD (United Nations Conference on Trade and Development). 1999. *World Commodity Survey, 1999–2000*. Geneva, Switzerland: UNCTAD.

USDA (United States Department of Agriculture). 2000. Situation and Outlook. *FAS (Foreign Agricultural Service) Online*. July. Available at http://www.fas.usda.gov/oilseeds/circular/2000/00-07/julsum.htm.

Voituriez, T. 2001. Palm Oil and the Crisis: A Macro View. In F. Gerard and F. Ruf, eds., *Agriculture in Crisis: People, Commodities, and Natural Resources in Indonesia, 1996–2000*. Montpellier, France: Centre de Cooperation Internationale et Recherche Agronomique pour le Developpement (CIRAD).

Wakker, Eric. 1998. *Lipsticks from the Rainforest. Palm Oil, Crisis, and Forest Loss in Indonesia—The Role of Germany*. November. World Wildlife Fund Germany.

———. 2000. Funding Forest Loss: Research Project to Determine the Involvement of Dutch Banks in Indonesia's Oil Palm Business. AIDEnvironment and the World Wildlife Fund.

World Rainforest Movement. 2001. *The Bitter Fruit of Oil Palm—Dispossession and Deforestation*. Moreton-in-Marsh, England: World Rainforest Movement.

greater uniformity of terrain. The taller varieties can cause soil erosion. Prolonged cropping inevitably causes soil degradation. Large areas that were once used to grow bananas tend to have been abandoned because the soils accumulate diseases or pests. Soil degradation resulted in these areas partly because the way they were originally cleared reduced soil fertility. Finally, prior to the use of chemical fertilizers (or where chemical fertilizers were considered too expensive), producers simply moved on to clear new areas. After some years or even decades of fallow, some lands have been brought back into banana cultivation once again. This happened in the Limón region of Costa Rica in 1970s and also with some farms in Honduras (Panfilo Tabora, personal communication).

Bananas grow rapidly. Fruit can be harvested the first year after planting root stock. Careful management can maintain a plantation for many years. In India, for example, some fields have been producing for more than 100 years. After the first year, crops are obtained from the offshoots of the original plantings. Both the number of offshoots and their production decline over time. Bananas can be picked green and will ripen after harvest. Bananas for export must be picked earlier than those for local consumption and will have more immature fruits.

Bananas are grown throughout the tropics. They are even less tolerant of frost than citrus, so they cannot be grown on the margins of temperate climates except when they are raised under some form of protection such as a greenhouse (as in Israel and Morocco). For much of the world, bananas are subsistence crops. They are high in carbohydrates and low in fat and proteins with a wide range of vitamins present in small amounts. In most areas of the wet tropics, bananas are interplanted with other food crops (e.g., rice, sorghum, chili peppers, sugarcane, cassava, and sweet potatoes) and tree crops (e.g., coffee, citrus, cocoa, and coconuts).

In plantations created for export production, bananas are not intercropped. The process of cultivating bananas for export includes the following operations. In most instances, virgin forests are cleared and all stumps are removed. The land is then plowed and contours or terraces are constructed as necessary depending on the slope of the land. A drainage system is established to drain water in the rainy season and to bring it in during the dry season. The appropriate varieties (depending on local soils, pests, and growing conditions) are selected, and cuttings are planted with the appropriate spacing. As the plantation begins to grow, weeding or weed controls are undertaken. As the plants grow, unwanted suckers are removed and pruning takes place to ensure a steady supply of bananas. Removal of suckers eliminates competition between stalks, which affects fruit size. Fertilization and applications of a range of pesticides are undertaken, often prophylactically to control diseases and pests. Plantations are irrigated as necessary.

In large-scale commercial operations, bananas are grown as a monocrop. They are planted on generally flat ground and on a grid. Because bananas are so susceptible to water fluctuations, deep trenches are usually cut through the fields to deliver water during the dry season and to drain it away from the roots during the wet season. Roads are laid out throughout the plantation as are overhead cables, which allow the product to be harvested, hung, and transported with minimal chance of

bruising the fruit. Increasingly, other types of trees are planted as windbreaks to reduce potential wind damage.

In large plantations, bananas are susceptible to a number of diseases and pests
(e.g., black sigatoka and nematodes) that require the use of several chemicals for control. Many of the chemicals used, however, are aimed not at diseases and pests but at
preventing external blemishes, which affect prices on international markets. In Costa Rica pesticide use on banana plantations has reached levels of up to 44 kilograms
per hectare per year. In 1987 banana cultivation used 35 percent of all insecticides
in Costa Rica and represented 35 percent of total producer costs (Astorga 1998).

Outside of export-oriented plantations, however, the use of pesticides is not a serious problem. When interplanted with other crops, bananas are relatively impervious to pests and other diseases.

In some of the smaller countries where bananas are produced for national and
even international markets, they are replanted every three to eight years. Replanting
is usually necessary because of declining yields as a result of poor root development,
which is related to compacted soil structure, poor drainage, reduced fertility, and increasing populations of soil nematodes. In addition, replanting helps to time harvests
for the seasons of greatest demand and highest prices. This practice is most common
on farms of 1 to 50 hectares (Stover and Simmonds 1987).

PROCESSING

Bananas produced for local markets are not processed. Stalks which contain a dozen
or more bunches (called hands) are cut from the banana plants and transported as
stalks to the point of sale, either local markets, corner stores, or even supermarkets.
At that point, the hands or bunches of bananas are removed from the central stalk
and sold individually. Most bananas sold in the world are never boxed or packaged
in any way.

By contrast there is a fair amount of processing and packaging for export bananas.
In almost all instances the processing is done on-farm. In plantations bananas are
brought to central processing plants along overhead, mechanized monorail transportation systems. A convoy of ten to fifty banana stalks (attached individually to
overhead hooks) is pulled by a machine operated by an individual and propelled by
a single-stroke engine.

At the processing plants, the individual hands or bunches of bananas are cut from
the central stalk and floated in a tank of running water. Tremendous amounts of water are used in flotation and rinsing operations. These operations are intended to remove pesticide residues on the peel as well as insects, spiders, or other foreign matter that might be in the bunches. The water used for this operation is more than 100
times that of the weight of the bananas.

The bananas are inspected, and any malformed or discolored bananas are removed. This is done manually, with oversight by representatives of the buyer in the
case of independent operations. The rejection rate varies from approximately 7 to 35

percent. Grading is subjective; it depends on how saturated the international market

percent. Grading is subjective; it depends on how saturated the international market is rather than the quality of the bananas per se. This is one of the ways that the large banana trading companies control volume and prices. Higher product rejection rates mean lower profits for producers.

In any case, the stalks (rachises) and reject bananas represent, on average, the same weight as the bananas that are exported. This waste is a disposal problem. When piled up it rots; when put in streams the organic matter consumes all the oxygen as it decomposes.

Because of the unique issues related to banana ripening, distance and time to market are key issues for bananas. Frequent shipments are imperative. The introduction of the refrigerated container is a precondition, today, of reaching more distant markets in a timely way. However, such containers impose minimum shipments of 20 metric tons or more if they are to be most efficient and competitive.

With refrigeration and by picking the fruit before it matures, there is a window of about five weeks between harvest and consumption. Without refrigeration, however, the time for consumption of the fruit is a matter of days. More recently, faster boats can carry banana cargoes to Europe from Latin America in eleven to fourteen days. These boats allow bananas to be picked closer to maturity, with better yields for the farms and with better flavor for consumers.

Historically, banana companies exerted pressure on producers through shipping. Delays in shipping could be even more of a disaster than high fruit rejection rates. In the United States, and Europe to a lesser extent, control is exerted through the internal distribution system as well as through the large ripening facilities. Banana companies have also used their control of the market to insist on packaging requirements (compartmentalized boxes, pallets, etc.) that reduce labor use in developed countries and make the movement of bananas more efficient. Packaging in Costa Rica represents some 33 percent of the total production costs of producers up to the point of loading the fruit on boats for export.

SUBSTITUTES

Internationally, bananas compete with an increasing number of fruits. Not only has the global area devoted to production of all fruit doubled in the past forty years, productivity has increased as well. In addition, improved refrigeration and transportation systems allow fruit to be transported farther and longer than ever before. Even so, bananas appear to be holding their own. They are still the highest income generators in the fresh vegetable department of grocery stores in the United States. Furthermore, bananas are considered an excellent food for babies and a convenient and well-accepted food for children. No other fruit has been able to dislodge bananas from this position.

In those countries where bananas are grown primarily for local consumption, the only other perennial crop with comparable per-hectare yields is oranges. However, nothing comes close to the combined caloric and mineral content of bananas, at least in those areas that have sufficient water to grow them. Furthermore, unlike most other starchy food sources such as root crops, bananas do not require cooking. Even in areas where they are cooked, they require less cooking and preparation time than substitutes such as corn, cassava, millet, or peanuts.

MARKET CHAIN

The market chain for bananas consists of growers, packers, transporters, cold storage and ripening companies, and retail outlets. These areas tend to be dominated by different players, but there is some merging of the overall functions. For example, most of the large fruit companies no longer produce most of their own fruit, although they still have some plantations. The risk of growing bananas has been left to individuals with supply contracts (and contracted prices) from the large fruit companies. The more oversupply there is of bananas, the shorter the contract.

In general, overall production per grower is increasing while the total number of growers is decreasing. Most growers have their own packing sheds, where bananas are inspected and packed into boxes, shrink-wrapped onto pallets, and loaded into refrigerated containers. The large fruit companies provide inspectors to make sure that growers' rejection rates are in line with company policy. As mentioned above, increasingly rejection is driven by oversupply rather than product quality. The large fruit companies usually take possession when the containers are loaded onto their ships. These same companies ship and warehouse the fruit in the consuming country. Once taken out of special refrigeration warehouses and containers, bananas ripen fully in five days and should be sold immediately. This is when the retailers receive their shipments.

Over the past forty years the real price of bananas has decreased by 39 percent, but profits have not necessarily declined. Productivity increased over the same period and production costs declined (FAO 2002). Consequently, profits in many areas have actually increased, although at this time producer profit margins in countries that have higher land and labor costs are quite low.

The data suggests that prices paid to producers of export bananas decreased by 10 percent between 1973 and 1983. Real income, however, increased over the same period. Prices paid to producers increased by 9 percent in the 1990s, even though the producers' share of the value generated in the market chain declined. Profits, not relative prices, are more important to producers. It was the relative increase in profits through the 1970s, 1980s, and early 1990s that generated the investments that are now flooding the market with bananas (FAO 2002).

In the 1990s prices slumped and the portion of the retail price that goes to the producer decreased. At the same time, the percentage of the price received by the retailer also declined. During the period in question, the consolidating, shipping, and distribution share of the banana price increased in both relative and absolute terms. Table 10.2 compares the percentage of the final retail price captured as the product moves from producer to retail. These figures are averages and only show

completely blemish-free bunches are exported. The presence of even one blemish on a bunch causes major loss in value or outright rejection. The result is a pattern of "precautionary" pesticide applications at every stage of the production process, whether they are needed or not. The idea is to prevent possible pest outbreaks rather than to treat them when and if they arise, even though spraying these chemicals when the target pests are not present kills more beneficial predators than pests. Pesticides are viewed as the best way to maximize harvests as well as to lower the risk of loss in the value of the harvest.

Export banana production and large-scale domestic banana production in countries such as Brazil depend on chemical controls. Some 286 different pesticides (fungicides, herbicides, and nematicides) have been authorized for use on bananas in Costa Rica either on-farm or in the packing sheds prior to shipping. A few of these are listed in Table 10.4. Most of the chemicals are produced in the United States, Switzerland, or Germany.

The use and cost of these chemical inputs is increasing. Pesticide use in banana production can reach 40 kilograms per hectare per year. In 1991 Costa Rica imported U.S.$56 million of pesticides, in 1994 $84 million, and in 1996 $100 million. In general, pesticides represent 20 to 35 percent of the total costs of banana production. The cost of fighting the disease black sigatoka alone can be as high as U.S.$1,000 to $1,200 per hectare per year.

In Belize, banana companies tend to mix all agrochemicals together and then spray them from a plane on a regular basis (usually about once a week, but sometimes more often). This spray is referred to by locals as a "toxic cocktail." It affects not only the bananas, but also the workers in the fields and their families who live in and alongside the fields, the biodiversity next to the plantation, and the water systems that flush the banana plantations on a regular basis. This latter point is very important. Bananas are planted on ridges with deep trenches cut between the rows of plants. The roots cannot survive in standing water, but the plants need water continuously. This means that water is a constant in banana plantations, and it is always flowing in or out to achieve the right balance.

The use of chemicals in the banana export industry is particularly important because, in most parts of the world, banana plantations have been established on the

TABLE 10.4. A Partial List of Pesticides Used in Banana Production in Costa Rica

Type of Pesticide and Chemical Class	Active Ingredient
Nematicides (organophosphates)	terbufos, cadusaphos, fenamphos, ethoprop
Nematicides (carbamates)	carbofuran, oxamyl
Insecticides (organophosphates)	chlorpyrifos
Herbicides (various)	paraquat, glyphosate
On-farm fungicides (various)	mancozeb, chlorothalonil, benomyl, tridemorph, propiconazole
Packing-house fungicides (various)	imazalil, thiabendazole, tridemorph, aluminium sulphate

Source: Astorga 1998.

TABLE 10.5. Agrochemicals Used by the Banana Industry in Belize

Chemical Input	Half-Life	Quantity Used	Loss via Leaching
Nitrogen (fertilizer)		809 MT/year	60–85%
Phosphorous (fertilizer)		256 MT/year	minimal
Potassium (fertilizer)		1,413 MT/year	60–85%
Calcium (lime, soil neutralizer)		1,760 MT/year	60–85%
Paraquat (herbicide)	500 days		
Mancozeb (fungicide)	70 days	10 MT/month; 31.9 MT/yr accumulated	
Ethoprophos (pesticide)	25 days (9% = 1yr)	1.87 MT/yr accumulated	

Source: Usher and Pulver 1994.

fertile, flat lands of coastal areas. This means that all the chemicals will more often than not have a relatively immediate impact on coastal wetlands as well as the inshore coastal areas and even nearby coral reefs. For example, it is well-documented that 60 to 85 percent of all fertilizer is lost via leaching and/or runoff (Usher and Pulver 1994). Nitrogen, potassium, and calcium are lost rapidly via leaching. In contrast, most phosphorous is attached to soil particles and is only leached if the soil sediment is washed off the plantations. Table 10.5 shows leaching losses for some applied nutrients as well as half-lives (the time required for half a substance introduced into an ecosystem to break down or be eliminated by natural forces) for some of the agrochemicals.

Solid Wastes

There are a number of residues and wastes that result from banana production and processing. These cause environmental problems. The volume of waste produced is at least equal to the volume of bananas produced. Twenty percent of the waste requires special treatment. Yet in Costa Rica the Ministry of Health found that 78 percent of plantations did not dispose of waste properly (Astorga 1998). A 1996 summary from Costa Rica, shown in Table 10.6, illustrates these issues.

There are two main types of solid wastes on banana plantations. The first are the organic remains of the bananas. These wastes include the bananas that are not of sufficient quality to export or even sell on the local market as well as the banana stalks; both are transported to the selection plants where the bananas are washed and boxed for sale. These wastes are created in such large quantities that they are simply thrown away rather than composted. They are often dumped at the edge of the plantation or in or nearby rivers, where their decomposition can consume the oxygen in the water and result in fish kills. The IUCN (the World Conservation Union, formerly known as the International Union for the Conservation of Nature) estimates for the organic waste generated by the banana industry in Costa Rica in 1995 supported the information reported in Table 10.6. The IUCN reported 283,217 metric

TABLE 10.6. Residues from Costa Rican
Banana Plantations

Type of Residue	MT/year
Polyethylene bags	4,406
Polyethylene packing material	2,171
Polypropylene twine	2,755
Fruit stems	225,000
Scrap bananas and rejects	278,000
Fertilizers	110,000
Nematicides	8,300

Source: LEAD International 1996.

tons of stalks and 225,525 metric tons of rejected bananas. These amounts were up from 1990 amounts that were, respectively, 152,798 and 121,672 metric tons.

The second important form of solid waste associated with bananas produced for export is plastic, including bags, rope, and pesticide containers. Each banana stalk is encased in a thin plastic bag treated with insecticide to keep insects and spiders off the fruit as it ripens. In addition, each stalk of fruit is attached to a pole by plastic twine so that its weight will not pull it over to touch the ground. Both these forms of plastic represent waste disposal problems.

Few programs exist to recycle these products. Because pesticide containers are contaminated, they must be handled separately. In 1995 the IUCN estimated that 4,510 metric tons of plastic bags and 4,832 metric tons of polyethylene rope were generated by Costa Rica's banana industry. These amounts were up from 1990 when the figures were 2,433 and 2,507 metric tons respectively.

Water Usage

The average water requirement for a mature banana crop is approximately 160 millimeters a month. In some countries this amount is exceeded seven months of the year as a result of rainfall. During the rainy season in some areas, more than a meter of water in excess of what bananas need must be drained from the fields. It is in this water that a lot of suspended solids and agrochemicals become part of the runoff.

In addition, a tremendous amount of water is used in the processing plants, where the bananas are floated prior to selecting and boxing them for export. Up to one hundred times the volume or weight of the bananas in water is used during the process of washing, selecting, and packaging bananas.

BETTER MANAGEMENT PRACTICES

The banana industry is faced with severe problems managing its impacts on habitat, soil, and water. The risks associated with continuing the current practices are serious. Land is being converted from forests; soils are being degraded beyond sustainable use; pesticides are accumulating in the environment; and rivers, streams, and coastal wetlands are being polluted. The Conservation Agriculture Network's "Standards for Banana Production" (2001) are a good start in the development of certification standards that will result in better management practices on the ground. A number of key, immediately practicable better practices are described below. More documentation and analysis are required to induce other producers to adopt these or similar practices or to discover their own ways of reducing impacts to more acceptable levels.

Manage Plantations for Continuous Cultivation

Recently, some banana plantation managers have been experimenting with replanting in areas that were previously abandoned. This is true of formerly abandoned banana farms in Panama (in the Changuinola area), Costa Rica (in the Limón area), and Honduras (in the Aguan watershed area). The adoption of the Valery banana variety that is resistant to the soilborne Panama wilt disease (a disease that ravaged banana production in plantations growing traditional varieties) has allowed these areas to be brought back into production, and this has reduced pressure on new areas (Panfilo Tabora, personal communication).

More importantly, this new variety allows producers to continue to cultivate the same areas after much-shortened fallow periods of only three years or even less in some cases. Due to the increasing cost of land suitable for banana plantations, more and more experimentation is taking place to find ways to undertake continuous cultivation.

Reduce Use of Agrochemical Inputs

In some areas, the application of agrochemicals is becoming more targeted and use per hectare is declining. Closer monitoring of nutrient imbalances, infestation rates, and the movement of disease vectors is a key technique used to determine the type of chemical to use as well as the best time for application. Farms in Costa Rica using this technique have reduced their pesticide spraying, for example, from forty-seven to as low as thirty-five times per year (Panfilo Tabora, personal communication).

Many chemicals applied to banana plantations are lost rapidly due to leaching. Smaller, more targeted applications spread out over the course of the year, for example, increase the efficiency of fertilizer use. However, the overall efficiency is still lower than it needs to be if the fertilizers are applied by themselves. One of the best ways to increase the efficiency of fertilizer use is to incorporate applications in organic material (either what is already on the field or mixed with compost) so that they bind and are released more gradually. This reduces leaching and makes the nutrients available in a slow release form over a longer period of time. Cover crops planted below the bananas are an efficient way of accomplishing this goal. Not only do they provide organic matter, but some species can also fix nitrogen that would otherwise have to be added to the soil. Different cover crops work better in banana plantations located in different parts of the world.

Another strategy is to invest in a more intense, enriched fallowing of degraded areas or areas that are at the end of the current production cycle. Enrichment planting with legumes for three to five years can rebuild the soil and decrease nematode problems. During the fallow, the cover crops are enriched with the nutrients that their deep roots bring up to the plants. The leaves become rich in potassium and phosphorus, which have been leached down in the soil over the years. The roots become rich in nitrogen, as a result of the nitrogen-fixing bacteria that live in root nodules on legumes. Fallowing also builds up organic matter in the form of litter on the surface, which acts as mulch and eventually increases soil organic matter as it decomposes and is incorporated back into the soil (Panfilo Tabora, personal communication).

There are also other techniques to deter the growth of specific diseases, including the use of resistant varieties as described in the previous section. In a three-year experiment on the commercial operation of an independent banana grower in Costa Rica, sprayings of beneficial microorganisms have increased yields, reduced foliar chemical spray frequency by 15 percent, and eliminated 75 percent of the nematicide applications. The microorganisms included lixiviates derived from compost and bokashi (fermented banana waste, described later in the section on reducing wastes) (Panfilo Tabora, personal communication). There are indications that this approach can be improved further. Experiments at EARTH's commercial banana farm have reduced foliar spray applications of pesticides by 30 percent. This overall approach has now become standard practice on plantations. EARTH researchers found that when beneficial microorganisms were incorporated into the spraying applications, the number of healthy leaves increased. Such applications have been followed by a marked drop in pest infestation (Panfilo Tabora, personal communication). However, with increased areas coming under cultivation, the total use of chemicals and the cumulative impacts are increasing even though per-hectare applications may be declining in some regions, so more work needs to be done in this area.

At EARTH University's packing house, there is a very deliberate separation of the fungal sprays to treat the banana cluster base (crown) from the rest of the water treatments used in processing. The fungal spray drips are collected and diverted to a dedicated settling pond where the chemicals degrade; the settling pond is covered so the chemicals cannot be further diluted. EARTH University has also begun to experiment with a naturally derived fungicidal compound that would allow for the elimination of fungal sprays altogether. The natural compound is based on citrus seed extracts and has been used for some two years. By all counts, both at the production end and with the buyers, it has performed well. Natural chemicals have increasingly become the norm (Panfilo Tabora, personal communication).

Through these and other strategies, EARTH University in Costa Rica has been able to reduce pesticide use by more than half. In Costa Rica it has also been observed that after some twenty to thirty years of continuous spraying of manganese (from the fungicide mancozeb), zinc, iron, and copper, these metals have accumulated to levels that are considered toxic to banana roots. This tends to stress or debilitate the roots and make them more susceptible to nematodes. Experiments at EARTH University's commercial farm suggest that applying organic matter to the soil can halve levels of these metals, and result in healthy and numerous roots (Panfilo Tabora, personal communication). A recent comparison of the chemicals used by the Chiquita company's Better Banana–certified farms in Costa Rica to a progressive but uncertified competitor suggest that the certification program is having positive impacts on the use of nearly all pesticides. This comparison is shown in Table 10.7. The study found that up to 2002, there was an overall trend to reduce the use of all chemicals with the exception of postharvest fungicides, which showed an increase.

Identify Appropriate Integrated Pest Management Programs to Reduce the Use of Pesticides

As explained below, there are several ways to reduce pesticide pollution by limiting the movement of these chemicals. However, the most effective way to reduce environmental contamination from harmful chemicals is to minimize the amount of the chemical that is applied. Currently, insecticide and nematicide applications are used as prophylactics, on a preventative basis. The introduction of a monitoring program would allow producers to apply chemicals only at those times when potential damage would justify the control measures.

An important aspect of any integrated pest management program is that the workers as well as the owners be educated about the problems of pesticide use and its alternatives. In fact, the education of people who use pesticides directly is one of the most effective ways to reduce total use. Through such programs, owners and workers both can begin to change their attitudes about the management and use of chemicals.

Chiquita reportedly has successfully employed IPM techniques that have reduced nematicide applications by half on 20,000 hectares of plantations (Rainforest Alliance 2000). However, the story is not totally in the numbers. Octavio Cuevas, a

TABLE 10.7. Comparison of Pesticide Use in Certified and Uncertified Banana Production in Costa Rica
(kilograms of active ingredient per hectare)

Type of Agrochemical	Uncertified Company	Better Banana Certified Company
Fungicides	60.4	44.8
Nematicides	11.1	6.3
Herbicides	2.2	0.8
Insecticides	0.5	0.1
Postharvest fungicides	0.1	0.2

Source: Rainforest Alliance 2000.

manager of 3,500 hectares of bananas in Colombia, is a good example. Two years ago, he believed that he needed chemicals to prevent insects from eating his plants and weeds from taking over the plantations. Now the farms he controls have nearly eliminated all classes of agrochemicals except fungicides. The ground is matted with vegetation that holds the soil and protects it from the elements. Butterfly larvae make latticework of banana leaves (without causing significant harm to yields), and frog larvae abound in the chemical-free canals. Weeds are controlled by hand, and experiments are underway to replace fertilizer with chicken manure. Meanwhile the banana production is setting records and the chemical bill is declining (Rainforest Alliance 2000).

Research needs to be undertaken on IPM specific to banana production. In this way, the most toxic chemicals could be targeted for reduced use or elimination altogether so that what is used is both more effective and less damaging. The successful use of bokashi and fermented organic matter at EARTH University (described earlier in the discussions on reducing agrochemical use and reducing wastes) suggests that it is increasingly possible and financially feasible to produce bananas with nontoxic forms of pest control.

Reduce Fertilizer Use

Most banana producers apply fertilizer regularly to their crops. They do this through calculations that are made in advance of actual needs. Little monitoring is undertaken to insure that the plants actually need or take up the fertilizer used. Setting up systems to gather information and monitor the use of and need for chemical fertilizers would reduce use over time. Such a system would require more labor, but it would likely reduce overall expenditures for fertilizers and might extend the life of the plantation.

The migration of fertilizer from banana plantations to other areas is a serious problem. There is no universal solution. However, progress can be made in addressing the problem by reducing the amount of fertilizer that is exposed on the surface of the land. Applying only the amount of fertilizer that the plants need reduces the excess amount in the environment that is available for leaching. There are now experiments with precision application in bananas that could help. Producers should also make sure that roots are healthy and numerous so as to reduce runoff. One strategy to stimulate root development is to apply humic acids, substances found in and extracted in liquid form from compost and well-decomposed organic matter. These can be purchased or made on the farm from composted wastes.

Another hidden environmental issue is the use of burned lime, also called quicklime (calcium oxide). In some areas, burned lime is used to provide calcium for neutralizing soil acidity on banana plantations. For example, in Belize, some 900 kilograms of burned lime are applied per hectare per year on banana plantations. Dolomitic lime is available from local mines, but it takes longer to break down in the soil. Burning the limestone converts it to a compound that breaks down very quickly, so it does a much faster job of neutralizing soil acidity than unburned limestone. If banana producers in places like Belize began to use some dolomitic lime when applying smaller amounts of burnt lime (which is absorbed more quickly), in three years they would be able to convert totally to dolomitic lime. This is important in a country like Belize because the fuelwood required to burn limestone results in deforestation. Another strategy to minimize or avoid the use of lime altogether is fallowing or green manuring. Acidity (low soil pH) is raised to near neutral in fallowed fields. If the fallows are well managed, they can also reduce contamination of and demand for natural resources at minimum cost (Panfilo Tabora, personal communication). There is, of course, the short-term opportunity cost while the land is taken out of production and devoted to production of green manure during the fallow, but this needs to be evaluated in light of the future reduction of input use, reduced downstream pollution/effluent liabilities, and improved long-term viability of the overall farming strategy.

An additional strategy for reducing fertilizer use is crop rotation in conjunction with fallowing. Crop rotation and fallowing can be profitable investments with returns that can rival net returns of bananas per hectare per year, based on savings in pesticides and fertilizers during the cropping years. Cover crops not only provide organic material to replenish the soil, they also reduce soil erosion by reducing exposure to sun and rain, which maintains populations of beneficial soil microorganisms and protects the structure of the soil.

Produce Packaging Materials on Site When Possible

There is a tremendous amount of packaging (boxes, liners, cover sheets, pallets, etc.) for exported bananas. Packaging represents about a third of all FOB costs (Free on Board, a standard shipping term implying that all costs have been paid and that the product is free and clear) of banana producers in Costa Rica. Wood pallets and corrugated boxes represent about 10 percent of the total weight of the cargo for banana shipments. This is an area where improved resource efficiency and the development of product substitutes could reduce substantially the direct and indirect impacts of the banana industry. Because these materials also represent a significant cost to the producers, there is a financial incentive to use them more efficiently.

Since most banana hauling boats are empty, or at least not full, on the return trip to plantation areas, pallets could be returned for multiple uses. Another way to improve performance in this area would be for banana producers to grow timber on areas of the farms that are being fallowed or that are marginal for bananas, but which might be optimal for timber production for making pallets. If leguminous species of trees were used, they could serve a double role by fixing nitrogen in the soil as well. Forested areas could be developed as havens for wildlife, even if they are not as diverse as natural habitat. In the future forested areas might also provide a separate stream of income from payments for carbon sequestration. In addition some of the pallets could be made from on-farm plastic wastes. This may also help to minimize the need for wood or fiber within the banana industry.

von Moltke, K. 1997. *Bananas.* Draft report for the World Wildlife Fund. Washington, D.C. 47 pages.

Worobetz, K. 2000. *The Growth of the Banana Industry in Costa Rica and Its Effects on Biodiversity.* Department of Biological Sciences, University of Alberta, Canada. Available at http://members.tripod.com/foro_emaus/Growth.htm.

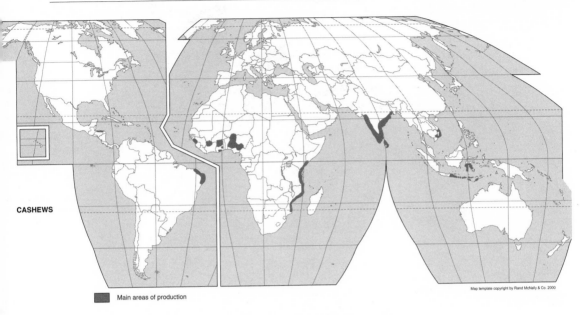

CASHEWS

Main areas of production

Map template copyright by Rand McNally & Co. 2000

CASHEWS *Anacardium occidentale*

PRODUCTION

Area under Cultivation	2.7 Million ha
Global Production	1.6 Million MT
Average Productivity	593 kg/ha
Producer Price	$425 per MT
Producer Production Value	$594 million

INTERNATIONAL TRADE

Share of World Production	62%
Exports	0.2 million MT
Average Price	$4,939 per MT
Value	$867 million

PRINCIPAL PRODUCING COUNTRIES/BLOCS (by weight)	India, Vietnam, Nigeria, Brazil, Tanzania, Indonesia, Côte d'Ivoire, Guinea-Bissau, Vietnam
PRINCIPAL EXPORTING COUNTRIES/BLOCS	India, Vietnam, Brazil, Tanzania, Guinea-Bissau, Côte d'Ivoire
PRINCIPAL IMPORTING COUNTRIES/BLOCS	United States, Netherlands, United Kingdom, Germany, Japan, Australia, Canada, France
MAJOR ENVIRONMENTAL IMPACTS	Conversion of natural forest in West, East, and southern Africa and in Brazil Plant toxicity tends to discourage other biodiversity in the same area
POTENTIAL TO IMPROVE	Good Better management practices are known Impacts and inputs are few and can be reduced further Organic and Fair Trade certification exists

---

Source: FAO 2003. All data for 2000.
Note: Production figure is in unshelled nuts and exports are in shelled nuts or kernels. The conversion rate used: 5 kg unshelled nuts = 1 kg shelled nuts.

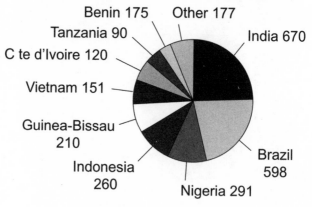

Area in Production (Mha)

- Benin 175
- Other 177
- India 670
- Tanzania 90
- C te d'Ivoire 120
- Brazil 598
- Vietnam 151
- Guinea-Bissau 210
- Indonesia 260
- Nigeria 291

OVERVIEW

The cashew is native to northeast Brazil. At the time of Portuguese colonization, cashew was a major food crop for Indians throughout the region. Large native stands of cashew trees were found throughout the northeast, especially on coastal lands but also well inland. The leaf of the cashew tree contains compounds that are toxic to other plants and animals. Leaf fall discourages the growth of other vegetation under the cashew tree. Also, the seed is surrounded by a concentrated caustic solution that burns the skin. This prevents wild animals from eating the seeds. These two characteristics encourage the growth of large stands of cashew trees that dominate the landscape.

Cashews can grow to be quite old. Unlike many trees, however, as they get older they tend to sprawl and branches touch the ground, take root, and become the base of other trunks that continue to sprawl, take root, and expand the tree farther and farther from the original base. The oldest and largest cashew tree now alive in Brazil is so old that it has spread over an entire hectare of land (Morton 1987). While little more than a foot in diameter at the main trunk, the tree is estimated to be more than a thousand years old.

Mature trees commonly have a canopy diameter of 12 meters. However, studies of roots show that while they often grow quite deep to tap into water sources, they also spread laterally at distances that are commonly twice that of the canopy.

The cashew is unusual because it flowers, sets fruit, and maintains a full leaf cover during the driest part of the year. Good yields occur when it is dry during the peak flowering period. About 14 percent of flowers are hermaphroditic; the remainder are male. About 70 percent of the hermaphroditic flowers fail to produce nuts. The crop is mainly cross-pollinated. Insects play an important pollinating role, as does the wind. The period from flowering to fruit fall is from fifty-five to seventy days.

In the sixteenth century, the Portuguese took cashew seeds to Africa, India, the Middle East, and other parts of the world. The value of the plant at that time was as cheap food (fruits and kernels), but more importantly it was a way to stop soil erosion in coastal areas. Cashews grow well in sandy soils and are extremely tolerant of saline soils. Cashews are grown throughout coastal areas in the tropics.

The trees produce nuts, fruits, gum, and charcoal. The only cashew product traded in any quantity internationally, however, is the nuts. The cashew fruit is as unusual as the rest of the tree. The shell of the nut is leathery and spongelike, not brittle, and contains a thick oil. The nut kernel is protected by an additional thin skin (the testa). The nuts are self-contained and are attached to the end of a fruit (called a cashew apple) that has many different uses, particularly in Brazil and in parts of Africa and Asia. There is no information available about the total production of fruits worldwide or the proportion of that total that is consumed.

PRODUCING COUNTRIES

The Food and Agriculture Organization of the United Nations (FAO) reports that there were nearly 2.7 million hectares planted to cashew trees in 2000. This data is somewhat incomplete, however. Brazil, for example, has large areas of natural cashew stands that are not captured in such statistics. In fact, it is estimated that most of Brazil's production comes from such areas. The FAO data indicate that the main producing countries by area are India (670,000 hectares), Brazil (598,490 hectares), Nigeria (291,000 hectares), Indonesia (260,000 hectares), Guinea-Bissau (210,000 hectares), Benin (175,000 hectares), Vietnam (151,000 hectares), Côte d'Ivoire (95,000 hectares), and Tanzania (90,000 hectares). According to the FAO these nine countries account for nearly 94 percent of all land used to grow cashews and just over 91 percent of all cashew nut production (FAO 2003).

In 1994 FAO statistics indicated that the main cashew producers by weight were Brazil (32 percent), India (24 percent), Mozambique (9 percent), Indonesia (6 percent), and Tanzania (5 percent). However, by 2000 India produced and processed more cashew nuts than any other country, as shown in Table 11.1. It was followed by Vietnam, Nigeria, Brazil, and Tanzania (FAO 2003). During the 1990s much of India's growth came from shelling the nuts produced in other countries, particularly in African countries. This happened because India's labor productivity costs are lower than Africa's. India is not only the biggest cashew producer, it is also the second largest consumer of cashew kernels. Indian cashews are known in the international market for their small size and overall shelling quality. In addition, though, they are know for their variable taste; this may result, in part at least, from the fact that nuts shelled in India have been grown in many different countries.

For the past 40 years, India and Brazil have dominated cashew production. In 2000, they still accounted for 47 percent of the global area planted to cashews and

TABLE 11.1. Main Cashew-Producing Countries by Area and Total Production, 2000

Country	Area Harvested (ha)	Cashew Nut Production (MT)
India	670,000	500,000
Brazil	598,490	114,467
Nigeria	291,000	184,000
Indonesia	260,000	90,400
Guinea-Bissau	210,000	72,725
Benin	175,000	26,000
Vietnam	151,000	270,400
Côte d'Ivore	125,000	78,000
Tanzania	90,000	121,200
Mozambique	50,000	57,894
World	2,742,167	1,600,002

Source: FAO 2003.

44 percent of total production. Along with Nigeria, Tanzania, Mozambique, and Kenya, they are the long-standing producers. In addition to these countries, however, there are a number of new producers that are increasingly important. These include Vietnam, Indonesia, Guinea-Bissau, Benin, Côte d'Ivoire and Ghana. In Vietnam, the value of cashew exports has surpassed tea to rank third among top agricultural exports, just after rice and coffee. Sri Lanka is another minor producer of cashews in Asia. Even though it is small in comparison to other countries, the crop is important, and there are some 30,000 small-scale cashew processors in Sri Lanka.

Cashews are an important commodity in Brazil. With a total area of some 700,000 hectares, the industry is responsible for annual revenues on the order of U.S.$200 million per year. In addition there are some 300,000 jobs created directly and indirectly by cashew production and the processing of nuts and juice (Porto and Paiva 2001).

Table 11.2 gives information on global cashew production based on how much production is exported and whether the exports are of unshelled or shelled nuts. While this data is incomplete (not all countries report it), it still gives an indication as to which countries consume most of their cashew nut production internally, which export the lower-value unshelled nuts, and which shell their own nuts to take advantage of the more valuable exports.

Range of Cashew Products

Several products can be produced from the cashew tree. These include nuts, fruits, nutshell liquid, and resin to name but a few. Of all these products only cashew nuts have significant trade internationally.

Cashew nuts are internationally the best-known product of the tree. Cashews are

TABLE 11.2. Global Production and Trade of Cashew Nuts, 2000

| Country | Cashew Nut Exports | | Total Production (MT) |
	Unshelled (MT)	Shelled (MT)	
Brazil		33,588	114,467
India	7,485	81,661	500,000
Kenya	511	87	12,500
Mozambique	–	4,700	57,894
Sri Lanka	28	91	4,610
Tanzania	94,482	–	121,200
Côte d'Ivoire	63,379	353	78,000
Guinea-Bissau	73,210	–	72,725
Indonesia	25,621	1,998	90,400
Vietnam	29,731	34,200	270,400
Nigeria	2,947	–	184,000
Benin	–	–	26,000
World	**229,189**	**139,438**	**1,600,002**

Source: FAO 2003.

TABLE 11.3. Chemical Composition of Shelled Cashew Nuts

Constituents	Percentage
Proteins	21.0
Fat	47.0
Moisture	5.9
Carbohydrates	22.0
Phosphorus	0.45
Calcium	0.05
Iron	5.0 mg/100g

Source: Davis 1999; CEPC 2003.

one of the most delicious and highly sought-after nuts. People seem to enjoy plain, roasted cashews the most. Cashews are also used to add flavor to a wide range of foods such as ice creams, sweets, chocolates, cookies, meat dishes, etc. There are hundreds of recipes that use cashew nuts. The cashew is a very nutritious food; as shown in Table 11.3 it is high in protein. It contains no harmful cholesterol and is rich in minerals and vitamins. It has as little as 1 percent soluble sugar. Thus, it can be safely consumed by those suffering from diabetes. When ground the cashews are a relatively easily digested form of protein and are recommended for people on chemotherapy or with HIV/AIDS. Cashews are useful for those with anemia since they are rich in iron. As with most nuts, however, some people are allergic to them. Cashew nut consumption may pose a health risk to persons sensitive to other nuts as well as to its botanical relatives poison ivy, poison oak, or poison sumac.

The cashew fruit (cashew apple) is rich in vitamins and amino acids. It can be used for making many typical fruit products such as jellies, jams, juice, wine, and liquor. In many parts of Africa the fruit is made into a wine and a spirit for sale on local markets. In Brazil the apple is used mostly to make juice, but it is also used in the manufacture of jams and alcoholic drinks. In Goa, India, the juice from the apple is fermented and then distilled into a cashew liquor called *feni*. In countries that utilize the cashew apples, the income from the sale of wine and liquor can be equal to the income from the sale of nuts (Jim LaFleur, personal communication).

The apple's juice has an antiscorbutic (antiscurvy) effect due to its high vitamin C content. It is used extensively in the cosmetic industry, as a substance capable of capturing free radicals. It is used in the preparation of shampoos, lotions, and scalp creams. The juice from cashew apples, when the tannin has not been removed, is prescribed as a remedy for sore throat and chronic dysentery in Cuba and Brazil. Fresh or distilled, it is a potent diuretic and is said to possess sudorific (sweat-inducing) properties. The juice is applied as a liniment to relieve the pain of rheumatism and neuralgia (Morton 1987; Grieve 1995).

Even when discarded the cashew apple serves as food for livestock or wild animals. At this time, tremendous volumes of cashew fruit are thrown away by the nut industry. In Brazil alone an estimated 400,000 metric tons of fruit are thrown away each year, and most of the fruit is discarded in India as well. By 1997, Vietnam was producing some 500,000 metric tons of cashew fruit that were not being used (Kinh et al. 1997). Since most of this fruit is de facto organic, it could be certified and used as a backup juice (replacing white grape) in the organic juice industry.

Cashew nutshell liquid is a natural resin that is extracted from the honeycomb structure of the cashew nutshell. It contains 90 percent anacardic acid and 10 percent cardol. Both are caustic and can contaminate the nuts and blister the skin of the shellers (Davis 1999). The liquid is a by-product of the cashew industry and a versatile industrial raw material. There is considerable potential for its utilization in the development of drugs, antioxidants, fungicides, and other chemicals. In the tropics the liquid from cashew shells is used in some medicines for the treatment of ailments such as scurvy, warts, ringworm, cancerous ulcers, and even elephantiasis. The oil is also used for treating timbers to make them termite-proof.

The liquid is now distilled to make a number of substances but the primary ones are cardanol and cardol. The major use is to make cashew friction particles for the brake lining industry. The liquid is also used to make resins, varnishes, paints, plastics, insecticides, preservatives, drying oil, epoxy, binders in automotive strip linings and brake linings, and heavy-duty coatings that have the ability to stick to poorly prepared surfaces. The next generation of products from cashew nutshell liquid have lower viscosity and lighter colors. Many will help epoxy and friction formulators to meet the demands of the next century (AP Horticulture 2003; Cardolite 2003).

In less sophisticated shelling operations, the cashew nutshell liquid, the shell and the oil cake are often used as conventional boiler fuel to reduce overall energy costs. The liquid and shell can also be sold to other industries (e.g., foundries or cement works) for similar uses. The oil cake is also a suitable fuel for generating gas for boilers and internal combustion engines.

BOX 11.1. OTHER PRODUCTS FROM THE CASHEW TREE

A number of minor products can be produced from different parts of the cashew tree.

- Gum from cashew fruit stems is used as a varnish for books and woodwork. It is said to protect them from insects and ants.
- Anacardic acid, a substance derived from oil in the nutshell, has antibiotic properties against gram-negative bacteria and is also used to treat leprosy and ringworm.
- The black juice of the nut and the milky juice from the tree after incision are made into an indelible marking-ink.
- The stems of the flowers give a milky juice which, when dried, is hard and black and is used as a varnish.
- The timber from the tree can be used in furniture making, boat building, packing cases, and in the production of charcoal. When farmers trim the trees to increase production by exposing branches to direct light, they can also turn the trimmings into charcoal.
- The leaves have some medicinal uses. An infusion of the leaves can be gargled for sore throat. In Indonesia older leaves are used to make poultices to treat burns and skin diseases.
- The gum from the tree, when dried, has properties similar to gum arabic and guar gum. Those products sell for more than $2,000 per metric ton as stabilizers and emulsifiers in a wide range of foods from beer to ice cream to salad dressings. In the age of fat-free food, such emulsifiers are increasingly important to bind ingredients. However, since many are allergic to cashews, using it as an ingredient in many different foods could pose health and liability issues.

CONSUMING COUNTRIES

The United States, the European Union, Japan, and the countries of the former USSR are the major importers of cashew nuts. Combined, they account for some 90 percent of global imports. The main European importers are Germany, the United Kingdom, the Netherlands, and France. The United States is the largest importer of shelled nuts, importing nearly half of all exported cashew kernels.

India dominates the cashew nut market. It is the largest producer and importer of unshelled nuts, the largest producer of shelled nuts, and the second largest consumer of shelled nuts. In the past the former USSR accounted for 25 percent of the global import market for processed cashew nuts, but that market has shriveled up since the breakup of the former Soviet Union.

Asia and Europe also purchase significant amounts of cashew nuts. The Japanese market is the most important market in Asia, but it accounts for only 4 to 5 percent of the international trade of shelled cashew nuts. Some 30 percent of Vietnamese production goes to the Chinese market, while 40 percent is exported to the U.S. market. The cashew nut share of the nut market in the European Union is in the range of 3 percent of the total nut imports.

PRODUCTION SYSTEMS

The cashew tree is evergreen. It grows up to 12 meters high and can have a spread of 25 meters. Its extensive root system allows it to tolerate a wide range of moisture levels and soil types. Commercial production, however, is undertaken on well-drained, sandy loam or red clay soils. Annual rainfall needs to be at least 889 millimeters and not more than 3,048 millimeters. Cashew trees are most frequently found in coastal areas (Intermediate Technology Development Group, no date).

Many varieties of cashew are cultivated. Some produce better nuts and some produce better fruits. Looking just at the fruits, for example, type K 10-2 has the best size and juice content. Apples of BLA-1 and Ansur 1-27 have high carbohydrate content. The fruit of type M 6/1 has the highest sugar content. K 27-1 has a high vitamin C (ascorbic acid) content. BLA-40 has low tannin content. M 10/4 has high protein content. Vengurla 37-3 has high content of other extractives, and BLA-273 has high crude fiber content. Depending on the end use of the nut or fruit a producer might select one variety over another.

Cashews are grown mostly in arid, coastal areas of India, Southeast Asia, and East and West Africa. On the dry Pacific side of Central America there are also many recently established plantations of cashews. In Brazil they are located as clumps all over the arid area of Rio Grande do Norte between Fortaleza and Natal. They also grow wild inland as far west as the Agreste, a transition zone of sandy soils between the humid Atlantic coastal forest zone and the dry interior. Similarly, cashews are also found in semiarid areas of Asia such as Indonesia. Most of the plantations are on clay soils, though in some parts of Brazil they are on sandy soils. In general, cashews are an excellent crop for deforested and degraded coastal areas. They grow between sea level and approximately 760 meters and grow poorly in high altitudes where the temperatures are too low.

In many countries cashews are grown as border trees on ranches and in orchards with very little care, but the trees always look very healthy. In plantations, they are planted on grids as monocultures. There is nearly always grass cover on the ground but not much more vegetation, as production areas tend to be somewhat saline as well as very dry and this does not encourage other vegetation. The trees are allowed to grow very tall, and there is little tree modification except for topping them off when they become too high.

Cashews are easy and inexpensive to produce. They can be grown in small plots by farmers, and with simple but regular plant protection measures they yield well above 10 kilograms per tree per year. Field maintenance of cashews requires less time and money than most other perennial crops and any annual crop. Some pruning is necessary to create light for the best flowering, fruit growth, and maturation as well as to facilitate nut collection and to allow for weed control.

The tree's main soil requirement is unimpeded drainage. Good yields can be achieved without the addition of fertilizers or manures. Cashews are often grown satisfactorily on infertile sands. Though trees will bear without added nutrients, fertilizers (such as urea, rock phosphate, and muriate of potash) are sometimes used, usually applied in circular trenches around the plants. The best growers avoid applying fertilizer during heavy monsoons (which wash away fertilizers) and also when the soil moisture is poor (as soil moisture promotes better absorption of nutrients).

Cashew trees are usually planted at stake, i.e., the seeds are planted directly in the field. (Cashews can also be propagated by vegetative means including grafting, air layering, or tissue culture.) The seeds are floated prior to planting; any that float are not considered viable and are not planted. Ideally trees should be planted as near the beginning of the rainy season as possible. In Asia cashew trees are usually planted as space permits among coconut palms, mango trees, banana plants, cassava, etc. However, they are also planted in pure stands in southern Tanzania. When cashews are planted at a close spacing, the trees must be thinned as soon as their canopies meet.

Once established, trees and fields need little care. In the first couple of years, lower branches and suckers are removed. Intercropping may be done during the first few years, with cotton, peanuts, or yams. Cashews grow quickly and start producing fruits in the second or third year. Full production is attained by the tenth year, and trees continue to bear until they are about thirty years old. When the trees are older, cattle and other livestock are allowed to graze among them to keep the weeds in check.

Harvesting is done when the fruit changes color from green to yellow, orange, red-orange or red. This can be done by picking with a pole-picker that has an attached net. This tends to be the preferred method of harvesting to ensure that the fruits are intact when sold as fresh fruit or into the juice market.

Nut producers prefer to simply let the fruits with attached nuts fall to the ground. They fall only when they are mature. However, once nuts fall, they must be picked up within a week or they will begin to discolor. The nuts are picked from the ground, separated from the fruit, washed, and dried under the sun. Small growers can store the nuts in a dark, dry place for some months in order to collect a sufficient volume to sell. The nuts are eventually sold to intermediaries.

Yields vary. In Tanzania, yields of cashew nuts from pure stands average about 590 kilograms per hectare. However, under optimal conditions yields in Tanzania can reach 1,100 kilograms per hectare. Studies have shown that with genetic selection, yields can reach as high as 2,200 kilograms per hectare (Agriculture News from Africa 2000).

The World Bank estimates that roughly 97 percent of cashews are harvested from wild growth in Brazil and small peasant holdings throughout the world. Almost all of East Africa's cashew crop is grown on small farms. The Bank estimates that only 3 percent come from monocrop cashew plantations (Rosengarten 1984, as cited in Davis 1999). Consequently, cashews are an extremely important income generator for small farmers and the rural poor. For most of these producers cashew is an attractive crop because it provides income even when completely neglected. Furthermore, on-farm or local processing can increase income. Thus, the crop provides value-added processing employment opportunities unlike many agricultural products, as well as foreign exchange earnings for the country.

Another important factor with cashews is that they can be produced on a wide range of soil types so that many producers can grow them. They also produce both fruit and nuts. Either or both can be sold for income or consumed on the farm depending on the needs of the producer. This makes them a very versatile source of income, akin to honey.

There are some constraints to cashew production, however. Some of these constraints have to do with achieving scales that would allow the processing to become more efficient. For example, in order to avoid complying with labor laws and save taxes in countries such as India and Brazil, many cashew processors keep the size of their units small so as to enjoy the status of small-scale industries. Extracting cashew kernels from their tough outer shell is a laborious job. The fact that shelling cashews is so labor-intensive has increased employment opportunities in both urban and rural areas for hand shelling. While this system creates more higher-value whole nuts, hand shelling is slower than shelling nuts with mechanical technology. The new technology, though reducing labor costs over time, results in much greater breakage. The main advantage of the new technology is that it tends to meet buyer requirements for hygiene, quality, quantity, and schedule.

There are other issues that affect the production and value of cashews. Some diseases affect cashew nuts. For example, an insect pest in Brazil has reduced the marketable nut harvest by some 50 percent. Many producers also do not know how to manage their groves by pruning, spacing, or weeding. Harvest and storage techniques are not always appropriate, and improper techniques can produce nuts of no or low value.

The fact that cashews are essentially a source of income for small farmers and landless labor is both the blessing and the curse of the industry. There is, as a consequence, an absence of market integration or coordination between producers, shellers, and exporters. This results primarily from the fact that for most farmers this is a rather small source of income compared to other cash or food crops. They do not see it as a crop that is worth a lot of effort. There are also often antagonistic relations between growers, shellers, and the industry as a whole. Many intermediaries are involved in the market chain and this tends to make it much less transparent in terms of price and product quality.

PROCESSING

Shelling cashew nuts is undertaken in large factory operations as well as decentralized, smaller-scale operations and even piecework systems connected to either of the other systems. There are hundreds of thousands of small-scale cashew processors. Guinea-Bissau has some 90,000 and Sri Lanka has 30,000, for example. Similarly, small-scale processors are responsible for approximately 95 percent of all shelled raw nuts sold into the market in Mozambique (Wandschneider and Garrido-Mirapeix 1999). In Brazil there are twenty-three large mechanical processing plants with capacity to shell 240,000 metric tons of nuts per year. There are also some 120 small, manual processing plants with a combined capacity to shell 20,000 metric tons of nuts per year (Porto and Paiva 2001).

The cashew harvest lasts for two months. Due to the protective liquid in the shell, nuts can be stored for up to a year before processing, although most processors feel that quality begins to deteriorate after six months. One of the main costs of processing facilities of any kind is the working capital to stockpile the nuts that will be shelled over the next six to nine months.

Whether small or large-scale commercial processing is to be undertaken, the nuts need to be dried and then stored in an aerated and dry environment. The initial drying usually takes place in the sun. The nuts are then sorted by size.

After the nuts are dried to 10 percent humidity, the rest of the processing can begin. In small-scale artisan processing systems the nuts are boiled in water for 40 minutes in order to separate the kernels from their shells and to loosen the paper-thin skins on the nuts. In larger commercial operations, the nuts are passed through a pressurized steam system, usually an autoclave, to do the same thing. In both instances, the shells are then removed.

Small-scale processing units treat the steamed or boiled nuts differently than larger systems, however. In the small-scale or artisanal system the nuts are again sun-dried and sorted by size. The reason the nuts are dried again is to shrink the nut from the shell after it has absorbed water from the boiling process. In large-scale shelling operations, the soaked nuts are partially dried in ovens to shrink them from the shell and reduce the drying time required before they can be shelled. Drum roasting and hot-oil roasting are also undertaken in larger factories prior to shelling to neutralize the liquid in the shell.

The next processing step in both systems is to slice the tough, leathery cashew shell off of one side of the nut and then separate the kernels from their shells. Cashew processing—the removal of the kernels from the tough shell and the skin— is more complicated than it appears because of the irregular size and shape of cashew nuts. In fact, the irregular size and shape make the mechanical shelling of perfect, whole nuts very difficult.

In more primitive shelling operations, now mostly for home or local consumption, the unshelled nuts are roasted in an open pan. In this process the nut begins to smoke. It is then removed from the fire, cooled and shelled immediately by hand. This process tends to scorch parts of the kernel, reducing its value and making it hard to sell on international markets. These nuts, however, have a distinct flavor that is preferred in the growing regions.

In all small-scale operations, and many large-scale ones too, the shelling process is done by hand at least for the larger nut sizes that are more valuable when sold as whole nuts. The biggest problem for the industry has been to find ways to undertake this process mechanically to lower labor costs. Some of the larger commercial shelling operations use mechanical shelling, either with automated conveyor belts that pass the nuts through blades to cut them open or using dry freezing to crack the nuts in some of the most capital-intensive operations. Unfortunately, mechanical shelling ends up with the majority of the shelled nuts being broken (e.g., 55 percent or more). Broken nuts have far less value. By comparison, experienced hand shellers can produce up to 90 percent or more whole kernels from the nuts they shell (Porto and Paiva 2001). Two people with specialized skills are required to work together to shell nuts by hand—one cuts the shells and the other peels the outer shell from the kernel. Two people can shell and peel about 15 kilograms of kernels per day (this is the main reason that cashew nuts are more expensive than other nuts). When shelling by hand, workers have to be careful because the liquid in the nut shell can burn the skin if proper measures are not taken. Normally this means using a vegetable oil or even rubber gloves to cover the hands at all times.

Another disadvantage of the larger processing plants is the tendency of the steam pressure system to cause discoloration and a deterioration of quality. Once nuts are heated, chemical changes can occur in the oils that tend to reduce their freshness and overall shelf life.

After shelling, the kernels are baked in ovens for 7 to 8 hours at 70 degrees Celsius to lower their moisture content to 4 percent. Reducing the humidity of the kernels increases their shelf life. Depending on the technology employed, drying can be undertaken either in the open sun, in solar furnaces, or in high-volume furnace dryers. Increasingly, shelling operations use the nutshells as fuel to power the high-volume driers. Once the moisture has been removed the nuts are cooled.

After the nuts are cool enough to work with, the paper-thin inner skin is removed. This skinning can be done either by hand or by machine. By hand, one person can remove the skin from about 10 to 12 kilograms of kernels per day (Intermediate Technology Development Group, no date). Breakage during the peeling can be as much as 30 percent.

Once skinned, the kernels are ready to be classified and packed. Classification is done by size of whole kernels, color, and physical integrity (e.g., wholes, halves, various-sized pieces, crumbs, dust). There are dozens of classifications of cashew nuts, from the largest whole nuts all the way to cashew powder. Each is recognized by traders and each has a different value. The overall high value of cashew nuts is sufficient to warrant such differentiation of the product.

Once sorted, cashew nuts are vacuum-packed in 20-kilogram Mylar bags inside cardboard boxes. Alternatively, some factories seal the nuts in 20-pound (9-kilogram) tin cans. Oxygen causes the quality of nuts to deteriorate, so whenever possible oxygen is removed from the containers by flushing them with nitrogen. The nuts are generally exported in large container lots. The total weight of a container is 15 to 17 metric tons. However, because all boxes or cans are clearly marked, nut sizes can be mixed within a single container depending on the purchase of the buyer.

No further processing is ever performed for international markets. However, in local markets many of the nuts are made into confections, pastries, and other products. This does not, however, take place at the shelling factories.

At large-scale processing facilities, another stage of processing occurs after shelling and before the kernels are dried. Cashew nutshell liquid is extracted by the expeller method. It is literally pressed mechanically to release the liquid from the honey combed nutshell. This liquid, which in the past has been treated as toxic waste, is now sold as a by-product with many different industrial uses (see above).

er, this was the exception rather than the rule. So long as the markets do not increase dramatically for cashew nuts, these impacts are likely to be within acceptable limits because producers will not be able to afford the inputs.

Habitat Conversion

With 97 percent of the total cashew crop produced by small farms or collected from the wild in Brazil, there is still very little large-scale habitat conversion associated with the production of cashews. Plantation establishment could pose environmental impacts, especially in drier, often more fragile and more marginal areas.

Agrochemical Use

There is very little weed control involved in cashew production, other than allowing livestock to graze beneath mature trees. As a result, herbicides are generally not used. Similarly, there are few fertilizers used in cashew production (other than urea rock phosphate and muriate of potash during initial planting), and these are applied very sparingly due to their cost. In fact, most farms do not receive any fertilization for years.

Few if any pesticides are used in cashew production in most areas. However, in some parts of Africa two species of *Helopeltis* can damage cashew trees by sucking juices from the leaves, young shoots, and inflorescences. These pests have been controlled by dieldrin sprays or dusting with benzene hexachloride (BHC), or DDT plus BHC (Agriculture News from Africa 2002). If the area for cultivating cashews is large, aerial sprays can be employed. This creates a potential for pesticide toxicity as the pesticide mixture is sprayed over a large area.

One reason for the minimal use of agrochemicals is that producers do not perceive cashews as an important profit-making crop. Rather, it is insurance for obtaining income without high costs during lean times. On many cashew farms the cashew is in fact the vegetation that provides the most food for wildlife; in many areas there are few sources of food for wildlife with as high nutritional content.

BETTER MANAGEMENT PRACTICES

Cashew trees are one of the few crops that generally have a more positive than negative environmental impact associated with their cultivation. This is because they tend not to be cultivated in plantations that require clearing large areas of land. The notable exception is in Brazil where some cooperatives, with government assistance, have planted thousands of hectares of the trees. Rather, cashews are most often planted in clumps and as border vegetation. Increasingly, they are volunteers that are self-seeded and tolerated by landowners. Cashews support and protect wildlife with their shade, their fruits, and their vertical and horizontal architecture. Even in large plantations, cashews are still a major food source for wildlife.

Cashew trees are very effective at retaining soil and stopping erosion, especially in coastal areas. As explained in the beginning of this chapter, this was why the Portuguese introduced cashew trees in many coastal areas of the tropical world that remain among the most important producers today. Along the coast of Orissa in India, cashew trees are still planted as shelterbelts and windbreaks. They stabilize sand dunes and protect the adjacent fertile agricultural land from drifting sand. More recently, cashews have proven useful as a species of choice for reforestation in degraded areas. They are one of the few trees that do well under such conditions and that generate both food and income for producers so that their survival is ensured.

In general, however, there are a number of important, practical ways that cashew producers, and the industry as a whole, could be made more efficient and profitable. These include:

1. Increasing yields by pruning, replanting, or "topping" existing trees with new grafts in order to increase production and control pests.
2. Bringing more areas into cashew cultivation, especially areas that are marginal, abandoned, or degraded.
3. Undertaking value-added hand processing at the farm or community level so that more of the value of the finished product accrues to the producer or the community.
4. Processing the cashew apple into juice, dried fruit, wine, liquor, and other products. The fruit could provide an excellent source of vitamins and minerals for many households that are short on both. Income from the sale of juice could also be significant given the large amount of fruit that is currently abandoned.
5. Increasing the ability of local producers and shellers to sort their nuts by standard grades so they are not penalized by buyers who, in turn, capture the value (often as much as 10 to 15 percent) of sorting the nuts by grades.
6. Improving local storage capacity and conditions so that cashews can be stored until hand shellers have a chance to process the entire crop. This would also allow the production to be sold onto the world market more gradually so that dumping would affect prices less.
7. Increasing transparency and competition among buyers to improve producer prices.

OUTLOOK

Cashew nuts are seen as a valued product by consumers in developed countries. Not only does this mean that consumers are willing to pay more for cashews than for other nuts, it also means that they are more likely to pay somewhat more for other cashew products as well, including fruit, honey, etc. In fact, premiums for organic and Fair Trade certified cashews attest to their consumer appeal.

Cashew nuts have maintained their value over the past forty years, unlike any other single commodity covered in this book. Without changing the basic price of cashew nuts, there are a number of ways that an increasing proportion of the ultimate retail value of the nuts could accrue to the producers. Recent efforts in Guinea-Bissau have shown that in-shell nut exports can be shelled locally, significantly increasing the income to the local economy. If these efforts can be replicated elsewhere, cashews have the potential of becoming the cornerstone of poverty reduction programs in many coastal tropical areas.

Such a strategy could have a positive impact on the environment as well. If cashew nut producers can make more from their nuts, this added income will take pressure off of other resources, in turn making their use more sustainable. This could have obvious impacts on other agricultural resources. In addition, many cashews are produced on coastal areas, so it is likely that increased income from nuts would take pressure off of wetland and marine resources as well.

REFERENCES

Agriculture News from Africa. 2000. Africa Cashew Articles. *Agriculture News from Africa.* Available at http://www.newafrica.com/cashew/articlepg.asp. Accessed February 2002.

AP Horticulture. 2003. *Cashew (Anacardium Occidentale).* Department of Horticulture, Government of Andhra Pradesh. Available at http://www.aphorticulture.com/Cashew.htm. Accessed June 2003.

Axtell, B. L. (prepared by). Researched by R. M. Fairman. 1992. Minor Oil Crops. FAO Agricultural Services Bulletin No. 94. Rome: UN Food and Agriculture Organization.

Bedi, B. M. 1971. Cashew Nut Dermatitis. *Indian Journal of Dermatology.* 16: 63–4.

Cardolite. 2003. *Concise History of the Commercialization of Cashew Nutshell Liquid.* Available at http://www.cardolite.com/www/cnsl_history.htm.

Cashew Export Promotion Council of India (CEPC). 2003. *Cashewnut—A Versatile Health Food.* Available at http://www.cashewindia.org/html/c0600frm.htm. Accessed 2003.

Centers for Disease Control. 1983. Dermatitis Associated with Cashew Nut Consumption—Pennsylvania. *Morbidity and Mortality Weekly Report.* Available at http://www.cdc.gov.

Davis, K. 1999. *Cashew.* ECHO Technical Note. Fort Myers, Florida: Educational Concerns for Hunger Organizations (ECHO). Available at http://echonet.org/tropicalag/technotes/Cashew.pdf

Falzetti, F. and J. C. Faure. 1985. Cashew Development Program of FAO in the World. In E. V. V. Bhaskara Rao and H. Hameed Khan, editors. *ISHS (International Society for Horticultural Science) Acta Horticulturae* 108, International Cashew Symposium. November 1, 1985, Cochin, India.

FAO (Food and Agriculture Organization of the United Nations). 2003. *FAOSTAT Statistics Database.* Rome: UN Food and Agriculture Organization. Available at http://apps.fao.org.

Garg, A. No date. *Indian Cashew Exports—Looking for New Areas of Cultivation.* Available at http://www.trade-india.com/master/product/cashewexports.html.

Grieve, M. 1995. Cashew. In *A Modern Herbal.* Arcata, California: Ed Greenwood. Original version published 1971. New York: Dover Publications. Available at http://botanical.com/botanical/mgmh/c/casnut29.html.

Intermediate Technology Development Group, Ltd. No date. *Cashew Nut Processing.* Technical Brief. Technical Information Services. The Schumacher Centre for Technology and Development. Available at http://www.itdg.org.

International Tree Nut Council. 2002. Cashew—Cashew Situation and Outlook. *Global Statistical Review.* January. Available at http://inc.treenuts.org/stats_cashew_jan02.html.

Kinh, L. V., V. V. Do, and D. D. Phuong. 1997. Chemical Composition of Cashew Apple and

Cashew Apple Waste Ensiled with Poultry Litter. *Livestock Research for Rural Development.* January. Volume 9. Cali, Colombia: Centro para la Investigación en Sistemas Sostenibles de Producción Agropecuaria (CIPAV). Available at http://www.cipav.org.co/lrrd/lrrd9/1/kinh91. htm.

Morton, J. 1987. Cashew Apple. *Fruits of Warm Climates.* Pp.239–240. Miami, FL: Julia Morton, distributed by Creative Resource Systems, Inc.

Murthy, B. G. K. and M. A. Sivasamban. 1985. Recent Trends in CSNL Utilization. In E. V. V. Bhaskara Rao and H. Hameed Khan, editors. *ISHS (International Society for Horticultural Science) Acta Horticulturae* 108, International Cashew Symposium. November 1, 1985, Cochin, India.

Palmer International, Inc. 2003. *Nature's Phenol, Cashew Nut Shell Liquid.* Available at http://www.palmerint.com/cnsl.htm. Accessed 2003.

Porto, M. C. M. and F. F. A. Paiva. 2001. *Cashew Nut Miniplants in Northeastern Brazil: A Successful Partnership.* Technical Workshop on Methodologies, Organization, and Management of Global Partnership Programs. Rome, Italy: Global Forum on Agricultural Research. 9–10 October 2001.

Quenum, B. M. 2001. *True Causes behind the Collapse of Mozambique's Cashew Nut Industry. Part 1: Cashew Nut Worldwide Market.* Available at http://businessafrica.hispeed.com/africabiz/cashew.

Rosengarten, F. 1984. *The Book of Edible Nuts.* New York: Walker and Company.

Russel, D. C. 1985. The Future of the Cashew Industry. In E. V. V. Bhaskara Rao and H. Hameed Khan, editors. *ISHS (International Society for Horticultural Science) Acta Horticulturae* 108, International Cashew Symposium. November 1, 1985, Cochin, India.

Smith, N., J. Williams, D. Plucknett, and J. Talbot. 1992. *Tropical Forests and Their Crops.* New York: Comstock Publishing.

Vilasa Chandran, T. and V. K. Damodaran. 1985. Physico-chemical Qualities of Cashew Apples of High-Yielding Types. In E. V. V. Bhaskara Rao and H. Hameed Khan, editors. *ISHS (International Society for Horticultural Science) Acta Horticulturae* 108, International Cashew Symposium. November 1, 1985, Cochin, India.

Wandschneider, T. S. and J. Garrido-Mirapeix. 1999. *Cash Cropping in Mozambique: Evolution and Prospects.* Food Security Unit Technical Papers No.2. A report by the European Commission's Food Security Unit Mozambique. Maputo, Mozambique.

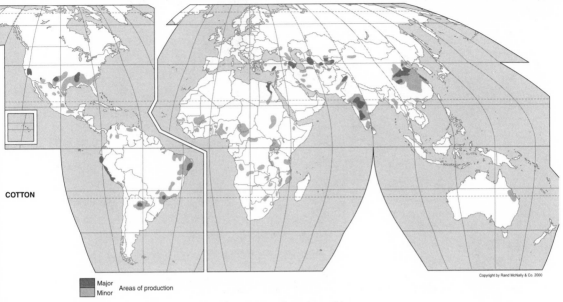

COTTON

Major
Minor   Areas of production

Copyright by Rand McNally & Co. 2000

COTTON *Gossypium hirsutum*

PRODUCTION		INTERNATIONAL TRADE	
Area under Production	32.7 million ha	Share of World Production	26%
Global Production	54.6 million MT (Seed)	Exports	7,663 million MT
	19.1 million MT (Lint)	Average Price	$1,020 per MT
Average Productivity	1,670 kg of seed/ha	Value	$7,818 million
	584 kg of lint/ha		
Producer Price	$616 per MT		
Producer Production Value	$33,644 million		

PRINCIPAL PRODUCING COUNTRIES/BLOCS (by weight)	China, United States, Pakistan, India, Uzbekistan, Turkey
PRINCIPAL EXPORTING COUNTRIES/BLOCS (of cotton lint)	Uzbekistan, Australia, United States, China, Greece
PRINCIPAL IMPORTING COUNTRIES/BLOCS (of cotton lint)	Turkey, Indonesia, Mexico, Thailand
MAJOR ENVIRONMENTAL IMPACTS	Habitat conversion Soil erosion and degradation Agrochemical use Water use and contamination
POTENTIAL TO IMPROVE	Poor Organic cotton exists but does not address water and some other sustainability issues BMPs have been identified, but reducing overall water and toxic chemical use will be difficult Genetic modification offers potential to reduce agrochemical use, but may cause other impacts if introduced

Source: FAO 2002. All data for 2000.

Total Lint Production (Million MT)

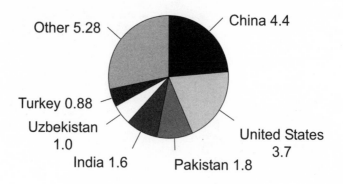

China 4.4

Other 5.28

Turkey 0.88

Uzbekistan 1.0

India 1.6

Pakistan 1.8

United States 3.7

OVERVIEW

The use of cotton has been dated to 3000 B.C. The word cotton is derived from the Arabic *qutton* or *kutn*, meaning the plant found in conquered lands, which refers to Alexander the Great's conquest of India. Cotton requires 180 frost-free days per crop. As a result, it is produced between 36 degrees south latitude and 46 degrees north latitude in tropical and subtropical climates.

Cotton achieved true "commodity" status in 1753 when Carolina cotton was listed on the London exchange. By 1861 cotton had become the single most important crop traded in the world, and more than 80 percent of it was grown in the southern United States. The surge in demand for cotton came from the industrial revolution, in particular from the expansion of the textile industry and the change from wool to cotton.

In the past several annual and perennial cultivars of cotton were grown. Each was adapted to different growing conditions and produced cotton of different-length fibers and natural colors. Over time the trend has been to breed whiter cotton with more and longer fiber. Most perennial, or tree, species of cotton have been abandoned because they cannot be produced or picked by machine, even though their long fiber is highly sought after. More recently some producers have begun to revive cotton varieties that have natural colors other than white to eliminate the dying process. Others have begun to produce organic cotton. Neither of these trends represents a significant share of either local or global markets.

Cotton is the largest money-making nonfood crop produced in the world. Its production and processing provide some or all of the cash income of over 250 million people worldwide, and employ almost 7 percent of all labor in developing countries. Nearly all activities associated with cotton production, processing, and manufacturing are becoming more concentrated in the hands of fewer companies and fewer countries. Cotton textiles constitute approximately half of all textiles (Banuri 1999).

PRODUCING COUNTRIES

Cotton is grown on farms in more than 100 countries. India (8.6 million hectares), the United States (5.3 million hectares), China (4.0 million hectares), Pakistan (2.9 million hectares), and Uzbekistan (1.4 million hectares) lead all countries with 68 percent of the world's total area planted to cotton. In 2000 the world production of cotton was 19.1 million metric tons of lint (FAO 2002).

Approximately 2.5 percent of the world's arable land is used to grow cotton (Banuri 1999). The total area devoted to the crop has changed little since the 1930s, but overall production has tripled and there have been significant shifts in where the production takes place (Soth 1999). For example, in the United States overall declines in area planted during the past seventy years were offset by increases in production.

Globally, yields averaged 1,670 kilograms of seed or 584 kilograms of lint per hectare in 2000. The most efficient cotton producers were Israel (which led all producers with average yields of 3,827 kilograms per hectare) followed by Syria, Mexico, and Spain, each of which produced at about double the global average per hectare. The major producers of cotton by weight (not area) are China, the United States, Pakistan, India, Uzbekistan, and Turkey (FAO 2002). About two-thirds of all cotton is now produced in less-developed countries, with China the biggest producer by far. In the 1980s, several African countries increased their production of cotton. For several of these, cotton ranks in the top two exports by value, as shown in Table 12.1. World exports of cotton lint are dominated by Uzbekistan, Australia, the United States, China, and Greece (FAO 2002).

TABLE 12.1. Cotton's Ranking of Total Exports by Value for Selected Countries, 1990–91

Leading Export	Second Largest Export	Third Largest Export
Benin	Sudan	Central African Republic
Burkina Faso	Togo	Egypt (fiber, yarn)
Chad	Zambia	Madagascar
Mali	Zimbabwe	Paraguay
Pakistan (yarn, cotton, fabric)		Syria
Uzbekistan		Tajikistan

Source: UNCTAD 1994.

CONSUMING COUNTRIES

Nearly two-thirds of all raw cotton production is used in domestic manufacturing. The remainder is exported. Globally, the main cotton-consuming countries are China, the United States, India, Pakistan, and the European Union. Global imports are dominated by Turkey, Indonesia, Mexico, Thailand, and China. In each of these countries, cotton is used primarily for textile manufacturing.

Cotton consumption in developed countries had declined to approximately 35 percent of overall fiber consumption by the early 1980s. After that point, cotton began to be seen as a nonsynthetic, comfortable, natural alternative to many other fibers. Within ten years, cotton had increased again to nearly half of all fiber consumption in developed countries. Conversely, as cotton use increased in developed countries, its percentage of total fiber use in less-developed countries began to decline. At the same time, less-developed countries began to produce more of the world's cotton, thread, and cloth for export.

Most producers of raw cotton now undertake value-added processing to increase the overall value of the crop. That trend is reflected in the fact that the ten largest cotton-producing countries consumed 50 percent of the global cotton output in 1986 and 77 percent of an even larger volume in 1996 (FAO 1977, as cited in IISD/WWF 1997 and Banuri 1999). Thus, they "consume" the raw cotton by manufacturing it to thread or even cloth for export.

PRODUCTION SYSTEMS

Cotton takes about five months to reach maturity. It is planted in rows about 1 meter apart. Planting, weeding, and even harvesting are increasingly undertaken by machine. Fertilizers are applied regularly. In addition, a number of different pesticides are used both as preventative measures and to treat specific pest infestations.

During the growing season, cotton produces flowers that turn into green seedpods, or bolls. Fibrous seed hairs grow in the closed boll. The fibers are from 2 to 4 centimeters long. The fibers can be picked by hand or by machine; today most are picked by machine. Once picked, the fibers are passed through a ginning machine to eliminate the seeds. They are then spun into yarn and dyed and woven to create different fabrics.

During the past century, cotton production has shifted from a labor-intensive industry to a capital-intensive one as machinery and chemical inputs have been substituted for labor. This has even occurred in developing countries. Even so, in countries like Pakistan, most cotton is still grown on small farms of less than 1 hectare.

Most of the increase in productivity has resulted from genetic improvements and green revolution technologies. The overall goal of genetic modification is to improve the basic characteristics of the plant. Through selective breeding programs, cotton now has enhanced fiber strength and fiber length, and a broader geographic range of production. Genetic work on cotton has also focused on insect- and disease-resistant cultivars. For example, work is being undertaken to change the shape and size of leaves that provide the nutrients for insects. Researchers are also attempting to increase the speed at which cotton ripens to limit the plant's vulnerability to insects and other pests as well as its overall water and input requirements. This work also extends the range over which cotton can be produced profitably.

Finally, through gene splicing, breeders are introducing insect-repellent genes into plants such as cotton (Banuri 1999). Recently, at least two transgenic cultivars of Bt (*Bacillus thuringiensis*) cotton have been developed. These cotton cultivars, produced in both the U.S. and China, produce low levels of insecticides that deter specific insect pests that attack cotton. Other cultivars of cotton are being developed that are fungicide-, herbicide- and pesticide-tolerant. Because pesticide applications are targeted to address specific pests, such varieties are intended to reduce overall pesticide use. Transgenic Bt cotton now accounts for most cotton area in the United States, and, globally, genetically modified cotton is one of the more widely planted genetically modified organisms (GMOs) at this time (Osgood 2002). The environmental impacts, positive or negative, of GM cotton are not yet well understood. However, one of the concerns about Bt cotton is that it produces and releases low levels of pesticides which may create resistance much the same way overuse of antibiotics does.

Cotton requires a substantial amount of water during the growing cycle. However, it is also very sensitive to rain (either rain that is excessive or that occurs when the cotton bolls are maturing) and humidity, which can encourage diseases. Consequently, most cotton is produced in more arid lands where humidity is not an issue and where water can be provided by irrigation as needed. Cotton is irrigated on 53 percent of all land where it is cultivated. More importantly, 73 percent of all cotton is produced on irrigated land.

To reduce insect infestation and to maintain soil nutrients, cotton is often rotated with other crops. If cotton is grown continuously on the same land, pest populations build up and agrochemicals must be used to control them. Cotton plants are susceptible to a large variety of pests and diseases that can lead to stunted growth, poor color, lower yields, and even the death of the plant. Cotton's main insect pests are bollworms, budworms, leaf worms, and weevils. Traditional pest-control methods were labor-intensive and included hand-picking pests, intercropping, crop rotation, and burning infected residues. Over the last 100 years, most of these methods have been abandoned in favor of chemical pesticides (Banuri 1999). The value of pesticide use in cotton alone is estimated at U.S.$2 to $3 billion annually. This is a significant proportion of production costs and is close to 10 percent of the annual value of the crop (Murray 1994, as cited in Banuri 1999).

PROCESSING

Cotton has several uses. In addition to the longer fiber that is used for thread and textiles, shorter cotton fibers (or linters) are used for cotton balls, tea bags, paper, or stuffing for sofas. The seed, which is 60 percent of the harvest by weight but only 10 to 25 percent of the value, is pressed to extract the oil. Cottonseed oil is used as vegetable oil and in margarine and other foods. The solid remainder is called cotton-seed cake and is used for cattle feed. Cottonseed, cottonseed oil, and cottonseed cake production are dominated by China, the United States, the former USSR, and India (FAO 2002).

A wide range of cotton products are exported from most cotton-producing countries. These range from cotton lint to other manufactured items. Table 12.2 outlines the cotton exports from Pakistan, the world's fifth largest cotton producer.

Cotton has one of the greatest environmental impacts of all agricultural commodities during its processing. The water and energy requirements during the processing and manufacturing of cotton textiles are tremendous. It can take up to 200 liters of water to produce, dye, and finish one kilogram of textiles (EPA 1996, as cited in Center for Design 2001). Globally, the textile industry is estimated to use 378 billion liters (100 billion gallons) of water each year. While these figures include all types of textiles, nearly half of all textiles are cotton and it is safe, therefore, to conclude that cotton uses a significant amount of water. Wastewater from textile production is often difficult to treat as it contains high concentrations of color, BOD, total organic carbon, dissolved solids and high content of toxic metals (chromium, copper, cobalt, lead, zinc, etc.) (Parekh 2003). An estimated 10 to 15 percent of 700,000 metric tons of dye is released globally each year in the effluent. In the United States, each surveyed textile factory in 1989 produced an average of 1,100 metric tons of solid waste each year (American Manufacturers Institute 1989, as cited in PPRIC 2003) with annual estimates for the industry in excess of 1,000,000 metric tons of solid waste each year. The industry also uses a tremendous amount of energy. One textile processing plant in Bulgaria uses 4,800 metric tons per year of heavy fuel oil and 7,300 MWh per year of electricity (Galatex 2003).

The manufacture of cotton textiles also has tremendous impacts through the use and flushing of dyes. It has been said, for example, that you can see what the next year's trends in clothing colors will be by looking at Hong Kong's harbor. Wastewater from dying can vary in chemical composition, making treatment difficult. In fact, one of the reasons the dying industry has largely moved out of the United States and Europe is because of these countries' stricter regulations for wastewater treatment.

TABLE 12.2. Exports of Cotton and Cotton Products from Pakistan, 1995–96

Item	Value (billions of rupees*)	Percentage of Total Production
Cotton lint	19.44	13.2
Cotton waste	0.17	0.1
Cotton yarn	54.05	36.8
Cotton cloth	43.28	29.5
Specialty items	2.14	1.5
Garments	27.64	18.8
Total cotton sector	146.73	100.0
Total exports	294.74	

Source: Banuri 1999.
* In 1995, U.S.$1 equaled approximately 34.3 Pakistan rupees.

Prior to 1750, more than three-quarters of all textiles were made of wool, about a fifth were made from flax, and the rest, just 4 percent in the year 1700, were made of cotton. Over the next century and a half, cotton came to dominate textiles globally, accounting for more than 85 percent of world fiber consumption by 1900. By 1999, it was the source of 48 percent of global textile production (Soth 1999).

The importance of cotton has declined during this century because of the increasing production and trade of cotton substitutes. Today cotton is around 48 percent of worldwide fiber use, while synthetics make up about 45 percent. The other fiber commodities include flax and wool as well as fibers derived from oil or wood pulp. For a number of uses these alternate fibers are good substitutes for cotton. Even so, overall fiber consumption is increasing and the production and use of cotton are increasing as well.

MARKET CHAIN

The cotton market chain can be divided into three different areas of activities — production, processing, and marketing. Each of the three areas is dominated by different players. Production tends to be controlled, albeit increasingly ineffectively, by government. In many countries the government controls research, extension, input supply (both what is allowed and its availability in some countries), and credit. Of course in some countries, like the United States, government subsidies are also important. On the other end of the market chain, apparel manufacture is controlled by the large retail chains that buy the clothes, representing a classic buyer-driven commodity chain. In the middle of the chain, however, thread, yarn and cloth manufacture are undertaken by a wide range of players that are not well organized or controlled either by the private sector or by government. Companies involved in yarn production, through manufacturers' associations, could lobby to influence government policies, but government does not comprehensively address this segment of the market chain in many countries.

Information about the market chain for cotton is hard to obtain. However, it appears that fifteen major trading companies dominate the market. These privately held companies are estimated to control between 85 and 90 percent of internationally traded cotton. There is a general suspicion that they use their influence on trade and price policies in ways that are detrimental to growers. For example, traders have been charged with using their control of traded cotton to convince mills not to buy directly from growers but rather to buy from established merchants (Morris 1991).

There is great variation in production even in a single country, as an analysis of the market chain in Pakistan illustrates. In that country, for example, there is a great deal of variation in size of production units; formality of the contractual connections between producers, processors, and the rest of the market; the nature of competition; and underlying cultural and governance systems. There are 1.3 million cotton farms, of which roughly half are smaller than 2 hectares (Banuri 1999). A vast

majority of cotton farms are operated as family farms by owners or tenants with limited literacy and access to technology. The main determinant of technological
change is the government, through its rather ineffective extension services.

At the other extreme are large-scale textile processors and small-scale garment
manufacturers, both influenced directly and indirectly by international corporations
that are clearly part of a much more formal and organized market chain. In the middle, in Pakistan, are large-scale spinning units and small-scale, informal weaving
units. The latter number in the tens of thousands, mostly operating as family enterprises, with virtually no governance (Banuri 1999).

MARKET TRENDS

Between 1961 and 2000 global cotton production increased by 100 percent while
trade increased by only 33 percent. This implies a relative increase in the use of domestic cotton in producer countries for value-added production at the local level.
During the same period, prices for raw cotton declined by 58.9 percent (FAO 2002).

One of the biggest and most influential players in the cotton market is China. In
1998 China single-handedly exacerbated already falling prices by putting more than
200,000 metric tons of cotton on the market. The country's stated goal was to buy
2 million metric tons of soybeans. The government's ultimate goal appears to have
been to redirect producers away from cotton area and toward grains, food crops, and
soybeans. Through this conscious market manipulation, more than 600,000
hectares were transferred from cotton to the production of other crops. Over three
growing seasons, the total land planted to cotton declined by 13 percent. However,
when China's stocks have worked their way through the textile industry, there is likely to be an increase in cotton prices, which will stimulate production in other countries.

The clear signals that should be sent by such events are often clouded in the reality of cotton trading. For example, trading is complicated in developing countries
where a large number of traders serve as intermediaries between producers and
processors. It is not in traders' interest to be candid and give producers complete or
up-to-date information about potential increases in cotton prices.

The cotton market was affected in the late 1990s by three different and unrelated
international financial crises: those in Southeast Asia, Russia, and Brazil. The Southeast Asian financial crisis had perhaps the largest impact on cotton markets. In
1998–99, Thai cotton spinners reduced their purchases by 35 percent, Taiwan cut its
purchases by 15 percent, the Philippines by 10 percent, and Indonesia by 40 percent. In the end, Indonesia's cotton demand fell by 23 percent to 350,000 metric
tons. Due to the uncertainty of economic recovery and the high cost of working capital, most Asian companies were unwilling to hold large stocks of cotton. Just-in-time
delivery became the norm. Only top-of-the-line cottons were purchased on a longer-
term basis (UNCTAD 1999).

The financial crisis in Russia also had a big impact on global cotton markets. In a
matter of months, demand decreased from 200,000 metric tons to 75,000 metric

tons. This in turn freed up 125,000 metric tons of cotton from Central Asia for the international market, which further depressed prices (UNCTAD 1999).

The response in Brazil was a little different. The Brazilian financial crisis brought the already depressed cotton market to its knees. The overall level of imports remained the same in weight and volume, but the grade and quality of cotton purchased deteriorated. Furthermore, Brazilian textile makers shifted their purchases to countries that are known to produce cheaper, inferior cotton (UNCTAD 1999).

These events raised spot prices for cotton above New York futures prices. Most manufacturers were only buying what they needed at the time or to fill orders that were in hand. Nothing was being done on speculation. The expectation was that prices would fall even further, so buyers were reluctant to enter the market. Ironically, even though manufacturers had less working capital, their expenditures for cotton went up when there should have been more cotton on the market. The increase in prices did not go to the producer, however. Instead it went to those buyers and distributors in developed countries who had access to working capital and who could hold product until it was needed. This situation lasted for most of the 1997–98 season (UNCTAD 1999).

There is another major factor that could affect the price of cotton globally and, consequently, where cotton is produced. The issue of concern to most people in the cotton industry is the strength of the Chinese currency. If the Chinese yuan were devalued, millions of spools of cotton yarn held in China would then become competitive on the global market. This would tend to push the price down. At that time, those holding higher-valued inventories would suddenly find their position eroded. This would push prices even lower. These factors could quickly cause prices paid to producers to plummet.

Agricultural policies also have an impact on cotton production and global market trends. In some countries, cotton is a strategic crop (a crop that is deemed to be extremely important to a country's security, for example, in the USSR where cottonseed oil was important for lubricating weapons). In others, agricultural policy is beginning to shift away from cotton and towards food production. These later shifts have led to declines in overall cotton production. Elsewhere, increasing water scarcity and tighter regulations of water management have reduced the availability of water for irrigated cotton production. As a consequence, some cotton producers may now be more interested in improved or more efficient water management systems for cotton. This could stimulate the identification and adoption of better management practices, but it is also likely to result in higher-priced cotton.

Declining commodity prices, increased energy costs, and uncertain economic conditions in many parts of the world have contributed to overall stagnation of cotton prices and consequently of production. For example, the real costs of irrigation projects have doubled in the last twenty-five years. One estimate suggests that in the 1980s alone real costs for irrigation rose between 70 percent and 116 percent (Serageldin 1996, as cited in Dinar 1998). Any technological changes that reduce the cost of irrigation or increase production from it, however, could make cotton and other irrigated crops more competitive. This would tend to expand irrigated areas as well as their environmental impacts. If anything, the increasing scarcity of fresh water on a global scale is likely to make producers invest in higher-priced, more efficient management in order to stay in business at all.

ENVIRONMENTAL IMPACTS OF PRODUCTION

While habitat conversion is a problem associated with cotton production, the most important production impacts are the use of agrochemicals (especially pesticides) and water. The quality of soil and water and the impact on biodiversity in and downstream from the fields are also major concerns. Finally, because of the high use of pesticides there are a number of human health concerns, both for farm workers and for nearby and downstream populations.

On the processing and manufacturing side, the use of industrial chemicals is of concern, especially those associated with dyeing textiles and finishing clothes. These chemicals affect not only the environment but also workers in the processing and apparel industries. Of particular concern is the use of carcinogenic dyes and chemicals, especially azo dyes.

At the producer level, the main environmental impacts from cotton production in order of importance include use of agrochemicals, water use, soil erosion and degradation, freshwater contamination, and habitat conversion and the associated loss of biodiversity. Each is discussed separately below.

Use of Agrochemicals

When produced with conventional agricultural practices, cotton generally requires the use of substantial amounts of fertilizers and pesticides. Globally, cotton accounts for 11 percent of all pesticides used each year, even though the area of production is only 2.4 percent of the world's arable land. With regard to the subset of insecticides, cotton producers use 25 percent of all insecticides used each year. In developing countries, estimates suggest that half of the total pesticides used on all crops are applied to cotton. Forty-six insecticides and acaricides (compounds used to control mites and ticks) comprise 90 percent of the total volume of all pesticides used on cotton. Five of these are classified as extremely hazardous, eight as highly hazardous, and twenty are moderately hazardous (Soth 1999).

The use of pesticides poses health risks to workers; to organisms in the soil; to migratory species such as insects, birds, and mammals; and to downstream freshwater species. Research on the cause of fish deaths in the United States showed that pesticides, even used with the proper application, harm freshwater ecosystems. Endosulfan is a pesticide that is classified as highly toxic. In August 1995 endosulfan-contaminated runoff from cotton fields in Alabama resulted in the death of more than 240,000 fish along a 25-kilometer stretch of river (PANUPS 1996). In another instance, gulls in Texas were killed 3 miles from cotton fields where parathion was sprayed when they ate insects that had been poisoned. Studies have estimated the human impact from pesticides used on cotton to be as high as 20,000 people killed

Runoff from cotton fields contaminates rivers, lakes, and wetlands with suspended solids, pesticides, fertilizers, and salts. These pollutants can affect biodiversity directly due to their toxicity or indirectly through long-term accumulation.

Underground aquifers can also be contaminated with chemicals, pesticides, or salts from cotton production. This draws into question any potential future uses of the water.

Habitat Conversion

Much of the land used to cultivate cotton has been in production for generations. This is true of areas in China, the United States, Egypt, Pakistan, India, and Brazil. However, other areas have been converted rather recently. The Pacific coastal plain from Mexico to Panama, for example, was converted from natural cover and slash-and-burn/fallow cultivation systems to permanent agriculture after 1950. By the late 1970s 400,000 hectares of Central American cotton fields were producing over a million bales of cotton annually, making it the third largest cotton-producing region after North America and the former Soviet Union. Virtually all the hardwood forests there were destroyed as were coastal savannas, evergreen forests, and coastal mangrove swamps. Only 2 percent of the original forests in the Central American cotton-production areas remain. As a result of labor concerns in the 1970s and declining yields (even with increased chemical inputs) in the 1980s, much of this area was converted from cotton production to pasture and beef production.

Cotton can indirectly cause the conversion of habitat as well. For example, the construction of dams to create reservoirs for irrigation water supplies can destroy considerable areas of riverine habitat and the species it supports as well as migratory species within river systems. In addition, the mechanization of cotton production, and its subsequent abandonment, in Central America displaced considerable numbers of landless laborers who then moved into highland, forested areas where they cleared land to produce subsistence crops.

BETTER MANAGEMENT PRACTICES

The current production of cotton is not only environmentally unsustainable, it undermines the necessary conditions for future cotton production. A tremendous amount of work will be required to bring cotton production into line with even minimally acceptable environmental standards. The strategy then must be to focus on reducing the most significant impacts. Toward this end, the overall goal of a conservation strategy for cotton should be to promote the sustainable production and use of cotton by minimizing the impacts of overall water withdrawal as well as pollution of freshwater ecosystems from cotton production (Soth 1999). Measuring the impact on freshwater ecosystems could serve as a useful evaluation for the adoption of better practices. For example, impacts on fresh water will be reduced if less water is taken from rivers during key times, if fewer agrochemicals are used (because of more effective targeting), and if less soil is lost from erosion.

In order to evaluate improvements, it is important to have specific, measurable targets both for the environmental impacts of production and the percentage of cotton that is produced using improved techniques. Some of the techniques will be the application of advanced irrigation technology and the use of more ecologically sound growing methods, such as organic farming or integrated pest management (IPM).

For farmers, the interest in sustainable cotton is direct. They stand to save water resources, maintain soil quality, maintain present and future incomes, and reduce health problems. It is also quite likely that they will actually save money by reducing expenditures for pesticides and other inputs.

For the rest of the cotton market chain, there is also direct interest in sustainable cotton production. Every business that buys and uses cotton—from yarn makers to weavers, textile manufacturers, and retail clothing stores—has an interest in a stable, sustainable supply of cotton.

The issue, then, is how to promote more sustainable cotton production within the overall constraints of the current regulatory structure as well as the overall cotton market chain. Producers in different parts of the world do not have to comply with the same regulations and consequently have different production costs. Furthermore, any additional regulatory changes in one country could put those producers at a disadvantage vis-à-vis unregulated producers in other countries. Changes to make production more sustainable also cost money in up-front investments. In addition, individual actors in a production chain respond to changing incentive structures which are often linked to overall governance. In the absence of effective governance, transition costs will be inequitably distributed and will vary for the different players in the market chain (e.g., they will not be the same for producers as for manufacturers, etc.). With cotton, the higher costs of the transition to sustainability appear to fall disproportionately on the producers and manufacturers. However, the benefits are more likely to accrue to mass retailers who have a comparative advantage (in labeling, packaging, advertising, and possibly even certification) in creating and taking advantage of consumer interest but relatively few actual costs in changing production systems and reducing impacts on the ground (Banuri 1999).

Switching to production systems that reduce environmental impacts will be impossible without the development of clean technologies. Alternative technologies exist for processing and are in the experimental stage of production, for example new technologies and management practices for organic and "green" cotton production. While technologies exist at the conceptual stage that have been implemented with specific producers (at least for IPM), the dissemination and application have not yet been undertaken, so the full range of feasible options is not yet identified (Banuri 1999).

Banuri (1999) suggests that any feasible program to encourage sustainable cotton production must intervene in the existing governance systems. These systems need either to be strengthened to facilitate the transition or to be transformed through investments and environmentally based investment screens, technical assistance, environmental certification programs or buyer screens, or government regulatory programs based upon better management practices. Whenever possible, these different approaches should be structured to send the same or complementary signals to the market chain so that they reinforce each other and increase the likelihood of success for each.

Most initiatives to reduce the environmental impacts of cotton production are not producer-neutral (they do not affect all producers equally). They will all tend to favor those participants with the strongest, most dominant positions in the market chain. This includes those who have adopted more modern production approaches; who operate at a relatively large scale; who have preferential access to credit, technology, and/or government resources; who have a near-monopoly over a portion of the market chain; or who are able to take their profits first from the consumer dollar (Banuri 1999).

Encourage Organic Production

Organic methods produce cotton without the use of synthetic fertilizers and pesticides. Instead they depend on natural processes to increase yields and disease resistance, partly through enhancing soil quality. Organic production is also the only internationally recognized, independently assessed certification or label for cotton production (Banuri 1999). By 1993 organic production was estimated at between 6,000 and 8,000 metric tons, or less than .04 percent of total global output for cotton. Some 75 percent of all production was in the United States.

Many, but not all, of the main environmental problems from cotton production could be addressed by switching to organic production. Organic standards for cotton have already been established and are available for review. However, organic standards do not set limits on the water that can be used to grow the crop, and this is the main problem with current cotton production. The water issue must be addressed to make organic cotton sustainable. In addition, while synthetic chemicals are not allowed in organic production, naturally occurring ones are. What this means is that a number of pesticides that include copper are allowed, even though they are toxic to soil organisms and other nontarget species.

There is also some evidence that organic cotton might not produce the volume of product that is desired for a wide range of reasons. In the United States, for example, interest in organic and naturally colored cotton in the late 1980s and early 1990s stimulated the establishment of whole new companies, product lines, and the on-farm certification of several producers. In the end, after several years of stable or in some cases increasing production levels, production began to decline (even with crop rotation) and prices increased dramatically. At this time it is not clear why these declines occurred or what it would take to correct them.

In the United States, keeping organic cotton production segregated was not a major problem at the farm level or even when the cotton was sold. However, keeping the cotton segregated throughout the different processing activities from ginning to spinning and weaving operations proved to be very difficult and expensive. All non-organic cotton had to be cleaned out of the operations. Because of the huge scale required to make these operations competitive within a global economy, it would be very costly and time-consuming to clean them out between runs of cotton that need to be segregated. There simply was not sufficient organic cotton to keep separate processing facilities in operation. As a consequence, the cost of spinning and weaving organic cotton was much more expensive than conventional cotton, and most manufacturers wanted nothing to do with it.

Even though organic cotton sales have declined, it appears that there is still consumer interest in the product. Such production can be encouraged through the purchasing policies of manufacturers and retailers that wish to be proactive; they can decide to give preference to organic cotton, and pay a premium for it, or only purchase organic products. Raw cotton is a tiny percentage of the cost of cotton textiles. Costs could be kept down by targeting producers in less developed countries where labor can be substituted for chemicals. Smaller spinning and weaving operations could be dedicated to organic cotton as production grows. Cutting-edge companies with high mark-ups might be willing to work on this approach, but it is doubtful. It is much more likely that interest in organic cotton will only grow considerably if a really large company decides to make a commitment to organic cotton that would stimulate the market accordingly.

Reduce Water Use

In general, improved irrigation systems and water management could reduce water losses to 15 percent or less from current levels of 60 percent on average (Ait Kadi 1993, as cited in Kirda 1999). In Israel, for example, water shortages have led to the development of very efficient drip irrigation systems. In such production systems, the total water used to produce a kilogram of cotton is far less than the 7,000 to 29,000 liters of water required to produce a kilogram of cotton with conventional means. Furthermore, drip irrigation systems produce the highest cotton yields of any cotton production systems in the world. Today, however, only 0.7 percent of irrigated areas globally use drip technology because of its high costs (Soth 1999).

Improved cultivation techniques also reduce water use. For example, conservation tillage reduces overall water use because crop residues are left on top of the soil, allowing them to act as water-conserving mulch. In Brazil a number of producers report that corn is grown in rotation with cotton and other crops because it provides more mulch. Similarly, pasture grasses are planted at the same time as corn, between the rows, to provide more biomass that will act as mulch and through their root systems help to build up the organic matter in the soil. Careful crop rotations reduce the need for pesticides and fungicides in addition to reducing water use.

Integrated pest management (IPM) for cotton builds on practices that farmers have used for centuries. These include adopting cultivars that are resistant to pests, altering the time of sowing and harvest to minimize exposure to pests, cultivating to reduce weeds, and removing crop residues. Pesticide use can be reduced by carefully monitoring pest levels and by targeting applications. The least toxic pesticide is chosen whenever possible; botanical pesticides such as neem and various tobacco extracts are also used. IPM reduces pests to "economically manageable" levels rather than aiming for complete eradication. Cultural practices such as crop rotation and intercropping are used to help keep down pest populations. Physical controls such as hand-killing pests and using pheromones to trap pests are also employed when possible so that fewer toxic chemicals are needed (Banuri 1999). IPM, however, does allow the use of standard chemical controls when necessary.

One study in India found that IPM resulted in higher cotton yields and a 28 percent decline of unit costs (Kishor 1992, as cited in de Vries 1995 and Banuri 1999). In short, using IPM for cotton has been found to be economically and environmentally beneficial. An added benefit is that IPM generates more employment. Still, most studies about the impacts for cotton are qualitative rather than quantitative, and few long-term studies have been undertaken. Applications of IPM for cotton have not worked on a broader scale (Banuri 1999), but there has also been little systematic attempt to apply the research on IPM that has been undertaken to date.

Rework Subsidies to Promote Conservation

Current U.S. price-support subsidies for cotton production account for as much as half of the income the 25,000 American cotton growers receive for their crop. This program insures that American cotton growers receive $0.70 per pound for their cotton, when the world price is only $0.40. Such subsidies have a direct impact on cotton production throughout the world. At the very least, they squeeze producers in countries that cannot subsidize production (or at least cannot subsidize it as much), forcing them to cut corners. Thus, U.S. government subsidies are matched by environmental subsidies in many less developed countries where producers are forced to cut corners to reduce their costs in order to compete with subsidized cotton production.

While subsidies may be inevitable, they should be used to achieve concrete conservation results. They could be used, for example, to retire the least productive lands or to require the adoption and use of improved practices, such as more efficient irrigation. They could also be used to wean producers from the use of the most toxic chemicals and reduce chemical use over time by subsidizing a switch to integrated pest management.

In a carrot-and-stick approach, policies could be developed to address pollution, toxic chemical use, water use and effluent issues. "Pollution taxes" could complement the subsidy approach described above while helping governments address the nonpoint-source pollution (the cumulative impact of cotton production in a region with many producers) caused by cotton and other forms of agricultural production. These types of policies would tend to push cotton producers and those who work with them to identify, refine, and adopt better management practices such as those described in this section.

OUTLOOK

Humans need fiber for clothing and other products. Other sources of plant-based fiber such as hemp, sisal, flax, and wood pulp do not at this time appear to be viable alternatives to cotton, either in terms of the quantity of material that could be produced or the overall environmental impact of production. Likewise, synthetic fibers do not appear to be viable alternatives, as their production (mostly from petrochemicals) also raises serious environmental issues.

Cotton seems to be unavoidable. However, no matter what the advertisers say, there is nothing "natural" about cotton. It uses too much water, too many pesticides, and produces too much pollution. The environmental impacts of cotton production must be reduced. The question is how. Producers have little slack, since most of the profits from cotton are not made at the producer end. However, one place to start may be with the fact that many of the ways to improve cotton production are more labor-intensive. Therefore, one of the best ways to start making cotton more sustainable is to eliminate subsidies and market barriers; this would promote production in those countries where labor costs are low enough to make possible the adoption of labor-intensive practices.

Another way to reduce cotton's impact significantly would be to begin to charge for pollution. Cotton production is one of the biggest sources of agricultural pollution, as it is the largest user of toxic chemicals in agriculture. Pollution is the Achilles' heel of the industry. However, cotton interests are entrenched in both developed and less developed countries; it was cotton and rice interests in California that ultimately created the political momentum to push through the U.S. farm bill in 2002. This will be one of the more difficult industries to change.

REFERENCES

Allen, W. 1994. Sustainable Cotton Production: A Niche Market or a Must Market? *Beltwide Cotton Conferences*. Vol. 1: 410–412. Meeting held January 5–8, San Diego, CA.

Banuri, T. 1999. Pakistan: Environmental Impact of Cotton Production and Trade. Prepared for United Nations Environment Programme by the Sustainable Development Policy Institute, Islamabad, and the Institute for Environmental Studies, Amsterdam. Draft.

Centre for Design. 2001. *Aiming for Sustainable Product Development: Textiles*. Melbourne, Australia: Centre for Design at RMIT University. Available at http://www.ecorecycle.vic.gov.au/asset/1/upload/Aiming_for_Sustainable_Product_Development_Textiles_(2001).pdf.

Cheeseright, P. 1994. Business and the Environment: A Costly Colour Run—Textile Dye Pollution is Turning Some of Britain's Rivers Purple. London: *The Financial Times*. May 11, p. 22.

de Vries, H. 1995. *An International Commodity Related Environmental Agreement for Cotton: An Appraisal.* Amsterdam: Vrije Universiteit, ICREA Research Team.

Dinar, A. 1998. Irrigated Agriculture and the Environment—Problems and Issues in Water Policy. In *Sustainable Management of Water in Agriculture: Issues and Policies.* Paris: OECD, p. 41–56.

EPA (United States Environmental Protection Agency). 1996. *Best Management Practices for Pollution Prevention in the Textile Industry,* RPA/625/R-96/004. September. Washington, D.C.: United States Environment Protection Authority, Office of Research and Development.

FAO (Food and Agriculture Organization of the United Nations). 2002. *FAOSTAT Statistics Database.* Rome: UN Food and Agriculture Organization. Available at http://apps.fao.org.

Galatex. 2003. *Energy and Water Conservation Program at a Textile Processing Plant in Bulgaria.* Varna, Bulgaria: Galatex A.D. Available at http://www.rec.org/ecolinks/bestpractices/PDF/bulgaria_galatex.pdf. Accessed 2003.

Gillham, F. 1995. Cotton Production Prospects for the Next Decade. World Bank Technical Paper Number 287. Washington, D.C.: The World Bank.

Gleick, P. H. (ed.). 1993. *Water in Crisis: A Guide to the World's Freshwater Resources.* New York: Oxford University Press.

IISD/WWF (International Institute for Sustainable Development and World Wildlife Fund). 1997. *The Cotton Industry: Towards an Environmentally Sustainable Commodity Chain.* Report Prepared for the Workshop on Cross-National Environmental Problem-Solving. School of International and Public Affairs, Columbia University.

Kirda, C., P. Moutonnet, and D. R. Nielsen. 1999. *Crop Yield Response to Deficit Irrigation.* Dordrecht: Kluwer Academic Publishers.

Klohn, W. E., B. G. Appelgren. 1998. Challenges in the Field of Water Resource Management in Agriculture. In *Sustainable Management of Water in Agriculture: Issues and Policies.* Paris: OECD, p. 31–39.

Merme, M. 1993. The Emerging Market of Green Organic and Naturally Coloured Cotton. Sloan 25 Masters Programme, London Business School. November.

Monsanto. 1999. Fact Sheet On Pesticide Use. Originally published at http://www.biotechknowledge.com/showlib_biotech.php32, Biotech Knowledge Center, Monsanto. Now available at http://journeytoforever.org/fyi_previous2.html, dated March 17, 2000.

Morris, D. 1991. *Cotton to 1996: Pressing a Natural Advantage.* Special Report No. 2145. London: The Economist Intelligence Unit.

Murray, D. L. 1994. *Cultivating Crisis: The Human Cost of Pesticides in Latin America.* Austin: University of Texas Press.

Osgood, D. 2002. Biotechnology and Related Issues for WWF-USA. Background document prepared for World Wildlife Fund. February 20.

PANUPS. 1996. *Endosulfan Responsible for Alabama Fish Kill.* Pesticide Action Network North America Update Service.

Parekh, B. K. 2003. *A Prospectus On: A New Process for Removal of Organic Pollutants from Textile Dyeing Wastewaters.* Lexington, KY: Center for Applied Energy Research. Available at http://www.caer.uky.edu/services/propect.htm. Accessed 2003.

PPRIC (Pollution Prevention Regional Information Center). 2003. *Pollution Prevention in Textiles.* Available at http://www.p2ric.org. Accessed 2003.

Serageldin, I. 1996. Irrigation and Sustainable Development for the 21st Century. Address given at the 16th Congress of the International Commission on Irrigation and Drainage. Cairo, September 15.

Soth, J. 1999. *The Impact of Cotton on Freshwater Resources and Ecosystems—A Preliminary Synthesis.* Fact Report (draft). C. Grasser and R. Salemo, eds. Zurich: World Wildlife Fund. 14 May.

UNCTAD (United Nations Conference on Trade and Development). 1999. *World Commodity Survey, 1999–2000*. Geneva, Switzerland: UNCTAD.

———. 1994. *Handbook of International Trade and Development Statistics, 1993*. Geneva, Switzerland: UNCTAD.

USDA (U.S. Department of Agriculture). 1993. *Cotton and Wool—Situation and Outlook Report*. CWS-73. Washington, D.C.: USDA. August.

Vaissayre, M. 2001. A Basic Outline of the Insect-Related Stickiness Problem and Its Management in Cotton. In Gourlot, J. P. and R. Frydrych, eds. *Proceedings: Improvement of the Marketability of Cotton Produced in Zones Affected by Stickiness*. Lille, France, July 4–7, 2001.

Watson, J. 1991. *Textiles and the Environment*. Special Report No. 2150. London: The Economist Intelligence Unit. April.

Woodburn Associates. 1995. *The Cotton Crop and Its Agrochemical Market*. Edinburgh, UK.

WWF (World Wildlife Fund). 1999. *Workshop Materials on Cotton and Freshwater*. Minutes and Conclusions. 1–2 July, 1999 at WWF Switzerland, Zurich.

Zilberman, D. 1998. The Impact of Agriculture on Water Quality. In: *Sustainable Management of Water in Agriculture: Issues and Policies*. Paris: OECD, p. 133–149.

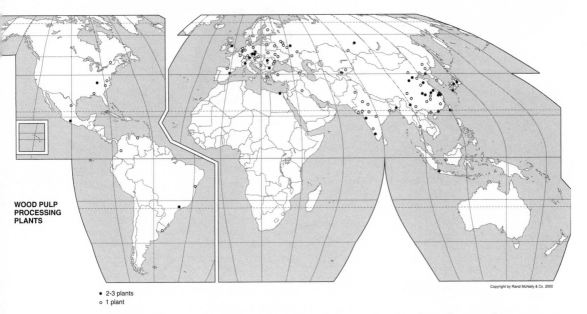

WOOD PULP PROCESSING PLANTS

- • 2-3 plants
- ○ 1 plant

Copyright by Rand McNally & Co. 2000

WOOD PULP   *Eucalyptus* species, *Acacia* species, *Casuarina* species, *Gmelina arborea*, and *Pinus* species

PRODUCTION		INTERNATIONAL TRADE	
Area under Cultivation	10 million ha	Share of World Production	12%
Global Production	323.0 million MT (Pulp)	Exports	37.8 million cubic meters
	189.0 million MT (Paper & board)	Average Price	$440–620 per MT
		Value	$20,720 million
Average Productivity	15 m³/ha/year		

PRINCIPAL PULP-PRODUCING COUNTRIES (by weight)	United States, Canada, China, Finland, Sweden, Japan, Brazil, Russia, Indonesia, Chile
PRINCIPAL PLANTATION-GROWING COUNTRIES	China, United States, Russia, India, Japan
PRINCIPAL EXPORTING COUNTRIES	Canada, United States, Brazil, Sweden, Chile, Finland, Russia, Indonesia, Portugal
PRINCIPAL IMPORTING COUNTRIES	United States, China, Germany, Italy, Japan, France
MAJOR ENVIRONMENTAL IMPACTS	Habitat conversion and degradation Pollution from agrochemical use and processing Burning during plantation establishment
POTENTIAL TO IMPROVE	Good Siting can reduce impact on biodiversity, natural habitat, and ecosystem functions Planting and management can maintain soil fertility and reduce impacts BMPs can reduce costs and increase benefits

Sources: FAO 2001a, 2002; PPI 2001; Cossalter and Pye-Smith 2003. All data for 2000.

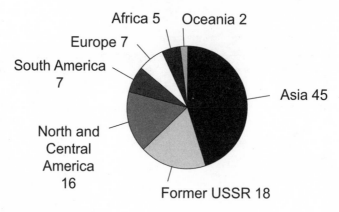

Percent of Global Tree Plantations

Africa 5   Oceania 2
Europe 7
South America 7
Asia 45
North and Central America 16
Former USSR 18

Source: Adams and Efransjah 2001.

OVERVIEW

Writing has long been associated with civilization. It is the basis for how we learn from others. The first known forms of written communication were petroglyphs and cave paintings from tens of thousands of years ago. Such communication, however, was not mobile. The Sumerians were the first to resolve this problem in about 4000 B.C. with the creation of clay tablets that, although heavy, could be transported. Since then wood, waxed boards, stone, ceramics, cloth, bark, bronze, silk, bamboo, and tree leaves have all been used to preserve the written word.

While there have been many other ways to write and communicate, paper has dominated the meteoric rise in communication and learning around the globe. The word paper is derived from papyrus (*Cyperus papyrus*), a plant found in African wetlands and particularly along the Nile River. Some 5,000 years ago the Egyptians created sheets of papyrus by peeling and then slicing papyrus stems into strips and laying them at right angles to form more sturdy mats. The strips were then pounded together and left in the sun to dry. The Egyptians, Greeks, and Romans used papyrus for record keeping, spiritual texts, and works of art. The ancient Greeks also used a kind of parchment made from animal skins to create sheets for writing, but this proved to be expensive as it took many sheep skins to produce a single copy of any major work.

Similar materials were invented independently around the world. During the second century A.D., the Mayans fashioned a similar pounded bark product for bookmaking in Central America. In the Pacific, Islanders made a paper-like bark cloth that was decorated and used for clothing and ritual objects.

The invention of paper as we know it today did not come for another three thousand years, and even then it was slow to spread around the world. Ts'ai Lin is generally credited with inventing paper in China in 104 A.D. The first paper was made by pounding the inner bark of the mulberry tree and hemp, rags, or bamboo fibers in water until they formed a fibrous, watery paste. The solution was then poured over a flat piece of coarsely woven cloth to let the water drain through. This left a thin layer of fibers on the cloth which, when dry, became sheets of paper. This substance was relatively cheap to make, lightweight, and portable, and it had a surface that was very good for writing. With the invention of paper, literature and the arts flourished in China. Equally important, paper allowed the kind of ongoing communication that made possible the governance of larger areas.

Papermaking was common in China before it spread to Vietnam, Tibet, Korea, Japan, Nepal, present-day Uzbekistan (the center of Tamerlaine's empire), Iraq, and Syria. By the tenth century Arabs had begun to substitute linen fiber for wood and bamboo. This created an even higher quality of paper. The first paper mill in Europe was built in Xátiva, Spain, by the Moors. When the European armies drove the Moors from the city in 1233, they captured the paper mill and learned the secrets of papermaking, which spread throughout Europe. Because of the Catholic Church's demand for paper, Italy quickly became the largest paper producer and exported the product throughout Europe. France and Germany followed suit, with Germany greatly improving the process and making the finest paper of the time.

Still, communication was complicated because all writing was done either by hand or by carved wood blocks. In 1450 Gutenberg invented moveable type and the printing press. This allowed documents to be reproduced quickly and in very large volumes, dramatically increasing the demand for paper.

As late as the nineteenth century paper was still made by hand. The first papermaking machine was invented in 1798, but no one would invest in the company to turn the prototype into a commercial reality. As the machinery improved, the major bottleneck was the supply of material from which to make paper. In 1850 Keller in Germany developed the first paper made from wood pulp, but the paper was of very poor quality. In 1852 Burgess perfected the use of wood pulp in papermaking in Germany by using chemicals to break down the fiber. This process was improved further so that by the end of the nineteenth century economical, high-quality mass-produced paper became a reality. Globally, paper production doubled to about 2.5 million metric tons per year by 1900. Newspaper, book, and magazine publishing flourished, as did literacy. Recently there have been many predictions that electronic media will create the paperless office; however, if anything it has resulted in increased paper use.

The shift to wood pulp as the major ingredient in papermaking, combined with the ever increasing demand for paper, put increasing pressure on forests. Forests are one of the few living natural resources that people still use in such large quantities. Forests cover 3.87 billion hectares, or 30 percent, of the Earth's land area. Only a small percentage of these forests are actually harvested commercially, however. By 2000 paper and paperboard products accounted for over 50 percent of the value of global forest product exports (FAO 2001b).

As demand for forest products, especially paper, has increased, there have been attempts to produce forest products from trees cultivated in plantations. The Food and Agriculture Organization of the United Nations (FAO) estimates that 5 percent of the world's forests are actually plantations rather than native vegetation. Plantations are tree "stands established by planting or seeding in the process of afforestation or reforestation" (FAO 2001a). Distinguishing between natural forests and tree plantations can be difficult, in terms of both calculating the amount of forest in a country and evaluating the habitat for biodiversity. Degraded native forests, stripped of economically important species, may provide no more ecosystem services than plantations. Conversely, abandoned plantations and those that are interplanted with native, long-maturing tree species such as mahogany or teak may contain considerable biodiversity and be more similar to natural forests than to other plantations.

PRODUCING COUNTRIES

According to the Food and Agriculture Organization of the United Nations (FAO), the area planted to plantations increased from 17.8 million hectares in 1980 to 43.6 million hectares in 1990 and 187 million hectares in 2000 (FAO 2001a). One-third of all tree plantations are in the tropics and two-thirds are in temperate and boreal areas. China, the United States, Russia, India, and Japan, each with more than 10 million hectares of plantations, account for more than 65 percent of global plantation area (Cossalter and Pye-Smith 2003). Few of these plantations are of fast-growing trees used for pulp.

The extent and distribution of fast-growing pulp plantations, as well as their productivity and length of crop cycle, are shown by species group in Table 13.1. By 2000, there were an estimated 10 million hectares of fast-growing plantations intended for pulp production with an additional 0.8 to 1.2 million hectares planted each year (Cossalter and Pye-Smith 2003). This trend is expected to continue well into the future. The sole purpose of these pulp plantations is to produce large volumes of uniform, small-diameter logs as quickly as possible and at competitive prices. Financially viable pulp wood plantations now are expected to yield at least 15 cubic meters of wood per hectare per year (Cossalter and Pye-Smith 2003).

As late as 1900, because of the widespread availability of natural forests, there was no real need to plant tree plantations. During the last century this began to change and prior to 1950, tree plantations were started in Europe, the United States, Australia, and New Zealand, and developing countries such as South Africa, India, Chile, Indonesia, and Brazil. In the 1950s, Japan, Korea, and China also undertook massive tree planting programs. Between 1965 and 1980, the area devoted to tropical tree plantations tripled (Cossalter and Pye-Smith 2003).

According to Adams and Efransjah (2001) Asia accounts for about 45 percent of global pulp plantations, followed by the countries of the former USSR (18 percent), North and Central America (16 percent), South America and Europe (7 percent each), Africa (5 percent), and Oceania (2 percent). In the 1990s Asia dominated the establishment of new industrial tree plantations, many of them for pulp and paper

TABLE 13.1. Fast-Growing Pulp Plantations by Species and Country

Species	Mean annual increment at an operational scale (m³/ha/year)	Time to reach maturity (years)	Estimated extent of fast-wood plantations ('000 ha)	Main producing countries (in decreasing order of importance)
Eucalyptus grandis and various eucalypt hybrids	15–40	5–15	3,700	Brazil, South Africa, Uruguay, India, Congo, Zimbabwe
Other tropical eucalypts	10–20	5–10	1,550	China, India, Thailand, Vietnam, Madagascar, Myanmar
Temperate eucalypts	5–18	10–15	1,900	Chile, Portugal, northwest Spain, Argentina, Uruguay, South Africa, Austria
Tropical acacias	15–30	7–10	1,400	Indonesia, China, Malaysia, Vietnam, India, Philippines, Thailand
Caribbean pines	8–20	10–18	300	Venezuela
Pinus patula and *P. elliottii*	15–25	15–18	100	Swaziland
Gmelina arborea	12–35	12–20	100	Costa Rica, Malaysia, Solomon Islands
Paraserianthes falcataria	15–35	12–20	200	Indonesia, Malaysia, Philippines
Poplars	11–30	7–15	900	China, India, USA, Central and Western Europe, Turkey

Source: Cossalter and Pye-Smith 2003.

production. By 1995 Asia contained 40 percent of the world's tree plantations and nearly 60 percent of all plantations established since 1985 (FAO 2001b).

Major countries with pulp plantations include Indonesia (with 4.3 million hectares thus far set aside for plantations, and 1.4 million hectares already planted), Brazil, South Africa, New Zealand, and Chile. In 1995 South African plantations produced 17.6 million cubic meters of timber, of which more than 35 percent was used for pulp. New Zealand produced the same amount of timber from plantations and used almost 30 percent for pulp (CCFM 2001). Brazil now contains the world's largest area of planted eucalyptus trees, although Australia contains more total trees as eucalyptus are native there (Mattoon 1998). Australia has an estimated 1.3 million hectares of tree plantations, and the government is proposing to triple the total area of plantations by 2020 (Ryan 2002a). Areas planted to trees for pulp are increasing in Malaysia, Vietnam, Thailand, Uruguay, Paraguay, Argentina, Venezuela, Colombia, Mexico, Congo, and Swaziland (World Rainforest Movement 1999). In temperate areas in China, Chile, Portugal, Spain, Argentina, Uruguay, and South Africa also have significant pulp plantations (Cossalter and Pye-Smith 2003).

CONSUMING COUNTRIES

In 1998 more than 40 percent of chemically produced paper-grade pulp shipments went to Western Europe, and more than 13.6 million metric tons of pulp were consumed in the countries of Western Europe (excluding Scandinavia). The United States was the second largest consumer of pulp, using over 6.4 million metric tons (PPI 2001).

The main consumers of paper products are not necessarily the same as the primary pulp consumers. The United States is by far the largest consumer of paper and paperboard products, using over 90 million metric tons in 2000. Following the United States are China (36 million metric tons), Japan (32 million metric tons), Germany (19 million metric tons), the United Kingdom (13 million metric tons), and France and Italy (11 million metric tons). Following these are Canada, Korea, and Spain, at roughly 7 million metric tons each (PPI 2001). While developing countries such as China could potentially represent large markets for paper products as standards of living rise, industrialized countries still accounted for over 75 percent of consumption in 1998. Furthermore, the FAO predicts that industrialized countries will be largely responsible for growth in consumption through 2010 (Mattoon 1998).

PRODUCTION SYSTEMS

Native forests initially supplied raw materials for the pulp and paper industry, but increasingly the industry is turning to monoculture plantations on a global scale to supply its mills. Old-growth and second-growth forests in North America's Pacific Northwest as well as Scandinavia, Chile, Indonesia, and elsewhere once served as important suppliers of raw materials to the paper industry. These regions have declined and continue to decline in importance as their forest resources have been depleted and as pressures from environmental groups and others to protect those forests have increased. The pulp and paper industry is currently shifting to other sources so that it can utilize previously unexploited natural forests; the industry is also planning for increasing reliance on industrial tree plantations. At present, industrial plantations supply an estimated 15 to 30 percent of the world supply of pulpwood (Mattoon 1998; World Rainforest Movement 1999), but the percentage is increasing rapidly.

Worldwide, plantations of fast-growing tree species are established for a number of reasons. Trees planted to provide people with fuelwood, to protect watersheds and halt soil erosion, or to offset carbon emissions can be considered "nonindustrial" plantations. By contrast, "industrial" plantations yield raw materials for commercial products, including timber for processing into wood and wood products as well as pulpwood for making paper and paper products. Plantations may be established for a single use such as pulp production, or multiple uses if trees are harvested at different ages to feed into different product streams. In addition to the timber, the byproducts (e.g., limbs, bark and live wood, cuttings, and trimmings) from industrial timber plantations as well as wild harvested wood are often fed into the pulp-manufacturing process. However, this chapter focuses only on pulpwood plantations, the industrial plantations of fast-growing tree species established to supply the pulp and paper industry with raw material.

Trees commonly grown in plantations include species of *Eucalyptus, Acacia, Casuarina, Pinus*, and *Gmelina arborea*. They are harvested on short rotations of six to nine years (pine rotations for timber are twenty to thirty years, but when grown for pulp it is harvested on shorter rotations, at least in the tropics). To give an estimate of the relative abundance in industrial plantations, the FAO (2001a) states that 10 percent of the world's 187 million hectares of plantations are eucalyptus species, 4 percent acacias, 20 percent pines, and 30 percent of all plantations were classified as "unspecified." Trees commonly grown on pulpwood plantations in Indonesia include *Acacia mangium* (which accounts for 80 percent of all plantations), *Acacia crassicarpa, Gmelina arborea*, and *Eucalyptus deglupta* (Barr 2001).

Pulp and paper plantations in the Northern Hemisphere are concentrated in Scandinavia and, to a lesser degree, in North America. Trees grown in northern regions are largely species of pine (*Pinus*), spruce (*Picea*), and fir (*Abies*) with long rotations of nearly thirty years. By contrast, more rapid tree growth in tropical climates allows tree rotations of as little as six years for eucalyptus (*Eucalyptus*) and acacia (*Acacia*). These species not only mature more rapidly, they also require less land to produce the same amount of pulp. Annual growth rates of 3 to 5 cubic meters per hectare in Canada and 10 cubic meters per hectare in the southern United States cannot compete with tropical yields of 25 cubic meters per hectare growing acacia in Indonesia and 30 to 40 cubic meters per hectare growing eucalyptus in Brazil (Mattoon 1998). Eucalyptus trees are often selected for cultivation for both their rapid growth and their pulp content; eucalyptus trees can grow as much as 30.5 meters (100 feet) in seven years. Commercial eucalyptus plantations in Brazil can produce an average of 40 to 50 cubic meters per hectare using some improved varieties. The potential, however, is thought to be as high as 100 cubic meters per hectare or greater (FAO 2001a; Strategic Environmental Associates 1996; WRI et al. 1998). Table 13.2 compares the productivity of two different forest plantation models; fast-growing eucalyptus in Brazil, and longer-rotation softwood in New Zealand (Cossalter and Pye-Smith 2003).

A number of factors combine with fast tree growth to encourage plantation establishment in tropical countries. Land is cheaper in developing countries, and fees to rent land in state forest reserves can run as low as U.S.$0.30 in Indonesia, $2.50 in Thailand, and $2.80 in Uganda (World Rainforest Movement 1999; Eraker 2000). A variety of other governmental subsidies may be available for establishing plantations, including direct subsidies, tax exemptions, special low-interest loans, and inexpensive labor. In addition, environmental regulations are often less stringent in developing countries, and as a result it is considerably easier and cheaper to establish plantations there.

Pulp is increasingly manufactured in the same countries where plantations are established. Pulp is a value-added product. Because wood represents 40 to 70 percent of the cost of making pulp, pulp produced in tropical areas is cheaper than that made in northern regions of the Northern Hemisphere (World Rainforest Movement 1999). The cost per metric ton of bleached hardwood pulp in Sweden is 185 percent higher than in Brazil (WRI et al. 1998).

Pulp is the most costly input in paper manufacturing, so less expensive pulp results in cheaper paper products. However, tropical countries still accounted for a relatively small proportion of overall global paper production. In 1999, tropical countries produced a large share of the world's raw pulp, but only 9 percent of paper and paperboard products (FAO 2001b).

Industrial tree plantations offer a number of advantages over natural forests. Plantations supply uniform raw material from a smaller area of land. The raw material from natural forests is varied, so pulp mill equipment must be able to handle different tree species of varying sizes. In Indonesia the pulpwood yields from natural mixed tropical hardwood stands are lower than those from *Acacia mangium* plantations by 50 to 150 cubic meters per hectare (Amec Simons Forest Industry Consulting 2001). Plantations are also more reliable and predictable pulp producers. They

TABLE 13.2. A Comparison of the Volume of Wood Produced in Two Plantation Models

Plantation Type	Area of operation (hectares)	Mean annual increment at an operational scale (m³/ha/year)	Time to reach maturity (years)	Wood produced per hectare (m³)
Fast-growing wood Aracruz Cellulose S.A.	180,000	43	6.5 to 7	After 4 rotations: 28 years, ± 1,000 m³
Longer-rotation softwood New Zealand average	1,650,000	20	25 to 30	After 1 rotation: 28 years, ± 560 m³

Source: Cossalter and Pye-Smith 2003.

are planted and available for harvest on regular cycles. Production can be geared toward the supplies needed to maintain the pulp mill at peak capacity.

Pulp plantations can be established under a variety of conditions. Cultivation requirements vary between the prominent species used. Pulp plantations can replace crops, grasslands, peatlands, and degraded or old-growth forests, as well as virtually any other land use. In Indonesia plantations have been established on areas of grassland, scrub or brush, peatlands, heavily logged and degraded forests, and forests in a more natural state (Amec Simons Forest Industry Consulting 2001).

Clearing methods depend on the type of vegetation on the area to be planted. Native forests are usually clear-cut prior to plantation establishment. In most instances, however, those forests are high-graded (harvested first for lumber-grade trees), and then the rest are cut and used in the pulp mills. Fire is often used in the clearing process to eliminate trees with no commercial value or trimmings left after harvesting timber or pulpwood. Bulldozing, plowing, and windrowing are intended to concentrate the remaining trees, branches, and roots into piles or windrows.

Seedlings are raised in nurseries in massive quantities. They are grown in black plastic bags filled with fertile potting soil and are normally watered by overhead spray. Seedlings are clones produced by tissue culture or other forms of vegetative propagation rather than from seeds. There is very little genetic diversity within the seedlings planted at any one time, although some variation occurs over time as the genetic stock is continuously improved. Seedlings are planted manually or mechanically, depending on the slope of the land and the amount of tree residue still in the area. Fertilizer is generally mixed into the planting holes for seedlings; soil amendments may be added as well to alter the pH. In peatlands, seedlings are sometimes planted higher than the surrounding area to avoid root damage from waterlogging.

As the trees start to grow, local plant species that sprout from seeds or roots are killed with herbicides or cleared manually or mechanically. Fertilizer is applied periodically, and pruning or thinning undertaken as needed to enhance growth. When plantation trees are harvested, they may be clear-cut or selectively felled, through either mechanical or manual labor (FAO 2001a). In general, it is more efficient to harvest all the trees in one area at a time than to cut them selectively. Some species will regenerate from their roots for as many as three cuttings before production declines to a point that it makes sense to replant the areas. Other species can only be harvested once and then must be replanted. Trees that can be harvested multiple times have lower overall production costs, everything else being equal.

Depending on the species, a production cycle can start anew after harvest without replanting. Some species of all of the main pulp plantation trees (species of *Eucalyptus, Acacia, Casuarina, Pinus,* and *Gmelina arborea*) resprout after being harvested. Aracruz Cellulose S.A. in Brazil replants eucalyptus after the second harvest (WRI et al. 1998), even though this is not always necessary.

Genetic manipulation and selective breeding have played key roles in altering productivity. Brazil's plantation industry is notable for its efforts to create genetic modifications that increase growth and/or pulp content. As a result, Brazil's eucalyptus plantations are considered the most productive in the world. Brazil's Aracruz

Cellulose S.A., for example, is considered the world's leader in genetic refinement and silvicultural activities, in addition to being the world's largest producer of bleached eucalyptus pulp (WRI et al. 1998).

Selective breeding and genetic improvements can increase pulp yields by up to 30 percent. Transportation costs are also reduced by increasing the per-hectare wood volume and cellulose content through genetic improvements. Less land is needed to supply a mill, and plantations can be located closer to pulp mills. By 1998, as much as half of the delivered cost of pulpwood at pulp mills was the cost of transportation. Any increased productivity that occurs closer to the mills reduces overall costs dramatically. Conversely, it also makes it more difficult for producers of pulp from natural forests to compete.

Wadsworth (1997) reviewed more than fifty years of forestry research about the potential deterioration of pulp yields from continual use of the same sites. Studies show that there is no significant decline in productivity for the second and third plantings of certain species (*Pinus radiata* and *P. patula*), while other species show a marked decline. Proving declining productivity is complex, and determining which factors cause that decline is even more so. For some species it is common for the second cutting to yield more than the first. Some have speculated that this is due to the benefits of an established root system; this appears to be the case for eucalyptus (Wadsworth 1997).

Wadsworth concludes that though there is clear evidence that habitat quality and soil quality of a natural tropical forest site decline after it is deforested, there is no conclusive evidence that successive plantations of timber crops inevitably result in further decline. He also writes that nutrient losses associated with repetitive harvests will eventually bring on such declines. Such declines, however, could also result from the loss of nutrient variety and a decline in the chemical composition of the litter and soil (e.g., one species of tree leading to less diverse soil, and erosion reducing the litter from the previous cover). Also, certain tree species appear to create more nutrient-rich soils. For example, soils that are created under oaks are apt to have more nutrients than those under pines (Wadsworth 1997).

While it is clear that plantations affect sites and under some conditions may cause deterioration, tree plantations have some potential for sustainable yields. The major causes of yield declines over time appear to be more related to management practices than to site deterioration. Careful harvesting techniques, the conservation of organic matter, and appropriate management of weeds and undergrowth can all minimize nutrient loss or damage to soil.

Theoretically, there should be no limit to the number of tree crops that can be produced on any given area. However, very little is known about how to produce unlimited crops. Too little research has been undertaken over a sufficient period of time to demonstrate the impacts of various production techniques. Production practices clearly affect the long-term productivity of any given area. Since companies typically own the land on which they have plantations, this motivates them to take measures to sustain the land's productivity (WRI et al. 1998).

Large-scale, industrial monoculture plantations of trees, like other forms of monoculture agriculture, face an increased risk of disease and pest problems as compared to mixed-species, diverse plantations, or natural forest stands (FAO 1999). While diseases and pests are certainly an issue, for the most part they have been relatively easily combated through careful selection of tree species and varieties plus chemical controls (McNabb 1994). The fewest disease outbreaks and highest yields are achieved when individual varieties are matched with the sites that are best suited for them. By matching traits to sites, producers can reduce their need for fertilizers, pesticides, and other inputs and activities—all of which not only have environmental impacts but also reduce profits.

Several viruses and other diseases are problematic on industrial plantations, though disease problems vary depending on the species. Various forms of heart rot, root rot, and rust diseases are the greatest danger to acacia plantations in Southeast Asia. Some diseases may do very little damage during the first planting of an area, but they can do far greater damage to subsequent crops. Another concern is that plantations of identical clones are more susceptible to diseases. By selecting varieties for desired traits and then using vegetative propagation to produce large numbers of identical trees for plantations, the natural disease resistance supplied by genetic variation is lost (Old 1997). For example, fungal diseases have been spreading through monoculture eucalyptus plantations in Vietnam (Lang 2002).

Birds, mammals, insects, and fungi can all be pests on pulp plantations. In some pine plantations in Chile, there has been a decrease in fox numbers; the resulting increase in rodents and rabbits has created pest problems (World Rainforest Movement 1999). Leaf-cutter ants have been problematic in eucalyptus plantations in Brazil, and companies now try to destroy ant nests before planting (McNabb 1994).

Well-run plantations are quite profitable. Aracruz Cellulose S.A., the world's largest producer of pulp from eucalyptus, is so efficient and has improved production of the species to such an extent that its pretax profit margins are more than 51 percent (WRI et al. 1998). Aracruz has increased production from 30 cubic meters to 45 cubic meters per hectare per year in only seven years. These yields compare favorably with Chile, where yields average only 20 cubic meters per hectare per year. When seedlings are ready for harvest, the trees are felled and trimmed and the undergrowth is knocked down. Aracruz has abandoned the use of chain saws and now uses mechanical harvesters that improve yields and efficiencies. When the logs are taken to the mills, the crowns are left on site, but the larger branches are often removed to make charcoal (WRI et al. 1998).

PROCESSING

Because the cost of wood pulp is to a large extent determined by the cost of transporting pulp logs to pulping mills, in many countries the two are increasingly located near each other. In fact, one of the main advantages of plantations is that the harvesting can feed the product into fixed mill sites much more cheaply than harvests from natural forests, which have longer cutting cycles. However, a hectare of natural forest in Indonesia, for example, produces more pulp at the time of the first cutting than a hectare of plantation trees. The difference is that the plantation pulp yields continue to be high after the initial cutting while natural forests take much longer to regenerate.

Processing of plantation trees begins in the field with harvest. Chain saws, heavy machinery, and transport vehicles are all required to cut down and remove trees. The harvested wood then passes through a debarker to remove most of the bark, which cannot be used for papermaking. The waste bark can be used as a soil amendment, mulch, or even burned as a fuel. Some companies remove bark in the field, not at the mill. Bark removal in the field is better for the environment. The bark eventually decomposes and returns nutrients to the soil, increasing the soil's productive potential; before decomposing, the shavings form a blanket on the ground that reduces erosion and helps protect soil from heavy machinery (WRI et al. 1998). The trunks are usually then passed through chipping machines, which chop the wood into 25-millimeter (1-inch) chips. At this point, the chips can be processed through either a digester or a refinery.

Wood consists of cellulose fibers stuck together with lignin that must be broken down to yield pulp. Wood pulp can be obtained chemically, by using chemicals to separate the fibers, or mechanically, by grinding the wood between stones or metal plates in the presence of water. Whole logs or wood chips can be used to make pulp.

In a digester, the wood chips are cooked using chemicals to remove the lignin. The by-products of the process are then used to provide energy for the mill and help with the recovery of the pulping chemicals. The resulting chemical wood pulp is then reconstituted with water to produce a coarse mixture.

Refiners use machines to grind the wood chips and separate the fibers. This process has a higher yield than chemical pulping, but it also has much higher energy costs. The mechanical wood pulp is then also reconstituted with water to make a coarse mixture. Once the pulp from all the different sources is made into a coarse wet mixture it can be blended into a single pulp. (Fiber sources may include waste paper, as described in the next section.) Blending at this stage is what will ultimately determine the final quality of the product. Dyes or other ingredients may be added at this stage to produce papermaking stock. The stock is then treated in order to separate and fray the wood fiber to the quality required for the final product. Finally, the product is run through a screen to remove any impurities such as chip remnants.

Mechanically and chemically produced wood pulp can be sold as a bulk commodity. However, Asian companies are increasingly processing pulp into paper as a way to add value to the pulp production. Brazil, by contrast, still exports a large volume of pulp to Europe and other countries.

SUBSTITUTES

There are at least three different product substitutes that affect the amount of virgin paper pulp used—other fiber substitutes, plastic, and electronic communication. As the availability of forest-based raw materials declines, producers have turned to innovative ways of expanding the fiber supply. The paper industry now uses shorter lengths, offcuts, residues, and waste (FAO 2001b). Nearly 60 percent of the fiber used in papermaking in 1998 came from virgin pulp; the rest came from recycled paper and nonwood fiber sources such as wheat straw. It is possible to increase the proportion of recycled paper—Germany, the United Kingdom, and Japan have increased it to 50 percent or more (Mattoon 1998). However, an added process to remove glues and inks must also be performed on waste paper for it to be used as a fiber source. In China 11.3 million metric tons of nonwood pulp were produced in 1999. Over two-thirds of the pulp produced in China is made from bamboo, bagasse, reed, rice straw, wheat straw, and other nonwood sources (Ryan 2002b). By contrast, in 1998 nonwood fiber made up less than 1 percent of total fiber for paper in the United States (Mattoon 1998). Innovations, new technologies, and expanding use of nonwood sources of fiber have the potential to decrease the industry's reliance on wood.

Plastic and other wraps have made considerable inroads in the use of paper for packaging, at least in developed countries and urban areas. This is true of boxes and overall packaging of dry goods as well as most of the packaging used for fresh meat and produce in grocery stores. Unfortunately, while plastic wrap takes up less space in landfills than paper that is not recycled, very little plastic packaging is recycled in most countries. In addition most plastic is made from petroleum, which is not a renewable resource.

The electronic era was introduced with considerable fanfare and seen by many as a way to create paperless offices. This has not been the case. In fact, electronic communications have actually increased per capita paper consumption, in both developed and developing countries.

MARKET CHAIN

In 1996 an estimated 12 to 21 percent of wood pulp was traded internationally, while the large majority was consumed near the source of manufacture (Strategic Environmental Associates 1996). The market for pulp is characterized by a lack of product differentiation, and pulp moves duty-free around the globe. In the 1990s world shipments of chemically produced paper-grade pulp grew at an average of 2.4 percent per year. However, the export growth was low from the traditionally high producers in North America and Nordic countries, and the market share of the rest of the world (primarily Brazil and Asia) grew as their shipments increased at an average annual growth of 5 percent (PPI 2001).

The pulp and paper industry is truly global, involving the worldwide trade of pulp, raw materials, and paper. By 2001 paper and paperboard products accounted for more than 50 percent of the value of global forest product exports (FAO 2001b). Five large companies produced almost 9 million metric tons of chemical market pulp (which includes chemically produced paper-grade, fluff, and dissolving/specialty pulps) (PPI 2001).

From 1990 to 2000, the share of total processed wood exported has increased, with 34 percent of wood-based panels, paper, and paperboard exported (up from 25 percent in 1990), and 20 percent of pulp exported (up from 16 percent in 1990). However, tropical timber products (from plantations and native forests) accounted for less than 10 percent of global pulp, paper, and paperboard products exported in 2000 (FAO 2001b).

In the United States, approximately 25 percent of the timber harvest is used for pulp production. This 25 percent accounts for 45 percent of the material used in pulp production; recycled material accounts for 30 percent; and residues from sawmills, veneer mills, etc. account for the final 25 percent of pulp sources (Strategic Environmental Associates 1996). Globally, a significant share of pulp comes from the wood chips that are the by-products of sawmills.

MARKET TRENDS

Between 1961 and 2000 the international trade of pulp increased from 9.8 million metric tons to 37.8 million metric tons. The average price corrected to 1990 values decreased from U.S.$480 per metric ton to $446 per metric ton, with a total real decline of 7 percent over the period in question. Given the nearly fourfold increase in supply, the price of pulp has been buoyed by a dramatic increase in demand.

The contribution of eucalyptus to world pulp supply is likely to increase, as it has become a more popular plantation species in recent years. In 1996 less than 1 percent of wood pulp came from tropical hardwoods such as eucalyptus (Strategic Environmental Associates 1996). Because the financial incentives to apply genetic manipulation and breeding can be strong, the use of these techniques is likely to increase. Genetically modified trees have also been developed. In the future, as long as consumers accept the products, they will be used to establish plantations as well. The increasing productivity of pulp plantations also affects global prices. For example, the Brazilian pulp giant Aracruz Cellulose S.A. increased the average yield by 50 percent in only eight years. Indonesian companies have done the same. Ultimately, genetics and climate will create the highest-producing plantations.

Pulpwood has a long growing cycle when compared to other agricultural crops. As a result, the pulp industry is highly cyclical (once an investment is made, the producer has to wait years to harvest) and driven by high capital investments in pulp mills and supply that can vary tremendously based on weather and other conditions. Consequently, the price of pulp is prone to dramatic shifts. For example, from 1993 to 1996 the price of a ton of benchmark pulp rose from U.S.$390 to $1,000 per metric ton and then fell back to less than $500 (World Rainforest Movement 1999). Overplanting of pulpwood trees can lead to a glut on the world market, which is beneficial for paper manufacturers and users but makes pulpwood cultivation less profitable and less attractive to producers. In addition, increased globalization (improvements in technology, communications, and transportation) means that pulp producers can sell to a greater range of buyers. As a result, the pulp cycle has more frequent, steeper, and longer-lasting price swings, and profit margins are low throughout the industry (WRI et al. 1998).

Mills with very large capacities were built in Indonesia in an attempt to avert the cyclical price declines of the pulp and paper market (Barr 2001). As expansion in tropical countries continues, northern countries struggle with more expensive raw materials, high production costs, and mills that are older and less efficient. British Columbia in Canada has been the most expensive pulp-manufacturing region in recent years due to high chip costs, labor, and other factors.

Current news releases announcing new pulp mills and plantations reflect the increase in the establishment of plantations used for pulp. At a time when plantations are increasing in number and area, new technology and innovations are reducing costs and increasing productivity. While the trends may be somewhat uneven globally, new technology will provide financial incentives to shift away from harvesting natural forests. Over the short term, such innovations increase the profit margin. Over the long term, they increase production and lower consumer prices (Sedjo 2001).

Another strategy to mitigate cyclical variation on world markets has been to integrate paper production with pulp production. Since the mid-1990s, global production of paper and paperboard increased steadily, and was not affected by the Asian financial crisis of 1997–98. However, prices decreased considerably in 1998 and 1999 before recovering in 2000 (FAO 2001b).

Harvest of fast-growing plantation species is expected to increase dramatically over the first half of the twenty-first century. As of 2000 fast-growing industrial tree plantations accounted for approximately 10 percent of the global industrial wood harvest. By 2050 they are predicted to account for 50 percent of the harvest, while industrial plantations of native species are likely to account for an additional 23 percent of the harvest. These fast-growing plantations are predicted to cover approximately 200 million hectares, or 6 to 7 percent of the world's forested area as of the year 2000 (Sedjo 2001).

Indonesia has attracted global attention for massively expanding its pulp and paper industry in the last decade. Indonesia's pulp and paper production increased sevenfold from 1987 to 1997. Indonesia currently produces 20 million cubic meters of pulp, which requires some 4.3 times that amount of wood. Indonesia's natural forests can no longer supply this volume. While Indonesia's forests continue to be converted for pulpwood at an alarming rate, the industry is looking towards plantations as the only way to maintain profitability and a constant supply of raw material. The government has provided generous subsidies for the establishment of plantations and, in 1997, allocated 4.3 million hectares to be cleared of natural forests and planted with plantations. These mixed tropical hardwood natural forests are being clear-cut and fed into pulp mills prior to and during the establishment of plantations. However, the 4.3 million hectares of land allocated greatly exceeds the area of land needed to support the pulp industry, and it appears the generous allocation of land for plantation establishment was done primarily to supply producers with ample supplies of mixed tropical hardwood forests (Barr 2001). It will be several years before the established plantations are supplying trees at their full capacity. In 1998–99 less than 8 percent of the 100 million cubic meters used by the pulp industry came from plantations (Barr 2001).

In developing countries governments have historically dominated the plantation sector of the pulp industry. This situation is changing in some countries, and the private sector (both individuals and corporations, often in partnership) is moving into the forest plantation industry. The private sector has assumed a major role in plantation development in Brazil, India, Indonesia, Malaysia, and Thailand. Issues of financial risk and economic viability are even more important when plantations are privately held. Governments are now making plans to "privatize" or sell their plantations to individuals or companies in Australia, Brazil, Chile, Indonesia, Malaysia, New Zealand, and South Africa. Out-grower schemes, in which private companies work in partnership with communities or small farmers, have also become more common. Foreign investments in plantations in Southeast Asia, Oceania, and South America are also increasing (FAO 2001b). These include commercial banks, investment funds, and pulp and paper companies from China, Japan, Europe, and North America.

ENVIRONMENTAL IMPACTS OF PRODUCTION

The environmental problems resulting from forest plantations are a subject of much debate. Proponents of plantations maintain that they are environmentally beneficial, allowing efficient production of forest products on a small area of land and therefore easing pressures on natural forests. Some argue that plantations have many of the ecological attributes of natural forests (such as similar leaf fall, soil percolation, and accumulation of organic matter) and are more beneficial ecologically than non-forested areas (Wadsworth 1997). By definition, however, forest plantations are large monoculture areas, usually of exotic species, and so contain far less biodiversity than natural forests. One of the major environmental issues of concern, then, is whether plantations are created on areas recently cleared of natural forests or on pasture or degraded agricultural land. Globally, most plantations have been created after the conversion of natural forest or logged-over and degraded forest areas.

Pulp plantations can have a number of negative environmental effects, primarily habitat conversion and deforestation, pollution from agrochemical inputs, and environmental degradation as soil quality and water cycles are altered. The increased burning associated with forest clearing is also a serious concern. These can be mitigated to some degree through good planning and vigilant management. The best standards of management are not always employed in developing countries (precisely where pulp and paper plantations are expanding most rapidly) because of constraints on resources and capacity as well as the lack of incentives and enforcement of existing laws and regulations that should affect such operations. Each environmental impact is discussed separately below.

Habitat Conversion and Deforestation

The establishment of extensive monoculture plantations results in a loss of biodiversity, irrespective of the vegetation type existing before the plantations were established. Indonesia's policy of clear-cutting native forests with outstanding biodiversity and then establishing plantations is an extreme example. Similarly, native or old-growth forests have been, and in some cases still are being, logged and chipped in Canada, the United States, Chile, and Tasmania; afterwards they are replaced by plantations. Conversion of natural forests to plantations accounts for 6 to 7 percent of all forest conversion each year (Cossalter and Pye-Smith 2003). According to some estimates, 15 percent of all plantations in the tropics were established on lands where natural forests were cleared immediately prior to planting the seedlings (Mattoon 1998). However, even if plantations replace degraded forest or grasslands, there is a loss of biodiversity when monoculture plantations of exotic, introduced species are planted.

The initial clearing is not the end of the process. After seedling trees are planted, other seedlings and sprouts of native vegetation are attacked aggressively to prevent them from competing with the desired species. This is done through hand or mechanical weeding or with the use of herbicides. Once tree plantations are established and branches extend to close the forest canopy, few other species will appear. In short, there is little biodiversity. Native flora and fauna may not be able to adapt to the new habitat. The intensive management of plantations means that epiphytes, parasites, and climbing flora common to tropical forests do not have an opportunity to develop. Soil flora and fauna also decline due to changes in soil composition and leaf litter. The use of agrochemicals affects leaf litter, and the lack of mature or dead trees results in less habitat available for fungi and insects. Crop pollinators and other ecosystem services offered by natural forests are no longer available. The few native insects and animals that do find a way to adapt to a specific niche within a plantation tend to increase exponentially because of the large size of industrial plantations. This often causes serious problems, resulting in the need for increased use of agrochemicals or, in severe cases, abandonment of the plantations. For example, the pine shoot moth (*Ryacionia buoliana*) proved such a problem for pine plantations in Uruguay that they were abandoned (World Rainforest Movement 1999).

Plantations are sometimes established on marginal or unsuitable lands, which may also increase environmental problems. In Indonesia, the Sinar Mas group intends to establish extensive plantations on peatlands, which will increase fire and environmental management risks. An independent audit concluded that these risks have not been adequately addressed (Amec Simons Forest Industry Consulting 2001). Furthermore, such lands are not as productive, so they will have financial implications that could affect the overall viability of the company.

Soil Erosion and Nutrient Loss

In plantations with intensive, short rotations, nutrients in the soil are depleted and soil becomes more acidic over time. Frequent management interventions, use of heavy equipment, and tree removal all disturb the soil, reduce organic matter, and increase erosion. Each of these impacts can contribute to other impacts as well. For example, soil erosion worsens the impacts of floods.

In Brazil, the policy of Aracruz Cellulose S.A. is to leave tree crowns and small branches on site after the harvest in order to help protect the soil (WRI et al. 1998). A study of fast-growing species in the tropics concluded that 70 to 80 percent of the nutrients in the tree were removed from the plantation when timber and bark were harvested. Removal of such large quantities of nutrients results in the need for large amounts of fertilizer to restore soil fertility. Leaving slash (branches and other residue) on the site after harvest could reduce nutrient loss by 25 percent. Leaving bark could reduce loss by another 5 to 10 percent. Extending the time of harvest also reduces nutrient loss, as shorter harvest-and-replanting cycles remove more nutrients. Harvesting *Gmelina arborea* every five to six years causes significantly more nutrient loss than harvesting every thirteen to fifteen years (Wadsworth 1997).

Increased Risk of Forest Fires

Severe forest fires burned around the world in 1997 and 1998, spurred on by El Niño–related drought conditions. In 1997–98 forest fires in Indonesia were extremely destructive. A total of 9.7 million hectares burned, with cost estimates ranging from U.S.\$4.5 billion to \$10 billion. An estimated 75 million people were affected by smoke or haze. Subsequent studies identified the use of fire to clear land for oil palm and pulpwood plantations as one of the main causes of these fires. Approximately 80 percent of the fires originated on industrial holdings (Mattoon 1998), and roughly 100,000 hectares of plantations burned in Kalimantan and Sumatra. Acacia and eucalyptus plantations are especially susceptible to fire because their leaves have a high oil content and young trees have thin bark that is not yet fire-resistant (Barr 2001). Most Indonesian companies have poor fire prevention and suppression practices. Fires were also widespread, although not as severe, in 1999–2000 (FAO 2001b).

Burning forests are a significant contributor to climate change. A recent study estimated that Indonesia's 1997 forest fires, most of which were started to clear land for agriculture, released between 0.81 and 2.57 gigatons of carbon (Page et al. 2002).

Changes in the Water Cycle

The environmental impacts of plantations on the water cycle are not fully understood. However, it is clear that changes occur. The amount of water falling on the soil is different after a plantation is established because the new trees and foliage are uniform; they do not have the same diversity of size and shape as the flora found in native forests. Runoff and absorption of rainfall also vary in response to factors such as the amount of leaf litter generated by the plantation and the type of humus produced by that litter as it decomposes. Heavy equipment compacts soil, which increases the speed and amount of runoff and reduces absorption. Some plantation trees use a large amount of water per hectare, although some studies have shown that they consume no more water than other herbaceous vegetation (Wadsworth

1997). In Kenya plantation softwoods managed on twenty-year rotations actually consume less water per hectare than natural forests, and only 10 percent more than perennial pasture grasses (Pereira 1967, as cited in Wadsworth 1997).

There is, however, a direct correlation between a species's rate of growth and total water consumption, so plantations that use fast-growing species such as eucalyptus have high water use. This high rate of water use can cause problems in areas surrounding the plantation, as less water is available for crops, freshwater ecosystems, and the generation of hydroelectric energy and other industrial activities.

Pollution from Agrochemical Inputs

Herbicides, pesticides, fungicides, and fertilizers are all used on tree plantations. Herbicides are used to remove native plants that return after land is first cleared. Monoculture plantations are susceptible to pest and disease outbreaks, which are typically controlled with insecticides and fungicides. However, some studies have suggested that eucalyptus plantations use fewer pesticides and fertilizers than crops such as corn, soybeans, or wheat (WRI et al. 1998).

The extent of chemical use varies depending on the company and region. The overall environmental impact depends not just on the quantity of chemicals used, but also on how and when they are used. Riocell S.A. in Brazil uses only the herbicide glyophosate (in targeted areas) and a single insecticide, which is widely applied (WRI et al. 1998). In New Zealand, however, more than thirty different pesticides have been used, including organochlorines (the class of highly toxic chemicals that includes DDT) (Mattoon 1998). Mechanized application, especially aerial spraying, can be inaccurate or drift and can result in the excessive use of chemicals.

Pollution from Processing Mills

No attempt will be made to address the full range of environmental impacts of pulp mills. Much is already known about this form of industrial pollution. However, since pulp processing mills are increasingly part of pulp plantations, it is important that their most significant impacts are mentioned here.

Pulp mills produce effluents that are high in solids, nitrogen, phosphorous, and organic compounds. In Europe the effluent released from pulp and paper mills has seen a reduction in its biochemical oxygen demand (BOD, a measure of pollution) by more than 70 percent since 1990. In Europe some 95 percent of pulp and paper mill effluents receive primary and secondary wastewater treatment (Confederation of European Paper Industries 2000). While there are improved methods for treating effluents, they have not been uniformly adopted or enforced through government regulations throughout the world. In addition to organic matter and other natural substances in the effluent, in many places high quantities of chlorine are used to bleach pulp to a uniform color as well as to improve binding, printability, and reproduction capacity; increase strength; and reduce yellowing over time. There is currently a shift away from chlorine gas in favor of other bleaching techniques such as chlorine dioxide and ozone, but this is not true everywhere.

Air pollution has also been a problem of paper mills. The use of lower-quality fuel sources in the past caused major emissions from pulp mills, including carbon dioxide, sulfur dioxide, nitrogen oxides, and particulate matter. Sulfur emissions also occur as a result of the pulping process itself. Sulfur dioxide has been reduced by switching from heavy fuel oil to sulfur-free and low-sulfur fuels, replacing fuel oil with natural gas, or controlling the production process more carefully. In addition, the use of natural gas produces less carbon dioxide than the fuel sources used previously (Confederation of European Paper Industries 2000).

The amount of water used in pulp and paper mills varies depending on the quality of the paper or paperboard produced as well as the size of the paper machine. In Western Europe the paper industry uses, on average, about 35 cubic meters of water per metric ton of pulp produced. However, the amount of water used can exceed 100 cubic meters per metric ton for high-quality grades of paper (Confederation of European Paper Industries 2000).

Social Impacts

Industrial pulp plantations can have a number of negative social impacts if not properly planned and managed. When governments designate land for the establishment of plantations, the land is normally described as vacant or unused when in fact it may be inhabited, utilized, or claimed by local people. Sometimes these people are ethnic minorities or indigenous groups not fully integrated into the mainstream economy. In addition to being displaced by tree plantations, local people can also be affected by the application of chemicals such as herbicides and pesticides. Conflicts between plantation companies and displaced or local people are commonplace in many areas of the world (Barr 2001; Eraker 2000; Mattoon 1998). Social conflicts are likely to continue to rise in tandem with increases in plantation area planted and increases in human populations (WRI et al. 1998).

Because plantations occupy such large areas, they often monopolize local employment opportunities and fix wages with little room for negotiation. The actual contribution of labor to production costs in forestry operations may be as high as 75 percent in some cases (WRI et al. 1998). This means that most pulp plantations are very concerned about labor and, in particular, how to bring the costs down.

Some pulp plantations and mills rely on local communities to provide a significant proportion of their raw material, either from plantations of their own or through legal or illegal harvesting from natural forests. Increasingly, companies are establishing systems similar to contract farming. Communities neighboring pulp mills are encouraged or even supported financially to plant fast-growing species to sell to the mill. If there are not two mills nearby, then there is no competition and wood prices tend to be set at levels that are highly advantageous to the buyer. Local outsourcing lowers a company's labor costs as well as its fixed investments in tree plantations.

There is increasing interest on the part of both investors and buyers in the overall environmental impacts of pulp plantations as well as what management is doing to reduce them. This is resulting in both investment screens and certification programs that are designed to reduce investment risks, improve product image, and increase the confidence of investors and buyers in the final product.

Investment screens are a mechanism to pressure pulp producers to improve environmental and social practices. Investors in the pulp and paper sector have been targeted on two fronts to adopt this approach: the risk bad investments pose to their corporate reputation and the financial loss from unsustainable business practices. Given the tiny profit margins of many pulp operations, banks that fund them are beginning to understand that the adoption of better practices is one way to insure that a pulp company will be more viable than its neighbor, all other things being equal. Increasingly, policies that explicitly consider environmental practices and social concerns help to guide finance decisions. Considering environmental and social issues can protect the corporate reputation and increase financial opportunities. Experiences from two companies in Brazil, Aracruz Cellulose S.A. and Riocell S.A., show that actively integrating social and environmental concerns into business strategies can create opportunities to increase efficiency and develop competitive advantages (WRI et al. 1998). Banks such as ABN Amro in the Netherlands have also developed specific pulp plantation investment strategies that require potential investments to be screened according to environmental and social criteria. Increasingly, such screens are seen as an integral part of the bottom-line analysis for a proposed investment and not just peripheral criteria evaluated afterward.

Forest Stewardship Council (FSC) certification of plantation forests offers another tool to promote better management practices. A number of pulp and paper companies have certified their forests and now can market their products as FSC-certified. The certification program now allows products with waste, recycled, or reused wood to be certified (FAO 2001b). Consumers, and in turn industrial buyers, are increasingly aware of environmental sustainability and social responsibility issues when making purchasing decisions. Some industrial buyers, particularly those purchasing pulp for well-known brands, will not buy any pulp from Indonesian acacia plantations because of their environmental problems. Some buyers have indicated that the situation might change if Indonesian acacia plantations obtain FSC certification. Such certification would indicate a major shift from current practices (Roberts 2002). This is a clear indication of the role some pulp buyers are willing to play to make the pulp industry more sustainable.

It is hard for companies to reduce their damaging environmental impacts, however, if they do not know what they are. Some countries require companies to report at regular intervals on soil erosion, suspended solids, and other water quality issues. In many cases, however, companies are measuring such environmental criteria themselves. This is seen as a scorecard, a way to know where they stand and monitor the impact of their performance. It is also important baseline data that can be used to show improved performance and quite possibly to reduce input costs. For example, some companies now recycle more than 90 percent of the chemicals used in the digestion phases of pulp processing. Companies have also been able to reduce the BOD levels of their effluent by as much as 90 percent and the levels of some of the toxic compounds even more (WRI et al. 1998).

Countless studies have been conducted on plantation management showing a number of practices that can increase productivity and at the same time increase the sustainability of plantations. These findings are summarized here.

Choose Appropriate Sites and Species

Where a plantation is sited can be the single largest contributor to its environmental impact. This is true both of where the entire site is located (e.g., on sloping land or in peat areas) as well as which areas within a plantation are planted and which are left in or returned to native vegetation.

Site selection can also affect plantation tree growth and productivity. Everything else being equal, the impact of a good or superior site on production can range from 28 percent to 139 percent more than the production from a mediocre site, depending on the tree species (Wadsworth 1997).

Some companies have found that by selecting species and varieties for specific sites and microenvironments yields can be increased. Aracruz Cellulose S.A. matches individual varieties with the sites best suited for them, reducing the need for fertilizer, pesticides, and other inputs or activities that have negative environmental impacts (WRI et al. 1998). More than 100 genetic varieties and clones of three major eucalyptus species and hybrids have been developed in order to find the varieties most suited to specific conditions of the sites being planted. Likewise, APP in Indonesia has developed eucalyptus clones that perform better in different circumstances.

Maintain Species Diversity

In some instances, yields have been increased by planting multiple species. In Hawaii, for example, eucalyptus planted with leguminous trees produced half again more dry weight than monocropped eucalyptus (Wadsworth 1997). In addition, mixed-species plantations can be better ecologically because they offer a better balance of soil nutrients and more variety of habitat for birds and other wildlife.

Most pulp plantations, however, are planted to a single species rather than a mix of trees. Even so, managers have found ways to reduce some of the negative impacts of large areas of monocrop tree plantations. Interspersing plantations with native reserves, especially if these are linked to form biological corridors, tends to maintain ecological balance and promote biodiversity within plantation estates. By supporting populations of natural predators and breaking up extensive monoculture stands, such areas help to control pests and diseases. Several plantations in Asia and the Americas have maintained 20 to 30 percent of their area in natural habitat.

In Brazil one company utilizes trees harvested from 162 different farms in 23 municipalities (WRI et al. 1998). While this increases transportation costs considerably, the company has decided to not concentrate the production of raw material in large plantations to avoid the environmental and social problems associated with such plantations.

Improve Planting and Replanting Techniques

While clearing forest areas for plantations, any plant material from the clearing process (such as felled trees and branches) should be pushed into rows that follow the contours of slopes on the site. In this way, the material forms barriers to minimize soil erosion and hold moisture; the barriers also retain organic matter and nutrients that are released over time as the materials decompose. Such windrows help protect soil even on nonsloping sites.

Areas that are being reclaimed from degraded agricultural land or pasture need to be planted to crops that build up the soil before any tree seedlings are planted. Careful choice of cover crops will increase overall organic matter, add nitrogen, and cover and hold the soil to reduce erosion.

Many companies plant rows of seedlings by hand in soil that is minimally prepared. They have found that minimizing the disturbance of soil during planting saves them money and reduces the need for fertilizers. Minimizing soil disturbance maintains microorganism communities in the soil, which in turn encourages healthier and more rapid root development. It also preserves the mulch and leaf litter on the surface, which protects the vitality of soil and retains moisture.

Reduce Agrochemical Use

Several ways have been identified to reduce the use of agrochemicals in the pulp industry. Often input use can be reduced relatively simply. The most important way plantations can reduce input use is to maintain organic matter on the surface of the land, which acts as a barrier to weed growth, holds moisture in the soil, and improves plant health. In addition, as the surface material decomposes it builds up levels of organic matter within the soil, which increases the soil's ability to retain nutrients and water. By minimizing the leaching of nutrients, soil organic matter can reduce the need for fertilizers.

Many plantation pulp companies are beginning to consider the soil an asset that must be not only conserved but also maintained. For that reason, they should invest in protecting the soil. They could do this in a number of ways that have been discussed above. Another way would be to invest in science and dedicated laboratories to develop management techniques to monitor and improve soils. Finally, there is far too little information exchanged between companies; considerable money is invested in reinventing the wheel, rather than in replicating practices already known to improve overall soil health.

Some of the more progressive companies now use spotters to determine if there are specific disease or pest issues. The data is collected, stored, and recalled by quadrant. It is used as the basis for deciding whether pests or diseases have reached a point that requires chemical control. As a result, interventions can be used only where needed rather than for the entire plantation or prophylactically through the life of the plantation. Over time, managers can anticipate problems in areas that are more susceptible to pests and diseases. If such patterns persist over time, managers have sufficient information to retire them if operational costs exceed revenues. During plantation establishment, herbicides tend to be used only on the areas directly adjacent to the seedlings, and then only as needed.

Watering huge plantations is expensive in terms of labor, equipment, and volume of water used, so there are strong economic incentives to avoid watering. In the past, many plantations were forced to water their trees once or more after they were planted and additionally during times of drought. However, most companies find that if they plant trees at the beginning of the rainy season, they no longer have to water by hand or set up expensive and wasteful irrigation systems.

Improve Harvesting Methods

High-technology harvesting equipment can increase the efficiency and productivity of harvest while leaving undergrowth and slash (harvesting residue) on the ground to protect the soil. Using machines with tires rather than metal tracks tends to reduce soil compaction as well.

Leaving bark and trimmings in the field is much better for the environment, as explained in the earlier discussion of processing. The bark and trimmings create a mulch that reduces soil erosion and builds soil organic matter. However, stripping the bark in the field costs more; it represents a 5.66 percent increase in the cost of forestry production compared to wood whose bark is removed at the factory. In the case of Riocell S.A. in Brazil, bark removal results in an annual additional cost of about U.S.$1.5 million, or an increase of 1.23 percent in total production cost (WRI et al. 1998). To evaluate the financial impact of debarking on the plantation, one would need to know whether leaving the bark in the plantation fields reduces the cost of nutrients or increases overall growth sufficiently to make up for the added cost. This is a relatively straightforward calculation for pulpwood plantations, but so far such calculations have not been done.

Calculating the costs and benefits of bark removal is further complicated by the fact that some processing plants use the removed bark as fuel. Vertically integrated companies must determine if the savings from the mulch that is left behind exceed the savings from using the material in the processing mill to reduce overall energy purchases. Some vertically integrated companies produce up to 80 percent of their total energy needs by burning bark and other by-products. Other companies convert bark and other processing waste into marketable by-products. More than 99 percent of the solid wastes from processing in one mill in Brazil is used to make fertilizer or soil amendments, or added to cement (WRI et al. 1998).

Eliminate Burning

Burning increases carbon dioxide, degrades soil, decreases soil nitrogen, and kills soil microorganisms. In addition it creates significant amounts of air pollution and has devastating effects on biodiversity. Increasingly, plantations have adopted no-burn policies, either for plantation establishment or for harvests or both. In some cases, as in Indonesia and Malaysia, this has been regulated by governments. In Brazil, Riocell S.A. was one of the first pulp mills to eliminate postharvest burning voluntarily (WRI et al. 1998). While the company's internal no-burn policy was not generally accepted in the early 1990s, companies now see the advantages. Producers have found that the retention of organic matter in the soil more than makes up for the additional costs by reducing the need for expensive fertilizers and irrigation. In fact, the benefits are felt for some time as the organic matter continues to reduce the cost of inputs and increase overall productivity.

OUTLOOK

Globally, tree plantations are assuming increasing importance as a source of pulp. Paper and paperboard products will increasingly come from raw materials cultivated in plantations. Improved genetics for plantation species will make harvesting pulp from natural stands less economically viable. This should have a very positive impact on natural forest habitat unless an overall decline in value encourages people to convert the habitat to other uses. Programs of payments for ecosystem services (e.g., carbon sequestration or watershed management) could also be developed to help maintain natural forests.

The area of industrial forest plantations will continue to increase. The economics of pulp production and processing are driving the industry to create larger plantations with attached pulp processing mills. The Philippines, Mexico, and China are all planning significant expansion of industrial plantations, with China aiming to establish 9.7 million hectares of forest plantations between 1996 and 2010 (FAO 2001b). Plantations will increase in importance to supply not just pulp and paper products, but also to produce engineered wood products such as laminated veneer lumber (LVL) and glue-laminated timbers. As the supply of large logs decreases globally, these composite materials will increase in importance (FAO 2001b). Over time, lumber may well become the most valuable product from plantations currently dedicated to pulp production.

This trend will provide pulp at increasingly low prices, but there are other impacts of the evolving system that are not nearly as obvious. For example, it is quite likely that huge areas will be planted to monocrop plantations in the name of efficiency. These areas are unlikely to include biological corridors to support biodiversity *unless* producers can be shown that such corridors make sense financially. This can happen either because producers recognize the advantage of leaving native vegetation in areas that are not economically viable for planting, or because buyers demand that producers meet basic biodiversity and ecosystem criteria as a cost of doing business. Similarly, unless it proves economically advantageous, such plantations are likely to be established by clearing existing forests rather than degraded pasture or agricultural land.

REFERENCES

Adams, M. and Efransjah. 2001. On the Conference Circuit: International Conference on Timber Plantation Development. 7–9 November 2000, Manila, Philippines. In *ITTO Newsletter*, vol. 11, no. 1, January.

Amec Simons Forest Industry Consulting. 2001. *APP Pulp Mills and Sinar Mas Group Forestry Companies: Preliminary Sustainable Wood Supply Assessment.* Vancouver: Amec.

Barr, C. 2001. *Banking on Sustainability: Structural Adjustment and Forest Reform in Post-Suharto Indonesia.* Bogor, Indonesia: Center for International Forestry Research (CIFOR) and Washington, D.C.: World Wildlife Fund.

CCFM (Canadian Council of Forest Ministers). 2001. Forest 2020 press release. Available at http://www.ccfm.org/forest2020/plantationsworld_e.html.

Confederation of European Paper Industries. 2000. *Environmental Report.* Brussels, Belgium: Confederation of European Paper Industries.

Cossalter, C. and C. Pye-Smith. 2003. *Fast-Wood Forestry—Myths and Realities.* Bogor Barat, Indonesia: Center for International Forestry Research.

Eraker, H. 2000. *Colonialism: Norwegian Tree Plantations, Carbon Credits, and Land Conflicts in Uganda.* Oslo, Norway: Norwatch.

FAO (Food and Agriculture Organization of the United Nations). 1999. *State of the World's Forests 1999.* Rome: UN Food and Agriculture Organization.

———. 2001a. *Global Forest Resources Assessment 2000.* FAO Forestry Paper 140. Rome: UN Food and Agriculture Organization.

———. 2001b. *State of the World's Forests 2001.* Rome: UN Food and Agriculture Organization.

———. 2002. *FAOSTAT Statistics Database.* Rome: UN Food and Agriculture Organization. Available at http://apps.fao.org.

Lang, C. 2002. *The Pulp Invasion: The International Pulp and Paper Industry in the Mekong Region.* Available at http://www.wrm.org.uy/countries/Asia/Vietnam.html.

Mattoon, A. T. 1998. Paper Forests. *Worldwatch*, March/April, pp. 20–28.

McNabb, K. 1994. Silvicultural Techniques for Short Rotation Eucalyptus Plantations in Brazil. Paper presented at the Mechanization in Short Rotation, Intensive Culture Forestry Conference, Mobile, AL, March 1–3.

Old, K. 1997. Collaborating to Protect Acacia Plantations. *On Wood, CSIRO (Commonwealth Scientific and Industrial Research Organisation) Forestry and Forest Products.* Volume 17 (Winter). Available at http://www.ffp.csiro.au/publicat/onwood/onwood17.htm.

PPI (Pulp & Paper International). 2001. *Annual Review 2001.* Volume 43(7). Available at http://www.paperloop.com/db_area/archive/ppi_mag/2001/0107/contents.htm.

Roberts, J. 2002. The Case for Acacia. *Pulp & Paper International.* May 2002.

Ryan, R. 2002a. *Asian Pulp and Paper Markets: Australia.* Paperloop, Inc. Available at http://www.paperloop.com/newsinfo/regional/asia_australasia/australia_pp_market.shtml.

———. 2002b. *Asian Pulp and Paper Markets: China.* Paperloop, Inc. Available at http://www.paperloop.com/newsinfo/regional/asia_australasia/china_pp_markets.html.

Sedjo, R. A. 2001. From Foraging to Cropping: The Transition to Plantation Forestry, and Implications for Wood Supply and Demand. *Unasylva* No. 204, Volume 52.

Strategic Environmental Associates. 1996. *Resource Efficiency Study: Resource Efficiency Gains Available from Fiber System and Technology Choices.* Underwood, WA: Strategic Environmental Associates.

Wadsworth, F. H. 1997. Forest Production for Tropical America. *Forest Service Agriculture Handbook 710.* Washington, D.C.: U.S. Department of Agriculture.

World Rainforest Movement. 1999. Pulpwood Plantations: A Growing Problem. A briefing paper of the World Rainforest Movement's Plantations Campaign. Montevideo, Uruguay: World Rainforest Movement.

World Resources Institute, Aracruz Celulose S.A., and Riocell S.A. 1998. Efficiency and Sustainability on Brazilian Pulp Plantations. Chapter 5 in *The Business of Sustainable Forestry,* Chicago: The John D. and Catherine T. MacArthur Foundation.

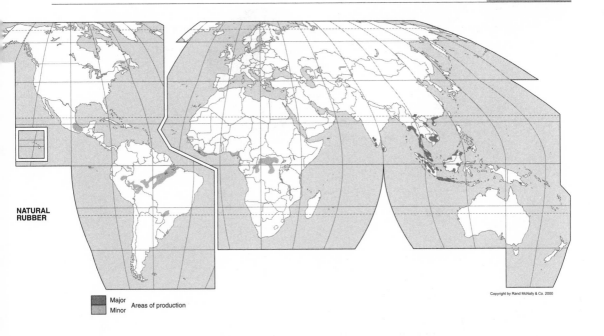

NATURAL RUBBER

Major
Minor — Areas of production

Copyright by Rand McNally & Co. 2000

RUBBER *Hevea brasiliensis*

PRODUCTION		INTERNATIONAL TRADE	
Area under Cultivation	7.7 million ha	Share of World Production	84%
Global Production	6.8 million MT	Exports	5.7 million MT
Average Productivity	888 kg/ha	Average Price	$680 per MT
Producer Price	$395 per MT	Value	$3,876 million
Producer Production Value	$2,688 million		

PRINCIPAL PRODUCING COUNTRIES/BLOCS (by weight)	Thailand, Indonesia, Malaysia, India, China, Vietnam, Sri Lanka
PRINCIPAL EXPORTING COUNTRIES/BLOCS	Thailand, Indonesia, Malaysia, Vietnam, Liberia, Côte d'Ivoire
PRINCIPAL IMPORTING COUNTRIES/BLOCS	United States, European Union, Japan, China
MAJOR ENVIRONMENTAL IMPACTS	Effluents from processing in the plantations and in processing plants Conversion of primary forest habitat is an issue mostly in China
POTENTIAL TO IMPROVE	Good Very little expansion planting BMPs are known and address most impacts Few inputs are being used Effluents are a problem but there are known ways to reduce them

Source: FAO 2002. All data for 2000.

Area in Production (MMha)

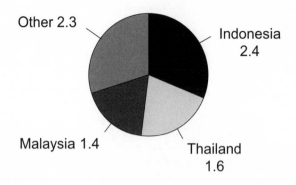

Other 2.3

Indonesia 2.4

Malaysia 1.4

Thailand 1.6

OVERVIEW

Rubber from *Hevea brasiliensis* dominates all other sources of natural rubber and is synonymous with what is now called rubber. Rubber was first known and used by Indians in the Brazilian Amazon. Exports of natural rubber collected in Brazil began in the eighteenth century. As far back as the early 1800s there were reports of rubber-covered slickers and boots being used by fishermen in the New England cod industry.

The development of vulcanized rubber in the late 1800s stimulated demand that led to the rubber boom. Instant millionaires were made in the Amazon, Indians were enslaved to gather rubber, and the poor from Brazil and all over the world were induced to move to the Amazon in the search for rubber. From 1890 to 1910 so much money was made that local elites sent their laundry to Europe where it could be done in clean water. An opera house was built in Manaus that rivaled any in the world. Tens of thousands of paving stones around the building were replaced with rubber "bricks" at the equivalent of $10 each so that carriages would be silent as they passed. European opera stars came to the Amazon, but many died of fevers and never left.

But the boom was not to last. Rubber was the object of one of the most publicized cases of alleged "biopiracy" in the world. In 1876 rubber seeds were taken from Brazil by Henry Wickham to Kew Gardens in England (smuggled or legally exported, depending on one's point of view). After addressing propagation problems, seedlings were shipped to British colonies in Asia, in particular Malaysia and Ceylon (now Sri Lanka) but also Indonesia. Production began in earnest around 1910, and the monopoly of wild Amazonian rubber was broken. The price plummeted.

By 1910 plantations had expanded tremendously; 245,000 hectares were being cultivated in Indonesia alone. Research in Indonesia during the early twentieth century led to the development of bud grafting, a propagation technique that greatly raised productivity. At this time rubber was still largely a plantation crop with only 8,100 hectares grown on small farms. However, with the new easy-to-learn propagation technology, that quickly changed. By 1940, 1.3 million hectares of rubber were grown by small-scale farmers compared to only 0.6 million hectares on plantations. By 1990 the balance had shifted even more with 2.6 million hectares grown by small-scale farmers and 0.5 million hectares on plantations (Burger and Smit 2001).

Only during World War II was wild Amazonian rubber highly sought again, and that was because the Japanese occupied all the rubber plantations in Southeast Asia. The Amazon was unable to provide the quantities of rubber necessary for the war effort. The Allies searched the Amazon for natural stands but also invested in research to develop synthetic substitutes. After the war production from the Amazon proved, once again, not to be competitive with rubber produced on plantations. After 1947 rubber ceased to be exported from the Amazon in commercial quantities.

By the 1980s plantation rubber production was in trouble. Synthetic rubber had eroded the market for natural rubber; today natural rubber makes up only 29 percent of the market. However, there are certain products that cannot be made with synthetic rubber. Airplane tires are 100 percent natural rubber, and automobile tires are 35 to 40 percent natural rubber. These two industries alone account for 70 percent of the natural rubber market. In the age of AIDS (acquired immunodeficiency syndrome), natural rubber is indispensable. Neither surgical gloves nor condoms can be made inexpensively from synthetic rubber. For the short term, anyway, natural rubber will have a market, although it is losing market share to synthetic rubber every year.

Rubber trees can be sold for timber once they have passed their productive life. The wood is a semihard, light-colored timber. It has a pleasant grain and can be used in wooden utensils, furniture, flooring, and chipboard making. Commercial exploitation has been rapid, and the timber currently commands a high value. In part the value is related to the ease of harvest associated with any plantation-grown tree. While any rubber trees can be sold for timber, plantation trees are easy to harvest and transport, many trees of harvestable age are located in a confined space, and plantations produce straight logs with few branches close to the ground.

PRODUCING COUNTRIES

By 2002 there were 7.7 million hectares of rubber in production, excluding the vast areas of natural rubber that are harvested in the Amazon. The countries with the most area planted to rubber trees include Indonesia (2.4 million hectares), Thailand (1.6 million hectares), and Malaysia (1.4 million hectares). These three producers account for 70 percent of all land planted to rubber trees and 67 percent of the 6.8 million metric tons produced annually. Côte d'Ivoire, Mexico, and the Philippines have the highest average yields at about double the global average of 888 kilograms per hectare. Thailand's average per-hectare yield is twice the level of either Indonesia or Malaysia (FAO 2002).

Thailand and Indonesia combined account for 57 percent of the world's supply of natural rubber. These two countries, together with Malaysia, India, China, Vietnam, and Sri Lanka, are the top seven producers and account for 80 percent of global production.

Brazil continues to produce rubber, but the vast majority of its rubber now comes from established plantations rather than from the wild. Most rubber plantations are monocultures, but some are intercropped with other species. All plantations in Brazil have been established outside of the Amazon in the states of Bahia, São Paolo, and Mato Grosso. These areas have shown higher productivity than natural rubber stands in the Amazon, and are outside of the range of the disease vectors found within the Amazon. Even with the plantations, Brazil has rarely exported rubber since shortly after the end of World War II. Ecuador, Guatemala, and Colombia also produce a small amount of rubber for local use (FAO 2002).

CONSUMING COUNTRIES

World consumption of rubber is dominated by the United States, China, the European Union (especially Germany), Japan, and India. Imports of natural rubber are dominated by the United States, the European Union, Japan, and China (FAO 2002).

PRODUCTION SYSTEMS

While rubber originated in the Amazon and the wild rubber trees of that region dominated trade in the nineteenth century, today all globally traded rubber is produced from planted trees. Plantations are established by clear-cutting tropical forests and then planting monocrop stands of rubber trees on a grid pattern to facilitate harvesting. On small-scale plantations, trees may be interplanted within agroforestry systems. After the initial planting within monocrop plantations, other vegetation is removed until the seed bank in the soil is exhausted or until the branches of the rubber trees extend to close the canopy and shade out other growth. Even though they originated in the Amazon, rubber trees do best where the water table is 1 to 1.2 meters or more below the surface (Goldthorpe 1993). This assures good soil aeration and the development of good root systems.

Planted trees are productive for thirty years or more. This means that virtually the entire rubber demand of the twentieth century was met by only three generations of rubber trees. As rubber prices have declined, mature or aged rubber plantations in Malaysia and Indonesia have been converted to other tree crops such as cocoa, pulp, or, more commonly, oil palm. The fact that former rubber plantations support new crops without intensive renovation suggests that the plantations did not cause a lot of soil erosion or soil degradation.

Most small farmers in countries like Indonesia use the traditional "jungle rubber" system of production. Smaller numbers of trees are planted in thinned natural forests or forests that are gradually converted to agroforestry orchards, depending on the amount of land owned.

Trees on plantations are planted in densities of 250 to 450 per hectare. Trees are tapped for their sap a couple of times each week. Productivity declines as trees get older, but if tapped properly the process does not threaten the tree. Some researchers in Brazil have suggested that tapping reduces seed productivity, however. While this is an important issue for wild trees, it is not important in plantations where all tapped trees have been planted.

Traditionally, some trees from previous plantings are left standing when plantations are cut down after 30 years or so or even when the rest of the plantation is cut because of declining rubber prices. These remnants are kept in reserve to meet immediate financial needs or to give producers an edge if rubber prices increase. Traditional trees take eight to fifteen years to mature before they can be tapped, and they are not as productive as new, input-intensive clonal varieties.

Through the 1990s, with the price of rubber generally declining and the price of food (mostly rice) increasing, small farmers were finding it increasingly difficult to cover their costs of living. Rubber came to supply only 75 to 90 percent of their income. Increasingly, up to 20 percent of their income was coming from paid labor on plantations of either oil palm or pulp (Penot and Ruf 2001).

In the 1970s rubber was one of the first perennial crops for which highly productive, vegetatively propagated planting materials became available to replace seed-grown stock. Bud-grafted, clonal varieties improved production and increased income, particularly to small farmers who relied on family labor. While only 15 percent of small farmers were using clonal varieties by the late 1990s, 86 percent were planning to plant or replant clonal rubber (Penot and Ruf 2001).

Clonal varieties offer several advantages over traditional varieties. They begin to produce within five years, tend to produce two to three times as much rubber, and generate 50 to 100 percent more net income (Gouyon 1999, as cited in Penot and Ruf 2001). While some clonal varieties are susceptible to leaf blights that reduce production by 30 to 50 percent, many of the clonal varieties perform better than traditional varieties on poorer soils, degraded areas, and areas with higher rainfall. The ability to use these varieties on degraded areas more than quadrupled the price of degraded land in parts of Indonesia between 1997 and 1999 (Penot and Ruf 2001).

At this time, many small farmers are diversifying their production. They are increasing their plantings of clonal rubber, but they are also planting oil palm. Producers do not see these crops as substitutes for one another. Rather, they are complementary aspects of an overall strategy to ensure reliable income. Most small farmers are also planting fruit trees. The fruit can be used both for consumption and for sale on local markets. Many of the small farmers cannot afford to plant input-intensive clonal rubber, so they are intensifying their agroforestry systems. If local roads and/or local fruit-processing facilities improve, then many small farmers are likely to increase their fruit plantings.

Over time, producers have learned how to plant and care for rubber plantations.

The oldest plantations in Asia are just now on their fourth generation. As production practices have come to more closely mimic natural forests (and with the absence of diseases native to the Amazon), production has risen from 250 kilograms per hectare per year to 2,500 kilograms per hectare per year (Goldthorpe 2003).

PROCESSING

Processing rubber begins with its harvest from the tree. Tapping rubber to collect the sap consists of making incisions in the bark, collecting the sap from the incision in a cup, and emptying the cup into a container. In plantations, new incisions are made about three times per week or some 120 times per year during the tapping season. Sap is collected every four to five hours throughout the period between incisions. The sap collected in the daytime is of higher quality and is coagulated by adding ammonia which maintains the higher-quality rubber. During the night the sap is exposed to bacteria that cause natural coagulation but create rubber of a lower quality. The collector visits each of twenty to thirty trees and pours the sap from each cup (about 500 milliliters) into a 15-liter container. When the container is full, a solution of ammonia (5 percent by volume) is added at the rate of 40 milliliters per liter of sap (Sonetra 2002).

The coagulated sap (latex) is then transferred to a tanker, which transports it to a rubber factory. At the factory, the latex is discharged into a holding tank. The latex contains about 25 to 30 percent "dry rubber"; water is added to the holding tank to dilute the latex to about 16 to 18 percent dry rubber content. The pH is usually about 6.6 to 6.9, and formic acid is added to reduce the pH to about 5 (Sonetra 2002).

On more rustic rubber plantations or small farms, one of the most important things producers must do to aid the processing of rubber is to add formic acid to the sap tapped from the trees to stimulate the coagulation of latex. Formic acid is one of the few costs to such producers. This process is sometimes referred to as prevulcanization. Once this coagulation has occurred, producers transport the treated latex either to the processing plants directly or, more often, to pickup points. The highest-quality rubber is treated with ammonia and then acid and processed within twenty-four hours of collection. This is one of the advantages of plantations: Not only are the collection and prevulcanization of rubber cheaper and easier to control, but it is also easier to transport the product to processing plants.

Further processing of rubber generally takes place off the plantation. The primary stage consists of processing latex and coagulum into sheets, crumb rubbers, or latex concentrate, and creates large quantities of effluent. In general 25 to 40 cubic meters of wastewater is produced for each metric ton of rubber produced. After the primary stage comes the process known as vulcanization. In 1839 Charles Goodyear invented this process, which uses sulfur, lead, or zinc oxide and heat to stabilize natural rubber by preventing it from turning brittle when cold and sticky when hot (Chapman 2002).

In 2000 the amount of natural rubber produced was 6.8 million metric tons (FAO 2002). Synthetic rubber production now amounts to more than 10 million metric tons per year, and so it exceeds the production of natural rubber. Most synthetics are petroleum-based. Because petroleum is readily available, most synthetics are cheaper for most applications than natural rubber. However, when synthetics must have the exact same elasticity and durability of rubber, then they are more expensive. This is why natural rubber still dominates some markets. Production of synthetic rubber is dominated by the United States, Japan, Russia, China, France, Germany, and Brazil. Consumption is dominated by these same countries (FAO 2002).

There are other plants that produce a form of latex than can be used for rubber. In fact, one of the major incentives for King Leopold of Belgium to occupy central Africa in the end of the nineteenth century was to coerce local residents to harvest wild latex from a long spongy vine of the *Landolphia* genus. In just over a decade an estimated 10 million Africans lost their lives either producing the rubber, being killed for not producing their quotas, or dying from the elements as they tried to escape King Leopold's occupying forces. Since the end of the nineteenth century, however, other rubber-producing plants have not proven as successful as plantation crops as *Hevea brasiliensis* (Hochschild 1999).

MARKET CHAIN

Rubber has one of the more simple market chains, and as a consequence the primary end users have periodically made efforts to vertically integrate rubber production. Henry Ford and others failed miserably in their attempts to establish rubber plantations in the Amazon during World War II. The Pirelli tire company had about the same amount of success in the Amazon. In West Africa, Firestone, Pirelli and others did successfully establish plantations, only to see them taken over or made unsafe as the countries were caught up in revolutionary movements in the latter part of the twentieth century.

In general, the trend is for small-scale farmers to produce more and more of the rubber in the world. The rubber is then sold to capital-intensive processing plants that have the capacity to handle the rubber produced from a very large region. After the rubber is processed and graded, it can either be sold or stored indefinitely before it is ultimately purchased and used by a manufacturer.

MARKET TRENDS

From 1961 to 2000 total natural rubber production increased by 221 percent, from 2.1 million metric tons to 6.8 million metric tons. Exports increased by 151 percent over the same time period. Prices declined by 82 percent (FAO 2002). The price of rubber has generally declined over the past fifty years. In 1995 the price was U.S.$1.60

per kilogram and by 1997 it had fallen to $1.30. Prices have continued to drop, so that by 1998 it was $0.60 and by 1999 it was only $0.55 per kilogram (Penot and Ruf 2001). By mid-2002 the price of rubber had bounced back to $0.87 per kilogram. While the currency collapse in Southeast Asia tended to protect rubber producers from price declines in the late 1990s (especially in Indonesia), the rising price of labor and rice made traditional rubber less attractive to plantation owners. As a result, many shifted their production to palm oil, cocoa, or pulp (Penot and Ruf 2001).

Penot and Ruf suggest that there are three main causes of price declines. First, world supply has increased by 4 percent while demand has risen only 3.5 percent. In addition, China, a major importer, has slowed its purchases after stepping up its domestic production. By the end of 1996 global stocks had recovered to some 2 million metric tons, which also depressed prices.

With production increasing faster than consumption, prices will continue to decline. Many of the increases in production have resulted from trees planted during the past twenty years. Among the factors leading producers to plant more trees are the growing awareness of AIDS and the speculation on the part of producers that this will spur increased markets for natural rubber through increased use of condoms and surgical gloves. Many of the trees in these recent plantations are only now becoming fully productive, so this is adding to the increases in production. A fair amount of planted rubber goes into production or is abandoned depending on the price of rubber. If prices continue to decline, some rubber plantations will be abandoned or converted to other crops.

While considerable effort and investment have been made to find substitutes for natural rubber, synthetic rubber cannot be fully substituted for natural rubber in many products at this time. However, given the size of the natural rubber market and the price of natural rubber relative to synthetic substitutes, it is likely that efforts to develop new substitutes will continue. Furthermore, as in the past, it is likely that substitutes will be found for an increasing number of uses to which only natural rubber can currently be put. This will reduce further the market for natural rubber.

ENVIRONMENTAL IMPACTS OF PRODUCTION

Rubber trees are long-lived. Because of the longevity of the trees and because synthetic substitutes have been developed for many of the products, expansion of rubber plantations has not been significant globally. The one notable exception is China, where natural habitat in the more tropical, southern part of the country was being cleared until very recently in order to establish rubber plantations. There have also been a number of quite large failed experiments to establish rubber plantations in the Amazon basin, but all of these efforts succumbed to disease after the native forests were destroyed.

The ongoing impacts of rubber production, then, are mostly linked to processing. Converting the liquid sap that is collected directly from the tree to latex produces considerable amounts of effluent. Some of the chemicals in the effluent are highly toxic. In addition, the conversion of sap to solid latex requires a fair amount of energy (either fuelwood or electricity) to separate it from the water after coagulation. Finally, the vulcanization of latex into rubber also releases effluents that are highly toxic in the environment. In many countries, the emission of effluents from rubber processing and vulcanizing plants is not well-regulated.

Habitat Conversion

A consequence of creating rubber plantations is the clearing of natural forests for the establishment of monocrop plantations. In addition, the timber is often stacked and burned. This results in a loss of the vast majority of forest species including those that live in the soil, which are exposed directly to sun and heat as well as rainfall and cannot survive the fluctuating heat and moisture levels. Soil exposure leads to erosion and the leaching of nutrients. Once rubber trees are planted, regrowth of any other vegetation is killed until the seeds in the soil are depleted or until the canopy is closed. Once rubber plantations are established they are recolonized by subsoil microorganisms as well as by small succulent and shrubby plants. While rubber plantations recreate some of the ecosystem functions of a natural forest, they harbor only a tiny proportion of the original biodiversity.

The area of the most active conversion of natural habitat to rubber plantations recently has been in China where rubber is considered a strategic crop (one that is so important that a country does not want to depend on others for it). Unfortunately, rubber is a tropical crop, and China does not have very much land that is suited for rubber cultivation. What is particularly unfortunate about this conversion is that much of China's land in tropical areas is quite hilly and subject to erosion. This leads to other environmental impacts, not only for China but also for those countries through which the Mekong River flows. For example, soil erosion alone has large impacts on drinking water, aquatic life, and siltation. In addition, the stripping of natural habitat tends to accentuate runoff during the rainy season as the water is no longer absorbed. This can contribute to flooding.

Pollution from Processing Rubber

One of the main environmental concerns with rubber production is the effluent from the initial stages of processing that most often occur in or near the plantations. The volume of effluent from rubber processing is twenty-five to forty times greater than the volume of rubber that is produced. There are two main types of effluents—the serum from the coagulation process and the water used to wash the rubber. The serum contains dissolved organic solids that readily oxidize and so create a significant biochemical oxygen demand (BOD) when they are dumped into water bodies. The washing effluent contains proteins, sugars, and other organic materials as well as inorganic chemicals. It also has high BOD, which can cause fish kills and harm other aquatic species in rivers and streams. In addition, some of the chemicals that remain in the sap after the latex is coagulated can be toxic (which is not surprising, as some serve the role of protecting the tree from pests).

The vulcanization of rubber is considered by people in the industry to be one of the most toxic industrial processes on the planet. Either lead or zinc oxide is used in the vulcanizing process. Even though zinc is probably the least toxic of the heavy metals, it is still quite toxic (even in very small doses) to invertebrates and many freshwater and marine species. These heavy metals can contaminate water bodies if the effluent is dumped into streams, and they are also released as rubber products are used or as they degrade. At this time, there is no way to reduce the heavy metals either in production effluents or in degraded products. Many people believe that the reason sneaker manufacturers moved to Southeast Asia is because of lower labor costs. In fact, it is probably equally important that the countries where shoes are now manufactured do not have stringent pollution control or worker health and safety measures.

The extent of heavy metal pollution from the degradation of rubber products is more than one might expect. It is estimated that more than 3,000 metric tons of zinc are released into the environment per year from tire wear alone. This represents about 25 percent of the anthropogenic release of zinc into surface waters (Chapman 2002). The European Union uses about 100,000 metric tons of zinc per year in the manufacture of rubber products. In order to reduce pollution from this manufacturing, the European Union has proposed standards of 1 to 3 milligrams per liter of zinc in effluent and 0.5 milligrams per cubic meter in stack emissions (Chapman 2002).

BETTER MANAGEMENT PRACTICES

At this time, most of the better management practices focus on increasing the productivity and life of existing rubber plantations. Several methods have been developed to maintain or increase soil quality. These include terracing steep hillsides, contouring on slopes, constructing bunds (earthen embankments constructed to reduce erosion), and installing silt pits. In addition, the use of ground cover, cover crops, and intercropping can all reduce soil erosion on rubber plantations, increase productivity, and reduce the need for costly inputs.

Most of the improvements to processing and wastewater management take place off farm and are more likely to occur when they are regulated by law. If standard end-of-the-pipe treatment measures are in place, effluents are not a problem. It is doubtful, however, that such treatments are common in processing plants in any of the less-developed countries that are the primary producers of natural rubber. At best, the effluent can be captured and put back onto the rubber plantations. This will reduce the pollution of freshwater ecosystems and thus reduce the damage to freshwater biodiversity. In general, processors find that they use fewer chemical inputs and water when they are required to ensure cleaner effluent. In the end, this saves them money.

Several techniques can be employed to reduce soil erosion. Each of these practices also helps to build organic matter, maintain soil nutrients and soil structure, retain water, and support microorganisms that benefit the maintenance, nutrient cycling, and building of soil. On steep, hilly terrain rubber trees should be planted on the contour to prevent soil erosion; this process is known as contouring. Terraces do an even better job of reducing erosion, but these require considerable investments to build (Goldthorpe 1993).

Soil erosion along terraces and on gentle slopes can be minimized by digging silt pits and constructing bunds. Silt pits trap the soil particles that are carried in runoff; they also hold some of the rainfall on site so it has time to sink into the ground. Bunds are earthen embankments that check the flow of water during heavy rains (Goldthorpe 1993). Planting bushy materials on the bunds can further minimize erosion after the bunds have settled.

Keeping the ground covered is one of the best ways to minimize erosion. Natural vegetation like ferns, grasses, and shrubs should be encouraged to rapidly cover the exposed soil surface during planting. In the absence of natural vegetation, rapidly-spreading creeping legumes can be sown as cover crops around the young rubber trees. Legumes increase nitrogen in the soil and reduce the need for chemical fertilizers. Equally important, they reduce erosion and exposure to the elements and increase organic matter.

Mulch around the base of rubber trees prevents soil exposure and holds nutrients and moisture, which is especially important during the establishment of plantations. Mulch also reduces chemical runoff. Mulch can be created from clearing the undergrowth in the plantations or from trimmings cut from the trees themselves. Mulch is most important during the early years of plantation establishment, before the canopy closes, when both of these sources are more plentiful.

Another way to reduce soil erosion after the planting stage is intercropping, growing other plants between the rubber trees. Intercropping has been used effectively with cacao and coffee in the Philippines, with tea and cacao in Indonesia, and with hearts of palm in Brazil. However, intercropping has not been widely practiced with rubber except by some integrated farms with multiple product lines. Most plants are shaded out by mature rubber trees. For about three months per year, however, rubber trees shed their leaves, leaving the understory with sufficient sunlight for other crops to grow. Short-lived legumes could be planted during this period to rejuvenate the soil provided there is enough moisture (often the trees lose their leaves during the dry season). Intercropping provides the additional benefit of supporting greater biodiversity, especially in plantations that have been cleared and replanted.

Research suggests that the biomass of the mature rubber plantation at 450 metric tons per hectare, while somewhat less than the biomass of 475 to 664 metric tons per hectare for Malaysian forests, compares favorably to the 295 to 475 metric tons per hectare for forests in Brazil, Papua New Guinea, and Thailand. Rubber plantations also perform well from the point of view of canopy cover and the production of leaf litter (Goldthorpe 1993). Sivanadyan and Moris (1992) conclude that a mature rubber plantation is a nutritionally self-sustaining ecosystem unlike other agricultural systems. Research in India has suggested that mature rubber plantations with closed canopies generate and recycle more nutrients and biomass each year than are harvested.

Rubber plantations can actually be useful for rehabilitating degraded agricultural areas and bringing them back into productive use. The leaf litter generated on rubber plantations provides organic matter that improves the physical properties of the soil (porosity, moisture absorption and retention) (Goldthorpe 1993). This could help to reduce agricultural conversion of natural habitat.

Improve Processing and Wastewater Management

There are several different practices that can be used to treat effluent from rubber factories prior to release or use as a soil amendment. In countries such as Malaysia treatment before release into natural waterways is required and increasingly enforced. A number of different treatments have been developed. For example, rubber factory effluents can be treated in an anaerobic pond system, in oxidation ditches, or in algae pond systems. However, the most common effluent treatment technology, which is the use of settling ponds, has a few drawbacks—not the least of which is that it takes sixty days (Goldthorpe 1993). That means that a considerable volume of water has to be held over time for treatment. Creating the treatment ponds requires a large area of land, construction expenses, and time.

Increasingly, effluent is tested and processing plants are required to reach certain levels of quality before they are allowed to release the material. Many rubber producers prefer to apply the effluent to their plantations as a soil amendment rather than to treat it to the level required before legally releasing it into rivers and streams. Good results have been reported from the experimental application of this effluent either through furrow irrigation by gravity, piped irrigation with sprinklers or trickle nozzles, or spray guns from tankers (Goldthorpe 1993).

Prior to application, however, the rubber particles need to be removed. This can be done through rubber traps or by allowing the effluent to sit for three days so the particles settle out. The effluent has a foul odor, but this can be mitigated by adding microorganisms that partially decompose the compounds. One experiment has shown that the effluent can be concentrated into a slurry with up to 60 percent solids or further concentrated into a powder. Both make effective fertilizers (Panfilo Tabora, personal communication). Apparently, such applications do not result in a build-up of toxic substances in the soil.

OUTLOOK

In the past, rubber has been an important cornerstone for industrial development because of its many uses, especially its overall importance to transportation (not only for tires but also for hoses for motors of all kinds). As a result it has been considered a strategic crop.

At this time, synthetic substitutes exist for the vast majority of the original uses of rubber. In other instances such as automobile tires, the proportion of natural rubber has been cut dramatically. Even so, there are some uses for which there are no affordable substitutes. The growth in these uses, to date at least, has been offset by the development of synthetic substitutes for other rubber uses. So long as these trends continue to offset each other, then rubber production will be sufficient to meet global demands. No major increases in rubber demand are expected at this time.

REFERENCES

Burger, K. and H. P. Smit. 2001. International Market Responses to the Asian Crisis for Rubber, Cocoa, and Coffee. In F. Gerard and F. Ruf, eds. *Agriculture in Crisis: People, Commodities and Natural Resources in Indonesia, 1996–2000.* Montpellier, France: Centre de Cooperation Internationale et Recherche Agronomique pour le Developpement (CIRAD).

Chapman, A. 2002. *Zinc in Rubber Compounds and the Environment.* Brickendonbury, United Kingdom: Tun Abdul Razak Research Centre.

FAO (Food and Agriculture Organization of the United Nations). 2002. *FAOSTAT Statistics Database.* Rome: UN Food and Agriculture Organization. Available at http://apps.fao.org.

Goldthorpe, C. C. 1993. *Natural Rubber and the Environment: A Review.* UNCTAD/COM/21. Geneva, Switzerland: United Nations Conference on Trade and Development (UNCTAD). 27 August.

Gouyon, A. 1999. Fire in the Rubber Jungle—Fire Prevention and Sustainable Tree Crop Development in South Sumatra. *International Forest Fire News (IFFN).* September (21):48–56. FAO/ECE/ILO Committee on Forest Technology, Management and Training. Available at http://www.fire.uni-freiburg.de/iffn/country/id/id_21.htm.

Hochschild, A. 1999. *King Leopold's Ghost.* New York: Houghton Mifflin Co.

Penot, E. and F. Ruf. 2001. Rubber Cushions the Smallholder: No Crisis, No Windfall. In F. Gerard and F. Ruf, eds. *Agriculture in Crisis: People, Commodities and Natural Resources in Indonesia, 1996–2000.* Montpellier, France: Centre de Cooperation Internationale et Recherche Agronomique pour le Developpement (CIRAD).

Sivanadyan, K. 1992. Natural Rubber Latex Serum Concentrate: Its Agronomic Potential. *Planters' Bulletin* 210:3–8. Kuala Lumpur, Malaysia.

Sivanadyan, K. and N. Moris. 1992. Consequence of Transforming Tropical Rain Forests to *Hevea* Plantations. *The Planter* 68(800):547–67. Kuala Lumpur, Malaysia.

Sonetra, S. 2002. *Waste Water from Rubber Processing as Fertilizer for Water Spinach and Forage Cassava.* MSc. Thesis. Phnom Penh, Cambodia: University of Tropical Agriculture Foundation–Royal University of Agriculture of Cambodia. Available at www.utafoundation. org/utacambod/msc99thes/sonecont.htm.

UNCTAD (United Nations Conference on Trade and Development). 1999. *World Commodity Survey, 1999–2000.* Geneva, Switzerland: UNCTAD.

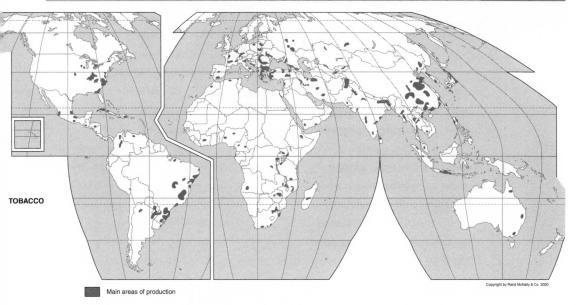

TOBACCO

Main areas of production

Copyright by Rand McNally & Co. 2000

TOBACCO *Nicotiana tabacum, N. rustica*

PRODUCTION		INTERNATIONAL TRADE	
Area under Cultivation	4.2 million ha	Share of World Production	83%
Global Production	6.8 million MT	Exports	5.6 million MT
Average Productivity	1,610 kg/ha	Average Price	$4,969 per MT
Producer Price	$2,985 per MT	Value	$28,017 million
Producer Production Value	$20,299 million		

PRINCIPAL PRODUCING COUNTRIES/BLOCS (by weight)	China, India, Brazil, United States, Zimbabwe, Turkey
PRINCIPAL EXPORTING COUNTRIES/BLOCS	Brazil, United States, Zimbabwe, China, Italy, Turkey
PRINCIPAL IMPORTING COUNTRIES/BLOCS	Russia, Germany, United States, Netherlands, United Kingdom, Japan
MAJOR ENVIRONMENTAL IMPACTS	Habitat conversion Soil erosion and degradation Agrochemical use Deforestation for drying tobacco Waste
POTENTIAL TO IMPROVE	Poor Most production shifting to developing countries where there is less ability to influence production practices Buyers have little interest in sustainable production Few alternatives to wood-based energy and paper use in many countries Focus is on stopping smoking, not making production sustainable

Source: FAO 2002. All data for 2000.

Area in Production (Mha)

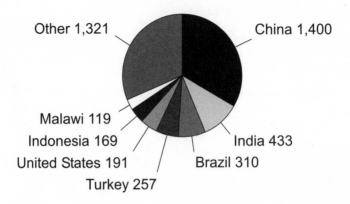

Other 1,321   China 1,400

Malawi 119
Indonesia 169   India 433
United States 191   Brazil 310
Turkey 257

OVERVIEW

Tobacco originated in the Americas, where its use is estimated to have begun around 6000 B.C. By the time of Columbus, tobacco was cultivated by Native Americans throughout the Americas and used for both ceremonies and medicine. Sailors on Columbus's first voyage were the first Europeans to use tobacco. By the middle of the sixteenth century, adventurers and diplomats (for example, Jean Nicot de Ville-main, France's ambassador to Portugal, after whom nicotine was named) promoted its use (Tobacco Free Kids 2001).

Initially, tobacco was obtained by trading with Native Americans and was used for pipe smoking, chewing, and snuff. The first crop cultivated by nonnatives was in Virginia in 1612 (Tobacco Free Kids 2001). Within seven years tobacco was Virginia's largest export. New, milder varieties of tobacco were found in Brazil and brought to the United States. Markets expanded and production expanded quickly to match it. Most tobacco was grown and harvested by slave labor.

Initially, tobacco was cured and cut for pipe smoking or ground into powder and used as snuff. Crude cigarettes have been around since the early seventeenth century. They became much more popular after the Civil War, however. At first, cigarettes were hand-rolled at the rate of three per minute. In 1880 the first cigarette-making machine was invented. It produced 200 cigarettes a minute (Tobacco Free Kids 2001). Today the fastest machines can make 160,000 per minute.

In 2000 global tobacco production was estimated at 6.8 million metric tons. The total area devoted to tobacco cultivation was about 4.2 million hectares. Tobacco is grown in more than 100 countries, including more than 80 developing countries (FAO 2002).

In 2000, China was the main producing country with 1.4 million hectares under cultivation. Other major producers included India (433,400 hectares), Brazil (309,989 hectares), Turkey (257,230 hectares), the United States (191,190 hectares), Indonesia (168,688 hectares), and Malawi (118,752 hectares). China accounted for nearly 34.3 percent of the global area planted to tobacco and 37.9 percent of total production. China, India, and Brazil produce more than half of the world's tobacco. The top seven producers account for more than 71 percent of land planted to tobacco as well as the same percentage of total production (FAO 2002).

Tobacco production increased by 59 percent between 1975 and 1997. For the most part, this increase occurred in developing countries. In those nations production increased by 128 percent during this time period while falling 31 percent in developed countries (Tobacco Free Kids 2001). More than 80 percent of the global tobacco crop is grown in the developing world.

Average tobacco yields globally are just over 1,610 kilograms per hectare per year. The United Arab Emirates produces yields that are 7.5 times the global average at 12,160 kilograms per hectare per year. Cyprus, Laos, Oman, and Uruguay each produce more than twice the global average per hectare per year.

In 2000, the leading tobacco exporters were Brazil, the United States, Zimbabwe, China, Italy, and Turkey. Exports amounted to about 29 percent of global production. Although the United States is a leading exporter (especially of high-quality, flue-cured tobacco), a number of developing countries have increased tobacco production considerably. For many of these nations tobacco is an indispensable source of foreign exchange, allowing them to import other commodities and consumer goods. For example, in Zimbabwe—the major tobacco producer in South Central Africa—nearly all tobacco (98 percent) is exported. About half of Thailand's tobacco crop is exported. In China, by contrast, tobacco consumption is growing so rapidly that most tobacco is grown to reduce the use of foreign exchange that is required to buy imported tobacco, and meet domestic demand.

For Malawi and Zimbabwe, tobacco is the leading export as ranked by value, as shown in Table 15.1. In these two countries, tobacco export earnings provide some 56 percent and 47 percent of export earnings, respectively. Indeed, tobacco provides 74 percent of foreign exchange earnings in Malawi and over 30 percent in Zimbabwe. In Brazil, the Dominican Republic, India, and Tanzania, tobacco amounts to about 5 percent of total agricultural exports.

TABLE 15.1. Tobacco's Ranking of Total Exports by Value
for Selected Countries, 1990–91

Leading Export	Second Largest Export	Third Largest Export
Malawi	Turkey	Malta
Zimbabwe		Zambia

Source: UNCTAD 1994.

CONSUMING COUNTRIES

China is the largest producer and consumer of tobacco in the world. It imports some 1 percent of its consumption. India is the second largest producer and consumer of tobacco in the world. In both instances, most of the tobacco consumed is grown locally.

Russia is the largest importer of tobacco in the world. It is followed closely by Germany, the United States, the Netherlands, the United Kingdom, and Japan (FAO 2002).

The United States imports some tobacco for special purposes. For example, Asian cigarette leaf tobacco is imported for blending, Puerto Rican tobacco for cigar filler, and leaf tobacco from Sumatra and Java for cigar wrappers. The United States also produces some of the tobacco it consumes; about two-thirds of the crop is grown in North Carolina and Kentucky (Brown and Williamson 2003).

Several factors have combined to reduce demand for tobacco. The anti-smoking campaigns in the United States appear to have permanently reduced demand in that country. By contrast, the Russian financial crisis has probably only temporarily reduced demand until the economy becomes stronger. The Chinese policy to encourage the use of tobacco land for other crops has caused an overall reduction in the area planted to tobacco. In 2000 Chinese production was down by 39 percent from the prior year, and the area devoted to tobacco was down by 25 percent to 1.5 million hectares (FAO 2002). However, demand in China is still increasing, as it is globally.

PRODUCTION SYSTEMS

The main tobacco species are *Nicotiana tabacum* and *N. rustica*. The latter is grown mainly in Asia. Smaller-leaved aromatic tobaccos tend to be grown for all purposes while larger, broad-leaved tobaccos are produced mostly for the cigar industry.

Tobacco production in the United States and some other developed countries has undergone great changes as technology has been developed so that increased mechanization dominates many aspects of production. Mechanized production in the United States has had other implications. The cost of production, for example, increased nearly 200 percent from 1980 to 1998, while the price of tobacco increased only 19 percent for flue-cured leaf. Put another way, the producer share of the price of a package of cigarettes declined from 7 cents to 2 cents (the tobacco companies' share increased from 37 to 49 cents in the same period) (Tobacco Free Kids 2001). In the meantime, because of the need to increase the scale of operations to survive financially, the number of U.S. tobacco farms declined by 32 percent while total production remained constant (NASS 1999).

Given the price of tobacco, investments in mechanization are not viable in most parts of the world. Thus, for most producers growing tobacco is still a back-breaking, labor-intensive job. Brazilian researchers estimated that some 3,000 person-hours per hectare per year are required to produce tobacco. And, as production shifts to developing countries, the number of child laborers increases. Globally, the industry estimates that 33 million people are engaged in tobacco cultivation, but the full-time equivalent employment levels are probably only a third of that. In addition, tobacco rarely accounts for more than 1 to 2 percent of total rural labor (Tobacco Free Kids 2001).

Labor requirements start with planting the seeds. While adult plants are hardy, seedlings are not. Seedbeds are created (either in greenhouses or glassed-in, enclosed areas in the fields) by carefully tilling the soil, sterilizing it with gas, heat, or pesticides to kill unwanted insects and weeds, and adding ashes to counteract the acidity of soil. Tobacco seeds are planted in the sterilized soil. They are tiny, with some 10,000 weighing only 1 gram. After spreading the seed on the beds, the soil is covered with hay or cloth to protect them. Within three to four months the seedlings have reached a height of 25 to 40 centimeters and can be transplanted into the fields, one at a time. Up to 25,000 seedlings are planted per hectare on ridges that are spaced about a meter apart.

When the plant reaches a specific height, the top is pinched off to improve leaf quality and quantity. For a period of five to six weeks, growers continue to remove new growth on the plant to ensure that the selected leaves grow to full size and mature evenly. The pinching of leaves from the plant causes it to ooze sap, which attracts a variety of insects that must be removed and killed. Some of the insects, like the hornworm, are camouflaged green, making them hard to find. On average, growers must tend between 250,000 and 400,000 individual leaves per hectare (Tobacco Free Kids 2001).

The process of harvesting can be extremely labor-intensive. If high-quality leaves are the goal, a technique called priming is used in which only three to four leaves are removed at a time starting at the bottom of the plant (Tobacco Free Kids 2001). The uppermost leaves, which have the highest levels of nicotine, are harvested last. Priming allows new leaves to form and mature over time. For lower-quality tobacco, all the leaves are harvested together with the stalk.

When grown as a monoculture (as it is in the most productive tobacco-producing regions), tobacco requires large amounts of fertilizers and other agrochemicals. One of the major problems with the crop is its susceptibility to water stress, diseases, insects, and soil chemistry (either deficiency of one or more nutrients or toxicity resulting from an excess of one or more nutrients). As a hardy, sun-loving plant, tobacco can tolerate a wide variation in rainfall patterns, provided they are not extreme, while still yielding a commercially viable crop.

Because of tobacco's susceptibility to stress and diseases, and also to take advantage of field conditions and to maximize yields, several varieties are planted in most fields. Ideally, the varieties that perform best would be chosen based on soil type, available moisture, and local pest problems such as nematodes and insects inside the field in question. However, given the increasing scale of production and the lack of seedling varieties at planting time in many parts of the world, it is impossible for even the most advanced farmers to take advantage of such precision-farming approaches to improve production on their land.

Soil and climatic differences cause significant variations in leaf characteristics. In turn, these differences greatly affect each leaf's suitability for use in the various manufactured forms of tobacco. Each geographic area produces a leaf with special characteristics that is particularly adapted for certain uses—cigarettes, cigars, pipe tobacco, chewing tobacco, or snuff.

In many developing nations, the soil is too poor and offers too few nutrients for many food crops. Most poor farmers cannot afford to use the fertilizers required to grow food crops such as cereals, grains, or beans, which, when sold on local markets, are not valuable enough to cover the cost of such inputs. The high value of tobacco relative to most food crops allows producers to afford the types of fertilizer and chemical inputs that allow the crop to be produced on poorer soils (Tobacco Free Kids 2001). Subsidized inputs (in cash or in kind) from the buyers don't hurt either. In the end, though, it depends on how much profit is left when the crops are sold and all debts have been paid. While tobacco is often a very attractive cropping alternative because of the amount of money it generates, many producers find that at the end of the season they have fewer net profits than expected.

One major trend with tobacco production in industrialized countries is the use of greenhouses in the production of healthy seedlings for transplant. This allows for lower usage of pesticides in the initial phases of production and for highly controlled application of fertilizers in the initial weeks of growth. The transplanting of healthy and vigorous plants in the primary phases of production is also thought to result in slightly higher resistance to stresses and disease.

Another major trend with tobacco production is contract farming. Under this system of production tobacco companies act as banks, extending credit to farmers at the beginning of the season in the form of seed, fertilizer, pesticides, and technical support. In return farmers pledge to sell their crop to the company at harvest. Company inspectors make regular visits to the farms to make sure their guidelines are followed. When the harvest is in and the time comes to pay the farmer, the leaf buyers determine the leaf grade and the price. As a result they often end up paying producers less than the original loans. In Brazil observers suggest that severe grading has reduced prices by 20 percent. There is some evidence that the companies decide prices among themselves to keep their costs lower (New York Times, as cited in Tobacco Free Kids 2001). This form of "debt bondage" is a way the tobacco companies are increasingly controlling producers (Tobacco Free Kids 2001). In Brazil, Universal and DIMON (two U.S.-based companies) contract with about half of all producers; most of the remainder have contracts with Souza Cruz, a subsidiary of British American Tobacco (Jones 1998b).

PROCESSING

Tobacco leaves are cured, fermented, and aged to develop aroma and reduce the harsh, rank odor and taste of fresh leaves. Fire, air, and more specialized flue curing are the most common ways to cure tobacco. Curing methods, along with environmental conditions and cultivation techniques, determine the characteristics of many of the grades.

Fire curing is used on 20 percent of all tobacco, primarily for cigarettes. The process dates from pre-Columbian times. It is undertaken most crudely by drying tobacco leaves over a fire where the smoke can cure them. Alternatively, fire curing can be undertaken in a structure that resembles a smokehouse, where the leaves are hung within a smoke-filled enclosure. Flue-cured tobacco, which constitutes just over half of all tobacco, is used to produce pipe tobacco, snuff, and chewing tobacco. Most globally traded production is flue-cured. It requires the leaves to be dried by radiant heat from flues or pipes connected to a furnace (Smith 1999).

Air curing accounts for 11 percent of all tobacco and is used mainly for cigars. In this process the leaves are hung in well-ventilated structures so that the air can circulate around them and dry them out. Any heat in this system is from the ambient air or from passive solar energy (i.e., heat from sunlight falling on the barn roof). Other tobacco—especially Turkish or other oriental tobaccos—is sun-dried. This process is used on 16 percent of all tobacco, for so-called oriental cigarettes (Smith 1999).

After curing, tobacco is allowed to age for six months to two years. The cured tobacco is graded, bunched, and stacked in piles called bulks or in closed containers for active fermentation and aging. Most commercial tobaccos are blends of several types, and flavorings are often added. These flavorings include maple syrup, chocolate, honey, vanilla, licorice, fruit extracts, and other sugars. Some concern has been expressed that many of these additives are used to help make cigarettes more palatable to children. After curing, tobacco is rolled into cigars, shredded for use in cigarettes and pipes, or processed for chewing or snuff. Because of the toxic chemicals in tobacco, any wastes that cannot be used to make other more valuable tobacco products are used to manufacture insecticides.

SUBSTITUTES

Tobacco is addictive. In fact, manufacturers have been accused of adding chemicals to tobacco to make it even more addictive. While many other products are equally addictive, none are true substitutes. Some research shows that alcohol consumption and tobacco consumption are often linked, and when linked it is much harder to stop either one (DeBenedittis 1999; Drobes 2002). However, most people trying to stop using tobacco products do not turn to other more addictive products. Most turn to food as an alternative and usually experience periods of pronounced weight gain, at least in the short term.

MARKET CHAIN

In recent decades, tobacco companies have shifted their purchasing to developing countries, where there is cheap labor and easy access to natural resources. Much effort and considerable money are invested to encourage farmers to grow tobacco. Companies have spent billions of dollars to build new factories, enter into joint ventures, and buy formerly state-owned factories. At the same time, these large corporations have been working closely with usually U.S.-based leaf companies (companies that source tobacco around the world, ensure quality control, and then sell to manufacturers) to expand the cultivation of lower-priced tobacco to supply the new factories. Philip Morris, British American Tobacco (BAT), and Japan Tobacco each own or lease manufacturing facilities in more than fifty countries. They purchase tobacco in dozens more (Tobacco Reporter 2001).

The world leaf market is dominated by three U.S.-based companies: DIMON, Standard Commercial, and Universal. These companies select, purchase, process, and sell tobacco leaf to the major manufacturers of tobacco products (Swoboda and Hamilton 1997). These companies are in charge of purchasing and quality control of tobacco for the main manufacturers. They also work with the major tobacco manufacturers to determine which countries will produce how much tobacco leaf and what kind. Much tobacco production is forward-financed by the manufacturers through the leaf companies that receive down payments from manufacturers to deliver tobacco of predetermined quality and quantity. The leaf companies in turn use that money as cash advances to growers. By helping to finance the farmers through the leaf companies, the manufacturers hope to reduce the risk of not being able to obtain product, the amount of time they are involved in the search for raw materials, and the overall price they will have to pay for product.

In Brazil DIMON pays about U.S.$100 million per year to provide tobacco farmers with fertilizer and other inputs. The company "agrees" to purchase the entire crop and in some cases even finances the curing barns as well (Jones 1998a). In Tanzania DIMON contracts with more than 30,000 tobacco growers by providing similar assistance (Tuinstra 1998a). In Poland, Philip Morris established a growers' fund for 18,000 tobacco producers to improve the quality of their crop, and BAT has given U.S.$3 million in no-interest loans to Polish farmers (Tobacco Free Kids 2001). Philip Morris has sent twelve American advisors to provide guidance to local tobacco growers in China. Producers in Argentina, Azerbaijan, India, Malaysia, Turkey, and Vietnam all receive similar loans, technical assistance, and infrastructure investments from leaf buyers (Tobacco Free Kids 2001).

Since a small number of leaf companies or their buyers set the prices, they leave farmers little choice, in any given year, but to accept the prices that are offered. In some cases, the buyers forward-contract to buy the crop through various grow-out schemes before it is even planted. In such cases, they stipulate not only which inputs can be used but also when, how much, and how often. In many instances, such contract farming operates on the notion of prevention of possible problems rather than treating them as they arise. In such circumstances, many producers also buy inputs from a state or private-sector buyer who ends up buying the tobacco at harvest. In the long run, most small-scale farmers never profit greatly from tobacco production. In addition, the inherent value of their land declines due to the terrible toll that tobacco takes on its nutrient and organic matter levels. Most tobacco farms have extremely low nutrient and organic matter levels and are easily farmed until they are no longer productive. Even for larger-scale commercial farmers, especially in the developed countries, most of the profits from growing tobacco come in the form of government subsidies.

In some developing countries, governments intervene in tobacco markets to manipulate the price paid to producers, usually to the detriment of the producers. In 2002 in Zimbabwe, for example, tobacco sold at auction for U.S.$1.65 per kilogram. This is a good price, but the farmers were not happy because the government grabbed most of the value by manipulating the exchange rate. The official exchange rate of the Zimbabwean dollar is 55 per U.S. dollar, despite inflation of more than 100 percent per year. On the black market, however, the local currency has only one-tenth that value. Farmers are not allowed to hold hard currency and must turn in their sales receipts to the government, at which point they are paid at the official exchange rate. Instead of allowing farmers to reap the real benefit of their crop, the government "gives" them a subsidy equal to some 80 percent of the "official" value of their crop. In the end, the government and certain officials pocket more than 80 percent of the actual value of the crop by holding on to the hard currency (*The Economist* 15 June 2002).

Regardless of the tobacco production and curing system employed, the profits from tobacco accrue largely to large multinational companies. As multinational cigarette companies increase their overseas manufacturing capacity, the leaf dealers have followed, setting up leaf procurement and processing facilities near the new factories. Today they operate in dozens of countries on five different continents. In a constant drive to increase profits, these companies regularly shift production from one country to another based on the concession they can negotiate with the local governments, irrespective of the impact this may have on local growers or economies.

Global tobacco companies based in the United States spend $11 million per day to advertise and promote cigarettes, to enhance their profits, and to protect their markets and market shares. Some promotional tactics that have proven effective in reaching potential young smokers around the world include the use of cartoon images, free cigarette giveaways, sponsorship of events that appeal especially to young people, and the use of cigarette logos on youth-oriented products. Intense tobacco marketing campaigns are increasingly common in developing countries such as Egypt, Indonesia, China, and Brazil. In these countries population size, low rates of smoking among women, increased disposable incomes available to larger numbers of landless workers, and the lack of legislative controls offer promising new markets to the tobacco industry (DeBenedittis 1999).

MARKET TRENDS

The massive increase in global tobacco production fueled by tobacco industry financing has resulted in a worldwide oversupply of tobacco and a decline in prices. From 1960 to 1989 the world price for flue-cured tobacco declined in real terms by 1.1 to 1.7 percent per year. This trend accelerated between 1985 and 2000, when the real price fell 37 percent to U.S.$1,221 per metric ton (Jacobs 2000, as cited in Tobacco Free Kids 2001).

There are some indications that the lack of competition in the markets hurts prices. Leaf companies have declined in number and are directly linked to the manufacturers. In Zimbabwe some 70 percent of the market is controlled by subsidiaries of DIMON and Universal (Tuinstra 1998b). An article in *Tobacco International*, a pro-business magazine, recently suggested that the small number of leaf buyers in Malawi has reduced competition to such a point that one company is buying about 50 percent of the local crop (Kille 1998). In 2000 prices declined by 14 percent. In 2001 they declined even further to as little as U.S.$0.10 per kilogram; the same leaves would have sold for between $1 and $2 per kilogram in 1999 (South Africa Broadcasting Corporation 2000, as cited in Tobacco Free Kids 2001).

Another factor that affects prices is currency value. When Brazil devalued its currency in 1999, tobacco prices in Argentina dropped from 15 to 25 percent because Brazilian leaf was suddenly cheaper for the buyers (Bennett 2000).

Companies are also increasing profits and reducing tobacco use per cigarette by figuring out new ways to process tobacco so that it increases in volume. Originally this was done with chlorofluorocarbons (CFCs), but since they were banned it is done with liquid carbon dioxide, nitrogen, or isopentane. In these processes the tobacco is soaked with liquid gas, which solidifies at atmospheric pressure. Hot gases are then pumped into the mix, causing the dry ice (solidified carbon dioxide) or other solidified gas to vaporize and thus puff the tobacco up by 60 to 100 percent (Tobacco Free Kids 2001). Expanded tobacco costs a little more to produce, but because less is used it pays for itself. In the near future ultralight cigarettes will contain 40 to 50 percent expanded tobacco, but only 20 percent for light cigarettes and 10 percent in full-flavor cigarettes (Tobacco Free Kids 2001).

Another factor that is likely to affect overall tobacco prices and trade globally are the proliferation of international trade agreements (such as the World Trade Organization and a dozen or so regional or hemispheric ones) which liberalize trade in goods and services. Cigarettes are also affected by the removal of trade barriers, which tends to introduce greater competition, lower prices (both to consumers and producers), and increased expenditures on advertising and promotion to stimulate demand. The World Bank (1999) reports that in four Asian countries that opened their markets in response to U.S. trade pressure during the 1980s (Japan, South Korea, Taiwan, and Thailand) consumption of cigarettes per person was almost 10 percent higher in 1991 than it would have been had the markets remained closed.

Similarly, as the United States and the European Union reduce their overall support for tobacco, either through subsidies or market protection, the crop is being grown in a number of developing countries where production costs are cheaper. The response among producers has been so overwhelming that global supply is now greater than demand. If anything, tobacco may be an indication of what will happen if global markets become more open through the reduction or elimination of production subsidies or market protection.

Per capita tobacco consumption in developed countries has stabilized with actual drops in the United States and increases in Europe. Overall use is expected to decline in both over the next century. However, tobacco use is increasing among teenagers and young women in both areas as well as among poorer males and females in both. In the United States, the tobacco industry loses close to 5,000 customers every day—including 3,500 who manage to quit and about 1,200 who die (INFACT 2003).

Globally, demand for tobacco continues to increase steadily. World consumption of tobacco increased by 7 percent from 1992 to 1995 and was expected to increase by an additional 7 percent by 2000. By 2025, the total number of smokers is expected to rise from its current 1.1 billion to some 1.6 billion (World Bank 1999). To meet the increases in tobacco consumption, world production of tobacco also increased by 7 percent from 1992 to 1995—with the largest increase in developing nations such as Indonesia, China, and Zimbabwe (FAO 2002). More recently, with the decline in tobacco prices due to oversupply, China has discouraged production in favor of other food and fiber crops, leaving the door open for increased imports (UNCTAD 1999).

Developing countries account for the greatest increases in consumption. By the mid-2020s it is predicted that only about 15 percent of the world's smokers will live in developed countries, as there will be a shift in the use of tobacco from developed to developing countries. In the developing world, per capita cigarette consumption has risen on average by more than 70 percent during the last twenty-five years (INFACT 2003). In Nepal and Haiti, per capita cigarette consumption in 1990 was over 240 percent higher than in 1970, while in Cameroon and China the relative increase was over 150 percent (Gajalakshmi et al. 2000). In these countries use is expected to increase slightly faster than the population over the next few decades.

Most tobacco companies have pinned their hopes for future profits on expanding markets in developing countries. The industry is targeting Africa, Asia, and Latin America. The tobacco industry has also been expanding into Eastern Europe and the former Soviet Union.

Nowhere is the potential of this strategy clearer than in China. The Chinese cigarette market is already three times the size of the market in the United States. Chinese cigarette consumption accounts for more than 30 percent of the world's 5.4 trillion cigarettes sold each year. Since international sales into the Chinese market amount to less than 1 percent of this market, the tobacco industry feels that there is immense potential for expansion there (Nelson 1993, as cited in O'Sullivan and Chapman 2000). As China discourages local tobacco production, this may well encourage an increase in imported tobacco products.

Tobacco advertising has been predicted to take a different shape in the coming years. Marketing campaigns are likely to target minorities and women. Women, especially those in developing countries, constitute a major untapped market for the cigarette companies. Companies will continue to target children, but the hooks used are likely to be much more subtle as a way to reduce legal actions. Greater emphasis will also be placed on developing niche markets. For example, special advertising may target groups that identify with fringe cultures. Also, drinkers will become a greater marketing niche.

In the foreseeable future, tobacco farmers are not likely to respond to calls to cease production of a commercially viable crop. They will continue to invest in production as long as consumer demand for tobacco products exists at current or even higher levels, and as long as they can be assured that their crops will sell quickly and profitably. As prudent businessmen, they will also invest in new opportunities as they arise.

ENVIRONMENTAL IMPACTS OF PRODUCTION

Tobacco is mostly known for its damaging social and health impacts. It is one of the largest causes of premature death worldwide and is the most common preventable cause of death in the world. Tobacco kills nearly 10,000 people every day worldwide. In twenty years that figure is expected to climb to some 30,000 deaths per day, or 10 million per year (Tobacco Free Kids 2001).

Tobacco is a known or a probable cause of some twenty-five diseases. These include lung cancer, heart disease, stroke, emphysema, and cancer in other parts of the body. In addition, secondhand smoke, or environmental tobacco smoke, poses a severe health risk to those exposed to it. For this reason, such people are known as "passive smokers." Secondhand smoke has been found to carry nitrosamines (potent cancer-causing agents) and glycoproteins (proteins that cause allergic reactions). Studies have shown that nonsmokers exposed to secondhand smoke can suffer significant damage to the functioning of their small airways. The International Agency for Research on Cancer of the World Health Organization has concluded that secondhand smoke causes cancer (BBC News 2002).

Smokers and those exposed to tobacco smoke are not the only people at risk. In addition to the health impacts associated with using tobacco products, agricultural workers (especially children) who weed and tend tobacco plantings or pick tobacco have been reported to experience "green tobacco sickness" (GTS). This is a type of nicotine poisoning caused by the absorption of nicotine through the skin. Absorption can be accentuated when those who work with tobacco do not wear gloves or protective clothing, as is common in most developing countries. Green tobacco sickness is considered a serious occupational hazard. It is characterized by symptoms that include nausea, vomiting, weakness, headache, dizziness, abdominal cramps, and difficulty in breathing, as well as fluctuations in blood pressure and heart rates. Research in North Carolina (the largest tobacco-producing state in the United States) suggests that 41 percent of all tobacco workers experience GTS at least one time during the tobacco harvest (Quandt et al. 2000).

The presence of large-scale tobacco production is often followed by tobacco processing plants in the same regions (Tobacco Free Kids 2001). In some areas, this has resulted in the use of child labor (for example, to roll cigars) as well as the introduction of smoking-related problems into the same areas. This particular problem has no easy fix, though it must be remedied to make tobacco an economically sustainable crop for the small farmer in developing countries. It must be addressed on a political level.

Because tobacco is implicated in such a large number of deaths annually, there has been relatively little research undertaken to promote the sustainability of a crop that ultimately kills its consumers. Tobacco as a crop, however, is responsible for damage to ancient forests, and it causes soil depletion through soil erosion and nutrient loss. Pollution also occurs from the extensive use of pesticides and fertilizers. These environmental impacts are discussed below.

Deforestation

A tremendous amount of wood is used to dry or cure tobacco. In southern Africa alone an estimated 200,000 hectares of woodlands are cut annually to support tobacco farming. This accounts for 12 percent of deforestation in the region. Most of the wood is used as fuel (69 percent), but wood is also used as poles for building curing barns and racks (15 percent) for hanging the leaves while they dry. The two most common methods of processing—fire curing and flue curing—both require fire or heat, though substitute fuels such as charcoal, coal, or oil can be used. One researcher estimates that 19.9 cubic meters of wood are used to cure every metric ton of tobacco in those areas where the energy comes from wood (Geist 1998). In addition, burning fuel to cure tobacco releases CO_2, which contributes to global warming.

In the United States, China, and Europe petroleum, coal, and natural gas are now common alternatives to wood. However, in most developing countries where increasing amounts of tobacco are harvested, producers still rely on the use of readily available and unregulated wood supplies from forests. Brazil, India, the Philippines, and most of Africa use wood for curing tobacco. A wood shortage is looming in Malawi and western Tanzania as a result of deforestation in the main tobacco-growing regions. Tobacco alone is estimated to account for 5 percent of Africa's total deforestation, and 20 percent of deforestation in Malawi (Geist 1999). The amount of deforestation attributed to the production and curing of tobacco is shown for several countries in Table 15.2.

Additional pressure on forests comes from the use of paper associated with wrapping, packaging, and advertising cigarettes. Modern cigarette-manufacturing

TABLE 15.2. Percentage of Total
Annual Deforestation Related to the
Production and Curing of Tobacco
in Selected Countries, 1990–1995
of Tobacco

Country	Deforestation (%)
South Korea	45.0
Uruguay	40.6
Bangladesh	30.6
Malawi	26.1
Jordan	25.2
Pakistan	19.0
Syria	18.2
China	17.8
Zimbabwe	15.9

Source: Geist 1999.

machines, for example, use more than 6 kilometers of cigarette-width paper per hour to manufacture cigarettes (Tobacco Free Kids 2001). The packaging of cigarettes and other tobacco products may require two to three times as much paper by weight as tobacco. It requires roughly 4 to 5 metric tons of wood from a forest to make 1 metric ton of paper, so every metric ton of tobacco product sold would require some 8 to 15 metric tons of wood for packaging material. Yet, according to the industry, wrapping and packaging account for only 16 percent of its overall use of forest products (Tobacco Free Kids 2002). If this estimate is correct, the industry uses 50 to 94 metric tons of wood per metric ton of tobacco, for a total of 340 million to 639 million metric tons of wood per year. Given the millions of copies of newspapers, periodicals, etc. sold every week throughout the world, it is quite likely that advertising is the largest single paper use of the tobacco industry.

Cigarettes that have not been extinguished properly also contribute to deforestation, posing a serious fire hazard. It is estimated that one-quarter to one-third of forest fires around the world are caused by careless smokers. If that estimate is accurate, then the amount of wood in forests that is burned by smokers could surpass the sums used by the entire tobacco industry. Apart from the human and property costs, such fires have huge impacts on forests, biodiversity, and watersheds.

When all these different impacts are totaled, it is clear that the tobacco industry is a major contributor to deforestation, which has serious ecological consequences including the loss of ecosystem functions and biodiversity as well as soil erosion and degradation.

Pollution from Pesticides and Herbicides

Tobacco is a delicate plant that is prone to many diseases and pests. Some management guides call for as many as sixteen applications of pesticides during the three-month growing period before the plants even leave the greenhouse (Goodland et al.

1984). The list of chemicals that are recommended includes some with toxicity levels that are quite high. In developing countries, chemicals that are commonly used include aldicarb (Temik), aldrin, butralin, endosulfan, chlorpyrifos, 1,3-dichloropropene (Telone), dieldrin, and DDT (Tobacco Free Kids 2001). A common soil fumigant that is used in tobacco production in developing countries is methyl bromide. It is also used to sterilize greenhouses (Watts 1998). This substance is a significant contributor to ozone depletion.

High doses of herbicides and pesticides can be dangerous to workers and to the environment. These chemicals can cause damage to eyes, skin, and internal organs, and are potentially carcinogenic and mutagenic. Exposure to these chemicals poses a considerably higher risk to children than adults, since exposure in the early years can lead to a greater risk of cancer as well as damage to the development of children's nervous and immune systems. This is particularly worrisome since more and more children are working in tobacco fields as production shifts into developing countries. The tobacco companies and leaf buyers often work with local schools to insure that school schedules allow children to work in the fields (Tobacco Free Kids 2001). Runoff and leaching of these chemicals pollute waterways, affecting people who use those as a water source and harming freshwater biodiversity as well. Aquifers can also become contaminated, which in turn contaminates any wells that tap into these aquifers.

Tobacco fields can provide both food and cover for wildlife. However, that puts the wildlife at risk from pesticides. Insecticides are usually highly toxic to wildlife, while most fungicides and herbicides are only slightly toxic. Birds made sick by insecticides may neglect their young, abandon their nests, and become more susceptible to predators and disease. Many pesticides and herbicides that are not highly toxic can still be harmful to wildlife by reducing the food and cover that they need to survive or by contaminating water supplies. Runoff can decrease the aquatic foods necessary to the survival of aquatic animals (Barry 1991).

Soil Degradation

Tobacco is a demanding crop that depletes soil nutrients faster than many other crops. This is particularly problematic where soils are already characterized by low nutrient content. When tobacco is cultivated on the same land repeatedly with minimal rotation with other crops, there is a tendency for the soil to become exhausted and for crop pests to become endemic. This is why continuous tobacco cultivation requires ever increasing inputs of pesticides and chemical fertilizers.

Solid Waste

Perhaps the least obvious way in which smoking impacts the environment is though tobacco-related waste and litter. In the United States, the tobacco industry ranks eighteenth among all industries in the production of chemical waste. Globally, the tobacco industry produced an estimated 2.26 million metric tons of manufacturing waste and 210 billion metric tons of chemical waste in 1995. Nicotine is an example of the industry's toxic waste. Globally, the industry produces about 300,000 metric tons of nicotine waste per year (Novotny and Zhao 1999).

Litter is another problem. In the 1999 Clean Up Australia activities, the most common type of rubbish collected was cigarette butts, especially filters. Tobacco waste accounted for 9 percent of the ten most common items found (Tobacco Free Kids 2001). In the United Kingdom cigarette butts account for some 40 percent of street litter. UK smokers alone throw away 200 million butts and 20 million cigarette packages every day (Tidy Britain Group 1995). The International Coastal Cleanup Project reported that cigarette butts were 20 percent of all litter items found (Novotny and Zhao 1999). The filters are not readily biodegradable and can take from eighteen months to five years to break down. As the tobacco in the butts degrades it releases toxins into the soil. With tens of trillions of cigarettes sold and with considerable tobacco remaining in each when it is discarded, this is a significant source of toxic waste. Cigarette wrapping and packaging also contribute to litter. Even when disposed of properly, a huge amount of cigarette wrapping and packaging enters the waste stream and must be disposed of each year.

BETTER MANAGEMENT STRATEGIES

Of all the commodities in this book, tobacco is unique in that it is totally unnecessary to human sustenance, clothing, or intellectual development. Given the overall health, environmental, and economic impacts of the product, any attempts to reduce its environmental impacts should be seen as short-term activities with an overall goal of eliminating production, at least for human consumption, altogether. Even so, studies suggest that the number of smokers globally will increase for at least the next generation.

Any shift away from tobacco production must be made with some care. For many farmers—especially small farmers—there are a number of real economic incentives to produce tobacco. While such farmers may eventually stop producing tobacco altogether at some point in the future, in the meantime they need to understand how to produce it with fewer impacts, if for no other reason than to leave as many agricultural production options open for themselves and their children as possible.

At this time, most tobacco is grown by small farmers in developing countries. Most of them grow tobacco as their single, or most important, cash crop. The incentives available from governments and multinationals allow them to make more money growing it than other crops. In developed countries, subsidies and market protection also make tobacco production attractive to producers. These programs will need to be shifted as well.

There are a number of ways to reduce tobacco use and its environmental impacts as well as the environmental problems caused by tobacco production, processing, and marketing. The following approaches can help to accomplish a number of these goals. The first step is to improve monitoring and evaluation of environmental issues associated with tobacco production. Economic alternatives to tobacco production need to be identified and promoted. For example, the existing tobacco contract farming schemes should be used for the production of other products. Tree planting programs need to be developed and implemented to offset the problems of deforestation associated with the use of wood by the tobacco industry. Health education programs as well as anti-smoking programs should be promoted globally, especially in areas of production. These should include information about the hazards of secondhand or environmental tobacco smoke, as well as strategies to protect people from exposure to tobacco smoke. Fiscal policies to discourage the use of tobacco, such as taxes that increase faster than the growth in income, should be developed and promoted. These will include eliminating incentives that maintain or promote tobacco use. They should also include elimination of tobacco advertising, promotion, and sponsorship.

Implement Soil Conservation Strategies

In the meantime, however, there are ways to produce tobacco that have fewer environmental impacts. In the Unites States, for example, no-till production is a workable option for farmers trying to save topsoil and soil moisture and to make tobacco production more sustainable. To replace tillage, herbicides are required to kill the existing vegetation before planting. Periodic pesticide use is also required to minimize pest outbreaks. However, because this technique maintains ground cover at all times, it minimizes pesticide runoff as well as soil erosion. In addition it maintains higher levels of soil organic matter, which significantly improves soil water retention. A disadvantage of the use of the no-till technique is the need for specialized equipment for irrigation and to maintain weed control.

Use Integrated Pest Management (IPM) and Other Conservation Strategies

A comprehensive guide to better management practices for flue-cured tobacco can be found through the North Carolina Center for IPM at North Carolina State University. The center has produced general works on pest management as well as more specific works on insect pest management and even specific manuals on the control of tobacco budworms and stink bugs.

Studies have also shown that many pest problems can be controlled, though not eliminated, by using resistant tobacco varieties. Although no varieties are resistant to all pests and diseases, a farmer can choose a plant with resistance to major local diseases and pests. This will reduce considerably the use of pesticides.

In general, it makes both financial and ecological sense to reduce the use of expensive inputs. The risk of pesticide toxicity to wildlife, for example, can be reduced considerably by judicious use of pesticides and other agrochemicals, which is a cornerstone of integrated pest management (IPM) practices.

Improve Efficiency of Curing Process

Recent surveys of twenty-three tobacco-growing countries, including Brazil, found that an average of 5.5 kilograms of wood were used to cure 1 kilogram of green tobacco where flue curing is practiced. The average fuelwood use has fallen steadily over the past decades due to the introduction of more efficient furnaces and improved barn and drying shed designs (ITGA 1997).

Innovations in the United States have also focused on curing and include the use of insulation and circulating fans to increase the heating efficiency of drying barns. Also new regulations have led to the development of drying systems that control the amounts of tobacco-specific nitrosamines (or TSNAs). These compounds are a by-product of the drying process and a chemical reaction between nitrogen compounds in combustion and nicotine in the tobacco leaves. The presence of TSNAs is thought to be responsible for many of the damaging effects of tobacco on human health. With the retrofitting of drying barns with heat converters or exchangers it is possible to eliminate most detectable TSNAs (PBS 2001).

OUTLOOK

The most important questions that will affect tobacco production and consumption in the future revolve around liability issues and, related to that, government regulation and taxation of the industry. So long as tobacco can be produced and sold legally, there will be people who will buy it. However, if governments develop regulations regarding advertising, the number of new smokers recruited each year may decline dramatically.

Liability issues will continue to grow in developed countries, with the United States taking the lead but soon followed by the European Union when the health care and associated costs of the industry to society are fully recognized and accounted for. In response to these pressures, tobacco companies are likely to sell off assets. Philip Morris sold Miller Brewing company and 16 percent of Kraft Foods in a move which analysts think will eventually result in the sale of both companies. RJR has been more aggressive about the sale of its assets. It has sold all of its international business interests as well as its interest in Nabisco and all its holdings (Tobacco.com 2003). While such sales might give the appearance of generating funds to cover liabilities, they allow companies to split off their more promising income-generating centers. In this way, investors could be sheltered from increasing liabilities in developed countries such as the United States while tobacco companies in the United States are allowed to go bankrupt. Some of the money generated by sales has been used to buy back stocks from shareholders.

While the future is uncertain for tobacco, it is not likely that the industry will disappear soon. Projections show the number of smokers, as well as total consumption, to be on the rise. Tobacco companies are looking to China and developing country markets to increase their overall sales and profits. It is not clear that the governments in those countries realize, or at the very least are prepared to address, the long-term social and economic costs of increased domestic tobacco consumption.

REFERENCES

Anonymous. 2002. Bonkers: How Zimbabwean Farmers Unwillingly Subsidise Their Government. *The Economist*. Harare, June 15.

Barry, M. 1991. The Influence of the US Tobacco Industry on the Health, Economy and Environment of Developing Countries. *New England Journal of Medicine*. March.

BBC News. 2002. Second-Hand Smoke 'Causes Cancer'. *BBC News World Edition*. June 19. Available at http://news.bbc.co.uk/2/hi/health/2053840.stm.

Bennett. C. 2000. Brazil's Success Hurts Neighboring Argentina. *Tobacco Reporter*. May.

Brigden, L. W. 2002. *Viewpoint: Tobacco Marketing—Where There's Smoke, There's Deception*. International Development Research Center (IDRC) Report. Available at http://www.idrc.ca/ritc/en/publications/publications6.html.

Brown and Williamson. 2003. *Economic Impact Information: From the Farmhouse to the Auction Warehouses*. Available at http://www.brownandwilliamson.com. Accessed 2003.

——. 2001. *Golden Leaf, Barren Harvest—The Costs of Tobacco Farming*. November. Washington, D.C.: Campaign for Tobacco Free Kids.

DeBenedittis, P. 1999. *Future Trends in Tobacco Marketing: Minorities and Niche Markets*. Peter D & Co., Inc. July 8. Available at http://www.medialiteracy.net/research/.

Drobes, D. J. 2002. *Concurrent Alcohol and Tobacco Dependence Mechanisms and Treatment*. Available at http://www.niaa.nih.gov/publications/arh26-2/136-142.htm.

Ellison, K. 1997. Tobacco Farming Central Shifts to South America. *Miami Herald*. Sunday June 29th.

FAO (Food and Agriculture Organization of the United Nations). 2002. *FAOSTAT Statistics Database*. Rome: UN Food and Agriculture Organization. Available at http://apps.fao.org.

Gajalakshmi, C. K., P. Jha, K. Ranson, and S. Nguyen. 2000. Global Patterns of Smoking and Smoking-Attributable Mortality. In: Jha, P. and F. Chaloupka, editors. 2000. *Tobacco Control in Developing Countries*. Oxford: Oxford University Press for the World Bank and World Health Organization. Available at http://www1.worldbank.org/tobacco/tcdc.asp.

Gehlbach, S. H., L. D. Perry, W. A. Williams, and J. S. Woodall. 1975. Nicotine Absorption by Workers Harvesting Green Tobacco. *The Lancet*. No. 1.

Geist, H. J. 1999. Tobacco: A Driving Force of Environmental Change in the Miombo Woodland Zone of Southern Africa. Paper presented at African Environment: Past and Present. Oxford University. July.

——. 1998. How Tobacco Farming Contributes to Tropical Deforestation. In Abedian et al., eds. *The Economics of Tobacco Control*. Cape Town, South Africa: Applied Fiscal Research Center.

Goodland, R. J. A., C. Watson and G. Ledec. 1984. *Environmental Management in Tropical Agriculture*. Boulder, CO: Westview Press.

Hammond, R. and A. Rowell. 2001. *Trust Us: We're the Tobacco Industry*. Campaign for Tobacco Free Kids and Action on Smoking and Health. Available at http://www.ash.org.uk/html/conduct/html/trustus.html.

INFACT. 2003. *Help Stop the Marketing of Tobacco To Children*. Boston, MA: INFACT. Available at http://www.infact.org/youth.html.

ITGA (International Tobacco Growers Association). 1997. ITGA Response to Panos' Report

"Tobacco Killing Brazil's Forests." November. Available at http://www.tobaccoleaf.org/Issues/panosbrazil.htm.

Jacobs, R. et al. 2000. The Supply Side Effects of Tobacco Control Policies. In Jha and Chaloupka, eds. *Tobacco Control in Developing Countries.* Oxford, UK: Oxford University Press.

Jones, C. 1998a. Harvesting 'Virginia Crude' Tobacco Keeps Rest of Farm Afloat. *Richmond Times-Dispatch.* 28 June.

———. 1998b. A Virginia Grower Sees Offshore Battle Brewing. *Richmond Times-Dispatch.* 8 June.

Kilborn, P. T. 1994. Along Tobacco Road, A Way of Life Withers. *New York Times.* 28 August.

Kille, T. 1998. Big Domination. *Tobacco International.* September.

National Agricultural Statistics Service (NASS). 1999. *1997 Census of Agriculture.* United States Department of Agriculture. Available at http://www.nass.usda.gov/census.

NIOSH (National Institute for Occupational Safety and Health). 1996. Southeast Center Studies Ways to Prevent Green Tobacco Sickness. *NIOSH Agricultural Health and Safety Center News.* 4 August.

Novotny, T. and F. Zhao. 1999. Consumption and Production Waste: Another Externality of Tobacco Use. *Tobacco Control* 8:75-80.

O'Sullivan, B. and S. Chapman. 2000. Eyes on the Prize: Transnational Tobacco Companies in China 1976–1997. *Tobacco Control.* September 9(3). Available at http://tc.bmjjournals.com/contents-by-date.0.shtml.

PBS (Public Broadcasting System). 2001. *Search for a Safe Cigarette.* October 2. Boston: WGBH Educational Foundation.

Quandt, S., T. A. Arcury, J. S. Preisser, D. Norton, C. Austin. 2000. Migrant Farmworkers and Green Tobacco Sickness. *American Journal of Internal Medicine.* February.

Smith, M. 1999. Types of Tobacco. Available at http://www.ces.ncsu.edu/pitt/ag/tobacco/tobtypes.html. Raleigh, NC: North Carolina State University Cooperative Extension.

South African Broadcasting Company (SABC). 2000. Malawian Riot Police Keep Calm in Tobacco Growers Protest. SABC. 18 April.

Swoboda, F. and M. Hamilton. 1997. The Largest Independent Tobacco Merchants are Based in VA, but Their Growth is Abroad. *Washington Post.* 7 July.

Tidy Britain Group. 1995. *Inkpen Litterbug Report.*

Tobacco Free Kids. 2001. *Tobacco & the Environment.* Available at http://tobaccofreekids.org/campaign/global/.

Tobacco Reporter. 2001. International Cigarette Manufacturers. March.

Tobacco.com. 2003. RJR and Philip Morris Prepare for Bankruptcy. April 28. Available at http://www.tobacco.org/resources/general/030427chemer.html.

Tuinstra, T. 1998a. African Potential. *Tobacco Reporter.* November.

———. 1998b. Pressure Building. *Tobacco Reporter.* June.

UNCTAD (United Nations Commission for Trade and Development). 1994. *Handbook of International Trade and Development Statistics, 1993.* Geneva, Switzerland: UNCTAD.

———. 1999. *World Commodity Survey, 1999–2000.* Geneva, Switzerland: UNCTAD.

van Hulten, M. 1996. *Tobacco: Disastrous for the Third World and the Environment.* May. CAN (Clean Air Now) News. Available at http://www.nietrokers.nl/e/hulten.html.

Watts, R. 1998. Tobacco: Time to Branch Out. *African Farming.* September–October.

World Bank. 1999. *Curbing the Epidemic: Governments and the Economics of Tobacco Control.* Washington, D.C.: The World Bank.

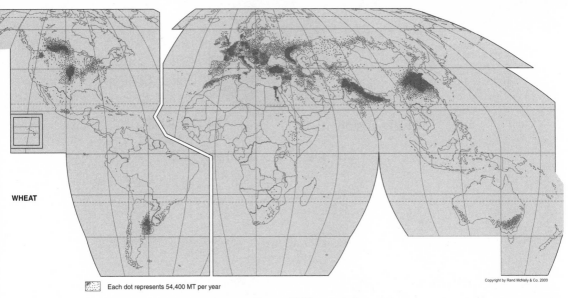

WHEAT

Each dot represents 54,400 MT per year

Copyright by Rand McNally & Co. 2000

WHEAT *Triticum* species

PRODUCTION		INTERNATIONAL TRADE	
Area under Cultivation	213.7 million ha	Share of World Production	22%
Global Production	585.0 million MT	Exports	129.0 million MT
Average Productivity	2,737 kg/ha	Average Price	$123 per MT
Producer Price	$140 per MT	Value	$15,897 million
Producer Production Value	$81,900 million		

PRINCIPAL PRODUCING COUNTRIES/BLOCS (by weight)	China, India, United States, Russia, Canada, Australia, Argentina
PRINCIPAL EXPORTING COUNTRIES/BLOCS	United States, France, Canada, Australia, Argentina
PRINCIPAL IMPORTING COUNTRIES/BLOCS	Brazil, Italy, Iran, Japan, Algeria, Egypt, Indonesia, Belgium, Morocco
MAJOR ENVIRONMENTAL IMPACTS	Habitat conversion Soil erosion and degradation Water and agrochemical use and pollution Burning of crop residue
POTENTIAL TO IMPROVE	Fair Breeding programs may take pressure off land when the wheat genome is better understood Inputs are higher per metric ton of production than many other crops Scale of production makes the adoption of many BMPs difficult Subsidies drive production and impacts in the U.S. and EU and globally as well

Source: FAO 2002. All data for 2000.

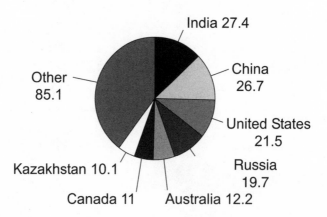

Area in Production (MMha)

India 27.4

China 26.7

United States 21.5

Russia 19.7

Australia 12.2

Canada 11

Kazakhstan 10.1

Other 85.1

OVERVIEW

Wild wheat, and its close relative barley, originated in Asia's Fertile Crescent. The archeological record suggests that people were collecting and eating wild wheat and barley as early as 17,000 B.C. Both bear their seeds on the tops of stalks that shatter when ripe, dropping the seeds to the ground where they germinate (Diamond 2002). This trait helps wild plants to reproduce, but makes it difficult for humans to gather the seeds. Though in the wild a mutation that causes the seeds not to drop can be lethal, this is attractive to human gatherers. Initially, it was unconsciously selected; stalks that did not drop their seeds were the only ones that could be harvested by humans after they ripened. After these seeds were gathered, they were probably dropped accidentally and later planted on purpose. In both ways the seeds became more dominant relative to their ancestors. It appears that wheat may have first been consciously planted as early as 8500 B.C. (Zohary and Hopf 2001; Lev-Yadun et al. 2000). Recognizable agriculture followed by at least 6000 B.C. The earliest evidence of people using bread wheat (the spelta variety, commonly known as spelt) is from 4700 B.C. in the Caucasus region between the Black and Caspian Seas (Zohary 2001).

Those who first domesticated foods such as wheat had a clear advantage over all but their closest neighbors. The surpluses agriculture produced freed parts of the population from hunting and gathering, which enabled them to develop the world's first metal tools, writing, armies, and eventually empires that allowed them to spread their food-producing technologies, among other things (Diamond 2002). As a result of these activities and subsequent domestication programs, wheat is now probably the most common plant on the planet, with rice a close second.

Early agriculturalists in Mesopotamia quickly found the production of cereals such as wheat to be unsustainable due to low and inconsistent rainfall, the soil erosion and degradation that followed deforestation, and salinization from irrigation (Diamond 2002). As the Mesopotamians destroyed their resource base, agriculture spread west into present-day Egypt and Israel and northwest into Europe. As farming shifted, so did political power. Three important precedents were established. First, agriculture was the basis not only of economic but also political power. Second, agriculture allowed and even encouraged population growth. With reliable food supplies and even surpluses, birth spacing could be reduced to one to two years. The fastest population growth rates presently occur in agricultural areas. And third, agriculture outstrips the ability of a region to regenerate nutrients that are the basis of productive and profitable agriculture, creating moving agricultural frontiers.

At least 5,000 years ago wheat had spread to North Africa (especially Egypt), Asia, and parts of Europe, areas where it has been in continuous cultivation ever since. The term ancient Egyptians used for wheat is thought to be *kamut*, which meant "soul of the Earth." Varieties of these traditional wheats are still in cultivation today in the United States.

PRODUCING COUNTRIES

The Food and Agriculture Organization of the United Nations (FAO) lists 122 wheat-producing countries. Globally, 213.7 million hectares of land were devoted to the cultivation of wheat in 2000. India (27.4 million hectares), China (26.7 million hectares), the United States (21.5 million hectares), and Russia (19.7 million hectares) dominate (FAO 2002). The four countries that have the most land in wheat production account for 45 percent of all land planted to wheat. The next nine countries with the largest amount of land devoted to wheat account for 34 percent of the global area devoted to wheat. Other significant producers with more than 5 million hectares under cultivation include Australia (12.2 million hectares), Canada (11.0 million hectares), Kazakhstan (10.1 million hectares), Turkey (8.7 million hectares), Pakistan (8.5 million hectares), Argentina (6.5 million hectares), France (5.3 million hectares), Ukraine (5.2 million hectares), and Iran (5.1 million hectares). Globally, more land is used to produce wheat than any other commodity. Many countries devote a quarter to half of all agricultural land (FAO 1996) to the production of this basic commodity, as shown in Table 16.1.

India, China, and Russia devote a tremendous amount of land to wheat but do not export significant quantities. The United States, France, Canada, Australia, and Argentina, by contrast, dominate wheat exports, accounting for 90 percent of all exports. Exports, however, account for only 10 to 20 percent of global production depending on the year (FAO 2002). Most countries attempt to produce enough wheat to meet internal demand, to be self-sufficient in the world's most basic foodstuff.

Global yields are low by comparison to rice because only one wheat crop can be grown each year. (In many areas other crops such as soybeans or hay are grown sequentially after the wheat is harvested.) Globally, average yields are about 2,737

TABLE 16.1. Percentage of Agricultural Land Devoted to Wheat, 1995

25–49 Percent		10–24 Percent	
Azerbaijan	Mongolia	Afghanistan	Macedonia
Belgium-Luxembourg	Morocco	Albania	Moldova
China	Nepal	Algeria	Netherlands
Czech Republic	Pakistan	Argentina	Paraguay
Denmark	Romania	Armenia	Poland
Egypt	Slovakia	Australia	Portugal
France	Switzerland	Austria	Russia
Greece	Syria	Bulgaria	Saudi Arabia
Iran	Tunisia	Canada	Slovenia
Iraq	Turkey	Georgia	South Africa
Israel	Turkmenistan	Germany	Spain
Italy	United Kingdom	Hungary	Tajikistan
Kazakhstan	Uzbekistan	India	Ukraine
		Jordan	United States
		Kyrgyzstan	Uruguay
		Lebanon	Yugoslavia

Source: FAO 1996.

kilograms per hectare per year. Several countries (Belgium, Denmark, Egypt, France, Germany, Ireland, Namibia, the Netherlands, New Zealand, Sweden, Switzerland, the United Kingdom, Zambia, and Zimbabwe) produce at levels that are more than double the global average. Of the largest producers, China has the highest average yields with 3,990 kilograms per hectare. Russia has the lowest with 1,350 kilograms per hectare (FAO 2002).

Global wheat production stands at about 585 million metric tons per year. This would be enough wheat to fill the cars of a train stretching 2.5 times around the Earth. The top four producers account for 47.8 percent of all production by volume (FAO 2002).

CONSUMING COUNTRIES

Foods that are made with wheat account for a major part of the diet for more than a third of the people on Earth. Given that most wheat is not exported, the main consumers of wheat are also the main producers. The main importers of wheat are Japan, Brazil, the countries of the former USSR, China, and Indonesia. Wheat imports are spread over most of the globe, however. The five leading importers account for only 28 percent of all wheat imports (FAO 2002).

Demand for wheat is increasing faster than population growth. According to Baenziger and Gill (2001) some two-thirds of the increased demand is based on population growth and one-third is due to increased income. What these statistics do not show is that wheat is increasingly used for animal feed rather than for direct human consumption. The direct consumption of wheat tends to provide a smaller portion of calories in the diet as incomes rise. For each 1 percent increase in real income, the demand for wheat increases by only 0.5 percent. Wheat's main attraction is that it can be used to make convenience foods that are ready to eat or quickly cooked for human consumption. These include a wide range of already-baked products and pasta, which is easily and quickly cooked.

PRODUCTION SYSTEMS

Wheat grows in a wide variety of climates and soils. However, good soil and moderate climate are both critical to good yields and healthy wheat crops. Wheat prefers fairly dry, mild climates. High-quality, disease-free seeds are also important, and of course wheat must be planted and harvested at the proper time if yields are to be optimal. Some longer-growing varieties of wheat are planted in the fall to give them a head start on the growing season, while others can only be planted in the spring.

Production systems changed very little in the first 6,000 years that wheat was cultivated. Wheat was originally sown, tended, harvested, threshed, and processed by hand. It was first irrigated some 6,000 years ago in the Tigris and Euphrates River area. In that region Sumerians first used animals to prepare fields and thresh the grain by walking on it. This was how wheat was produced until recently.

Changes in production began when mills were developed for processing, with wind or water power driving the millstones to grind the grain. Subsequently, animal-drawn and eventually self-propelled machines were introduced to plant, harvest, and thresh the grain in the field. Even so, in much of the world wheat is still planted and harvested by hand, with animals used only to plow the fields.

The mechanization of wheat farming seems to have driven the mechanization of agriculture in general. In 1834 McCormick invented the reaper, a horse-drawn machine that increased the harvest rate of wheat fourfold, from 0.8 hectares per day by hand to 3.2 by machine. Some fifty years later the Case steam-powered threshing machine revolutionized the ability of farmers to thresh grain in the field, even though all the grain still had to be shocked and hauled to the stationary threshing machines. Self-propelled machines were subsequently developed that cut and threshed as they moved through the field; they retained the grain while depositing the straw back on the field.

For nearly five millennia increases in wheat production were accomplished by increasing the acreage planted to wheat. In many parts of the world this was accomplished by plowing huge expanses of prairies, steppes, and pampas and turning under the native vegetation. This process destroyed the habitat of animals and literally turned upside down the soil habitat where most of the biodiversity and biomass resided.

Today wheat is the most widely grown crop in the world, but the area devoted to its cultivation has been fairly constant since the 1960s. Productivity has more than doubled since the mid-1960s. But per-hectare increases in yields through breeding programs for wheat have not kept pace with productivity gains in soybeans, corn, or rice. As a consequence, the area devoted to wheat has remained relatively stable while the area devoted to the other crops has increased because farmers find it more rewarding to grow other crops that give higher financial returns. New varieties of other crops (corn and sorghum, in particular) have been developed that are more drought-resistant, so these crops can now be grown in areas where wheat, barley, rye, or oats were grown in the past. Finally, wheat is mostly consumed by humans. But while demand is increasing faster than population growth, yield increases have exceeded demand for human consumption. Unless wheat becomes a major source of animal feed (and there are some indications of that trend, with 30 percent of wheat production now used for animal feed in the United States), the amount of land devoted to wheat is not likely to change. In fact, increases in yields may actually reduce the total amount of land devoted to the crop.

Between 1966 and 2000 wheat yields increased in the United States (from 1,890 to 2,780 kilograms per hectare) and Australia (from 1,050 to 1,820 kilograms per hectare). Yields in France started higher than other producers in 1966 and continued to climb at higher rates, from 3,260 kilograms per hectare to 7,490 kilograms per hectare. However, the most significant increases in yield have occurred in developing countries. China and India, for example, have shown some of the most significant improvements in yields over the same period. In China per-hectare yields have increased more than fourfold in the same period, from 850 to 3,990 kilograms per hectare. China has by far the highest average yields of any major wheat producer, yields that are nearly half again higher than those in the United States. In India, while the increased yields have not been so great they have still increased threefold (from 900 to 2,700 kilograms per hectare). India's yields are now nearly equal to those in the United States (Baenziger and Gill 2001).

Wheat is grown on some 4 percent of all the land area of the continental United States. About 17 percent of the 45,000 farmers who cultivate wheat produce two-thirds of the total crop. Wheat production in the United States is very chemical-intensive. According to the U.S. Department of Agriculture, wheat producers apply more than 7.3 million kilograms of pesticides to their fields every year, at average rates ranging from 1 to 3 kilograms per hectare. About half of winter wheat fields in the United States are sprayed with herbicides, while some 95 percent of fields of the spring and durum wheat varieties are sprayed. The most common pesticides applied include the herbicides 2,4-D, dicamba, MCPA (4-chloro-2-methylphenoxyacetic acid), metsulfuron, tribenuron, and chlorsulfuron (USDA 2000, as cited in Kimbrell 2002).

Nitrogen is the nutrient most frequently applied in fertilizer for wheat production in the United States. About 85 percent of the U.S. winter wheat crop receives nitrogen applications and about 10 percent requires potassium as well. Similar application levels apply to spring and durum wheat, but a larger proportion of those crops also receive applications of phosphate (Kimbrell 2002). Nitrogen, potassium, and phosphate are the three nutrients found in most fertilizers, as they are the ones most likely to be needed by growing plants.

Another major input for wheat production is fossil fuels. Wheat requires greater use of fossil fuels than corn, sorghum, or soybeans per metric ton of production because productivity is lower and there are few methods of conservation tillage that work for wheat. As a consequence, more machinery is used per ton of production than for other cereals or grains.

Wheat production can generally be categorized as intensive or extensive based on the use of machinery, agrochemicals, and water per hectare and per metric ton of product. Intensive systems require more inputs and generally result in higher production per hectare. Higher levels of production are found in more humid environments. In more arid areas wheat is produced at the margins of other agropastoral production systems and production is generally less intensive. As a rule of thumb, the more intensive the use of inputs the greater the environmental impact per area (but not necessarily per metric ton of production) for areas already in production. Once wheat production systems have been established, the overall environmental impact of production depends primarily on the inherent characteristics of the land, soils, and climate where the wheat is grown as well as the intensity, methods, and seasonality with which inputs are used (Runge 1994).

Table 16.2 compares the impacts of three different production systems—intensive in high-income countries, extensive in high-income countries, and intensive in low-income countries. The table describes and compares the inputs used, crop rotations employed, environmental impacts, and BMPs (Runge 1994).

Like other agricultural crops, wheat suffers from pests. Rust is the most destructive wheat disease. Caused by fungi, it attacks the leaves and stems and can significantly reduce yields. Grasshoppers and locusts are two of the more common insect pests. They contribute to annual losses of some 10 percent per year in the United States and sometimes much more in other producing countries. Weeds, too, are a common problem, robbing wheat of both moisture and nutrients.

Increased wheat yields in many developing countries have come largely as a result of public-sector investments, either from governments or through bilateral or multilateral agencies. In the United States, 41.4 percent of investments in wheat have come through the private sector (Frey 1996, as cited in Baenziger and Gill 2001). In Europe, due to highly flexible rotations and strong seed organizations, wheat breeding is done primarily by the private sector. Seed companies protect these markets by providing a full line of seeds, including wheat (Baenziger and Gill 2001).

Globally there are more than 30,000 known varieties of wheat. These fall into six main classifications—hard red winter wheat, soft red winter wheat, hard red spring wheat, hard white winter wheat, soft white wheat, and durum wheat. Of the 30,000 known varieties, only 1,500 or so bred for commercial production dominate what is grown today. Most of these varieties (80 percent or so) were developed by companies, universities, or research laboratories in the United States. Pioneer HiBred International alone produces more than 3,000 new genetic combinations annually, which are field-tested on 40,000 trial plots (Kimbrell 2002).While most of these varieties are not commercially viable, wheat development is still largely an American business. The other creators of commercial wheat varieties include Canada (with seventy-nine varieties to date), Mexico (forty-eight), Russia (twenty-six), and Australia (fifteen). China and India, the world's leading wheat producers, have produced a combined total of only five varieties (Kimbrell 2002). Unfortunately, while

TABLE 16.2. Comparison of Different Wheat Production Systems

	Intensive in High-Income Countries	Extensive in High-Income Countries	Intensive in Low-Income Countries
Examples	Paris Basin	Western Canada, Australia	India
Inputs used	Machinery Higher fertilizer use Higher pesticide use Some irrigation	Machinery Lower fertilizer use Lower pesticide use	New varieties Fertilizers Pesticides Irrigation
Crop rotations	Sugar beets/wheat/barley or corn/wheat/barley	Clover (in Australia)	
Environmental impacts	Leaching of nitrogen Nitrate and pesticide residues Water pollution	Soil erosion (wind and rain) Land degradation	Degradation Soil erosion (wind and rain) Salinization Alkalinization Deforestation Shifting cultivation
Improved practices	Better rotations Reduced fertilizer and pesticide use	Conservation tillage	Reduced fertilizer and pesticide use

Source: Adapted from Runge 1994.

it is easy to create nearly endless numbers of varieties through manipulation, very few of them are commercially viable, and farmers tend to rely on a small number of varieties that they plant for a shorter time.

Cultivated wheat is polyploid; it has more than two sets of chromosomes. Wheat used for bread is a hexaploid (has six times the original chromosome number), while the durum or hard wheat that is used for pasta is tetraploid (has four sets of chromosomes). Wheat has extensive germ plasm resources within the cultivated and related species gene pools. Because wheat is polyploid, it tolerates chromosome loss, deletions, and additions much better than plants with only two sets of chromosomes (Baenziger and Gill 2001). The origin of wheat provides for its extensive germ plasm resources. The hexaploid bread wheat genome, for example, is comprised of some 16 billion base pairs (chemical compounds that make up genes).

Improvements in wheat yields have come partly from technology and partly from improved genetics. Technological improvements have revolved around the use of agrochemicals, water, mechanization, and developing and communicating better information about production. Breeding efforts for wheat, including those from the green revolution, have focused on a few key traits (Baenziger and Gill 2001):

- Semidwarfing genes, which focus the plants' energy on producing more grain and shorter stalks that are more easily harvested and less susceptible to wind and storm damage;
- Day-length insensitivity, which allows for broader adaptation of wheat so that it can be produced in different parts of the world; and
- Systematic international germ plasm exchange, which means that new varieties and traits can be experimented with and can be crossed more easily with local varieties.

In the future, many of these same traits will drive genetic advances in wheat breeding. International germ plasm exchanges will certainly be the basis for most wheat breeding programs. However, a better understanding of wheat genomics and genetic manipulation will drive the breeding methodology because of the tools now available to plant breeders. And, finally, improvements in related technologies such as mechanization and better on-farm management practices, coupled with better information, will continue to generate yield improvements in the future (Baenziger and Gill 2001).

The most impressive gains will be made in wheat when the wheat genome and the variations are more fully understood. It will be important to understand how wheat is dispersed in the gene-rich regions where it is found in the wild and where many different varieties are cultivated. In particular, it will be important to understand which types of genes are in those regions, what particular environmental factors they are most or least adapted to, and what proportion of overall wheat genes (or genes of its relatives) are in those regions (Baenziger and Gill 2001). Finally, understanding which genes are associated with which characteristics—for example, yields, disease resistance, drought tolerance—will also reduce the costs of and improve the performance of genetic manipulation. All of these suggestions can be done with the same technology as those used to create transgenic varieties, but without the introduction of any genetic material from unrelated species. There has been work, for example, to create a transgenic Bt wheat which would produce its own chemicals as a defense against soil-based pests.

New conditions will require adaptations in the varieties being grown. The reliance of producers on only a few varieties of wheat can have severe repercussions. In 1996, for example, a fungal disease known as Karnal bunt swept though the wheat belt of the United States, ruining more than half of the crop and leading to the quarantine of more than 116,000 hectares. As producers come to rely on a smaller and smaller number of varieties, such disease susceptibility can be expected to become more common. To date the development of new wheat varieties has largely been seen as a legitimate government expenditure, since wheat is a basic food and thus its development can be seen as a food security issue. Certainly, the development of new varieties has benefited producers and consumers the world over. It is hard to argue that this is not a legitimate social expenditure for the global consumers of wheat. However, when the technology is developed with government funding, it is a legitimate question as to whether it should be given to seed companies who not only profit from it but can also patent the technology.

PROCESSING

Wheat processing commences with harvesting and threshing. It includes drying the grain, removing the husks and any other foreign matter, and removing the bran, ending up with the final milling of the grain into white flour. The development of water- and wind-driven mills to grind wheat into flour changed agriculture fundamentally. Mills were expensive to build, maintain, and operate. Consequently, they were usually owned by a very large landowner, the church, or royalty. Each farmer was charged a fee (often one-sixteenth of the total grain milled) to mill the wheat into flour. Over time, smaller mills were destroyed or made illegal as a way to force people to use the royal or church-owned mills. In the end this forced families (including farmers) to buy flour rather than make it themselves (CyberSpace Farm 2002a).

Technological developments have allowed cleaner, more uniform flour to be produced using much less labor. By 1870 flour mills in the United States were far more automated and required only three workers to produce a much larger volume of flour than ever before. Over time mills have become larger in size, fewer in number, and more specialized. For example, in 1873 there were 23,000 mills in the United States grinding wheat, corn, rye, and feed grains. By 1993 there were only 205 wheat mills, and Kansas milled almost 10 percent of all the flour in the United States (CyberSpace Farm 2002b). Most countries have their own milling operations, whether they produce wheat or not. It is easier to ship and store wheat as grain than it is as flour.

Processing wheat is energy-intensive and produces a lot of dust and organic matter as by-products. Over time each of these by-products has been captured and sold into other product lines. Most wheat is milled into white flour, which does not contain the germ or bran. When white flour is produced, the wheat germ and wheat bran are sold for other uses. Whole wheat flour contains both wheat bran and white flour. The flour milling process is normally integrated into the packaging process and the marketing of the different products. Therefore, flour milling companies control more of the sale price than the retailer; these companies also have more control over the price paid for grain than the producer of the raw grains.

Most of the processing of the grain involves dry product. Drying improves the storage life of grains and makes them easier to process. Energy (usually natural gas or electricity) is used to dry grain, either on the farm or in nearby grain elevators, to keep the wheat from spoiling. In addition, considerable energy (in the form of gasoline and diesel) is required to transport the product.

PRODUCT SUBSTITUTES

For human food, there are many substitutes for wheat, but few of them are as well accepted around the world. Oats, barley, and rye are substitutes, but total consumption and consequently the areas in production of all have declined over the past 100 years. Corn, rice, or manioc flour is preferred in parts of Latin America, Asia, and Africa. As incomes increase in these areas, however, consumption tends to switch, at least in part, to wheat.

As societies have become wealthier an increasing portion of their calories has come from meat and vegetable oils. From a strictly caloric point of view, these too must be seen as product substitutes.

Wheat is used increasingly in animal feed at least in the United States, the European Union, and Australia. This is true both for terrestrial animals and increasingly for aquaculture feeding operations as well.

MARKET CHAIN

The market systems for wheat are similar to those for corn and soybeans and have changed over time in the same way. In the nineteenth century most wheat was bagged at the farm and sold in the same bag all the way to the mill that processed it into flour. With the creation of commodity markets, wheat was graded when it first entered the market and was sold to a trading company. Over time it became easier, more efficient, and more competitive to store, transport, and process fewer varieties as each had slightly different characteristics and required separate handling and storage facilities before, during, and after processing.

Wheat is given one of six grades when it is brought to central grain elevators. Grading depends on the weight and the quality of wheat. Wheat of similar grade and the same variety is stored in the same place. Thus, wheat from one area is substitutable for wheat from another.

At this time, most commercial production is sold to trading companies that have large holding elevators both in the areas of production as well as in or near large grain markets. These grain elevators dominate the landscape. Wheat is transported by truck, rail, or barge to centers where it is made into flour or exported to other countries. Wheat that is exported is independently graded and inspected by the government.

Some of the grain trading companies also make flour. Most, however, sell their grain to flour companies. The flour mills are quite large in order to take advantage of economies of scale. If anything, the wheat market chain is even more concentrated (e.g., involves fewer companies) than that of other cereal grains as wheat processing is mostly to make flour and very little of that is sold directly to consumers. Normally, flour is held and sold, as needed, to the manufacturers who make it into food products.

MARKET TRENDS

Between 1961 and 2000 total production of wheat increased 163 percent, the amount of wheat traded internationally increased 179 percent, and the price of wheat declined by 61 percent (FAO 2002). In 1961 total global wheat consumption for all purposes was 225 million metric tons, of which 9 percent (19 million MT) was fed to livestock. Both total consumption and feed consumption rose steadily over the next three decades. By 1992 total wheat consumption had more than doubled to 562 million metric tons. In addition to increases in demand for human consumption, by 1992, wheat fed to animals had risen more than fivefold, to 107 million metric tons, accounting for 19 percent of total consumption worldwide.

Several factors have affected the world wheat market. Russia had a 40 percent drop in production in 1998 that buoyed the world market, but Russia's devaluation of its currency made imports expensive so other cereals and starches were substituted whenever possible. In addition, the price of pork has collapsed in the United States and Europe. While this first affected corn and soybean prices, wheat prices quickly fell as well since all are used in feed (UNCTAD 1999). It is clear from this example that the price of wheat is linked more to meat prices in developed countries than to production and consumption in any single country.

When Cargill (the second largest grain trading company) bought Continental (the fourth largest), the new grain trading company had a 17.2 million metric ton (631 million–bushel) capacity. That made it the single largest grain-trading company in the world. Archer Daniels Midland (ADM) had been the largest with a capacity of 16.6 million metric tons (610 million bushels). Monopolies such as this tend to lower prices paid to producers, as there are fewer buyers for producers to choose from. An increasing number of producers, for example, have only one buyer for their product. While markets may be more transparent than ever before, a lack of competition tends to allow the big grain traders to pass a lot of their costs on to the producers as well as to the buyers. And since fewer and fewer companies dominate the global grain trade, these same conditions are true internationally as well. In fact, the one thing that the large grain trading companies have in common is that they are not publicly held companies. This means that their profit margins are not public. Even if all the different producers and buyers could organize themselves, there would be less room to negotiate price from either the supply or the demand side.

ENVIRONMENTAL IMPACTS OF PRODUCTION

The FAO has developed a qualitative assessment of the environmental impacts of various commodities (Runge 1994). This assessment includes wheat in a more general category of grains. One of the main environmental impacts of grains such as wheat is the destruction of natural habitat due to field enlargement or creating new fields following the abandonment of degraded areas. This in turn causes the displacement of indigenous species and biodiversity loss, due especially to the promotion of high-yielding varieties. Other major environmental problems are soil erosion, soil degradation, and water use and pollution. The increasing use of manufactured fertilizers and pesticides and creation of dust from dry milling also cause environmental problems.

Despite these problems, wheat has a fairly positive role because much of it is planted in the late fall, winter, or very early spring depending on the location and the variety. Due to its "off-season" nature it provides ground cover and reduces soil erosion. It requires less fertilizer and pesticides per hectare than many other crops (but often more per ton produced), and most wheat is not irrigated.

A broad range of pesticides is used on wheat including herbicides, soil fumigants, insecticides, and fungicides. While the quantities used per hectare are not large, because wheat is grown on so much land the total quantities of chemicals used are quite large. More importantly, these chemicals end up in runoff and freshwater systems if soil is not protected from erosion and if there is insufficient organic matter and mulch to bind and hold them.

Habitat Conversion

Given that wheat is grown on more land than any other single crop, its historical impact on the environment has been considerable. Habitat conversion for cultivation has been a very large environmental impact from wheat production in the past. The production of wheat, in fact, has created many of the current agropastoral landscapes that are seen as "natural" in Europe, northern Asia and the United States.

Habitat conversion continues to this day in such areas as the western United States, where subsidies make wheat production profitable in marginal areas that could not be farmed profitably otherwise. Many of the areas now under cultivation include some of the few remnants of blue stem prairie that had never before been plowed. While globally grasslands and savannas are not as biodiverse as many other terrestrial ecoregions, they are nonetheless unique. Furthermore, they provide essential ecosystem services (e.g., overall water retention and runoff and carbon sequestration) that biodiversity in other regions depends upon.

The expansion of wheat production into fragile ecosystems of the Northern and Southern Hemispheres would threaten remaining biodiversity. While the consumption of wheat is certainly increasing in tropical countries, most production is still in the colder temperate regions because tropical wheat remains a poor producer compared to that produced in temperate zones. Intensive breeding programs could certainly change this in the future. If that happens, it could become possible and even profitable to grow wheat on the savannas of such places as Africa.

Most of the temperate areas that are suited for wheat production are already under cultivation. The remaining areas include savannas with fewer water resources. To become productive these areas will require huge investments in irrigation infrastructure that will drastically change ecology and habitat.

Soil Erosion and Degradation

Because landscapes and subsurface geology vary widely, the impacts of grain production on land resources also vary. Soils vary tremendously within a given field as well as over broader landscapes. Soil characteristics such as depth, particle size, nutrient composition, organic matter content, and physical chemistry are all altered by wheat production. These characteristics, which determine the capacity of soil to produce grain, must be restored or rebuilt over time if an area is to remain productive.

It is clear from historical records that wheat production can dramatically affect soil conditions and nutrient cycles. Cultivation of wheat has led to converting previously fertile areas into dust bowls and deserts, as in North Africa in the first millennium and parts of North America in the early twentieth century. As populations increase, agricultural technology changes, and human-induced soil degradation increases, the land devoted to wheat production increases or at the very least shifts to new agricultural frontiers.

Continuous cropping of wheat and other grains often depletes nutrients even in the most fertile soils. Replacement can occur through fallowing, crop rotation, leaving crop residue on the surface, applying animal manure, or using manufactured chemical fertilizers. Applying too much nitrogen or any other nutrient can cause imbalances that harm plants as well as the environment, especially if other essential nutrients (such as potassium) are limiting yields or if the applied element is already in sufficient supply in the soil. There are several methods of application, but the best are those that put the nutrient in the soil rather than on top of it. Nutrients applied to the surface are more easily leached away. These nutrients, if allowed to leave the field through runoff or leaching, will induce plant growth (including algal blooms) at their new location.

Water Use and Pollution

There are several impacts of wheat production on water resources. First, because many countries see wheat as a strategic food crop and want to be self-sufficient in basic food production, the crop is often grown in areas with insufficient rainfall to support it. Wheat is the second most irrigated crop cultivated globally. Irrigation always brings with it the problem of soil salinization, which at low levels reduces yields but

rotating crops, and using polyculture production systems are often abandoned. This simplification of production practices tends to accelerate the degradation of the soil and the land resource base in general.

Forms of production that are more conserving or restoring of the soil can still be practiced if producers are convinced that they are economically viable. Fallowing (in which land is removed from wheat cultivation and left to pasture or growing soil-building cover crops) provides a good example. Most nutrient depletion occurs in the upper level of the soil where the roots of the wheat are concentrated. Enriched fallows of some seven years (where plantings are chosen to increase fertility) can build up soil fertility. For example, legumes grown to fix nitrogen and deep-rooted cover crops to scavenge for native or leached potassium, phosphorus and other trace elements and pull them to the surface make it easier for short-rooted crops such as wheat to utilize the nutrients. In addition to reducing the need for fertilizers, such extended fallows also increase the soil's organic matter tremendously and restore soil biodiversity and soil structure. These, too, have a positive impact on future production. Thus, fallowing is both a soil conservation and rejuvenation measure that can help to return soils to their former vitality.

Fallowing also provides temporary habitats and increases foraging areas for wildlife. Many producers in developed countries have found that encouraging wildlife in such areas allows them to make money by selling hunting rights, mostly for game birds but also for some mammals. Depending on the price of wheat, such hunting fees can represent 5 to 25 percent of net income. The major problem with this strategy, at least in the United States, is that most wheat-producing areas are located far from urban areas and few accommodations for hunters are available. By contrast, in developing countries, attracted wildlife can be an important source of protein for farm families.

Another possible strategy to reduce the financial impact of an extended fallow is through government-sponsored programs. In the United States, the Conservation Reserve Program (CRP) provides financial support for ten years. Producers can sign up part of their land for CRP to fallow those areas. Undertaking this approach on a partial basis would allow land to rotate into and out of production. Even if this is not the intended use of the CRP or other subsidy programs, adapting it in this way could help restore productivity in large areas of wheat cultivation in the United States and other areas as well. In Brazil, a similar system that does not involve government support involves cropping for three to five years followed by the use of the same land for pasture for seven to ten years.

Unfortunately, agricultural credit and investment systems do not incorporate these features as regular measures for consideration. Fallowing and crop rotations are not seen by financial institutions as being good investments, even though they improve the resource base and reduce future costs for expensive fertilizers and pesticides. In addition, many agricultural extension agents often do not understand the value of such systems and denigrate them to local farmers. This reality is gradually shifting, however, and soil and environmental rehabilitation strategies are increasingly seen as worthwhile investments.

Avoid Burning Crop Residue

Wheat is grown in many parts of the world with relatively short growing seasons. As a consequence, the stubble and crop residues do not have much time to decompose. Even when plowed into the soil, straw can take several years to decompose, especially in climates with short, dry summers and long winters. As a consequence, burning is a common practice in many parts of the world where there is an insufficient market for straw. Burning wheat stubble not only destroys valuable organic matter but also causes air pollution and releases carbon dioxide into the environment, therefore becoming an environmental problem.

It is possible, however, to use stubble to improve the soil. If slow decomposition is an issue, it can be inoculated with decomposing microorganisms before it is plowed under (Panfilo Tabora, personal communication). As the wheat stubble breaks down it is converted into organic matter in the soil, which then binds with water, fertilizers, and pesticides, reducing the runoff from all of these.

Increasingly, farmers in developed countries are selling their straw to particle- and fiberboard manufacturers, who use it to make engineered timber. By the early 1990s some wheat producers were making 15 to 25 percent of their net income from the sale of straw. At that time they were lamenting the dominance of shorter wheat varieties, which, while more storm- and drought-resistant, produce less straw.

Improve Water Quality and Reduce Water Pollution

Wheat producers can reduce the overall impacts of production by managing their use of water more efficiently. While the major wheat producers do not irrigate, this is not the case in parts of Europe, the Middle East, and Asia where irrigation is most commonly undertaken by gravity-fed flooding of fields. This is not an efficient use of water. Moreover, whether wheat's water needs are met by rainfall or irrigation, water will be used more efficiently by the plant if there is more organic matter in and on the soil to absorb it and release it more gradually. The elimination of burning would help to build up organic matter as would various forms of conservation tillage. Such practices reduce effluent runoff and pollution.

As the scale of wheat production has increased, more and larger machinery has become the norm. However, per hectare wheat production is less productive than most other cereal grains so farmers tend to farm more land to make the same or even less money. Under these circumstances, it is difficult for producers using large machinery to protect lower-lying areas of fields that are most affected by runoff. For most producers, it is easier to cultivate such areas during planting than to leave them in grass which reduces erosion but requires far more precision to engage and disengage both the machinery to till the soil prior to planting as well as the planter itself. The maintenance or reestablishment of grass-covered waterways in wheat fields would reduce dramatically overall runoff and pollution issues. Similarly, terraces would reduce runoff and erosion, but depending on the amount of land moved, could make it difficult to maneuver large machinery. Also, the price of wheat may simply not be high enough to justify such earth works.

OUTLOOK

Global consumption of wheat and wheat products is increasing only at about the rate of production increases. An increasing portion, nearly 20 percent, of all wheat is being fed to animals. Because of increased yields, the amount of land devoted to wheat production is stable to declining. These trends are not expected to change.

A major factor that could affect the overall consumption of wheat is the market for other substitutes, either cereals or starches. For example, rice production has peaked in many areas, so with populations continuing to increase it appears that consumers may switch to wheat or other foods out of necessity. Corn, barley, sorghum, and millet are also important substitutes for wheat that, depending on shifting cultural values, may have increased or decreased levels of consumer preference in the future. Finally, potatoes and cassava are also both increasing as sources of calories in the diet globally.

Recent decisions to abandon Bt and other GMO research on wheat imply that there will not likely be major changes in production practices or overall production levels in the near future. But this could change in the future. Another outstanding issue is whether the U.S. government in particular or other governments more generally will continue to invest in genetic research and breeding programs that could increase wheat yields to new levels or expand even further the conditions under which cultivation can be undertaken. In all likelihood, neither the production nor consumption of wheat is likely to change a great deal in the near future.

REFERENCES

Associated Press. 1998. Pesticide Deforming the Frogs? *Channel 4000*. March 23. Available at http://www.channel4000.com/news/stories/news-980323-062043.html.

Baenziger, P. S. and K. S. Gill. 2001. *Give Us This Day Our Daily Bread: The Future for Wheat*. Paper presented at the American Association for the Advancement of Science Annual Meeting, San Francisco, CA.

Betz, V. M. 2002. Identification of the Initial Site of Einkorn Wheat Cultivation. *Athena Review*. 12(1).

Bonjean, A. and W. Angus. 2001. *The World Wheat Book—A History of Wheat Breeding*. Andover, UK: Intercept.

Cabello, G., M. Valenzuela, A. Vilaxa, V. Duran, I. Rudolph, N. Hrepic, and G. Calaf. 2001. A Rate Mammary Tumor Model Induced by the Organophosphorous Pesticides Parathion and Malathion, Possibly Through Acetylcholinesterase Inhibition. *Environmental Health Perspectives*. 109(5): 471–479.

Chapman, G. P. 1992. *Grass Evolution and Domestication*. Cambridge, England: Cambridge University Press.

CyberSpace Farm. 2002a. *Planet Wheat: A Short History of Wheat*. Kansas Women Involved in

Farm Economics (WIFE), sponsor; High Plains Journal, host. Available at http://www. cyberspaceag.com/wheathistory.html. Accessed August 2002.

——. 2002b. *Planet Wheat: A Short History of Flour Milling*. Kansas Women Involved in Farm Economics (WIFE), sponsor; High Plains Journal, host. Available at http://www. cyberspaceag.com/flourmillinghistory.html. Accessed August 2002.

Danielson, S. 2002. Pesticides, Parasite May Cause Frog Deformities. *National Geographic News*. July 9. Available at http://news.nationalgeographic.com/news/2002/07/0709_020709_ deformedfrogs.html.

Diamond, J. 2002. Evolution, Consequences and Future of Plant and Animal Domestication. *Nature* 418(8 August): 700–707. London: Macmillan.

FAO (Food and Agriculture Organization of the United Nations). 2002. *FAOSTAT Statistics Database*. Rome: UN Food and Agriculture Organization. Available at http://apps.fao.org.

——. 1996. *Production Yearbook 1995. Vol. 49*. FAO Statistical Series No. 130. Rome: UN Food and Agriculture Organization.

Frey, K. J. 1996. *National Plant Breeding Study 1*. Special Report 98. Ames, Iowa: Iowa State University.

Kimbrell, A., Editor. 2002. *Fatal Harvest: The Tragedy of Industrial Agriculture*. Washington, D.C.: Island Press.

Lev-Yadun, S., A. Gopher, and S. Abbo. 2000. The Cradle of Agriculture. *Science* 288, 1602–1603.

Mills, P. K. 1998. Correlation Analysis of Pesticide Use Data and Cancer Incidence Rates in California Counties. *Archives of Environmental Health*. 53(6): 410–413.

Milwaukee Journal Sentinel. 1999. Pesticide Linked to Deformed Frogs, Studies Indicate. *Milwaukee Journal Sentinel*. October 6. Milwaukee, WI.

Reif, J. S. 1999. Companion Dogs as Sentinels for the Carcinogenic Effects of Pesticides. *Pesticides, People & Nature*. 1(1): 7–14.

Robertson, H. 2002. *Triticum (Wheat Genus)*. Cape Town, South Africa: Iziko Museums of Cape Town. Available at http://www.museums.org.za/bio/plants/poaceae/triticum.htm.

Runge, F. 1994. "The Grains Sector and the Environment: Basic Issues and Implications for Trade." Draft discussion paper prepared for the Intergovernmental Group on Grains of the UN Food and Agriculture Organization.

Smil, Vaklav. 2000. *Feeding the World—A Challenge for the Twenty-First Century*. Cambridge, Mass: MIT Press.

UNCTAD (United Nations Commission on Trade and Development). 1999. *World Commodity Survey, 1999–2000*. Geneva, Switzerland: UNCTAD.

USDA (United States Department of Agriculture). 2000. *Agricultural Chemical Usage Field Crops Summary 2000*. Washington, D.C.: USDA.

Wilford, J. N. 1997. New Clues Show Where People Made the Great Leap to Agriculture. *The New York Times*. November 18.

Zohary, D., and M. Hopf. 2001. *Domestication of Plants in the Old World (3rd ed.)*. Oxford: Oxford University Press.

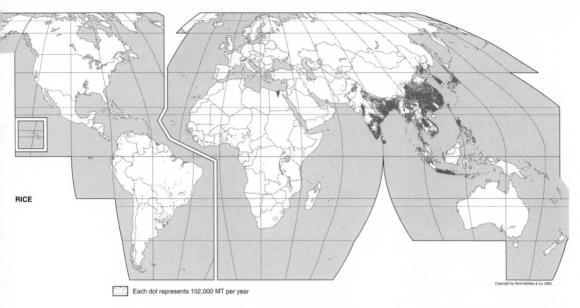

RICE

Each dot represents 102,000 MT per year

Copyright by Rand McNally & Co. 2000

RICE *Oryza sativa*

PRODUCTION
Area under Cultivation	154.1 million ha
Global Production	600.6 million MT
Average Productivity	3,897 kg/ha
Producer Price	$176 per MT
Producer Production Value	$105,970 million

INTERNATIONAL TRADE
Share of World Production	4%
Exports	23.6 million MT
Average Price	$279 per MT
Value	$6,460 million

PRINCIPAL PRODUCING COUNTRIES/BLOCS (by weight)
China, India, Indonesia, Bangladesh, Vietnam, Thailand

PRINCIPAL EXPORTING COUNTRIES/BLOCS
Thailand, Vietnam, China, United States, Pakistan, India

PRINCIPAL IMPORTING COUNTRIES/BLOCS
Indonesia, Côte d'Ivoire, Iraq, Iran, Saudi Arabia, Nigeria, Brazil, Japan

MAJOR ENVIRONMENTAL IMPACTS
Habitat conversion and biodiversity loss
Soil erosion and degradation
Agrochemical use
Water use and pollution
Production of greenhouse gases

POTENTIAL TO IMPROVE
Good
BMPs are known and cost-effective
Rice is a food security issue, governments will help improve
    production
Genetics (better understood for rice than for other crops) will lead
    to improved production and reduced impacts

Source: FAO 2002. All data for 2000.

Area in Production (MMha)

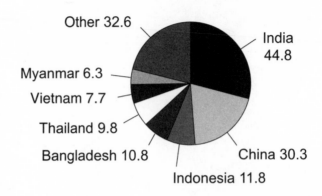

Other 32.6

Myanmar 6.3

Vietnam 7.7

Thailand 9.8

Bangladesh 10.8

India 44.8

China 30.3

Indonesia 11.8

OVERVIEW

For most of the world, rice is not only the staff of life but quite literally the stuff of life as well. In many parts of Asia the word for rice literally means food. The Chinese greet each other not by inquiring vaguely "How are you?" but by saying "Have you eaten yet?" It is understood that to eat is to eat rice. A person in China with job security is said to have "an iron rice bowl," while one who has been fired has had his "rice bowl broken." Khush (2000) has noted similar indications of the importance of rice in many Asian countries. In Thai, *khao* means both rice and food. The Emperor in Japan is considered the living embodiment of the ripened rice plant. In Bali it is believed that the god Vishnu caused the Earth to give birth to rice, and the god Indra taught people how to grow and eat it. In the Philippines rice is used at all important rituals. In fact, to show their mourning, relatives do not eat rice for several months after a death in the family (Khush 2000). Rice is not just food in such countries; it is culture.

Traditionally, hundreds or even thousands of rice varieties have had local cultural significance in rice-growing regions. Growing urban populations, however, have forced an increasing emphasis on producing large quantities of a few rice varieties, rather than smaller quantities of many varieties. The result of the "green revolution," is that these new high-yield varieties make more rice available at a lower cost. They also require increased inputs, contrasting sharply with previous cultivation techniques.

It is the demand for rice as a cheap food that is winning out over traditional varieties. Globally, the population of rice consumers is increasing at the rate of 1.7 percent per year compared to overall population growth of 1.3 percent (Khush 2000). Half of the world's projected population of 8 billion in 2025 will be rice consumers. Today, more than a billion rice consumers live in poverty. They have limited access to food and cannot afford to pay any more for rice. All things being equal, to feed these people means that rice production needs to increase by 35 to 40 percent by 2025 yet maintain a stable or even lower price (Khush 2000).

PRODUCING COUNTRIES

The Food and Agriculture Organization of the United Nations (FAO) (2002) reports that 114 countries produced rice in 2000. The total amount of land in rice is approximately 154.1 million hectares. More than 90 percent of the world's rice is grown and consumed in Asia. It is grown on about 11 percent of the world's cultivated land (Khush 2000).

The two main rice producers in the world, by area, are India with 44.8 million hectares under cultivation and China with 30.3 million hectares. Also important as major producers are Indonesia (11.8 million hectares), Bangladesh (10.8 million hectares), Thailand (9.8 million hectares), Vietnam (7.7 million hectares), and Myanmar (6.3 million hectares). These seven countries represent 78.3 percent of the land planted to rice and nearly 81 percent of the 600.6 million metric tons of rice produced globally in 2000. Other significant producers include the Philippines, Brazil, Pakistan, Nigeria, Cambodia, Nepal, the United States, Japan, Madagascar, and Korea (FAO 2002).

Total annual production was estimated at 600.6 million metric tons globally in 2000. This was a world record for production. The main producers by weight are China, India, Indonesia, Bangladesh, Vietnam, Thailand, Myanmar, and Japan (FAO 2002).

The FAO (2002) reported that the global average yield in 2000 was 3,897 kilograms per hectare per year. Australia, Egypt, and Greece were the only countries whose average yield exceeded twice that of the rest of the world.

CONSUMING COUNTRIES

Rice provides 23 percent of all the calories consumed by the world's population. This is more than wheat (17 percent) and corn/maize (9 percent). For the low-income populations in Asian countries, however, rice accounts for more than 50 percent of caloric intake. In such countries, the average adult eats about 160 kilograms of milled rice each year, which amounts to about 0.5 kilograms each day (Khush 2000). In the United States, by contrast, rice accounts for less than 2 percent of the per capita caloric food intake; the average American consumes less than 10 kilograms each year.

Most rice is consumed in the producing country. In 2000 about 23.6 million metric tons, or only 6 percent of global production, was exported. The main exporters are Thailand, Vietnam, China, the United States, Pakistan, and India (FAO 2002).

The largest importers are Indonesia, Côte d'Ivoire, Iraq, Iran, Saudi Arabia, Nigeria, Brazil, and Japan (FAO 2002). Depending on the year, China can also be a significant importer. Indonesia imported more rice in 1998 than any other country in history. As a result, the government began to encourage farmers to produce substitutes such as corn, cassava, and sweet potatoes. With the exception of Middle Eastern countries, rice imports shift from country to country each year as local production is reduced by adverse weather and imports are required to supply food needs. Middle Eastern countries must import nearly every year, as local supplies never satisfy demand.

The important thing to remember about rice (and other important sources of calories as well) is that its significance does not derive from feeding subsistence farmers. Rice and other major calorie sources are seen as essential because they feed the urban poor. Put another way, rice feeds the cities of Asia and keeps them politically stable at the same time. The policies of rice-consuming countries are aimed at keeping rice available and affordable.

PRODUCTION SYSTEMS

Rice is grown under a wide range of conditions. It is grown from Australia to northern China and from below sea level in India to 3,000 meters above sea level in Nepal (Khush 2000). Broadly speaking, there are two main kinds of rice: upland and lowland. Upland rice is characterized by cultivation that utilizes rainfall instead of irrigation. Pesticides, fertilizers, and improved seed varieties may all be used in the production of upland rice. Lowland rice tends to be produced on flat areas with irrigation and an entire package of inputs.

Upland rice is grown on open, often rolling lands. It is rain-fed, and there is no standing water in the rice fields. Depending on the year, it can suffer from prolonged water shortages. Yields average only 1 to 1.5 metric tons per hectare. About 12 percent of all rice land is cultivated as upland rice (Khush 2000).

Lowland rice is further divided into three types: irrigated, flood-prone, and nonirrigated. Most of the land planted to rice—about 55 percent—is irrigated and grown in paddies, and this area produces some 75 percent of all the world's rice. The twentieth-century improvements described below as the "green revolution" dominate irrigated rice production. Because of this technology the average yields for irrigated rice have been raised to 5.5 metric tons per hectare (Khush 2000).

Like irrigated rice, the nonirrigated form of lowland rice is grown in paddies, but it depends on rainfall rather than irrigation. Such fields suffer from both rainfall shortages and excesses. Average yields for this form of cultivation vary from 2 to 2.5 metric tons per hectare. About 25 percent of the world's rice is cultivated this way. Flood-prone rice is grown in the river deltas of Southeast and South Asia. It is produced in standing water of up to a meter. Yields vary from 1.5 to 2 metric tons per hectare. About 8 percent of rice land is cultivated in flood-prone areas (Khush 2000).

Rice cultivation has a long history in Asia. Rice has been grown in paddies since antiquity. The paddies are formed by creating earthen embankments (bunds or dams) around their edges to hold water. The bottom of the fields are made imper-meable by puddling, a process in which wet clay is made into an impermeable paste. While this is a good system for producing rice, it makes it very difficult for producers to switch to other crops.

Until recently, Asian rice farmers produced rice cheaply using a buffalo for pud-dling (unlike tractors, the buffalo is heavy enough to compact the clay but not heavy enough to break through the pan) and manual labor for transplanting, weeding, and harvesting. They were able to produce good yields with few inputs other than water and the sediment it carried. A single hectare could support a family of seven. This picture remained unchanged for centuries; Asia's population and rice production developed at the same speed (Witte et al. 1993).

However, all that changed during the twentieth century. The population began to expand more rapidly, and there was insufficient land for expanding cultivation to support more people. The situation was aggravated by the fact that urban areas be-gan to grow rapidly, and rural populations began to abandon rice production and move to the cities. For these reasons, rice production was no longer able to keep pace with demand. As a result, new cultivars and production systems were needed (Witte et al. 1993).

The new cultivars and production systems developed in the twentieth century have come to be known collectively as the green revolution. Rice is probably the best known of the green revolution crops. Green revolution technology consisted of genetic improvements and improved management practices involving use of pesti-cides and fertilizers, which, when combined, produced higher yields. High-yielding cultivars can produce up to 10 metric tons per hectare per crop, but scientists hoped for averages of 5.5 metric tons per hectare per crop (Witte et al. 1993). With irriga-tion, and the other green revolution inputs such as fertilizer, it was possible to pro-duce up to three crops per year.

Improved rice varieties and technologies developed at the International Rice Re-search Institute (IRRI) in the Philippines led to major increases in yields. Increases resulted from a variety of factors. Breeding programs increased production and de-veloped shorter rice cultivars, which can be fertilized to produce more without falling down, mature more quickly, and require fewer inputs such as water. Their en-ergy is focused on grain production.

In the 1970s this technology package spread throughout Asia. Government de-velopment programs were designed to supply seeds, fertilizers, insecticides, and even water along with the necessary credit to allow farmers to purchase them. Credit was often subsidized to insure that the programs were adopted by farmers. Governments also invested in irrigation schemes, transportation and roads, processing and storage facilities, and extension advice and supervision to make sure that farmers used the green revolution packages correctly (Witte et al. 1993).

Most of the major rice-growing countries became self-sufficient in rice as a result of the green revolution. Since the first high-yielding rice cultivar was introduced in 1966, yields increased from 2.1 to 3.7 metric tons per hectare per crop in 1999. Total production increased from 257 million metric tons to 600 million metric tons in

1999, an increase of 134 percent, while the area cultivated increased only 23 percent, from 126 to 155 million hectares (Khush 2000). Moreover, as of 2002 the cost of production was about 20 to 30 percent lower for high-yielding cultivars than for traditional ones. The decline in rice production costs and increased overall production have also meant declines in prices, which have benefited the urban poor and landless rural laborers.

The switch to high-yielding rice cultivars has contributed to a growth in income of rural landless workers. High-yielding varieties require more labor per unit of land to apply the necessary fertilizers and pesticides as well as to harvest and process the increased output (Khush 2000). Increased employment for rice production also has a multiplier effect, leading to increased overall employment in trade, transportation, construction, and services in rural areas. The growth of many Asian economies is in part related to growth in agricultural income and its distribution, which expands markets for nonfarm goods.

Despite the major increases in rice production from the green revolution, there are a number of reasons for caution. As a result of changes in U.S. export policy world rice prices plummeted in the 1980s, which affected domestic prices in Asia. Also, the associated lower production costs may not last forever. Green revolution rice has made farmers dependent on fertilizers and pesticides, whose price has tended to increase over time. Due to general economic growth, the price of labor is increasing in many rice-producing areas even as rural poverty increases. And pesticide poisoning is becoming more common for people and the environment.

More importantly, there are fundamental questions regarding whether modern rice production technology is sustainable. As early as 1990 Pingali (as cited in Witte et al. 1993) suggested that green revolution irrigated rice was beginning to show declining yields. The rate of yield decline at the International Rice Research Institute in both the wet and the dry seasons was 1.28 percent per year, a rate that can seriously undermine the 2 percent per year growth rates achieved through plant breeding during the green revolution. Globally, such yield declines continue to be a concern. The main causes of production decline appear to be forms of environmental degradation, including increased populations of pests and diseases as well as depletion of soil micronutrients and changes in soil chemistry due to intensive cropping and low-quality irrigation water.

Research in the Philippines suggests that initially only a third of farmers had the management skills to adapt the green revolution technology to their farms. Over time, the ability to understand and manipulate the new technologies on the farm has become even more important. A fairly sophisticated level of understanding is required both to implement the current technology and new developments and to solve problems. Producers need to understand tools such as irrigation dynamics and fluctuations, labor or capital availability, technologies to incorporate fertilizers more efficiently, and integrated pest management. As Pingali, Moya, and Velasco wrote, "further productivity gains in the post–green revolution era will come from more efficient use of existing inputs to exploit genetic potential of existing varieties. These 'second generation technologies' are more knowledge intensive and location specif-

ic than the modern seed-fertility technology that was characteristic of the green revolution." (1990, as cited in Witte et al. 1993).

It is essential to maintain the current high yields, which will require finding ways
to make current production more sustainable. And since there is very little new land
to convert to productive rice farming, yields must even be increased on existing areas. This job will be difficult, however, as some of the best rice lands are being lost,
water is being diverted to nonfarm uses, and labor is moving to cities (Khush 2000).
The way out of this conundrum is through genetic work on cultivars with higher
yield potential under adverse conditions as well as better management practices.
The improved practices will need to focus on integrated nutrient and pest management as well as better water and soil management.

Because rice is such an important staple in many parts of the world, it is a strategic crop. As such, its producers benefit from many subsidies. There are systems of
subsidized credit so producers can purchase inputs. In addition to more general subsidies, many countries provide water to rice farmers for free, and several others provide water at far below the actual costs (Witte et al. 1993). Fertilizer, by contrast, is
generally taxed rather than subsidized.

Rice production is also affected by a number of other macroeconomic conditions. Bourgeois and Gouyon (2001) offer several insights into how macroeconomic
conditions affected both production and profitability before, during, and after the recent Indonesian financial crisis. In the early 1990s, prior to the crisis, the incomes of
landowners (producers) grew more slowly than those of urban families, and the income of agricultural laborers actually fell in real terms. Urban areas became more
attractive through higher wages, and landowners found it harder to recruit laborers.
As a consequence, producers adopted labor-saving practices and technologies, including mechanized rice planting methods (Naylor 1992, as cited in Bourgeois and
Gouyon 2001). This increased rural inequity, as not all farmers could afford to invest
in or benefit from mechanization. Similarly, landowner farmers—as opposed to
renters or sharecroppers—tend to have higher incomes because they do not share
the profits from their production. Because of all these factors, small farmers, landless
farmers, and agricultural workers had low or zero growth in productivity and declining real incomes prior to the financial crisis.

The economic crisis that began in 1997 had a number of effects on farmers. First,
the cost of inputs increased, as these are largely imported, so farmers could no longer
invest in fertilizers or pesticides. Without fertilizers, production declined. Potassium
in fertilizer boosts pest resistance, so lack of fertilizer also lowered the rice's resistance to pests. Second, because food was scarce, farmers tended to plant rice three
times in a year rather than only once or twice, which was a traditional way of limiting the impacts of pests. Finally, farmers planted whenever they could rather than at
the same time as their neighbors. This meant that pests could move from one field to
another, gathering impact and numbers as they went (Bourgeois and Gouyon 2001).

This problem was accentuated for poor farmers who had to sell their crop at harvest to get immediate cash rather than store it until the price increased. Problems
were compounded when they went to buy expensive inputs for the next growing

season (Bourgeois and Gouyon 2001). In the end, many farmers shifted to export crops such as cocoa, coffee, shrimp, and spices, where prices in U.S. dollars helped to buoy their income. Depending on the price of rice, such farmers may or may not return to rice cultivation.

What is clear from both these examples is that during periods of economic boom and bust, producers of basic foodstuffs find their economic position slipping. Many personal needs as well as agricultural inputs are more expensive because they are imported or the price is set by international trade. Producers are forced to sell more of their production, more quickly, to buy what they need. Simply put, the price of foodstuffs does not adequately compensate those who produce them.

PROCESSING

Each type of rice has its own unique characteristics and texture. While there are thousands of rice varieties, grown in either paddies or upland plots, consumers buy them based on whether they are white, brown, red, or black; how much gluten they contain; whether they are scented; and whether they are long, medium, or short grain. Each type requires, or is the result of, somewhat different processing. Within each of these types there are a number of different grades.

The main processing steps for most types of rice include cleaning, husking, separating, milling, grading, and bagging. During cleaning, all foreign objects are removed. These include hay, straw, stones, branches, and—from paddy-grown rice—snail shells. During husking, the excessive husks are cleaned and rubbed off the rice grain. Any remaining husks are separated from brown rice by blowing them from the grain in a process very similar to winnowing other grains. The rice separator catches any unhusked grains remaining in brown rice by applying a difference in gravitational pull and surface friction. The rice with the husk removed is denser and tends to separate from that with husk still on it. The unhusked paddy is then returned to the husking process until its husks are separated. During milling, the outer bran layer is removed from the brown rice and then the bran is separated by air ventilation. White rice is what remains after the bran is removed from brown rice. This process usually takes two to three cycles, depending on the degree of milling required.

Once the clean rice has been milled, it is ready for grading. Grading is the process by which milled rice is separated by the integrity of the rice: whole grain, long broken rice (where the broken grain is 75 percent or less of the original grain) and broken. Rice is also graded on transparency (e.g., chalky), color damage (e.g., from insects, heat or water) or foreign matter (e.g., other plant seeds). In addition, the moisture level is measured. Most grading is done mechanically. Finished rice is stored separately according to its grade. After grading, the rice is bagged and is then ready for delivery. Bagging is normally done in units of 1, 5, 10, 15, 25, 45, 50, and 100 kilograms. In some larger markets, rice is sold in bulk by the metric ton. In general, the larger units are defined by the number of 100-kilogram bags. For example, a truckload of rice in many parts of Asia is 140 sacks of 100-kilogram bags, or the equivalent in 1-kilogram, 5-kilogram or 15-kilogram bags.

A by-product of processing is the rice bran, which is removed from the rice kernel in the process of milling white rice. This product is sometimes more expensive by weight than the polished rice itself because of its many uses as a medicinal supplement (e.g., to lower cholesterol and blood pressure and to control some forms of diabetes). It is also a highly sought-after feed ingredient due to its high content of amino acids and proteins.

Most rice is sold in its raw form after the husk and bran have been removed. It is then cooked and eaten directly by people with little more processing. However, rice is also manufactured into a number of different products. These include crackers, noodles, flour, various canned products, milk substitutes, vinegar, and wine. Some rice and residue from processing is used for animal feed.

PRODUCT SUBSTITUTES

As a source of carbohydrate in many diets, rice is simply one of many possible sources of food for energy. Wheat, corn, sorghum, millet, barley, rye, and other grains are all substitutes. Sweet potato, cassava, and many tubers can be substitutes as well. Tubers, in particular, are more versatile than rice because they can be planted in many types of soils and do not need major landscape transformations (e.g., irrigation works, terracing, etc.) to be produced.

It is important that the politics of food production be shifted to include discussions about crops that have fewer environmental impacts. The main advantage of rice is its versatility in storage and preparation and therefore its convenience for urban consumers. However, on farms and for rural areas, other substitutes are more easily grown and may be more desirable. While substitutes for rice may not store as well or have the same versatility, in rural areas this is not necessarily as important a consideration. Furthermore, a mix of other crops can lead to a more varied, and therefore healthier, diet.

MARKET CHAIN

The market chain for rice involves the producer, middlemen, rice mills, brokers, wholesalers, retail shops, and consumers. The price of rice at any point in the chain depends on the total amount of rice available (both from production and in storage) and the relative quality and volume.

The international commerce of rice is controlled by large-scale rice milling facilities. These large, sunk investments allow companies to achieve economies of scale. These companies tend to dominate the milling in the primary export countries (e.g., Thailand, Vietnam, China, Pakistan, and the United States). Once exported, the rice trade tends to be controlled by the same companies that dominate the international grain trade. Even so, rice is only a small part of grain trade.

For domestic markets and consumption, small mills in many rural areas can be financially viable. These can handle the produce of as little as 200 hectares. Since few

small farmers, by definition, produce 200 hectares of rice, most mills in areas dominated by small producers tend to be owned by cooperatives or by the government.
Recently there has been a trend in many countries for government-owned mills to
be privatized.

Larger mills, by contrast, tend to be integrated with larger production areas as
well as more vertically integrated into the market chain. Production from such
plants tends to be linked directly to storage, packaging, and distribution systems.
These large-scale operations are more common in Thailand, Brazil, Colombia, the
United States, and increasingly in Vietnam as well. The destination of rice produced
in such systems tends to be for urban consumption and/or for export. In other parts
of the world rice distribution is local, neighborhood commerce. Thus, the channels
in the market are diverse and based on the capacity for investments along the chain.

MARKET TRENDS

Between 1961 and 2000 rice production increased globally by 179 percent and the
quantity of rice traded internationally increased by 267 percent (FAO 2002). During
the same period, however, the value of rice traded internationally decreased by 61
percent. The increased availability of rice and a decline in the cost of production
have contributed to an estimated 40 percent decline in real prices of rice since 1960
on domestic markets. These price declines have benefited the urban poor and the
rural landless.

The International Rice Research Institute recently released a report which states
that if Asia is to satisfy demand for its growing population over the next thirty years, it
will have to increase production by 40 percent or 200 million metric tons by 2030.
Such increases will not come from improved genetics alone. Increased production
of that magnitude is likely to trigger expansion of cultivation, resulting in larger and
larger areas of natural habitat being converted both to upland and paddy-grown rice.

ENVIRONMENTAL IMPACTS

Many of the environmental problems from rice production result specifically from
the green revolution rice production technology. This technology has caused significant reductions in biodiversity within rice fields, particularly for paddy-grown rice.
However, it also has increased use of fertilizers and pesticides, which in turn has increased pollution of streams, rivers, and groundwater systems through runoff from
fields. Rice production also generates more greenhouse gases than any other major
agricultural crop. Each of these problems is discussed separately.

Biodiversity Loss within Existing Production Areas

Green revolution rice production technologies have increased production per
hectare as well as the number of crops that can be grown successively each year.

Clearly these production gains have reduced the habitat conversion that would have had to take place to produce as much rice using the traditional production systems prior to the development of the technology. Even so, green revolution production methods have tended to reduce quite significantly the amount of biodiversity that exists within the production system.

Traditionally, paddy fields are home to many species. Kenmore (1991) writes that "Rice ecosystems often have more than 700 animal species per hectare in highly intensified fields in the Philippines and over 1,000 so far described in Asian species of higher trophic level predators and parasitoids." The application of ever-increasing quantities of pesticides and synthetic fertilizers, however, has led to the disappearance of much of this biodiversity, including the beneficial nitrogen-fixing algae whose absence leads to greater dependence on synthetic nitrogen fertilizers. As the agrochemicals affect the microbial life, they also affect the entire food chain that depends on them. Paddies are no longer habitable by the dozens and dozens of species that different farmers harvested for food. In the end, the loss of water reptiles, fish, frogs, and snails deprives people of an important food source.

Current rice production is a monoculture activity undertaken in irrigated paddies. In these production systems, rice varieties are selected on the singular basis of productivity and are further interbred to maximize that trait. This approach has tended to shrink the gene base, and one of the characteristics that is disappearing is pest resistance, which often existed in traditional rice strains. This means that farms increasingly rely on pesticides to do what rice plants were capable of doing in the past.

Pollution from Fertilizers

Fertilizers affect water quality. For example, nitrogen absorption is quite low in rice production. Estimates from the Philippines suggest that only 30 percent of the applied urea is effectively utilized in rice cultivation. The bulk is lost through volatilization and denitrification. Nitrogen from synthetic fertilizers such as urea is oxidized (through nitrification) into nitrate, which in turn is converted to volatile gaseous forms and lost through denitrification. Losses in the form of ammonia are high, contributing to eutrophication of the paddy water, with a resulting high daytime paddy water alkalinity (Witte et al. 1993).

Phosphorus absorption is even lower. Estimates suggest that not more than 10 to 15 percent of phosphorus added to the soil is absorbed by the crop (Witte et al. 1993). The rest is often transformed to insoluble forms (a process known as phosphorus fixation), and only under certain conditions can these forms be made available to the crop.

Inefficient fertilizer use not only costs the farmer money and lowers profits, it also has a polluting effect downstream. Most of these impacts have not been quantified. It is known, however, that nutrient-rich waters coming from agricultural areas in China are a major cause of the frequent red tides along the coast.

Perhaps most important, the repeated and increasing use of synthetic fertilizers also alters the microbial balance that converts organic matter and dissolved minerals in the soil into forms that the rice plant can use. Over time, the reliance of farmers on synthetic fertilizers tends to lead not only to a slow degradation of soil fertility but also to a reduced ability of the soil to absorb chemical inputs.

Pollution from Pesticide Use

Pesticides are perhaps one of the most important environmental problems posed by rice cultivation as a result of both their overuse and misuse (Witte et al. 1993). Pesticides disrupt healthy ecological processes, as noted above. Equally important, pesticide poisoning is a health issue for both farmers and workers.

Modern rice production uses insecticides, herbicides, molluscicides, and to a small extent fungicides. In the major rice-producing countries of Asia, more agrochemicals are used on rice than on all other crops combined. In the Philippines, for example, 47 percent of all insecticides and 82 percent of all herbicides were used on rice (Witte et al. 1993). In the late 1980s and early 1990s pesticides that had been banned in other countries were still being used in Thailand and the Philippines. These pesticides include chlordane, DBCP, DDT, dinoseb, HCH (hexachlorohexane, better known as lindane), hexachlorobenzene, methyl parathion, mercury compounds, and PCP (pentachlorophenol). In the Philippines four pesticides (monocrotophos, methyl parathion, azinphos-methyl, and endosulfan) constituted 70 percent of the pesticides used in rice cultivation in the early 1990s (Witte et al. 1993).

Another impact of the increased use of agrochemical inputs is that many bioaccumulate. This means that people absorb chemicals not only from the rice, but also from other plants, and in concentrated doses from any animals that accumulate the chemicals from what they eat. One survey found that organochlorine insecticides were present in low levels in half of the blood samples taken from Filipino farmers. A report prepared for the Institute of Agricultural Economics in Hanover, Germany, estimates that nearly 40,000 farmers in Thailand suffer from various degrees of pesticide poisoning every year, and that the associated health costs amount to more than U.S.$300,000 per year. The external costs of health care, monitoring, research, regulation, and extension amount to as much as U.S.$127.7 million per year in Thailand alone (*Rice Today* 2002). Studies in Thailand have shown that pesticide residues exist in more than 90 percent of samples of soil, river sediment, fish, and shellfish (*Rice Today* 2002).

One of the problems of restricting or prohibiting the use of pesticides within a country, much less between countries when the products are traded, is to sort out the exact names of the pesticides that are used. Pesticides are often sold under brand names without reference to the chemical compounds included in them. One chemical, for example, is marketed under 296 trade names, another under 274 (*Rice Today* 2002). This makes transparency for users as well as monitoring by governments very difficult.

There is much speculation about the impact of climate change on the ability of agriculture to feed more people. However, agriculture itself can have a significant effect on global warming through the release of greenhouse gases. Continuously flooded rice fields, in particular, release methane to the atmosphere (Wassmann et al. 2000). One estimate places methane emissions from rice at some 10 to 15 percent of total global methane emissions (Neue 1993, as cited in Wang et al. 2000). Other estimates suggest that the contribution of rice paddies to global rates of methane emissions ranges from 5 to 30 percent (Minami and Neue 1993, as cited in Witte et al. 1993). If there is an increase in yields or areas harvested, methane emissions would increase as well.

The processing of rice in large dehuskers leads to the accumulation of large amounts of rice husk waste. This waste is normally burned to reduce the volume and therefore the disposal problem. The burning of waste releases both carbon dioxide and carbon monoxide into the environment.

Water Use

Irrigated rice requires about 1,200 millimeters of water per crop. This amounts to some 5,000 liters per kilogram of rice produced. In some areas water use for rice cultivation causes salinization of soils, making the land less fertile. Rice is a large and inefficient consumer of water, even by today's agricultural standards. The impacts of the total water withdrawals on biodiversity and ecosystem functions are not well studied. For example, it is not known whether taking water for rice cultivation and reducing flooding during rainy seasons is better or worse for biodiversity than taking water from river systems during the dry season.

The provision of water for rice production causes collateral damage as well. Many dams have been constructed to provide water for the irrigation of rice. These dams prevent migration within freshwater ecosystems and as a result reduce biodiversity. Dams and irrigation systems also increase disease vectors by providing breeding grounds for mosquitoes and other hosts who transmit the diseases where they did not exist before. These can include organisms causing bilharzia (schistosomiasis), malaria, and even diarrhea.

BETTER MANAGEMENT PRACTICES

There are many different kinds of rice production systems. Fortunately, there are ways to reduce the environmental problems associated with each. However, it now appears that some forms of rice cultivation may be far more productive and yet have fewer innate environmental impacts that would need to be addressed. Since many of these systems are particularly appropriate for smaller producers, they should be investigated and supported not only by those interested in the environment, but also by those interested in food security and poverty reduction.

As with other commodities, it is clear that improved efficiency of input use for rice production can increase yields while reducing costs. There are many ways to improve the efficiency of resource use. These include reducing pesticide and fertilizer use, improving water management, reducing effluents and soil erosion, and improving overall soil management. Many of these improved practices can also reduce the greenhouse gas emissions associated with rice production. Some of the better practices will, in addition, increase wildlife habitat and perhaps even increase producer income streams, as through the sale of hunting permits.

Finally, it is clear that better practices aimed at optimizing certain impacts may in fact contribute negatively to others. Each producer will have to determine which are the most important impacts to be reduced and what are the best ways to accomplish this.

Develop Innovative Production Systems

Rice can be cultivated without paddies and possibly still give yields that are superior to those obtained from paddy culture. It can be irrigated with drip irrigation or overhead sprays in areas where adequate and timely rainfall is not reliable. Rice can be grown in ways similar to vegetable crops and can even be mixed with vegetables and tubers in polyculture systems. New multicrop polyculture technologies could eliminate the need to convert broad areas of land for flooding, which can destroy habitat and local biodiversity in large regions, uses scarce water resources inefficiently, and can even fail. This perspective is virtually absent from the major rice research institutions, though in fact this "technology" already exists. In Indonesia, Japan, China, and India these types of farming have existed in very old systems for centuries. They can provide insights as well as alternatives to high-input paddy rice cultivation that offer fewer negative environmental impacts.

For example, organic and low-input forms of conventional rice production in Japan and Thailand already demonstrate very high yields. Some produce yields of more than 10 metric tons per hectare per crop and show signs of increasing even more (Hawken et al. 1999). This indicates that environmentally sound practices can produce the improved yields of rice that will be required to feed expanding populations. Other work suggests that additional improvements in rice production can be obtained by interplanting or sequence-planting rice with soybeans, field beans, or other legumes that improve soil fertility and soil biodiversity to improve plant vigor and resilience (Panfilo Tabora, personal communication). These polycultural cropping systems are also vital sources of protein for farmers, both from the plants produced and the wildlife attracted and harbored.

Rice systems contain some of the best-understood community relationships in the tropics. What is not well understood is the relationship of this biodiversity to ecological processes that either increase the viability of lower-input rice production systems or the provision of other marketable items or food for rice cultivators, such as the edible frogs, snails, and fish that have now largely disappeared because of pesticide use.

In China, ecological farming (farming based on the principles of organics combined with modern science and technology to improve yields and quality as well as input use efficiency) is practiced on about 5 percent or less of rice land. Even so the results are interesting. When rice is combined with azolla (an aquatic fern that has symbiotic relationships with nitrogen-fixing cyanobacteria) cultivation, azolla inhibits weed growth and then can be cut as a green manure for the next rice crop. In combining rice cultivation with fish or duck production, 37 to 84 percent of weeds are consumed by either the fish or the ducks. In addition, trials showed that there was a slight increase in soil organic matter with combined rice and fish production, plus increases of 16.4 percent, 50 percent, and 9.5 percent in soil levels of phosphorus, potassium, and nitrogen respectively. Levels of dissolved oxygen also improved. Because of the nutrient-rich wastes produced by fish and other animals there are lower quantities of fertilizers and pesticides in the effluent because less of each is used (Chen et al. 1993). Under some types of ecological farming, on 0.2 hectares of land some 1,800 kilograms of rice were produced as well as 130 kilograms of fish. Under traditional green revolution production systems only 1,668 kilograms of rice were produced under normal conditions (Chen et al. 1993).

Paddy rice production obliges changes in topographies that not only include clear-cutting but also changes in hydrology. Such changes are harmful to wildlife. In addition these changes are irreversible without considerable effort and investments when farmers decide to produce different crops. Investments are needed to develop or document technologies for rice production that are based on respecting the topographies and natural features of agricultural sites.

Reduce Pesticide Use

The effect of pesticides on estuaries, rivers, and fragile coastal zones are all reflected in the reduction of fish catch and aquatic biodiversity, as well as species that depend on aquatic biodiversity for food. Reductions in pesticide use will reduce the damage from agriculture on all downstream biodiversity. Data from Thailand and the Philippines suggests that integrated pest management (IPM) can reduce the use of insecticides on rice by 75 to 95 percent. Furthermore, there is no need to use the most toxic categories of pesticides to achieve the same or better results (Witte et al. 1993).

The first recorded implementation of IPM for rice on a massive scale was in Indonesia, where 50,000 farmers were trained in IPM techniques in 1990. Training of farmers was accompanied by the banning of fifty-seven trade formulations of rice insecticides and the introduction of pest-resistant rice varieties. The acreage previously affected by pests such as the brown plant hopper decreased from over 200,000 hectares to below 25,000. In addition, pesticide production in Indonesia dropped from 55,000 metric tons per year to 25,000 metric tons, while rice production increased from 28 million metric tons to 30 million metric tons (Kenmore 1991). It is not clear how much the program cost or whether it was cost-effective.

In the Philippines, by contrast, the costs of IPM training are well documented.

IPM program costs are estimated at 230 million pesos (the Philippine currency) per year over five years. Costs per trained farmer are expected to be 500 pesos for the training component only, or 1,150 pesos including management, monitoring, evaluation, and administration. This compares to reduced pesticide costs of approximately 448 pesos per hectare per crop. If two crops of rice are grown annually, this results in a cost recovery in less than one year for the average rice farmer with 1.6 hectares. Another way to look at the expense of the program is that it represents 0.18 percent of the proposed budget for the country's Rice Development Plan. Fertilizer assistance, by comparison, required 12 percent of the same budget (Witte et al. 1993).

In Vietnam, an IPM program reduced insecticide use in the Mekong Delta by an estimated 72 percent. What's more, the number of farmers who believed that insecticides would bring higher yields fell from 83 percent before the IPM campaign to just 13 percent after (*Rice Today* 2002).

In China, researchers found that interplanting disease-resistant hybrid rice reduced the severity of the disease known as rice blast by 94 percent and improved the yield of the highly valued glutinous variety by 89 percent (Zhu et al., as cited in *Rice Today* 2002).

Increase Efficiency of Fertilizer Usage

Synthetic fertilizers are expensive, and the volume used in rice production is quite high. Strategies should be pursued to reduce their use while maintaining yields. This will improve producer profits. There are several ways to increase the efficiency of fertilizer usage. Nitrogen utilization rates, for example, can be improved when it is incorporated into the soil rather than sprayed on the field. This prevents it from volatilizing. Another way to reduce the use of nitrogen by 30 percent or more is to incorporate the rice straw into the soils (Witte et al. 1993). By recycling rice straw through composting and mulching (e.g., basic organic fertilizer strategies), system "leakage" can be reduced considerably, thus minimizing long-term nutrient depletion.

A strategy based on nutrient management practices such as green manuring, the use of azolla, and recycling or composting crop and household wastes can restore soil fertility. The rapid cycle of building up and breaking down organic matter is what builds soil fertility. Such a strategy can limit problems of waste disposal as well as fertilizer costs (Witte et al. 1993). A more efficient use of fertilizer can reduce overall use and can be coupled with the better management practices described above.

Reduce Effluents

Allowing water to stay longer in the rice fields may be a simple way to reduce farm agrochemicals in runoff. Scientists at the Texas Agricultural Experiment Station have determined how many days water should stand in rice fields to allow breakdown of chemicals to safe levels. For example, some 22.7 kilograms per hectare (20 pounds per acre) of chemicals is the normal chemical content of water runoff from rice fields managed traditionally. If the water is left on the field for five to seven days longer, the level is reduced to 6.8 kilograms per hectare (6 pounds per acre), which is considered safe. In addition, the chemical fertilizers will be available to the roots of the next crop.

Synthetic fertilizers affect water quality, altering the microbial balance that is key to the conversion of organic matter and dissolved minerals into usable form. Water in rice paddies or in settlement ponds can be treated with microbial organic matter that is inoculated with beneficial microorganisms to reestablish its balance either for improved efficiency in the pond or prior to release from the pond.

Improve Water Management

New systems of rice cultivation allow producers to conserve much of the water that was used to cultivate rice thirty years ago. In Australia, for example, more accurate laser siting and leveling of irrigated fields have reduced water use by some 25 percent. Improved control of water movement on and off the land reduces the opportunity for rice pond water to enter the water table from rice fields. Other ways to reduce water use include growing shorter varieties with shorter seasons (meaning they ripen earlier). In Australia, such measures have reduced water use by 30 percent per hectare over the past ten years and increased rice yields 60 percent per water used in the same period.

In California some farmers have begun to employ new recirculating irrigation systems plus automated shutoff valves that conserve up to two-thirds of the water requirements of thirty years ago. Rice fields that are tilled in the fall and left open to drain freely after each winter rain lose thirty times more soil than rice fields where the stubble is left standing and water is allowed to collect (Ducks Unlimited 2002). Holding soil in the ponds makes downstream freshwater systems cleaner and consequently better wildlife and fisheries habitat. In addition, minimal fall tillage and ponding of winter rainfall promote the decomposition of rice straw and build organic matter in the soil. More thoroughly decomposed rice straw means less effort and expenditures at planting time. Maintaining standing water in the winter also appears to suppress germination and growth of winter weeds and thus reduces the work needed for spring field preparation.

Control Erosion and Improve Soil Management

The presence of organic matter in the soil is a key to minimizing soil erosion. However, rice stubble is normally burned or removed because it does not decompose quickly in areas where multiple crops are grown each year. When stubble is incorporated into the soil, its decomposition can release methane gas that damages the roots of subsequent rice crops and reduces productivity. Organic matter can be integrated back into rice fields through the use of effective microorganisms that contribute to a rapid decomposition of the stubble while at the same time trapping the methane and ammonia gases from the decomposition of the stubble and converting them into substances that are useful for plant growth. This system has been used in Japan and China, but it is still not widely adopted (Panfilo Tabora, personal communication).

Reduce Greenhouse Gases

There are several ways to reduce emissions of methane, a greenhouse gas produced in rice paddies. Research has shown that transplanting thirty-day-old seedlings, direct seeding on wet soil, and direct seeding on dry soil reduced methane emissions by 5 percent, 13 percent, and 37 percent, respectively, when compared to transplanting eight-day-old seedlings. Plowing also affects greenhouse gas production. For example, methane emissions following fall plowing were 26 percent less than they were following spring plowing (Ko and Kang 2000).

In addition to reducing rice production's overall contribution to the generation of greenhouse gases, rice fields can sequester some 10 metric tons of carbon per hectare per growing season—but only if crop residue is kept in the soil (Rice Producers of California 2003). If crop residue is burned instead of kept in the soil, the carbon is lost and rice becomes a net contributor to CO_2 production.

Burning rice husks at processing plants is not only harmful environmentally, it is a waste of resources. Rice husks have tremendous value in many greenhouse operations and are in fact bought and transported great distances as a valuable raw material for soil amendments.

Manage Rice Fields as Wildlife Habitat

Rice fields provide considerable food for waterfowl. Even so, this function could be improved considerably with a few rather small changes in management. Where game species are abundant, payments for hunting can increase producer income considerably. In California, several management changes have improved habitat for wildlife. Earlier flooding of the fields with existing nutrient-rich water has been shown to improve wetland food production for wildlife. Planting and harvesting rice later in the year makes waste grain available to waterfowl when they are migrating. Finally, managing some areas next to rice fields as habitat with natural grasses and sedges provides both cover and food for waterfowl.

Rice fields make excellent stopover points for migratory birds. In California some 95 percent of all wetlands have been lost, greatly reducing available stopover points. In northern California 500,000 acres of rice fields provide roosting grounds as well as food. The California Rice Straw Burning Reduction Act of 1992 has forced many rice farmers to use winter flooding of rice fields to assist in the decomposition of waste rice straw. This winter flooding in turn has helped provide winter habitat for millions of migratory birds and other wetland species. The fields are a resting ground for an astonishing 3 to 5 million migrating waterfowl every year on the Pacific Flyway and are home to over 141 species of birds, 28 species of animals, and 24 species of amphibians and reptiles. Thirty of these species are listed as endangered, threatened, or species of concern (California Rice Commission 2001).

OUTLOOK

Rice is the most important food crop for people cultivated at this time. A quarter of all the calories consumed by people come from rice, and this figure is 50 percent in Asia. As such, rice is also a strategic crop, one that is extremely sensitive politically. Production in most rice-producing areas is stable or even declining. If all things were equal this would bode ill for those politicians in countries where rice is the staple. But, all things are not equal. In some of the largest rice-consuming countries of the world (China, India, and Indonesia) populations are shifting to cities, where cheap food is not only a food security issue—it is a political survival issue as well. Politicians will be forced to ensure that urban populations have ready supplies of acceptably priced food or they will no longer be in office.

There are ways to produce more rice and to produce it more efficiently. Fortunately, many of these practices would also be more profitable, especially for small farmers. Given that small farmers make up a large proportion of the people that are fleeing the countryside for the cities, it should be possible to find ways to induce them to stay and farm profitably. While this is such an obvious solution, it ultimately will depend on getting the policies and incentives right. This needs to be coordinated not only at a national and local level and with a wide range of different players, but also internationally. Development assistance, research, and even credit need to be used in such a way that they encourage more rational rice production while achieving the scale and efficiency to provide for the needs of urban residents as well. It is not clear if all of this will happen in time.

REFERENCES

Bourgeois, R. and A. Gouyon. 2001. From El Niño to Krismon: How Rice Farmers in Java Coped with a Multiple Crisis. In F. Gerard and F. Ruf. Eds. *Agriculture in Crisis: People, Commodities, and Natural Resources in Indonesia, 1996–2000.* Montpellier, France: CIRAD.

California Rice Commission. 2001. *California Rice Commission—Environment.* Available at http://www.calrice.org/environment/environment.html.

Chen, D., B. Zhong, B. Shen, X. Li, and G. Yu. 1993. *Case Study on Eco-Farming in China with Special Emphasis on Rice.* Geneva, Switzerland: United Nations Conference on Trade and Development. UNCTAD/COM/19. August 27.

Childs, N. 2002. *Rice Outlook—U.S. 2001 Rice Crop Pegged at Record 213 Million Hundredweight.* Electronic outlook report from the Economic Research Service. Washington, D.C.: U.S. Department of Agriculture. January 14. RCS 0102. Available at http://www.ers.usda.gov.

———. 2002b. *Rice Outlook—U.S. 2001/02 Rice Imports Projected at Record 12.5 Million Hundredweight.* Electronic outlook report from the Economic Research Service. Washington, D.C.: U.S. Department of Agriculture. January 14. RCS 0202. Available at http://www.ers.usda.gov.

———. 2002c. *Rice Outlook—U.S. 2001/02 Total Supplies Projected at Record 255 Million Hundredweight.* Electronic outlook report from the Economic Research Service. Washington, D.C.: U.S. Department of Agriculture. January 14. RCS 0302. Available at http://www.ers.usda.gov.

Ducks Unlimited. 2002. *Rice Industry Caring for the Environment.* Available at http://www.ducks.org/conservation/rice_soil_conservation.asp.

FAO (Food and Agriculture Organization of the United Nations). 2002. *FAOSTAT Statistics Database.* Rome: UN Food and Agriculture Organization. Available at http://apps.fao.org.

Greenland, D. J. 1997. *The Sustainability of Rice Farming.* CAB International and IRRI (International Rice Research Institute).

Hawken, P., A. Lovins, and L. H. Lovins. 1999. *Natural Capitalism: Creating the Next Industrial Revolution.* Boston: Little, Brown and Company.

Kenmore, P. E. 1991. *Indonesia's Integrated Pest Management—A Model for Asia.* Manila: FAO Inter-Country Program for Integrated Pest Control in Rice in South and Southeast Asia.

Khush, G. S. 2000. Rice: How Much It Takes to Feed Half the World's Population. Paper presented at the 2000 meeting of the American Association for the Advancement of Science (AAAS). San Francisco, CA.

Ko, J. Y. and H. W. Kang. 2000. The Effects of Cultural Practices on Methane Emission from Rice Fields. *Nutrient Cycling in Agroecosystems.* 58:311–314. The Netherlands: Kluwar Academic Publishers.

Naylor, R. 1992. Labour-Saving Technologies in the Javanese Rice Economy: Recent Developments and a Look into the 1990s. *Bulletin of Indonesian Economic Studies* 28(3):71–91.

Pingali, P. L. 1992. Diversifying Asian Rice-Farming Systems: A Deterministic Paradigm. In Barghouti, S., L. Garbux, and D. Umali. 1992. *Trends in Agricultural Diversification: Regional Perspectives.* World Bank Technical Paper No. 180. Washington, D.C.: World Bank.

Rice Producers of California. 2003. Environmental Issues. Available at http://www.riceproducers.com/environment.html. Accessed 2003.

Rice Today. 2002. Pesticide Misuse. *Rice Today.* 1(1) April:10–11. International Rice Research Institute. Available at http://www.irri.org.

Runge, F. 1994. *The Grains Sector and the Environment: Basic Issues and Implications for Trade.* Rome: U.N. Food and Agriculture Organization.

Smil, V. 2000. *Feeding the World—A Challenge for the Twenty-First Century.* Cambridge, MA: MIT Press.

USA Rice Federation. 2001. *Environmental Advantages of Rice.* Rice Online Environmental Information. Available at http://www.riceonline.com/environ.htm. Arlington, VA: USA Rice Federation.

Wang, Z. Y., Y. C. Xu, Z. Li, Y. X. Guo, R. Wassman, H. U. Neue, R. S. Lantin, L. V. Buendia, Y. P. Ding, and Z. Z. Wang. 2000. A Four-Year Record of Methane Emissions from Irrigated Rice Fields in the Beijing Region of China. *Nutrient Cycling in Agroecosystems.* 58:53–63. The Netherlands: Kluwar Academic Publishers.

Wassmann, R., H. U. Neue, R. S. Lantin, L. V. Buendia, and H. Rennenberg. 2000. Characterization of Methane Emissions from Rice Fields in Asia. *Nutrient Cycling in Agroecosystems.* 58:1–12. The Netherlands: Kluwar Academic Publishers.

Witte, R., B. van Elzakker and J. D. van Mansvelt. 1993. *Rice and the Environment: Environmental Impact of Rice Production, Policy Review and Options for Sustainable Rice Development in Thailand and the Philippines.* UNCTAD/COM/22.

Zhu, Y., H. Chen, J. Fan, Y. Want, Y. Li, J. Chen, J. X. Fan, S. Yang, L. Hu, H. Leung, T. W. Mew, P. S. Teng, Z. Wang, and C. C. Mundt. 2000. Genetic Diversity and Disease Control in Rice. *Nature* 406 (August 17): 718–722. London: Macmillan.

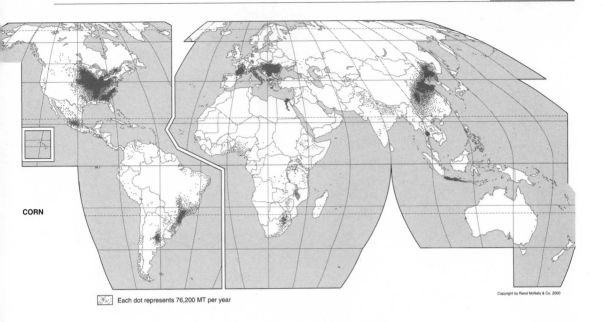

CORN

Each dot represents 76,200 MT per year

Copyright by Rand McNally & Co. 2000

CORN (MAIZE) *Zea mays*

PRODUCTION

Area under Cultivation	138.7 million ha
Global Production	593.0 million MT
Average Productivity	4,274 kg/ha
Producer Price	$111 per MT
Producer Production Value	$65,837 million

INTERNATIONAL TRADE

Share of World Production	14%
Exports	81.8 million MT
Average Price	$107 per MT
Value	$8,733 million

PRINCIPAL PRODUCING COUNTRIES/BLOCS (by weight)

United States, China, Brazil, Mexico, Argentina, France, India

PRINCIPAL EXPORTING COUNTRIES/BLOCS

United States, Argentina, China, France

PRINCIPAL IMPORTING COUNTRIES/BLOCS

Japan, South Korea, Mexico, Egypt, China, Spain, Malaysia

MAJOR ENVIRONMENTAL IMPACTS

Habitat conversion
Soil erosion and degradation
Agrochemical inputs
Water use and pollution

POTENTIAL TO IMPROVE

Good for commercial and subsistence agriculture
BMPs are known that reduce impacts for a wide range of production systems
BMPs save money and increase profits
Conservation tillage reduces inputs and impacts
Plant breeding can reduce impacts

Source: FAO 2002. All data for 2000.

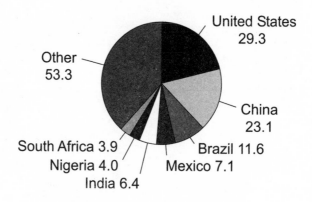

Area in Production (MMha)

United States
29.3

Other
53.3

China
23.1

Brazil 11.6

South Africa 3.9

Mexico 7.1

Nigeria 4.0

India 6.4

OVERVIEW

Corn (known as maize in much of the world) was domesticated by indigenous peoples some 7,000 to 10,000 years ago in Mexico or Central America. It is still consumed throughout the region, usually in the form of tortillas, and very often at three meals a day. Sometimes corn with a little salt is the entire meal. At the time of European conquest, corn was produced throughout North and South America in longitudes from 58 degrees north to 40 degrees south in virtually all areas with sufficient growing seasons and rainfall. The Europeans first took corn home to plant and then spread it throughout their colonies.

For many Indian societies in North and South America, corn was a sacred food. Its cultivation and consumption were the stuff of rituals. Corn was not planted, harvested, or eaten the first time each year without paying proper respect to the gods that provided it. Different varieties were often cultivated. Some were used every day, others only for special occasions. Some societies preferred yellow, others white, still others red or blue varieties. Some 30,000 varieties are thought to have existed. Corn allowed many indigenous societies throughout the Americas to develop surpluses of food that could be stored, thus freeing their time to do other things. Many of the first indigenous villages and settlements as groups ceased being nomads both in order to cultivate corn and because of the surplus it provided.

Over time, however, corn has lost its luster as a "choice" food in many parts of the world. While it is still a major source of food for many people on the planet, particularly poor people, as soon as they can afford it, corn consumers abandon it in favor of a diet containing more fruit, vegetables, and protein. At this time, corn is used not as a "gift of the gods" food but rather as animal feed.

PRODUCING COUNTRIES

The Food and Agriculture Organization of the United Nations (FAO 2002) reports that 158 countries produced corn in 2000. Seven countries account for nearly 61.6 percent of all land planted to corn and 73.5 percent of the 593 million metric tons of corn produced in 2000, as shown in Table 18.1.

The main producers by weight are the United States and China. These two countries account for 60.4 percent of the corn produced globally. Only 13.8 percent of corn produced each year is traded internationally. The main exporters are the United States and Argentina, with the United States accounting for half to two-thirds of all corn exports each year.

The United States grows nearly 40 percent of the world's corn. More than half of that production comes from only 20 percent of corn producers. Production is also concentrated in only a few states; the six leading corn-producing states account for more than 80 percent of the corn produced in the United States annually. Iowa alone is responsible for 9 percent of global corn production (Kimbrell 2002).

Geographically, corn is the most widely grown cereal crop in the world. Globally, average yields are approximately 4,274 kilograms per hectare per year. Kuwait has the highest yields for corn in the world, averaging more than 17,400 kilograms per hectare per year, or more than four times the global average, but such production comes at a very high price in terms of irrigation and other inputs. Belgium, Chile, France, Germany, Greece, Israel, Italy, Jordan, New Zealand, Qatar, Spain, Switzerland, and the United States all produce corn at twice the yields per hectare of the global average.

Corn production has been increasing steadily throughout the world. From 1974 to 1994, for example, global production increased by 86 percent. Though the area under corn cultivation is less than that for wheat or rice, corn produces more tons of

TABLE 18.1. Major Corn-Producing Countries in 2000

	Area Cultivated (million ha)	Total Production (million MT)
United States	29.3	251.9
China	23.1	106.2
Brazil	11.6	31.9
Mexico	7.1	17.6
India	6.4	11.6
Nigeria	4.0	5.6
South Africa	3.9	10.9
Global total	138.7	593.0
Percentage of total	61.6	73.5

Source: FAO 2002.

food than any other single crop. Unlike many other food crops, most corn is used for animal feed and not for direct human consumption. The exception to this occurs in parts of Mexico and Central America and Africa. Table 18.2 shows which countries devote the most land to corn production, as a percentage of available agricultural land (FAO 1996).

Corn production is increasing in parts of the world where it was not significant before. In the past three decades, for example, corn production has more than doubled in the Middle East and Asia. Those regions combined now produce twice as much corn as Latin America, the genetic home of corn. In Africa demand exceeds supply, and corn is rapidly replacing other basic food crops for direct human consumption (FAO 2002).

CONSUMING COUNTRIES

Most corn is consumed in the country of origin. The main corn-consuming countries are the United States, China, Brazil, and Mexico. The main importing countries, by contrast, are Japan, South Korea, Mexico, Egypt, China, Spain, and Malaysia. Japan purchases nearly 20 percent of the corn traded internationally, while South Korea purchases about 11 percent. In both cases these countries use corn to feed animals to meet increasing demand for animal protein. Due to economic growth over the past fifty years, both countries' consumption of animal protein has exceeded their ability to produce sufficient feed grain. While it might be more efficient to import meat rather than feed, there is a cultural preference for fresh meat (FAO 2002).

TABLE 18.2. Percentage of Agricultural Land Devoted to Corn, 1994

50 Percent or More	25–49 percent	10–24 percent
El Salvador	Belize	Angola
Guatemala	Benin	Austria
Malawi	Bhutan	Bolivia
Philippines	Brazil	Bulgaria
São Tomé/Príncipe	Côte d'Ivoire	Burundi
Tanzania	Ecuador	China
Zimbabwe	Egypt	Colombia
	Haiti	Congo
	Honduras	Ghana
	Kenya	Guinea
	Lesotho	Hungary
	Mexico	Indonesia
	Mozambique	Nicaragua
	Nepal	Panama
	North Korea	Togo
	Somalia	United States
	South Africa	Venezuela
		Zambia

Source: FAO 1996.

Many policies can affect the relationship between corn-producing and corn-consuming countries. As a result of the North American Free Trade Agreement (NAFTA), roughly one-quarter of corn consumed in Mexico, generally considered the birthplace of corn, is produced in the United States. Small-scale producers in Mexico (some 90 percent of corn farmers, cultivating less than 2 hectares each) cannot compete against the mechanized and subsidized larger American producers (Weiner 2002). Under NAFTA, limits on the amount of corn the United States can export to Mexico will be eliminated by 2008, and this will further reduce the Mexican-produced share of that market. Mexico had long subsidized the price of corn as a social net for the rural poor and as a tool to discourage rural-to-urban migration. Under NAFTA Mexico agreed to let the price of corn fall to the level of the international price. However, the international price of corn is greatly influenced by the United States, which subsidizes corn producers.

Corn is used for many different purposes. Over the last fifty years, however, corn has increasingly been used to feed animals. Of all cereal grains, corn provides the highest levels of conversion of dry feed to meat, milk, and eggs, making it the feed ingredient of choice in formulated feeds (Runge and Stuart 1998).

Roughly two-thirds of global corn production is consumed by animals, 20 percent directly by humans, 8 percent in industrial use for food and nonfood products, and 6 percent is used as seed (Runge and Stuart 1998). In addition to its primary use as a feed grain, corn by-products include corn oil, cornstarch, corn syrup, and a few thousand other products. When used as a source of hydrocarbons, corn-based products can be substitutes in any of several chemical processes. Corn can be used to make ethanol as a substitute for petroleum-based fuels or plastics.

PRODUCTION SYSTEMS

The productivity of corn depends on the climate, the fertility of the soil, and the availability of water. Hybrid seeds, fertilizers, pesticides, and improved production methods have also combined to increase productivity as well as the range where corn can be planted. In the United States, production has increased from an average yield of 1,250 kilograms per hectare in 1900 (Runge and Stuart 1998) to more than 8,591 kilograms per hectare by 2000 (FAO 2002). Over the past forty-four years, for example, corn yields have increased on average 49.78 kilograms per year (USDA 1996, as cited in Runge and Stuart 1998).

Plant breeding and hybrids account for 58 percent of yield gains. Hybrid seeds allow corn to be grown in different geographical areas, provide resistance to different corn pests, and permit good yields with different inputs. Mechanical production improvements (e.g., soil preparation, cultivation, no-till, and/or irrigation) as well as herbicides have allowed individual producers to cultivate ever increasing areas while reducing labor needs. Pesticides have increased yields by 23 percent by permitting earlier planting dates and season-long weed control. Improvements in fertilizers and their application are responsible for 19 percent of yield increases since 1958 (Runge and Stuart 1998).

Virtually all corn growers in the United States use agrochemical inputs. Synthetic nitrogen fertilizers are applied to 98 percent of the crop. Nitrogen can be applied as manure, ammonium nitrate, anhydrous ammonia, or urea as well as in compound with other nutrients such as potassium or phosphorus. In cooler regions, nitrogen is applied alongside seeds to start and stimulate growth. Most nitrogen is applied either "pre-plant" or beside rows so that it does not "burn" the crop. Phosphate (the most common form of phosphorus) is added to 85 percent of the crop, and potash (the most common form of potassium) to 73 percent. In addition, some 97 percent of cornfields receive herbicide treatments—the most common is atrazine—and a third are sprayed with insecticides (Kimbrell 2002). Commercial producers throughout the world use similar inputs, and small farmers are using increasing quantities of inputs as well.

In the United States more than 90 percent of fields are sprayed with herbicide whether the corn is grown as a continuous crop or in rotation with other crops or fallow. By contrast, insecticides are used on some 60 percent of corn grown continuously on the same field, but less than 25 percent of acres receive insecticides when the crop is grown in rotation. In fact when grown in rotation with small grains, less than 5 percent of corn acres are sprayed with insecticide (Runge and Stuart 1998).

A number of factors are beginning to reduce yields. As a result of continuous planting in some areas, the crop is susceptible to pests and soil erosion, resulting in a 23 percent decline in production. Elsewhere, a similar decline is thought to have resulted from several other factors. As the number of genetic varieties of corn planted is reduced, seed corn is increasingly susceptible to diseases and insect pests. For example, in 1970 U.S. corn production was decimated by corn blight. More than $1 billion of corn was lost, and in some places yields were reduced by as much as 50 percent (Kimbrell 2002). Pests continue to adapt more quickly than new technologies, like GMOs, can be developed. In addition, the average input costs have been increasing somewhat in real prices over the last four decades. While fertilizer costs have gone down, the costs of other chemicals have gone up somewhat. Energy costs have increased the most, by more than 50 percent (Runge and Stuart 1998). Considering overall declines in the real price of corn, this has led to greater producer interest in reducing operating costs through a more efficient use of inputs.

Recent breeding improvements of corn have allowed the expansion of the cultivated area by reducing the number of days to harvest, reducing the sunlight and mean temperature requirements, changing the disease resistance, and reducing the overall fertility or water requirements. Almost all of these improvements were done with conventional breeding programs rather than through the development of transgenic varieties. In general such changes have allowed the spread of corn production into colder climates with shorter growing seasons, which is the focus of the United States and European corporate plant breeders.

In the past in dedicated corn production areas, corn was planted after plowing, disking, and harrowing the fields to pulverize the soil. In some cases plowing was done in the fall as a way to allow farmers to plant earlier in the year, to reduce the workload in the spring or to begin the decomposition of the stubble from the year before. Now it is widely known that fall plowing is not good. It leads to soil erosion from wind and water, kills many beneficial soil organisms, and degrades soil structure (Runge and Stuart 1998).

Conservation tillage and no-till planting have become much more common within the last fifteen years or so in the United States, Brazil, and Europe. Planters deposit seeds in fields that have been only lightly tilled or not tilled at all. Crop residues are purposely left on the surface as a mulch that gradually decomposes and adds organic matter to the soil. The organic matter holds water and reduces erosion and the leaching of fertilizers. A critical constraint to the adoption of such practices is that the farm implements required are different from traditional ones. Therefore the front-end investments needed may be beyond the capacity of undercapitalized producers (Runge and Stuart 1998).

In some parts of the world traditional multicropping and nonmechanized production of corn still dominate the landscape. Increasingly, however, producers plant corn mechanically in monocrop stands. Today corn is the staple food crop for 350 million people living primarily in Latin America and Africa. Corn production is expanding most rapidly in Africa, where it is becoming a basic subsistence crop that also has strong markets for any surpluses.

Globally about half of the corn produced each year is from hybrid seeds, which must be purchased each year. The rest is from seeds retained each year by farmers. Corn planted from hybrids is more productive, so less land is needed for hybrid corn than for traditional varieties. Most land devoted to hybrid corn is planted as a monocrop using machinery and other inputs. The mechanized production of corn tends to take place on fields that are less hilly and marginal. However, because the mechanized production takes place on such large areas and with few waterways retained, it still can cause considerable erosion.

Much nonhybrid corn is planted on highly erodible soils. It is usually interplanted and/or sequentially planted with other crops. Hand cultivation that takes into account the nuances of the lay of the land and quality of the soil in conjunction with polycultural planting systems can reduce the overall erosion from the production of traditional corn varieties. While it is not clear which type of production causes the most soil erosion in absolute terms, it is possible that the traditional production systems create more erosion per kilogram of total production from all crops than more specialized, higher-input production systems.

There are some outstanding questions with regard to the impacts of the different corn production systems. For example, are larger producers, with more animals and more hectares of cropland, able to integrate and manage residual waste streams better than smaller ones? Does the combination of animal and cropping agriculture offer more opportunities for this integration than concentrated and specialized animal and/or cash grain farming alone? While smaller operations may generate fewer wastes in total, the per-hectare generation of wastes on such operations may be larger than those of larger farms capable of greater efficiencies. Opportunities for reuse, such as spreading manure on fields or more precise fertilizer application equipment, may require a larger scale to be economical (Runge and Stuart 1998).

PROCESSING

There are no significant environmental problems on or near farms posed by the processing of corn or corn products. With the exception of parts of Central America where so much corn is consumed directly by humans, processing takes place well away from corn-producing areas and is generally regulated by the laws and regulations that apply to any industry in the country in question.

When fed to animals, corn is normally cracked or ground into flour to aid digestion. While this process produces dust and uses energy, there are few other issues involved in such processing. Such processing is generally done close to feeding sites so that transportation is not a major issue.

Technological innovations over the past twenty-five years or so continue to increase the versatility of corn, not only as food and animal feed but also as an ingredient in a large variety of industrial products and processes. While corn's uses as ethanol, corn starch, and corn syrup are well known, less well known are its uses as a source for xanthan gum, vitamin C, biodegradable packing materials, lactic acid, corncob fuels, automobile paint, plastics, tires, chewing gum, foot powder, surgical dressings, adhesives, and whiskey (Runge and Stuart 1998).

The value of corn is ultimately determined by the demand for the end uses to which its products can be directed. For example, 25.4 kilograms of corn can be used to extract 14.5 kilograms (32 pounds) of cornstarch, or 14.5 kilograms (32 pounds) of corn sweeteners, or 9.5 liters (2.5 gallons) of ethanol. In addition, the remaining by-products can be used to make 5.2 kilograms (11.4 pounds) of gluten feed (at 20 percent protein) and 1.4 kilograms (3 pounds) of gluten meal (at 60 percent protein), plus 0.73 kilograms (1.6 pounds) of corn oil (National Corn Growers Association 1997, as cited in Runge and Stuart 1998).

SUBSTITUTES

Because corn has so many uses, it has substitutes and is a substitute for many products produced from other crops. Similarly, depending on prices or climatic conditions corn is often grown in rotation with other crops such as soybeans.

Ground corn is used in feeds for cows, pigs, and chickens. It is mixed in varying concentrations to either substitute for or to complement soybean meal, sorghum, wheat, fish meal, and cassava.

Sorghum is one of the main substitutes for corn used in animal feed. Sorghum is an excellent feed for animals and is used increasingly for chicken production, especially in the United States (see sorghum chapter). Because sorghum grows in slightly drier conditions and on poorer soils, it is an excellent crop substitute in many countries. In many less-developed countries, particularly in Africa, sorghum is used directly for human consumption, so there is already familiarity with it and the substitution process proceeds more quickly. Seed companies in the United States have developed strains of sorghum that are adapted to drought and to growing conditions in many parts of the world. These companies see production emphasis shifting to sorghum as the world becomes a drier place and as human populations expand into more arid regions.

Soybean meal and cake, cottonseed cake, and wheat are all used to complement or to substitute for corn in animal rations depending on overall price and on the nutritional levels required. (See the chapters on soybeans, cotton, and wheat for more complete discussions of these crops.)

There are several vegetable oil substitutes for corn oil. These include palm, palm kernel, coconut, cottonseed, soybean, sunflower, and canola (rapeseed) oil to name but a few. Developed countries are experiencing a shift from the first five oils listed above, which contain more saturated fats, toward less-saturated oils such as canola and olive oil. Olive oil has a limited market because of its price, so canola production has been increasing rapidly to provide a low-cost alternative. The FAO reported that globally more than 25 million hectares were planted to canola in 1999. The largest producers were China (7.5 million hectares), India (6.0 million hectares), and Canada (4.9 million hectares). These three producers accounted for more than 70 percent of all land devoted to canola production and more than 60 percent of all production (FAO 2002).

Finally, ethanol produced from corn is a substitute for gasoline. The U.S. Renewable Fuels Act of 2000 required that a certain level of all gasoline sold in the United States be from renewable sources. This could increase the yearly demand for ethanol from 6.8 billion liters (1.8 billion gallons) to more than 20.4 billion liters (5.4 billion gallons) per year by the year 2010 (American Corn Growers Association 2000a).

MARKET TRENDS

Between 1960 and 2000 corn production increased by 189 percent while the amount of corn traded internationally increased by 484 percent. Corn yields increased by 121 percent during the same period, and prices declined by 57.9 percent (FAO 2002).

In 1900 farmers received 70 cents of every food dollar spent. Today they receive less than 5 cents, and that figure is falling, especially for manufactured products. For example, a farmer receives less than 2 cents for the corn in a $4.00 box of corn flakes. Over the past twenty-five years the prices paid to farmers have remained relatively constant, while consumer food prices have increased by nearly 250 percent (American Corn Growers Association, 2002a).

One of the main issues currently affecting the market for corn is the production and sale of transgenic or genetically modified corn, such as "Bt corn" because it produces *Bacillus thuringiensis* (Bt), a bacterium that acts as a pesticide. At issue is whether Bt corn should be produced, and if so, whether it or products containing it should be labeled. The issue of ingredient labeling is complicated since corn is used in thousands of products.

Ultimately, however, consumers and governments will decide what consumers need to know from packaging. Producers are beginning to catch on as well. As Gary Goldberg, the chief executive officer of the American Corn Growers Association (ACGA), testified before the U.S. Department of Agriculture (USDA) in April 2000,

For agricultural producers, this debate over GMOs [genetically modified organisms] is not a safety, environmental, or health issue. It is an economic issue. GMOs have become an albatross around their necks, catching them in the middle of a debate between chemical companies, seed dealers, grain exporters, foreign and domestic customers, and U.S. and foreign government officials. Simply put, can farmers afford to grow a crop they may not have a market for? If we had to categorize this debate in one word it is "uncertainty." The uncertainty of not knowing whether our foreign customers will continue to purchase our products because we are dictating to them what they should buy; uncertainty over the issue of segregation and the responsibility for farmers to segregate on the farm, adding considerable expense; uncertainty over liability and who is liable for cross-pollination and contamination; and the uncertainty over corporate concentration and whether only a small handful of companies will control the production and distribution of seeds. These are the issues that concern farmers and they are the reasons that so many farmers have made the conscious decision to reject GMOs for this [2000] planting season.

Goldberg reported that the ACGA, in its own independent survey of members, found that there was a 16 percent reduction in Bt corn planted in 2000 compared to the previous year. A USDA survey found a 25 percent reduction in acres planted to Bt corn for 2000 compared with the previous year.

In short, a minority of players in the United States, and their political and corporate supporters, are jeopardizing the marketplace for conventional corn producers. The issue is really about markets. In 1997–98 the United States exported 2 million metric tons of corn to Europe. In 1998–99, corn exports to Europe totaled only 137,000 metric tons. U.S. competitors (Argentina, Brazil, and China) not planting GMOs captured the difference. Once again, the issue is complicated. Argentine corn producers are beginning to plant more Bt corn; Brazilians are reportedly planting a great deal of Bt corn illegally; and China may be exporting non-GM corn, but it is importing lower cost transgenic Bt corn (and transgenic soybeans as well).

Contrary to corporate claims, it is not clear that GMOs actually net more income for producers because seed and technology fees are higher than for conventional corn. On the other hand, producers believe they are saving money on reduced herbicide and insecticide use. They also see higher yields and production efficiencies.

The yield issue is interesting. To date, it appears that Bt corn yields may be 10 percent higher than non-Bt corn yields (American Corn Growers Association 2000b). If all farmers in the United States had planted Bt corn in 1999, the 10 percent increase would have added 238,760,000 kilograms to the U.S. inventory. In all likelihood this would have pushed the average price down to less than 4 cents per kilogram. What financial impact would this have had on farmers who are already finding it hard to survive financially? This situation has the feel of the commodity treadmill that producers throughout the world know so well. Farmers are also concerned that the technology is monopolized by a few companies that will be able to raise prices after producers have converted to the technology and are unable to convert back to other seed varieties.

One thing is clear: Many consumers, particularly outside the United States, do not want to purchase Bt corn or products that contain it. Consequently, on-farm segregation is becoming an important issue. Japanese purchasers (in 1999 the largest purchaser of United States corn with 398.9 million kilograms) have already warned American producers that they must segregate their crop if they want to continue to sell to Japan (American Corn Growers Association 2000c). This will be very difficult. In addition, many grain elevators are not physically prepared to handle two distinct grain flows. Whoever segregates the crop is likely to incur substantial costs for testing and certification as well as the time to clean out the combine and grain augers every time the product flow shifts from Bt corn to non-genetically modified corn.

Concerns about Bt corn are also found in the United States. Gerber and Heinz baby foods, Wild Oats supermarkets, Seagram's, IAMs pet foods, Genuardi's Family Markets, and Frito-Lay have moved to stop the purchase of Bt corn either in response to or in anticipation of consumer concerns about the product (American Corn Growers Association 2002b).

An increasing number of consumers have turned to organic corn products to ease their concerns about food quality in general and to avoid genetically modified products in particular. Certified organic products guarantee that the chain of custody must be traceable from the producer to the consumer. This is much harder to do for traditional versus Bt corn. This is one of the reasons the price of organic corn is high, often twice the price of conventional corn.

But is organic corn the solution to concerns about genetically modified, Bt corn or even the more sustainable production of traditional corn? It is not clear that the higher price for organic corn offsets its lower production levels and necessary crop rotations with lower-valued crops. While organic may make sense as a long-term strategy, it may not make sense in the short to medium term. It might also be more difficult to produce organic corn if a producer has existing debts for machinery or land. Thus, transitions to organic production may be very difficult to absorb financially.

Finally, it is now clear that for corn (and other genetically modified crops) pollen can drift very large distances and contaminate even organically grown crops. Some researchers estimate that due to pollen drift very little organic seed, much less organic produce, exists in countries such as the United States.

ENVIRONMENTAL IMPACTS OF PRODUCTION

Most of the environmental impacts associated with corn production occur at the farm level. The most important problems are habitat conversion, soil erosion and overall degradation, leaching of agrochemicals, and pollution of fresh water and groundwater.

Habitat Conversion

Corn is produced in a wide range of settings. In developed countries, corn tends to be produced on the same fields year after year or in rotation with other crops. In these countries any expansion of corn production tends to correspond with a decline in areas used for other crops. There is little habitat conversion for the production of corn at this time, although that was very common during the last 200 years in areas like the United States. Even where major habitat conversion took place decades or even generations ago, corn production is still changing the landscape. As machinery gets bigger, fields get bigger. This means the loss of fencerows and hedges that often were a safe haven for biodiversity. Larger machinery makes it more difficult to disengage to avoid grass-covered waterways. Tilling these areas in the past thirty years has led to increased erosion.

In developing countries corn production is gaining at the expense of other crops. Production is also becoming increasingly mechanized. In addition, however, habitat conversion for corn cultivation is also occurring. This is true of the planned colonization schemes in the greater Amazon region as well as the more generalized displacement of people in Central and South America (e.g., when labor-intensive crops such as cotton, coffee, and other crops were abandoned in favor of cattle). Perhaps the greatest impact of corn production on natural habitats at this time is occurring in Central and Southern Africa, where corn production is expanding more rapidly than anywhere in the world (due to government subsidies) and demand still far exceeds supply.

Soil Erosion and Degradation

Studies in the United States have shown that environmental susceptibility to erosion may or may not be related to overall productivity (Larson et al. 1988). In corn-producing areas of Minnesota, soils vulnerable to erosion and those low in productivity were often not the same lands. In fact, these two types of land were not correlated. However, land that is vulnerable to erosion eventually loses productivity.

A study of soil erosion in the corn belt areas of Iowa, Minnesota, Wisconsin, and Illinois indicates that erosion rates have declined. In 1932 erosion rates were more than 37 metric tons per hectare per year when corn production amounted to only 2.75 metric tons per hectare per year. By 1982 average erosion rates were down to 19.5 metric tons per hectare per year. By 1992, after 18 percent of all arable cropland had been taken out of production (including the most highly erodible areas) through the Conservation Reserve Program (CRP), erosion in the United States was estimated at 14 metric tons per hectare per year while corn production was about 8.6 metric tons per hectare per year (Runge and Stuart 1998).

Just as there is no correlation between a soil's susceptibility to erosion and its fertility, there is no correlation between row crop cultivation and erosion. Row crops have increased considerably in the United States, for example, from 1930 to the present, precisely when erosion declined. The correlation is between mechanical

row crop cultivation and erosion. No-till and conservation tillage row crops have reduced erosion dramatically. Erosion rates, however, still appear to be beyond replacement values and consequently are unacceptably high.

Declines in soil erosion result primarily from investments in conservation measures that include terraces, strip cropping, crop rotations, windbreaks, and switching
to conservation tillage (reduced tillage and no-till cultivation). By 1994 no-till farming techniques were practiced on about 12 percent of row crop production. Mulch
tillage (in which crop residue is left on the soil surface) and ridge tillage (in which
crop residue is collected in valleys alongside ridges of soil that are planted) were
practiced on another 26 percent of planted crops in the United States. This compared with 3 percent in 1984 and zero in 1930. Yet not all reductions in erosion result from producers' practices. Between 1985 and 1992 the U.S. government's CRP
program paid producers not to cultivate the most highly degradable areas (Runge
and Stuart 1998). This is probably the single most important cause for declining soil
erosion.

There have not been widely accepted studies on global soil erosion rates. Anecdotal evidence suggests that erosion is increasing in many areas even though producers know that it will destroy their ability to produce over time. For many, there is
no other option. They do not know any alternatives.

Use of Agrochemical Inputs

A range of pesticides is used widely on corn in the United States and elsewhere. Between 1964 (when Rachel Carson wrote *Silent Spring*) and 1991, herbicide use on
corn in the United States grew from just under 12 kilograms to just less than 100
kilograms of active ingredients per hectare (Runge and Stuart 1998).

In 1992 nearly 70 percent of the area planted to corn in the United States was
treated with the herbicide atrazine (Ribaudo 1993). Cornfield herbicides (like
atrazine) as a class of farm chemicals accounted for 47 percent of total agricultural
pesticide use in the nation in the early 1990s. Weed suppressants were applied to
about 95 percent of all corn acres (USDA 1991, 1992). Atrazine persists in soil, however, and moves in surface and ground water. Atrazine and nitrogen are thought to be
one of the main contributors to the "dead zone" in the Gulf of Mexico and a major
polluter of underground water supplies throughout the corn belt of the United States.

During the same period the rates of application of insecticides more than doubled from less than 1.5 kilograms per hectare to slightly less than 3.5 kilograms per
hectare (Runge and Stuart 1998). The important point here, however, is that the
composition of the insecticides changed dramatically away from some of the most
hazardous chemicals.

Fertilizers are used more commonly in the production of corn than they were 50
years ago. About 95 percent of all corn planted in the United States receives supplemental nitrogen. Fertilizers containing phosphate and potash are used on 75 to 80
percent of all corn. In all, corn accounts for almost half of total fertilizer use in
the United States (by comparison, wheat has 14 percent and soybeans 6 percent).

Fertilizer use is one of the main causes of water pollution and eutrophication in the United States.

By 1990 the U.S. Environmental Protection Agency estimated that roughly 650 different active ingredients were for sale as pesticides, down from 1,400 previously. However, about fifteen to twenty new materials are added each year in the United States (Runge and Stuart 1998).

Worst-case estimates of the impacts of eliminating the use of all agrochemicals suggest yield declines on the order of 53 percent for crops such as corn and 37 percent for soybeans (Knutson et al. 1990). However, according to Ayer and Conklin (1990) it is likely that these estimates "tend to underestimate the ability of producers to substitute other methods of pest and disease control, such as crop rotations, and to more carefully time and apply new and existing chemicals based on when and where they are most needed." Integrated pest management (IPM), cropping management, and precision agriculture are important new approaches, but current technology continues to rely on existing pesticides. Analysis also shows that reductions in the use of pesticides will affect the profits of farms of different sizes in different ways. There is good evidence that pesticides used on medium and especially larger farms are being substituted for other inputs such as labor and mechanical weeding (Runge and Stuart 1998).

A bigger issue, however, is pest resistance. By 2000, producers in the U.S. were using twenty times more pesticides and losing twice as much of their crop to pests as they were in 1950. This was one of the attractions of Bt corn. It produced its own insecticides. However, by as early as 1997, eight insect pests in the U. S. had become resistant to Bt (Conway 1997, as cited in Hawken et al. 1999).

Water Use

Because corn requires large amounts of water, it poses risks of crop failure for producers. One way to avoid the risk is to irrigate the crop. It is expensive to set up irrigation systems on the off chance that water will be needed in a bad year, however. Instead, irrigation systems tend to be set up in areas that are not suited to produce corn with the rainfall that is normally available. Unfortunately, ongoing irrigation is expensive and requires large amounts of water in areas where it draws on scarce water and energy supplies. While most corn grown throughout the world is not irrigated, in the United States corn is the second largest consumer of irrigation water (after alfalfa hay). Over half of all U.S. corn irrigation takes place in the state of Nebraska, where, in parts of the state, the crop could not be grown profitably without irrigation. Unfortunately, the Ogallala Aquifer that supplies all the water for this irrigation is fossilized water. This means that the aquifer has a limestone cap, and the water being drawn from it is not being replaced. Eventually the water will run out, but before that the energy costs of bringing it to the surface may become too expensive to produce corn profitably.

Corn production is increasing throughout Asia. In China corn is displacing other crops such as wheat and even rice. Increasingly, corn is also being grown with irrigation.

Water Pollution

Agriculture was acknowledged recently as the major source of surface water quality impairment in the United States. The USDA's Economic Research Service in 1994 found that agriculture contributed to water quality problems in 72 percent of impaired stretches of river, 56 percent of lakes and 43 percent of estuaries (Runge and Stuart 1998; Faeth 1996; USDA 1994).

Pollution from the Midwestern agricultural states (where corn is the major crop) contributes to an offshore "dead" zone in the Gulf of Mexico. The U.S. Geological Survey (USGS) investigated the sources of pollutants causing the dead zone. They concluded that 70 percent of the nitrogen delivered to the Gulf came from above the confluence of the Ohio and Mississippi Rivers (Alexander et al. 1995). An estimated 90 percent came from nonpoint sources, primarily agricultural runoff and atmospheric deposition (EPA 2001).

Corn production has had a significant impact on groundwater quality from the agrochemical inputs used in production. The impacts are not well documented. However, half of the U.S. population uses groundwater for its main source of drinking water. It is the sole source for many rural communities.

Cross-Pollination and Contamination by GM Corn

The issues of cross-pollination and contamination by genetically modified corn, and the associated liability issues, will probably drive some farmers away from transgenic seeds. Seed companies have come to assume that corn pollen drifts no more than 185 meters (600 feet). However, Neil Harl, an agricultural economist from Iowa State, has said that pollen can drift over 8 kilometers (5 miles) (American Corn Growers Association 2001). Who is right? Perhaps more importantly, who will be liable? If the seed industry is confident that the distance is 185 meters (600 feet), then they should assume liability for any crossing or contamination beyond that.

GMOs are expected to be a continuous source of concern, but it is their uncontrolled introduction and proliferation that are especially disturbing. Perhaps one of the most important issues is the potential impact of GMOs on microorganisms, upon which so much of sustainable agriculture will depend. Can GMOs change landscapes? Can GMOs pose hazards to the fragile habitats and to wildlife itself? These, too, may ultimately be questions about liability.

BETTER MANAGEMENT PRACTICES

Given what is already known about reducing the environmental impacts of corn production, it should be possible for conservation strategies to reduce production impacts significantly as well as to increase long-term profits. However, the strategies will have to be site-specific and tailored to different types of production. What will work with capital-intensive, market-oriented producers will be quite different than what will work most effectively with subsistence producers and small farmers who sell surpluses into local markets. Since both types of producers can have significant impacts and since both types of production could be improved, it would be wise to determine which type of production is most common in a biodiverse area before proceeding.

Understanding better management practices (BMPs) for corn cultivation should be the cornerstone of any strategy to reduce the impacts of production. Research on BMPs should identify not only the practices and their social and environmental impacts but also their financial implications. For example, integrated pest management generally improves profits while reducing pesticide applications. Many if not most BMPs are being identified, adopted, and promoted primarily to solve a problem for producers or due to market-based incentives to lower producer costs. Reduced environmental impacts are added benefits. For example, the overall reductions in erosion in the United States were not accomplished by growing less corn or other crops. Rather, they came about through investments in a variety of BMPs that include a wide range of conservation measures (Runge and Stuart 1998).

For many producers, however, the critical constraint to the adoption of BMPs may well be that many farm implements are expensive and require up-front investments. Well-capitalized producers must first amortize their existing investments. By contrast, undercapitalized producers may neither be able to adopt new, better machinery or abandon older, obsolete technology for which they have not yet paid.

BMPs could also provide guidance for both what is important to measure and how one might measure nonpoint-source pollution, nutrient balances, ground and surface water contamination, as well as other specific measures such as nitrogen use and runoff. In addition, they could be the basis for identifying targets and policies aimed at changing incentives as well as producer practices. All too often, policies are changed after a problem has been discovered rather than implemented earlier in the process in order to prevent problems.

Ideally, BMPs could be the basis for a certification program. However, they would have to be evaluated with measurable targets and indicators to monitor progress in achieving them. Such practices and the measurable impacts could serve both as the target and the yardstick by which to develop and measure policies.

Government can encourage the adoption of BMPs whether the goal is to change the use of specific chemicals or to change land use patterns. The CRP program is a case in point. Taking some of the most highly erodible land in the United States out of production had two impacts: soil erosion was reduced immediately, and biodiversity increased almost immediately. Government does not need to buy land to ensure conservation. It is far cheaper to buy conservation easements. For example, a program in the state of Minnesota has purchased permanent conservation easements along threatened watersheds to protect critical wetland habitat (Larson et al. 1988).

Finally, it is conceivable that global markets could push for the development of perennial corn varieties that could produce multiple corn crops in tropical areas while at the same time reducing their overall environmental impacts. Given that there would be less private sector interest in such varieties, development would need to be supported by governments. Such developments would certainly not only reduce the overall impacts of corn production but also change what are presently understood to be the better practices.

Adopt Conservation Tillage

Many BMPs save producers time and money. Conservation tillage saves on fuel, labor, and depreciation of farm equipment while improving soil structure and fertility (USDA 1995). Producers reduce mechanical cultivation, and the savings are substantial. A farm of 400 hectares saves 450 person-hours per year, more than U.S.$800 for machinery use and wear and tear, and U.S.$3,250 in fuel savings per year (CTIC 1997). Producers seem to be interested in switching to conservation tillage or no-till production if they do not have huge existing capital investments in machinery and if their land tends to be erodible.

Recent research regarding the impacts of conservation tillage suggests that it can have a significant positive impact on soil biodiversity while reducing erosion and increasing water retention in soil. Studies indicate that it increases earthworm populations two to three times above those in conventional fields, bringing the associated benefits of improved water infiltration, better crop rooting, and increased soil fertility (Scardena 1996). One of the key remaining questions about conservation tillage is: What are its overall impacts regarding herbicide use? One form, no-till, substitutes herbicide use for mechanical cultivation to prepare fields for planting and to control weeds.

BMPs are not found only in high-input forms of agriculture in developed countries. In the humid tropics, corn is cultivated on steep slopes and more marginal areas in the Andes, Belize, and mountainous areas of Mexico. Use of conservation tillage methods there is reducing impacts and increasing producers' yields and income. In Belize, for example, producers are now planting native nitrogen-fixing legumes simultaneously with corn. These plants overtake the corn after it is harvested, protecting the soil from exposure, reducing erosion, fixing nitrogen, and actually reseeding themselves for the next season's crop.

Increase Organic Matter in the Soil

The key to sustainable agriculture is the maintenance or rehabilitation of the soil. Organic matter in the soil is perhaps the single most important issue, as many other factors stem from it. For example, organic matter in the soil can trap or detain major water pollutants and chemicals.

Overall water use in corn production is related directly to the poor water retention of soils. The main cause for reduced water retention is the depletion of organic matter in the soil where corn is produced. Thus, any buildup of organic matter in the soil will result in a net water savings. The buildup in organic matter will also result in a net savings in fertilizer and pesticide use, because organic matter also reduces nutrient leaching and makes plants more vigorous and thus more pest-resistant. This means that organic debris from the harvest should be the returned to the fields rather than removed for any other purpose. There are now techniques, including adding decomposer microorganisms, that can make crop residues decompose very quickly into organic matter within the field itself. Conservation tillage, described above, is one of the ways to increase soil organic matter in corn production.

Use Microorganisms to Break Down Waste and Excess Nutrients

With organic matter present, microbial biodiversity can begin to decontaminate pollutants in the soil. It may be possible to undertake limited bioremediation using effective microorganisms in freshwater systems, but care should be taken before releasing alien microorganisms into such environments. They do, however, have the ability to convert nitrates into other forms of nitrogen that could become useful nutrients for other organisms. Pesticides can also be acted upon by microorganisms to change their toxicities or to detoxify them entirely. Sulfur and iron radicals can be converted by these organisms to become useful substances for other life forms.

Reduce Use of Fertilizers and Pesticides

In the cases of both pesticides and fertilizers, application timing and methods can greatly influence total use levels. More precise application of pesticides, and their increased efficiency, has resulted in continual increases in yields in the 1990s even as total pounds applied of active ingredients per area of land cultivated has fallen. Lin et al. (1995) report that switching from preplanting to after-planting applications and from broadcast to band applications can reduce nitrogen use by 4.5 to 64.8 kilograms per hectare.

The recommended use of nitrogen, for corn and for other crops, is often based on monocrop trials rather than on crop rotation systems of production. Researchers have found that with typical crop rotations, nitrogen applications can be reduced as much as 34 kilograms per hectare (30 pounds per acre) below the manufacturer's recommended rate (Vanotti and Bundy 1994, as cited in Runge and Stuart 1998). Not only is this a savings to farmers, it means that less fertilizer will be put into the ecosystem to pollute it.

In a study of continuous corn cultivation in Iowa, it was found that the herbicide atrazine was less concentrated in drain tiles under fields if it was applied in narrower bands than if it was spread over the entire crop. Banding allowed producers to apply one-third less (Kanwar and Baker 1994).

Biological substitutes for chemical pesticides (e.g., sex-linked insect attractors) can also be effective. These should be added to the options of producers around the world as well as encouraged by the various groups that work with them. As such measures become more common, their costs are far cheaper too. Microorganisms, for example, have proven to be a valuable disease control solution that is becoming cheaper as well. PlantShield (*Trichoderma harzianum*) and Mycostop (*Streptomyces griseoviridis*), manufactured by BioWorks, Inc. and Kemira Agro, respectively, are labeled for control of a number of vegetable crop diseases (Reid 2002). All these approaches can be developed further and applied more broadly. The application of some of these approaches can even reduce the use of some chemical inputs by making them more effective by reducing pest resistance to them.

Use Crop Rotation

The same crop cannot be grown continuously on the same piece of land year after year without causing serious damage and serious problems. The most successful low-cost, long-term corn production systems include crop rotation. Such rotations should include not only cash crops but also nitrogen-fixing legumes and high-biomass-producing plants as well so as to introduce more organic matter into the soil. Rotations are normally three to four years, but in some cases with organic producers, corn rotations can be five or six years and involve corn, soybeans, oats, hay, and fallow. In these systems corn is produced only every five years. The overall profitability of such operations depends not only on the value of all the different crops, averaged over the rotation, but also on the reduced input costs associated with continual monocropping of the same crop.

Crop rotations can be shortened if clover or other legume cover crops are sown when oats or wheat are planted and then allowed to grow for the remainder of that crop year, effectively getting two crops in one year. In the tropics, two to three sequential crops (e.g., those grown in the same year) are normally part of multiyear rotations.

Such rotations are not only undertaken by organic producers. In Brazil corn producers regularly have corn-soybean-cotton rotations. Within any given year, grass or other off-season crops are planted to increase the biomass. These crops are sprayed with weed killer, but the total amount of pesticides and fungicides used on such crops is half that of traditional producers in the area. In addition, these production systems are allowing smart, forward-looking producers to buy up degraded pasture land and rehabilitate it within four to five years. In effect, they are making as much money rehabilitating land as they do from growing marketable products.

OUTLOOK

The projections for increased consumption of animal protein, particularly in developing countries, imply an increase in demand for animal feed. At this time, no other feed is as important as corn for increasing the production of beef, pork, or poultry. Since most of the demand is coming from developing countries, it is possible, if not likely, that many producers in developing countries will abandon the production of traditional food crops in an attempt to capture some of the increasing market for corn for animal feed. With current price supports and subsidies for U.S. corn, it will be hard for producers in most developing countries to compete even on domestic markets in their own countries, unless their countries erect market barriers of their own.

The one possible bright spot for producers in this scenario is that current demand projections indicate that developing countries will need supplies of corn that are equal to the total current annual U.S. production. However, this most likely means that increased corn production may well reduce the production of traditional food crops. This will have a disproportionate impact on those who cannot afford to eat meat or other animal proteins on a regular basis.

Corn is grown throughout the world in both temperate and tropical areas, coastal areas and highlands. As a consequence, the production of corn causes a wide range of environmental impacts. These can include soil erosion and degradation as well as intense use of pesticides, fertilizers, and water. In all likelihood global corn production will increase dramatically as demand increases for animal feed grains. At the same time, increasing efforts will be made to reduce pollution from corn production, first in developed countries but later in developing countries as well. It is not obvious that the attempts to reduce the impacts of corn production will be sufficient or timely enough to offset rapidly expanding production.

REFERENCES

Alexander, R., R. Smith, and G. Schwarz. 1995. The Regional Transport of Point and Nonpoint-Source Nitrogen to the Gulf of Mexico. *Proceedings of the First Gulf of Mexico Hypoxia Management Conference*, Kenner, LA, Dec. 5–6.

American Corn Growers Association (citing Gary Goldberg, Chief Executive Officer of the ACGA). 2000a. *ACGA Commends Senators Daschle and Lugar for Legislation to Grow Demand for Ethanol*. ACGA News and Views. Available at http://www.acga.org/news/2000/051700.htm.

———. 2000b. *Corn Growers Assert that Increased Corn Yields Is the Wrong Reason to Plant Genetically Modified Corn*. ACGA News and Views. Available at http://www.acga.org/news/2000/071900.htm.

———. 2000c. Corn Growers Will Need to Segregate Crop to Keep Huge Japanese Corn Market. News from the American Corn Growers Association.

———. 2001. Letter from the President of the American Corn Growers Association to EPA's Christine Todd Whitman regarding the negative market impacts to farmers of Bt corn. August 26. Available at http://www.biotech-info.net/ACGA_letter.html.

———. 2002a. Consumer Food Prices vs. Farm Prices Over the Past 25 Years. Washington, DC: American Corn Growers Association. Available at http://www.acga.org.

———. 2002b. *Genetically Modified Crops: Questions & Answers*. GMO Brochure Section 2. Available at http://www.acga.org/programs/GMOBrochure/02.htm.

Ayer, J. E. and N. Conklin. 1990. Economics of Agricultural Chemicals: Flawed Methodology and Conflict of Interest Quagmires. *Choices* (Fourth Quarter).

Conservation Technology Information Center (CTIC). 1997. 14 Benefits for U.S. Farmers and the Environment through Conservation Tillage. West Lafayette, IN: CTIC.

Conway, G. R. 1997. *The Doubly Green Revolution*. London: Penguin.

EPA (Environmental Protection Agency). 2001. Action Plan Proposes Goal to Reduce Gulf of Mexico Dead Zone. *Nonpoint Source News—Notes*. Issue 65 (June).

Faeth, P. 1996. *Make it or Break it: Sustainability and the U.S. Agricultural Sector*. Washington, D.C.: World Resources Institute.

FAO (Food and Agriculture Organization of the United Nations). 2002. *FAOSTAT Statistics Database*. Rome: UN Food and Agriculture Organization. Available at http://apps.fao.org.

———. 1996. *Production Yearbook 1995*. Vol. 49. FAO Statistical Series No. 130. Rome: UN Food and Agriculture Organization.

Goldberg, G. 2000. Testimony before the USDA Advisory Committee on Agricultural Biotechnology. Washington, D.C., 27 April.

Hawken, P., A. Lovins, and L. H. Lovins. 1999. *Natural Capitalism: Creating the Next Industrial Revolution*. Boston, MA: Little, Brown and Company.

Kanwar, R. D. and J. L. Baker. 1994. Tillage and Chemical Management Effects on Groundwater Quality. In *Agricultural Research to Protect Water Quality*. Ankeny, IA: Soil and Water Conservation Society.

Kimbrell, A. Editor. 2002. *Fatal Harvest: The Tragedy of Industrial Agriculture*. Washington D.C.: Island Press.

Knutson, R. D., C. R. Taylor, J. B. Penson, Jr., and E. G. Smith. 1990. Economic Impacts of Reduced Chemical Use. *Choices* (Fourth Quarter).

Larson, G. A., G. Roloff, and W. E. Larson. 1988. A New Approach to Marginal Agricultural Land Classification. *The Journal of Soil and Water Conservation* 43:1(January–February).

Lin, B. H., M. Padgitt, L. Bull, H. Delvo, D. Shank, and H. Taylor. 1995. Pesticide and Fertilizer Use and Trends in U. S. Agriculture. Natural Resources and Environment Division. Agricultural Economic Report No. 717. Washington, D.C.: USDA–Economic Research Service.

Nadal, A. 2000. The Environmental and Social Impacts of Economic Liberalization on Corn Production in Mexico. A Study Commissioned by Oxfam GB and World Wildlife Fund International. September. Draft.

National Corn Growers Association. 1997. *The World of Corn*. St. Louis, MO: National Corn Growers Association.

Reid, J. 2002. Biological Disease Control in Veg Crops. *Ag Corner*. May 8. Ithaca, NY: Cornell Cooperative Extension.

Ribaudo, M. 1993. Atrazine and Water Quality: Issues, Regulation and Economics. Agricultural Resources: Cropland, Water and Conservation. USDA-ERS. AR-30. Washington, D.C.: U.S. Department of Agriculture.

Runge, F. 1994. *The Grains Sector and the Environment: Basic Issues and Implications for Trade*. Rome: UN Food and Agriculture Organization. November 1. 49 pages.

Runge, C. F. and K. Stuart. 1998. *The History, Trade and Environmental Consequences of Corn (Maize) Production in the United States*. Report prepared for World Wildlife Fund, Washington, D.C. March 1. 208 pages.

Scardena, D. Editor. 1996. *Profitable Midwest No-Till Soybean Production*. North Central Region Extension Publication No. 580. Columbus, OH: Ohio State University Extension.

Smil, V. 2000. *Feeding the World—A Challenge for the Twenty-First Century*. Cambridge, MA: MIT Press.

UNCTAD (United Nations Conference on Trade and Development.) 1999. *World Commodity Survey, 1999–2000*. Geneva, Switzerland: UNCTAD.

———. 1994. *Handbook of International Trade and Development Statistics, 1993*. Geneva, Switzerland: UNCTAD.

USDA (U.S. Department of Agriculture). 1991. Agricultural Resources: Inputs. Situation and Outlook. AR-24. Washington, D.C.: USDA-Economic Research Service.

———. 1992. RTD Updates: Tillage Systems. Resources and Technology Division. Washington, D.C.: USDA–Economic Research Service.

———. 1994. RTD Updates: Chemical Use Practices. Data updates from the Research and Technology Division. Washington, D.C.: USDA–Economic Research Service.

———. 1995. Tillage and Cropping on HEL (Highly Erodible Land). AREI Updates. Number 6. National Resources and Environment Division. Washington, D.C.: USDA–Economic Research Service.

———. 1996. *Corn Yearbook*. Washington, D.C.: Agricultural Statistics Board, National Agricultural Statistics Service.

Vanotti, M. B. and L. G. Bundy. 1994. Corn Nitrogen Recommendations Based on Yield Response Data. *Journal of Production Agriculture* 7(2).

Weiner, T. 2002. In Corn's Cradle, U.S. Imports Bury Family Farms. *The New York Times*. February 26.

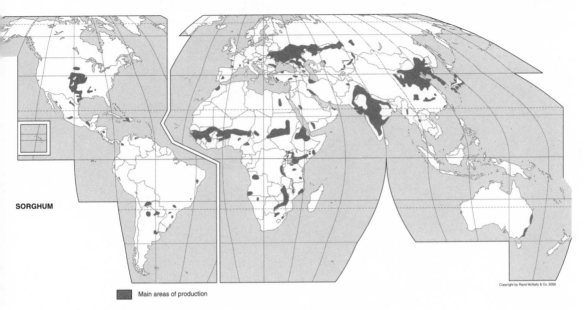

SORGHUM

Main areas of production

Copyright by Rand McNally & Co. 2000

SORGHUM *Sorghum bicolor*

PRODUCTION		INTERNATIONAL TRADE	
Area under Cultivation	42.0 million ha	Share of World Production	13%
Global Production	58.0 million MT	Exports	7.7 million MT
Average Productivity	1,381 kg/ha	Average Price	$98 per MT
Producer Price	$90 per MT	Value	$754 million
Producer Production Value	$5,222 million		

PRINCIPAL PRODUCING COUNTRIES/BLOCS (by weight)	United States, India, Nigeria, Mexico, Argentina, China, Sudan, Australia
PRINCIPAL EXPORTING COUNTRIES/BLOCS	United States, Argentina, France, Sudan
PRINCIPAL IMPORTING COUNTRIES/BLOCS	Mexico, Japan, Spain, Israel, Italy, Chile
MAJOR ENVIRONMENTAL IMPACTS	Habitat conversion Soil erosion and degradation Agrochemical use Poisoning in herbivorous animals Fire hazards
POTENTIAL TO IMPROVE	Fair BMPs are known and reduce input use and runoff in developed countries New varieties reduce soil degradation Many producers are poor and not well integrated into the market or extension systems Pests are a major problem for poor farmers

Source: FAO 2002. All data for 2000.

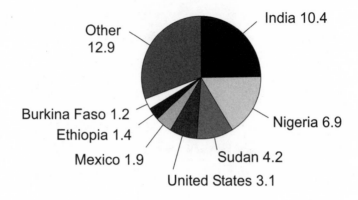

Area in Production (MMha)

India 10.4

Other 12.9

Burkina Faso 1.2

Ethiopia 1.4

Mexico 1.9

United States 3.1

Sudan 4.2

Nigeria 6.9

OVERVIEW

Sorghum is a grass in the same family as corn and sugarcane. It was domesticated about 5,000 years ago, and nearly all of its genetic material comes from varieties that were originally cultivated in Africa. Today, sorghum is a food staple in many parts of Africa and Asia. One of the main advantages of sorghum is that it is very drought-tolerant. Along with millet, it is planted by African farmers in dry and marginal areas as a hedge against famine.

Just as corn, cocoa, rubber, potatoes, and other crops were taken from the Americas to Europe, Africa, and Asia, sorghum was brought to the colonies from Africa. It is thought that it was brought from Africa to the Caribbean, where it was grown to feed slaves (DeWalt and Barkin 1987). It subsequently spread to Central, South, and North America. By the nineteenth century, sorghum was grown over much of the Great Plains as a drought-tolerant animal feed.

At the beginning of the twentieth century, the U.S. Department of Agriculture and the Texas Agricultural Experiment Station were working together in Chillicothe, Texas, to introduce and test different varieties of sorghum. During this period researchers bred and selected new lines of sorghum that were higher yielding, mechanically harvestable, resistant to disease, and adapted to a wider range of climatic conditions (DeWalt and Barkin 1987). In the 1960s and 1970s the DeKalb Seed Company began to develop lines of high-yield sorghum. They intended for Great Plains farmers to adopt these lines when the climate changed, or when the aquifers that they were using to irrigate corn dried up. That has not yet happened; instead, the new varieties were exported to much of Central and South America, Africa, and Asia.

PRODUCING COUNTRIES

Sorghum is cultivated to produce grain and silage. The United States, India, Nigeria, Mexico, Argentina, China, and Sudan are the largest producers of sorghum grain. This grain is used both for human food and for animal feed. From 1999 to 2001 the United States produced 20 to 25 percent of all sorghum grain globally. The top seven sorghum grain producers account for 77 percent of global production (see Table 19.1) (FAO 2002).

In other countries, especially in Asia, sorghum is produced primarily for silage. India is the largest producer of sorghum that is cut green for silage. It is used for feeding livestock, especially cows. In other countries, the grain is harvested and the stalks are fed to livestock.

The total area under production for sorghum in any given country is a function of whether the land is used for grain or fodder, as well as the overall efficiency of the production methods employed. Production efficiency, in turn, depends on general productivity, climatic and soil conditions, and whether the crop is grown in polyculture or monoculture systems.

India, with nearly 25 percent of the global land devoted to sorghum, leads all producers in area under cultivation. Nigeria, Sudan, the United States, Mexico, Ethiopia, and Burkina Faso also devote considerable land to the crop. In fact, these seven sorghum producers account for nearly 70 percent of the global area under

TABLE 19.1. Sorghum Production and Cultivation by Country, 2000

Country	Production (MT)	Area Cultivated (ha)
United States	11,951,910	3,126,630
India	8,862,700	10,397,900
Nigeria	7,711,000	6,885,000
Mexico	5,842,308	1,899,201
Argentina	3,350,513	723,600
China	2,608,456	894,860
Sudan	2,488,000	4,194,960
Australia	2,115,912	622,267
Ethiopia	1,548,720	1,359,190
Burkina Faso	1,016,275	1,225,223
Egypt	941,188	162,597
Brazil	779,608	523,970
Tanzania	664,200	638,700
Venezuela	581,526	286,697
World	57,964,600	41,964,377

Source: FAO 2002.

sorghum cultivation. Many countries devote at least a quarter of their agricultural land to sorghum, and Niger devotes at least 50 percent, as shown in Table 19.2.

In 2000 only 13 percent of all sorghum grain was exported. The United States accounted for more than 80 percent of all exports and exported some 55 percent of all domestic production, as shown in Table 19.3. Only France exported a larger proportion of total production.

CONSUMING COUNTRIES

Sorghum is consumed by people in most of the developing countries that produce it. Per capita consumption of sorghum is highest in Africa, especially in Burkina Faso, Ethiopia, Mali, Niger, Nigeria, Sudan, and Tanzania. In India sorghum is mostly used as animal fodder, with the exception of rural parts of Maharashtra where the grain is used as a food staple.

Sorghum grain is also used as animal feed in many countries, and nearly all international trade in sorghum is for animal feed. This is true in the United States, Mexico, Brazil, Japan, South Korea, and some European countries. In Europe, sorghum is mostly fed to poultry and pigs, but some is used for cattle as well.

TABLE 19.2. Percentage of Agricultural Land Devoted to Sorghum Production, 1994

50 Percent or More	25–49 Percent	10–24 Percent
Mauritania	Botswana	Chad
Niger	Burkina Faso	El Salvador
Somalia	Eritrea	Ghana
	Mali	Haiti
	Sudan	Lesotho
	Tanzania	Mozambique
	Yemen	Nigeria

Source: FAO 1996.

TABLE 19.3. Sorghum Export Statistics, 2000

Country	Export (MT)	Percentage of Crop Exported
United States	6,577,186	55.0
Argentina	770,324	23.0
France	229,006	61.6
Sudan	105,318	4.2
China	16,760	0.6
Australia	10,732	0.5
Italy	6,325	2.9
Venezuela	2,986	0.5
Nigeria	2,476	0.03
Burkina Faso	1,525	0.1
World	7,669,185	13.2

Source: FAO 2002.

Mexico is the main importer of sorghum grain, followed by Japan. Together, they accounted for just over 85 percent of global imports in 2000 (FAO 2002). Globally, most exported grain is used as animal feed.

PRODUCTION SYSTEMS

Sorghum is grown widely both for food and animal feed. Roughly 90 percent of the land planted to sorghum globally is in developing countries, mainly in Africa and Asia, where more than 70 percent of the crop is used for food. Most of the areas planted to sorghum are lands that are marginal for agriculture, i.e., they are subject to low and irregular rainfall and drought.

Broadly speaking, there are two different production systems. Intensive, commercialized production is concentrated mainly in the developed world and parts of Latin America and the Caribbean. Most of this production is used for animal feed. This production is characterized by monocrop cultivation and the use of hybrid seeds, fertilizer, and improved water management technologies. Commercial sorghum is fully mechanized. Seeds are planted in rows, and the plants are cultivated by machine or, in the case of no-till production, sprayed with herbicides. The crop is harvested by machine. Yields average about 3 to 5 metric tons per hectare. While this type of commercial production amounts to less than 15 percent of the area planted to sorghum, it produces more than 40 percent of global output. Furthermore, about 40 percent of this commercial production is traded on international feed markets (FAO and ICRISAT 1996).

Most of the sorghum produced around the world, however, is produced for subsistence or local food markets. For many households the goal is to produce enough sorghum to feed themselves, and depending on the year, many fail to achieve this goal. The production systems are extensive and characterized by few inputs. Most of the seeds are planted by hand, sown either by broadcasting the seeds or planting them in hills. Most subsistence and small-scale production of sorghum is in polyculture systems. The impacts of weed and insect pests are diminished by growing the plant in association with other crops. In some instances, improved seed varieties are being planted (especially in Asia), but for the most part these systems are far less intensive than those used for commercial production. Improved seeds are seen as a relatively cheap investment with a good chance of improving overall production. But for most cash-poor subsistence producers and small farmers, investments in other inputs compete directly with those for schooling or health care, both of which are seen as having higher potential returns on investment. Consequently, the use of fertilizer is limited, and very few producers have adopted moisture conservation techniques. As a consequence, yields average only 0.5 to 1.0 metric tons per hectare in most areas (FAO and ICRISAT 1996).

In Africa and Asia land holdings are small, and farmers have little equipment to use in production. While cultivation techniques on these small farms are relatively simple and generally follow traditional practices, there is an increased use of chemical inputs and mechanization. Some larger farms (as in Zimbabwe) are mechanized and more productive. In general, however, implements tend to be for hand use or are used with animal power. Only the larger farms use tractors and tractor-drawn equipment. Because sorghum is drought-resistant, it tends to be cultivated mostly under rain-fed conditions with little or no irrigation. Low prices for sorghum also tend to discourage capital-intensive inputs such as irrigation. In Sudan, for example, 90 percent of the crop is grown under rain-fed conditions. In most of India, where sorghum is cultivated as a silage crop to be cut and fed to dairy cows, it is also produced in traditional ways with few purchased inputs.

Whether grown commercially or for subsistence, sorghum production is limited by the same factors: birds, insects, diseases, weeds, and drought. Nearly 150 insect pests attack sorghum. In addition, sorghum is a host for more than 100 plant pathogens, including fungi, bacteria, viruses, and nematodes. The most serious diseases are grain mold, anthracnose, ergot (honeydew disease), root and stalk rots, and downy mildew. Globally, striga is the most common weed, and it has wiped out production in some areas.

From research undertaken in different parts of the world, it is clear that sorghum can produce more than 10 metric tons per hectare under ideal conditions (ICRISAT 2000). At this time, subsistence producers achieve, on average, only 5 to 10 percent of that amount. Higher yields will most likely be achieved through improved production practices rather than through increased or improved inputs. If sorghum is to achieve its potential contribution to improving food supplies and reducing poverty in marginal rural areas, the factors limiting its production in different circumstances must be identified and overcome.

Unfortunately, production trends appear to be headed in the opposite direction. In many poorer regions of the world, sustainability of sorghum is becoming an issue of great concern. Fallow periods are being shortened and more marginal lands are being brought into production. Marginal lands, often farmed by poorer producers, are cultivated with little or no fertilizer and few improved soil management techniques. As a consequence, soil degradation is increasing (FAO and ICRISAT 1996).

There are some noteworthy historical trends in sorghum production. The area planted to sorghum worldwide gradually increased from 1900 to 1985 when area planted (50 million ha) and total production (77.5 million MT) both peaked (FAO 2002). Production increases during that period, however, did not keep pace with population growth. This was probably due to the availability of other cereals as well as the replacement of manual labor first with draft animals and then with machines. Also, sorghum hybrids were not developed as quickly as those for corn. Even so, sorghum was well adapted to meet the basic food and animal feed needs in some areas of the world.

The introduction of sorghum hybrids, in conjunction with the genetic gains made in the poultry industry, made sorghum an increasingly attractive alternative to more expensive cereals such as corn and soybeans used for animal feed. Also, sorghum required fewer inputs than other cereal grains. While sorghum production for human food has been declining, much of the increased production has been aimed at poultry and, subsequently, pork feed. By the year 2000 more than 41.7 million hectares globally were planted to sorghum (FAO 2002).

Sorghum production in the United States mirrors this trend. From the late nineteenth century until the 1930s, the area planted to sorghum declined. This was probably due to increased efficiency of production with tractors rather than a declining demand for sorghum. During this period considerable sorghum was also cut green for silage and to make molasses. By 1950 only 4 million hectares were planted to sorghum. The area under production peaked again in the 1960s and 1970s and then began to decline. The declines in U.S. area cultivated appear to have occurred for a number of reasons. Hybrids proved to be far more productive than traditional varieties, and when fertilizers and other chemicals were thrown in production increased even more. These factors reduced the need for larger areas planted to maintain or increase total production.

The single most important technological change in sorghum cultivation since the 1950s and 1960s has been the development and use of hybrid seeds. Hybrids are now used widely throughout the world. As a result, productivity and uniformity in maturity and grain quality have increased. Hybrids have also encouraged mechanization and increased the use of fertilizers and other purchased inputs. In India hybrids are planted on about 55 percent of the total area. Yields on those areas have doubled over the past thirty years (FAO and ICRISAT 1996).

Most cultivated sorghum belongs to the species *Sorghum bicolor*. This species has more than five different races. Information regarding the specific varieties produced is difficult to obtain. The major races and the distribution of their cultivation is as follows:

- Bicolor: Most African, Asian, and American countries
- Guinea: African savannas and South Asia
- Caudatum: Northwest Nigeria, Chad, Sudan, Ethiopia, and Uganda
- Kafir: Southern Africa and northern Nigeria
- Durra: Sudan and India

There are several constraints that have affected productivity of sorghum. Most are related to production and include poor seedling vigor, damage from birds and other pests, harvest inefficiency, and lack of access to improved seeds, fertilizers, and pesticides (especially in developing countries). Once the crop is harvested, there are other problems. These include postharvest losses, storage complications, difficulties in marketing the produce, and fluctuations in demand based on availability and price of other cereal substitutes for food or animal feed.

Other factors have limited sorghum production in the United States. For example, comparable increases in productivity of corn and soybeans have allowed both to remain competitive with sorghum in the market. The water shortages that were expected to affect corn production in the western Midwest and Great Plains were offset by the development of more drought-resistant corn varieties and by the use of artisanal water from aquifers for hub irrigation systems. Finally, U.S. government policies that had allowed and even encouraged marginal land to be planted to grain sorghum were ended in favor of taking that land out of production and putting it into the Conservation Reserve Program (CRP). Thus, farmers were paid not to plant marginal and highly erodible land, and some of that land was not fit to grow much besides sorghum.

This has not been the case in every country, however. The adoption of sorghum both for food and feed in Mexico has been striking, particularly for a country where corn has been the staple for thousands of years. The hybrid sorghum, developed in the 1950s, was rapidly adopted by farmers. Sorghum was more productive than corn, and required less labor. By 1982 sorghum provided 74 percent of the raw material used in animal feed in Mexico (*Boletín Interno*, September 29, 1982, as cited in DeWalt and Barkin 1987). The net profit per hectare from sorghum was more than 4.5 times that from corn in 1983 (DeWalt and Barkin 1987).

However, as the Mexican and U.S. economies have become more integrated through the North American Free Trade Agreement (NAFTA), U.S. agricultural policies have had a greater impact on production in Mexico. In particular, U.S. subsidies for corn and other feed grains have affected prices in Mexico. Similarly, beef, pork, and poultry prices in the United States also influence prices in Mexico. This has dampened the profits from crops like sorghum, and farmers have begun to switch to higher-value, more labor-intensive agricultural products where they might have an advantage in the U.S. market.

PROCESSING

Sorghum is processed for human consumption to use either directly in food or in the production of beverages. In processing for food items, sorghum is either boiled and eaten, or ground into flour to be used as an ingredient in other foods. A number of traditional foods are made from sorghum in the areas where it has become a food staple. These foods include fermented and unfermented breads, stiff porridge, thin porridge, and steamed foods. Sorghum flour is used in various baked products such as hearth and flat breads, cakes, muffins, cookies, biscuits, tortillas, etc. Sorghum flour can be used as a substitute for any other whole-grain flour. In Southeast Asia and Nigeria sorghum flour is used to make noodles, pasta, and related products. In Botswana sorghum is processed to make weaning foods.

Sorghum is also used in the production of alcoholic and nonalcoholic beverages. Nonalcoholic beverages and dried malt extracts are produced from sorghum in Nigeria. Instant beer powder, ground malt, beer, and rice-product substitutes are made in South Africa. Sorghum-based beer and malt are also made in Mexico. Sorghum wine is produced in China and Taiwan.

Sorghum varieties that have higher sugar content in the stalks, known as sweet sorghum (as opposed to grain sorghum), are used as sugarcane substitutes. In the southern United States, the juice that is extracted from the stalks of sweet sorghum is used to make sorghum syrup, a substitute for maple syrup and blackstrap molasses.

Sorghum is also processed for use in livestock feed and for pet food. It is a major ingredient in chicken, pig, and cattle feeds throughout the Americas. In most cases the grain is pelleted or hammered in specialized mills. This material is then mixed with other feed ingredients depending on the species that is being fed and its age.

There are a few industrial uses of sorghum as well. Specially processed sorghum flour or meal (e.g., partially decorticated and acid-modified) is used as a low-cost adhesive. Similar sorghum-based materials are used in the manufacture of wallboard, as binding materials for fillers, and ore-refining materials. Sorghum is also used to make petroleum substitutes (U.S. Grains Council 2001). In the United States, Kansas and Nebraska are the leaders in the production of ethanol from sorghum.

SUBSTITUTES

Millet, corn, rice, wheat and other cereals, and soybeans are the principle substitutes for sorghum in food as well as in animal feed. Pearl millet is the main alternative to sorghum in many parts of the world because it has similar climate and soil requirements. In fact, when drought is expected, many farmers plant millet because that crop matures more quickly and requires less water than sorghum. Corn is another alternative crop to sorghum. In parts of Africa corn is grown for human consumption. However, in most parts of the world corn and sorghum are produced for animal feed, particularly for poultry. Sorghum is more drought-tolerant than corn.

In many countries that do not have specific production or price policies for sorghum, its production is affected by such policies for corn, rice, wheat, or other cereals. With the liberalization of cereal and grain markets in many countries, prices have tended to vary between different parts of a country as well as throughout the year. These variations will affect farmers' willingness to plant sorghum as well as other crops (FAO and ICRISAT 1996).

MARKET CHAIN

In most Asian and African countries sorghum has traditionally been a subsistence crop, with only small volumes entering a more formal market chain. As a consequence, markets are poorly developed. In most countries where sorghum is a food crop, markets tend to be local and regional with much of the product consumed within a very short distance of the producer.

Increasingly, however, sorghum is bought and sold commercially. These transactions tend to be dominated by people with sufficient capital to hold the product after harvest in order to sell it later in the year. Initially this occurs mostly in rural markets near areas of production or between neighboring households. Marketing channels between producers and the major urban centers are poorly developed but are becoming better organized. The factors that limit the development of more efficient markets within the African subsistence-oriented production sectors, for example, include the overall limited and/or variable trade volumes due to scattered and irregular supplies and the large distances and high transportation costs to get the sorghum to domestic markets. Such markets exhibit huge price fluctuations that result from short-term imbalances between supply and demand that peak during harvest. In addition, there are no specific price or production policies to promote stable markets, especially in Africa and Asia.

Because sorghum is an important food crop in many countries, more formal marketing arrangements have been effective in some parts of the world. For instance, the Sorghum Board of South Africa handles 60 percent of the crop each year. The Board ensures stability of supply and maintains its status as buyer of last resort. In practice, the Board has become the primary buyer rather than letting the grain enter the private sector, which traditionally buys sorghum from producers at low prices, holds the crop for a few months, and then charges much higher prices to consumers.

In other parts of the world where sorghum is produced primarily for animal feed, the market chain is similar to those for soybeans and for other grains such as corn, wheat, and rice. In general, livestock producers do not produce sorghum. As a consequence, there is still considerable space for sorghum buyers and traders. In fact, larger-scale producers of sorghum for animal feed are fully integrated into well-functioning markets. Most developed country producers sell all their production into such markets. In some instances, farmers feed the grain they produce to their own animals, which they later sell, but increasingly feeding is undertaken by another, specialized set of producers. The sorghum market faces another important challenge—the grain industry today is a global market filled with intense competition among a variety of potential sources of animal feed that are increasingly evaluated in terms of their energy and overall dietary content as well as their price, availability, and storage life. So far, it has not been demonstrated that sorghum has a higher feed value than other feed grains.

The market infrastructure in Asia is relatively well developed for both human and animal consumption of sorghum. This is especially the case in areas with high population density such as India and China.

In Australia the sorghum industry has undergone a major change with regard to its main traditional market, Japan. The changes in Australia probably anticipate similar changes in other surplus grain producing countries. Demand has shifted away from exporting grain to the more traditional international markets (such as Japan) toward keeping it on the more dynamic and expanding Australian domestic markets. At this time Australia is using sorghum as an animal feed to produce meat for sale on international markets. In short, the country is adding value to its grain, in the form of meat, rather than selling it directly to other countries where it is fed to animals. As shipping basic grains and disposing of animal waste from intensive feeding operations in many developed countries become more expensive, this trend of meat exports is likely to continue. This is especially likely if the European Union relaxes its policies that protect its own livestock producers.

MARKET TRENDS

World production of sorghum increased 40 percent from 1961 to 2000. International trade has increased by 83 percent, from 4.2 million metric tons to 7.7 million metric tons over the same forty years. Prices, on the other hand, have decreased 59 percent since 1961. From 1961 to 2000 average yields have increased 54.7 percent per hectare.

World sorghum production is projected to grow at 1.2 percent per year to about 74 million tons in the year 2005. This notwithstanding, population growth will outpace production increases, particularly in countries where sorghum is a vital food security crop. Food sorghum consumption will grow by about 15 percent by 2005, driven by a 39 percent increase in demand in Africa. By contrast, food utilization of sorghum in Asia is expected to drop by 8 percent, a continuation of the current trend. Other grains will be substituted for sorghum as human food, while sorghum production destined for animal feed will continue to increase.

Several other factors also affect sorghum market trends. For example, futures and options, transportation and logistics, farm policies, and private insurance continue to change the marketing and overall management of sorghum. Increasingly, issues such as identity-preserved hybrids and contract production will continue to add to producers' marketing options and strategies. Ultimately, the key to marketing sorghum depends on market supply and demand and information about the two.

Improvements in transportation and market infrastructure as well as more basic information are required in many parts of the world to encourage more intensified sorghum production. Broadening sorghum markets so that it can substitute for other grains in more global markets could also encourage production and increase returns to producers by reducing their dependence on local markets. Developing supplies of high-producing new varieties would help producers increase yields over time, as would improved input markets to ensure that agrochemicals are more readily available, cheaper, and more efficiently used.

ENVIRONMENTAL IMPACTS OF PRODUCTION

The primary environmental problems from sorghum production arise, ironically, from its unique plant characteristics. Sorghum is a drought-tolerant crop. Breeding has made it even more so. As a consequence, production is expanding into more marginal areas. In addition, the sorghum plant has certain chemical defenses that can poison animals or water supplies if the crop or effluent from it is mishandled. Other environmental problems from sorghum production include habitat conversion and habitat degradation, soil erosion and degradation, use of agrochemicals, and fire hazards.

Habitat Conversion and Degradation

The cultivation of sorghum poses severe threats to the integrity of many ecoregions around the world, including some of the most fragile. Because of its ability to survive and produce with less water and poorer soils than most other commodities, sorghum is grown on some of the world's most delicate land. In the Rift Valley of Eastern Africa, for example, land is increasingly being converted to crops such as sorghum, millet, and irrigated rice. These are now the most cultivated crops in Malawi.

Habitat conversion for cultivation drastically changes the composition of the local flora and fauna. The farming activities of rural people throughout this region are destroying and fragmenting large areas of natural habitat. This is the most important conservation issue in the area and sorghum is right in the middle of it.

In Asia and Africa, the long-term, continuous cultivation of sorghum leads to the infestation of fields by striga, a parasitic weed. Heavy infestations by this weed can leave the land unfit for cultivation. Striga can lead to declining soil fertility, and more importantly it can become so invasive that it strangles sorghum plants and makes farming unviable. In short, the areas with greatest infestations are abandoned (FAO and ICRISAT 1996).

One of the ways to reduce the impacts of striga is to plow and plant very early in the planting season before rains have started and the weed has become established. If planting is delayed for any reason, then striga becomes established and production will be severely affected. In the early 1980s, Ethiopian troops attacked peasants in the northern part of the country with the intent to disrupt food production. Their tactic delayed planting and striga became established in the fields. This was one of the major causes of the famine of 1984–85 (Clay et al. 1985).

Soil Erosion and Degradation

Sorghum causes erosion on slopes as shallow as 4 degrees (Buxton et al. 1997), and produces more erosion than most other cultivated crops (Thurman 1996).

The high potential for soil erosion tends to restrict sorghum production to soils with little slope. When grown for silage, most sorghum foliage is removed from the field when harvested. Soil is then at greater risk for water and wind erosion. In addition, nitrogen leaching from the soil can occur from the time sorghum is harvested until the next crop is planted and established. This can be a period of more than six months in areas with pronounced dry seasons or winters.

When grown as a grain, these issues are present but not as pronounced. Sorghum produces considerable foliage, which, if left in the field, acts as an effective mulch, especially if chopped at harvest. No-till practices can also reduce the impact of sorghum production on soil erosion while increasing soil health. Meyer et al. (1999) found that on research erosion plots, no-till practices reduced soil erosion for sorghum by more than 80 percent.

As a result of population pressures in most African countries, fallow periods are being shortened and more marginal lands are being brought into cultivation. These marginal lands are farmed with little or no fertilizer, and the cultivation methods used (e.g., planting rows without regard to incline) can be a major cause of soil degradation. In addition, changes in climate such as lower rainfall levels and higher temperatures lead to periodic drought, which makes cultivation riskier. These factors force farmers in parts of Africa to adopt inappropriate production practices. The net result is production practices that are unsustainable in the long term coupled with declining productivity in the short term (FAO and ICRISAT 1996).

Farmers have long observed that sorghum is "hard" on the soil. Grain sorghum is a soil-depleting crop because its fibrous root system is deep (relative to other crops such as corn or millet) and extensive; it fills the rhizosphere and uses up the nutrients present in the soil (Ross and Webster 1970). Still, these are precisely the characteristics that make the plant drought-resistant and able to tolerate marginal soil.

Use of Agrochemical Inputs

One of the main environmental impacts from commercial sorghum production is the use of agrochemical inputs. Sorghum is one of the most chemically dependent of all agricultural crops (Thurman 1996). Many types of fertilizers and pesticides are used in sorghum cultivation and storage, but since most storage occurs off-farm, only the impacts related to cultivation are discussed below.

On the cultivation side, a number of pesticides are commonly used on sorghum, including organophosphates such as chlorpyrifos, dimethoate, and malathion. Carbamates like carbaryl and carbofuran are increasingly used as well, and are highly toxic. Pyrethroids are also used and this is important because these substances are extremely toxic to fish and other aquatic life. Lindane is the only organochlorine that is used; it is registered for use as a seed treatment in sorghum. Fungicides are also used in sorghum fields as foliar sprays and for seed treatment. Mancozeb, a fungicide that is commonly used, is toxic to fish. Birds, insects, mammals, and reptiles are exposed to these agrochemicals.

Small grains such as wheat and sorghum provide food and cover for many wildlife species. Wildlife are directly exposed to pesticides when they eat plants or seeds with chemical residues or when they swallow the pesticide granules or water that is contaminated by them. They are exposed indirectly when they eat insects or other animals exposed to or killed by pesticides. Pesticides washed by rain into streams, ponds, or other wetlands can harm aquatic animals (Rollins et al. 1997).

Herbicides are used to control weeds both in and adjacent to sorghum fields. Globally, atrazine is the most commonly used herbicide in sorghum cultivation, alone or in combination with other preplant herbicides like alachlor, metolachlor, 2,4-D, dimethenamid, paraquat, etc. The greatest risk to wildlife from herbicides is the effects they can have on wildlife habitat, as wildlife rely on the trees, brush, grass, and weeds in and near fields for food and cover (Rollins et al. 1997).

The application of fertilizers and other chemicals can also affect soil microorganisms and hence soil fertility. Heterotrophic bacteria (those responsible for converting dead plant material into organic matter) in surface soil are several times more abundant in unfertilized soil than in fertilized soil (Barber and Matocha 1994).

Poisoning Animals

Defensive chemicals produced by the sorghum plant itself can be toxic to domestic animals and wildlife. The sorghum plant can be toxic to herbivores if it is eaten prior to flowering. Sorghum, sorghum hybrids, and related plants contain high levels of prussic acid (hydrogen cyanide). Prussic acid poisoning can occur when livestock is pastured on sorghum or when wild herbivores eat sorghum by mistake or during periods when other food is scarce. These characteristics are common with grain sorghum as well as related plants including Sudan grass, sorghum-Sudan grass hybrids, Johnson grass, and sweet sorghum. The presence of naturally occurring polyphenol chemicals in the sorghum plant, especially in varieties with purple undercoats, has also been found to affect the health of wild birds.

Fire Hazards

A wild species of sorghum, *Sorghum halepense,* has been known to create a fire hazard in natural ecosystems. The biomass yield of grain sorghum is quite high. It has been measured at about 35 metric tons per hectare in southern Europe. This biomass is often burned by farmers and can be a fire threat at that time.

Sorghum Silage Effluent

Effluent from sorghum silage has a very high biochemical oxygen demand (BOD). If this effluent is allowed to enter natural watercourses, it can reduce oxygen levels so dramatically that it can kill many freshwater organisms. Thus, effluent from sorghum silage is an environmental hazard when allowed to enter a watercourse.

BETTER MANAGEMENT PRACTICES

There are a number of strategies by which sorghum farmers can reduce the environmental impacts of production. Several are similar to those that have been discussed for other crops, including building the soil, reducing pesticide use, treating effluent, and developing systems of carbon payments.

It is possible both to reduce the impacts of sorghum cultivation on the environment and to benefit producers financially, but it requires good planning and proper execution. In general, the adoption of better practices can reduce the impact of sorghum production by maintaining productivity on existing lands indefinitely and reducing runoff and pollution. Several different, complementary practices can help to reduce the impacts.

Build the Soil

The use of cover crops and crop rotation help build and maintain the soil. The introduction of organic matter and green manure build and maintain productivity. In addition, these same treatments reduce the already-low fertilizer and water needs of sorghum as the organic matter binds more effectively with the nutrients and water available naturally.

Sorghum production reduces soil vitality through the direct exposure of soil to sunlight during preparation for planting and the period of time it takes the crop to form a canopy over the soil. No-till sorghum or interplanting sorghum with cover crops has been found to decrease runoff more than other cultivation methods (Meyer et al. 1999). This approach can reduce soil erosion by more than 80 percent and decrease runoff by at least 10 percent.

Soil erosion from wind and water is one of the most obvious impacts of sorghum production. In many parts of the world the better management practices are basic. Producers should not plant up and down the hillsides. Rather, they should plant horizontally, running rows perpendicular to the slope. Similarly, if wind tends to come from one direction, then rows should be planted perpendicular to the direction of the wind so that the tall rows act as miniature windbreaks.

Another BMP to reduce soil degradation is continuous farming. This refers to crops that are grown in sequence within a year, in contrast to crop rotation that occurs over a number of years. Farmers in many parts of the world are finding that planting an off-season ground cover (especially a legume), even if it does not provide a marketable crop, improves subsequent yields of other crops. The more biomass produced by the ground cover, the higher the subsequent yields.

Soil erosion can also be reduced by chopping and leaving large amounts of crop residue from grain sorghum on the field after the harvest. If the residue is mulched, this is an even more effective way to control soil erosion (McCarthy et al. 1993). Sorghum crop residue benefits the soil by assisting moisture retention. This, in turn, maintains beneficial organisms near the surface of the soil that assist with the recycling of nutrients into forms that are available to plants.

In the United States, a continuous cropping system of winter wheat–fallow–grain sorghum–fallow has been shown to conserve soil and water resources (Getachew et al. 1997). Annual plants such as sorghum, corn, and millet can make effective vegetation barriers if they are in place during critical erosion periods (James and Croissant 1994). Planting a sorghum crop in the summer-autumn period can also help to prevent soil erosion.

An additional benefit of this sequence planting is that sorghum provides food for wildlife during a period of scarcity. On the Great Plains in the United States, producers often plant sorghum to attract wildlife, which then also attracts either ecotourists or hunters (Rollins et al. 1997). Using this system, ranchers can generate up to 25 to 50 percent of their total income from the sale of hunting permits (Greg Simmods, personal communication). Since this has gone on for many years, it appears that the hunting off-take levels must be reasonably sustainable.

Reduce Pesticide Use

Another increasingly important set of impacts from sorghum cultivation are problems related to the indiscriminate and often unnecessary use of pesticides and herbicides during production. Most of these impacts can be prevented, or at the very least reduced, by the judicious use of agrochemicals. Using such chemicals to address specific problems, rather than prophylactically or indiscriminately, will not only reduce the environmental impacts, but will also make the producers more financially viable. In addition, it will make them less dependent on chemical solutions for addressing production problems.

For example, there are effective measures for addressing the problem of the striga weed pest. As mentioned above, early planting tends to reduce the impacts of striga. Deeper plowing tends to reduce its impacts as well. Maintaining ground cover and seeding at optimal times also help to reduce the impact of the weed. If carried out over a number of years, these approaches can help to reduce or even eliminate striga. Recently, newer varieties of sorghum have been introduced that are more resistant to striga infestation, thus preventing degradation of the soil that occurs when striga takes over a field.

Develop Payments for Carbon Sequestration

With biomass production of up to 35 metric tons per hectare per year, sorghum has tremendous potential to increase biomass within crop rotation and no-till management systems. Such dense biomass residue also has the potential to act as a carbon sink. This can increase the organic matter in the soil, thereby increasing productivity while reducing chemical and water requirements of subsequent crops. In addition, it could potentially be the source of carbon sequestration payments for farmers.

Manage Silage to Avoid Toxicity

Improved feed management can reduce the risk to animals of poisoning. Cutting silage at the appropriate growth stage and feeding appropriate amounts to livestock can prevent prussic acid or cyanide poisoning in animals.

Treat Effluent from Silage

The effluent from sorghum silage should not be allowed to enter water bodies. It is a very effective soil additive and should be reintroduced to fields unless that is not feasible financially. Many silage storage facilities, however, would permit producers to capture liquid as it builds up or drains from silos or other silage storage facilities and return it to the fields in liquid form. If the material will drain into water systems then it needs to be treated first. This can be accomplished simply through containment until the organic matter decomposes, reducing BOD. This process can be accelerated through inoculation with effective microorganisms. All these treatments would reduce or prevent environmental effluent problems.

OUTLOOK

Sorghum is one of the five main food grains of the world. To date, most of the attention on the crop has focused on increasing its acceptability and availability as a feed grain. However, much of the information collected during these efforts could also be used to improve sorghum production for subsistence and small-scale farmers. For such producers sorghum production has not come close to achieving its potential as a food crop. If sorghum could be produced consistently and as close to its known production potential as corn, rice, or soybeans, the result would be to increase food and income to some of the most marginalized people on Earth. In addition, intensifying production in areas already cultivated could take pressure off of some of the most marginal, drier, tropical habitats in Africa and Asia.

Also in its favor, the sorghum plant produces more biomass per hectare than any other cultivated cereal grain. This means that no-till or conservation tillage sorghum, without burning, can build organic matter in soil faster than almost any other food crop. It also means that if markets ever develop for carbon sequestration for agricultural crops, sorghum (probably along with corn) would be one of the most interesting to consider.

However, it is not clear that sorghum, if left to conventional international markets alone to determine its fate, will be sufficiently interesting as an animal feed. While it has considerable production gains that can still be achieved, and while it does well in drier climates (which may become more common in the future), it is not clear that sorghum is as nutritious or as palatable to animals as other animal feeds. Also, to achieve its potential, sorghum will require considerable expenditures in research and development, and it is not clear who would fund it.

REFERENCES

Barber, K. L. and J. E. Matocha. 1994. Rotational Cropping Sequence and Fertilization Effects on Soil Microbial Populations. *International Sorghum and Millet Newsletter*. 35(126).

Buxton, D. R., I. C. Anderson, and A. Hallam. 1997. *Performance of Sweet and Forage Sorghum Grown in Monoculture, Double-Cropped with Winter Rye, or in Rotation with Soybean and Maize*. Agricultural Research Service. Washington, D.C: U.S. Department of Agriculture.

Clay, J. W., B. K. Holcomb, and P. Niggli. 1985. *Politics and the Ethiopian Famine 1984–85*. Cambridge, MA: Cultural Survival.

DeWalt, B. R. and D. Barkin. 1987. Seeds of Change—The Effects of Hybrid Sorghum and Agricultural Modernization in Mexico. In H. R. Bernard and P. Pelto, eds. *Technology and Social Change*. Second Edition. Prospect Heights, IL: Waveland Press.

FAO (Food and Agriculture Organization of the United Nations). 2002. *FAOSTAT Statistics Database*. Rome: UN Food and Agriculture Organization. Available at http://apps.fao.org.

———. 1996. *Production Yearbook 1995*. Vol. 49. FAO Statistical Series No. 130. Rome: UN Food and Agriculture Organization.

———. 1996. *A Joint Study. The World Sorghum and Millet Economies—Facts, Trends, and Outlook*. Rome: UN Food and Agriculture Organization.

Getachew A., P. W. Unger, and O. R. Jones. 1997. Tillage and Cropping System Effects on Selected Conditions of a Soil Cropped to Grain Sorghum for Twelve Years. *Communications in Soil Science and Plant Analysis*. 28(1/2):63–71.

Holmes, P. A. 1992. The Marketing of Grain Sorghum in Australia. *AIAS Occasional Publication* 68:375–380. Melbourne, Australia: Australian Institute of Agricultural Science.

ICRISAT (International Crops Research Institute for the Semi-Arid Tropics). 2000. Genetic Resources and Enhancement Program Annual Report. Patancheru, India: ICRISAT.

James, T. A. and R. L. Croissant. 1994. Controlling Soil Erosion from Wind. Crops Online Fact

Sheet No. 0.518. Fort Collins, CO: Colorado State University Co-operative Extension. Available at http://www.ext.colostate.edu/pubs/crops/00518.pdf .

Maiti, R. K. 1996. *Sorghum Science.* New Delhi, India: Oxford and IBH Publishing Co.

McCarthy, J. R., D. L. Pfost, and H. D. Currence. 1993. *Conservation Tillage and Residue Management to Reduce Soil Erosion.* Agricultural publication G1650. Columbia, MO: MU Extension, University of Missouri–Columbia. Available at http://muextension.missouri.edu/explore/agguides/agengin/g01650.htm.

Meyer, L., S. Dabney, C. Murphree, W. Harmon, and E. Grissinger. 1999. Crop Production Systems to Control Erosion and Reduce Runoff from Upland Silty Soils. *Transactions of the ASAE (American Society of Agricultural Engineers).* 42(6)November/December:645–1652.

Rollins, D., T. W. Fuchs, and J. Winn. 1997. *Reducing Pesticide Risks to Wildlife in Small Grains and Sorghum.* Texas Agricultural Extension Service. College Station: The Texas A&M University.

Ross, W. M. and O. J. Webster. 1970. *Culture and Use of Grain Sorghum.* USDA/ARS Agriculture Handbook 385. Washington, D.C.: U.S. Department of Agriculture.

Runge, C. F. 1994. *The Grains Sector and the Environment: Basic Issues and Implications for Trade.* Rome: UN Food and Agriculture Organization. November 1. 49 pages.

Skinner, P. R. 1996. Sorghum Marketing and Utilization in South Africa. In *Proceedings of the SADC/ICRISAT Regional Sorghum and Pearl Millet Workshop,* Gaborone, Botswana, 25–29 July 1994. Patancheru, India: International Crops Research Institute for the Semi-Arid Tropics. 315–319.

———. 1997. Maryland Grain Sorghum Hybrid Performance Trials. Unpublished.

———. 1998. Agronomy Mimeo 20. In C. W. Smith and R. A. Frederiksen, eds. *Sorghum—Origin, History, Technology, and Production.* New York: John Wiley and Sons, Inc.

Thurman, W. N. 1996. *Assessing the Environmental Impact of Farm Policies.* AEI Book Summary. American Enterprise Institute for Public Policy Research. Available at www.aei.org/bs/bs6793.htm. Washington, D.C.: AEI.

UNCTAD (United Nations Commission for Trade and Development). 1994. *Handbook of International Trade and Development Statistics, 1993.* Geneva, Switzerland: UNCTAD.

U.S. Grains Council. 2001. Value Enhanced Grains. Washington, D.C.: U.S. Grains Council. Available at http://www.vegrains.org.

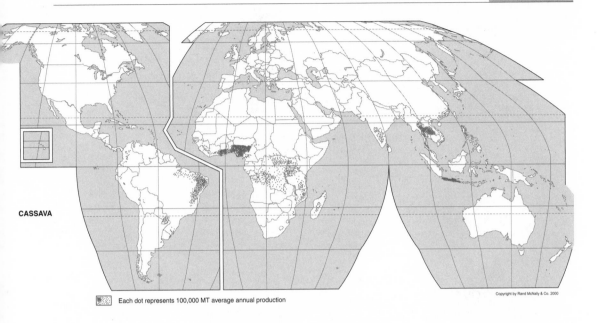

CASSAVA

Each dot represents 100,000 MT average annual production

Copyright by Rand McNally & Co. 2000

CASSAVA *Manihot esculenta*

PRODUCTION

Area under Cultivation	16.8 million ha
Global Production	178.6 million MT
Average Productivity	10,611 kg/ha
Producer Price	$69 per MT
Producer Production Value	$12,182 million

INTERNATIONAL TRADE

Share of World Production	9%
Exports	15.8 million MT
Average Price	$30 per MT
Value	$472 million

PRINCIPAL PRODUCING COUNTRIES/BLOCS (by weight)

Nigeria, Brazil, Thailand, Indonesia, Dem. Republic of Congo

PRINCIPAL EXPORTING COUNTRIES/BLOCS

Thailand, Indonesia, Vietnam, Costa Rica, China, Brazil

PRINCIPAL IMPORTING COUNTRIES/BLOCS

Netherlands, Spain, China, Belgium, Indonesia, South Korea, Portugal, Japan, Malaysia, Germany

MAJOR ENVIRONMENTAL IMPACTS

Habitat conversion
Soil erosion and degradation

POTENTIAL TO IMPROVE

Fair
Cassava is produced on marginal areas with big impacts
Most poor producers have few options, few have land titles
Polyculture (e.g., multiple food crops grown at the same time) production reduces cassava's impacts

Source: FAO 2002. All data for 2000.

Area in Production (MMha)

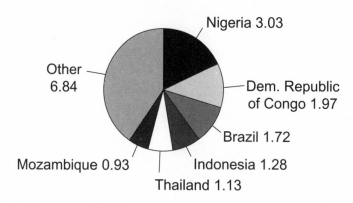

OVERVIEW

Cassava is the only food staple cultivated widely throughout the world that is poisonous to consume prior to processing. Cassava was discovered and domesticated by Indians in the Amazon. It was a food staple for lowland Indians long before the arrival of corn. Several varieties were cultivated and there are indigenous villages where dozens of varieties are still planted.

With the arrival of the Europeans, cassava was taken to many different parts of the world. In 1558 cassava was already reported on the margins of the Zaire River. Today it is consumed in many parts of Africa, and Nigeria is the main producer and consumer. Cassava was introduced to Asia during the seventeenth century, and today Thailand is the main exporter of the product.

Cassava is the highest producer of carbohydrates per hectare among staple food crops. According to the Food and Agriculture Organization of the United Nations (2002), it is the fourth most important food crop in developing countries after rice, corn (maize), and wheat. Two parts of the cassava plant are used for human consumption—the starchy roots and the leaves. Though the root is the main crop, in most cassava-growing countries in Africa the leaves are also consumed as a green vegetable, which provides protein and vitamins A and B. In Brazil the leaves are only used occasionally for special dishes. More often the leaves are used in dairy cattle rations.

As a food crop, cassava has some inherent characteristics that make it especially attractive to small farmers throughout the world. These characteristics include:

- *Multiplicity of end uses:* Cassava roots are rich in carbohydrates, especially starch, and can be consumed in a number of different ways.
- *Food security:* It is available all the year round, making it preferable to other, more seasonal staples such as grains, peas, and beans. Its roots can remain in the ground for several years after they mature.
- *Sturdy, tolerant, and pest-resistant:* Compared to grains, cassava is more tolerant of low soil fertility. In addition it is more resistant to drought, pests, and diseases.

These attributes combined with socioeconomic considerations have made cassava a leading crop in poverty alleviation programs. Many poor or landless farmers only have access to marginal land which is well suited to the production of cassava. The rural poor often need food at periods of the year when other crops do not produce and when there is little paid labor in rural areas. Cassava lends itself well to such conditions (Dostie et al. 1999). These attributes are also what gives cassava its final advantage over many other basic food crops—it is very cheap for poor people to buy.

PRODUCING COUNTRIES

In 2000 the total area planted to cassava was 16.8 million hectares and total production was 178.6 million metric tons. Nigeria was the largest producer with 32 million metric tons followed by Brazil (23 million metric tons), Thailand (19 million metric tons), Indonesia (16 million metric tons), and the Democratic Republic of Congo (16 million metric tons); these are the top five producers of cassava in the world (FAO 2002). An overview of cassava production is given in Table 20.1. There are two

TABLE 20.1. Production and Export of Cassava, 2000 (in metric tons)

Country	Production	Total Exported	Percent Exported
Nigeria	32,010,000	—	0.0
Brazil	23,335,974	56,666	0.24
Thailand	19,064,000	13,438,310	70.5
Indonesia	16,089,100	444,226	2.76
India	6,800,000	5,232	0.08
Dem. Republic of Congo	15,959,000	545	0.003
China	3,800,933	58,598	1.54
Paraguay	2,719,410	—	0.0
Madagascar	2,463,360	50	0.002
Vietnam	1,986,300	337,642	17.0
Colombia	1,792,380	938	0.05
Philippines	1,765,710	1,615	0.09
Malaysia	380,000	6,818	1.79
Costa Rica	67,402	146,537	217.4
World	178,567,247	15,755,728	8.82

Source: FAO 2002.

complicating issues that affect estimates of total cassava production. Most cassava is interplanted with other crops. In addition, most cassava is consumed by the producer or sold in local markets. It is thus very hard to track either the total global area cultivated or the volume produced.

Table 20.1 shows the major cassava producers as well as the major exporters. Thailand is the main exporter globally, exporting some three-quarters of all its production and more than 85 percent of global exports. Other smaller exporters include Indonesia, Vietnam, Costa Rica, China, and Brazil. While there is some year-to-year variation, Thailand is always by far the largest cassava exporter globally (FAO 2002).

CONSUMING COUNTRIES

Nigeria, Brazil, Indonesia, the Democratic Republic of Congo, Ghana, Tanzania, India, Thailand, Mozambique, Uganda, and Angola grow and consume more than 70 percent of all cassava grown each year (FAO 2002).

The main importing countries are the Netherlands, Spain, China, Belgium, Indonesia, South Korea, Portugal, Japan, Malaysia, and Germany (FAO 2002), as shown in Table 20.2. A few alternative markets have developed for cassava such as in the former USSR, but most of the international cassava trade continues to depend heavily on European imports, where it is used for animal feed.

PRODUCTION SYSTEMS

Cassava is a perennial woody shrub that grows from 1 to 3 meters tall. It can grow from 30 degrees north to 30 degrees south of the equator and from sea level to an altitude of 2,000 meters. The ideal temperature for cassava is about 20 degrees Cel-

TABLE 20.2. World Trade in
Cassava, 2000

Country	2000 (million MT)
World Exports	15.8
Thailand	13.4
Indonesia	0.4
Others	2.0
World Imports	17.4
Netherlands	3.4
Spain	3.3
China	2.8
Belgium	1.9
Indonesia	1.0
South Korea	0.7
Portugal	0.7
Japan	0.6
Malaysia	0.4
Germany	0.4
Others	2.2

Source: FAO 2002.

sius. On good soils with adequate rains, production can reach 30 metric tons per hectare. Tubers can reach 0.5 meters in length and 10 centimeters in diameter.

Cassava is propagated through cuttings 20 to 30 centimeters long taken from the woody stems. Spacing is usually 1 to 1.5 meters between plants. During the first year beans, corn, tobacco, or other annual crops are grown between the young cassava plants.

Cassava is produced under diverse ecological conditions and production systems. It has the ability to withstand poor environmental conditions such as low rainfall and infertile soil. In Africa cassava is mostly grown on small farms, usually intercropped with vegetables, plantation crops (such as coconut, oil palm, or coffee), yam, sweet potato, bananas, melon, corn, rice, and peanuts or other legumes. Only a few countries, notably Thailand and Brazil, produce cassava primarily as a single crop over relatively extensive areas of land. Monocropped cassava is rare in the rest of the world. Monocrop cultivation is sometimes associated with food production but more often with the production of animal feed or alcohol.

Land clearing and soil preparation demand high inputs of hand labor due to the vigorous growth of native vegetation in the lowland humid tropics ecosystem. Weed control is the most labor-intensive activity after crop establishment. Preemergence herbicides like fluometuron, diuron, and alachlor are currently recommended for weed control in cassava. Paraquat has also been recommended for postemergence application as a complement to hand, machete, or hoe weeding.

The application of fertilizer and pesticides remains limited among small-scale farmers due to the high cost and lack of availability. While the crop grows well with little or no fertilization, it responds well to fertilizer application on infertile soils. For the most part, small farmers use their own labor to remove pests. While there are pests that can affect production to a high degree, the pesticides to address them are not cheap and are rarely used.

There are several constraints for cassava production. In Africa, for example, pests and diseases cause yield losses as high as 50 percent (Ross 2002). One source suggests that improved pest and disease management and better processing methods could increase cassava production in Africa by 150 percent (Ross 2002). The production of cassava is also dependent on the supply of good-quality stem cuttings. The success rate for these cuttings is very low compared to the germination rate of grain crops. In addition, cassava stem cuttings are bulky and highly perishable; they tend to dry up within a few days. Moreover, they are costly to cut, handle, and transport.

There are several varieties of cassava. They tend to fall into two groups—the sweet and the bitter types. The bitter types contain higher concentrations of cyanogenic glucosides than the sweet types. Because cassava is propagated from cuttings, the rate of multiplication of new, improved cassava varieties is slow. This also retards their adoption, so there are relatively few different cultivars. Furthermore, because propagation from a producer's own cuttings is free, many are reluctant to spend money for improved varieties. To the extent that cultivars exist, they tend to be mostly regional. Some of the various cultivars that are commonly grown in countries throughout the world are listed below:

- Brazil: IM-158, IM-168, IM-175, BGM-021, IAC-12–829, IAC-576–70, Aipim, Pioniera, and Gigante
- China: SC 205 and Colombia CM 4031–2
- Colombia: Manihoica P-12, CG 1141–1, CM 3306–4, ICA-Sebucan, ICA-Catumare, and Manihoica P-13
- Cuba: CMC-40, Señorita, CEMSA 5–19, CEMSA 74–6329, and Jaguey Dulce
- Indonesia: Adira 1 and Adira 4
- Nigeria: TMS 30572 and 4(2)1425
- Paraguay: Meza-I
- Philippines: Kalabao, Golden Yellow, Colombia CM 323–52, Lakan 1, Datu 1, and Sultan 1
- Thailand: Rayong 1, Rayong 2, and Rayong 3

As a root crop, cassava requires considerable labor to harvest. Owing to their highly perishable nature, cassava roots require immediate or early transport to the marketplace, and the transportation costs involved are quite high relative to the value of the commodity because of the high water content of the roots. This remains a major constraint in cassava production for markets.

PROCESSING

Processing cassava for food for humans can be partially mechanized, but it still requires considerable labor. This makes it time-consuming given that the relative value of cassava is quite low by comparison to other food crops. In Brazil and other areas the wet, ground cassava meal is dried over a griddle to turn it into a flour that can then be stored. This takes a fair amount of time and firewood. Finally, many cassava varieties contain cyanogenic glucosides, and inadequate processing can lead to toxicity and even cyanide poisoning.

Avoidance of rapid postharvest deterioration and reduction of cyanide levels are traditionally the main reasons for processing cassava into different food products. Roots are processed into a wide variety of granules, pastes, flours, etc. The roots of sweet cassava, with only a third the levels of prussic acid compared to bitter cassava, can be consumed freshly boiled or raw. Cassava is used in the production of food and animal feed, and it also has industrial uses. In 1993, 58 percent of the world production was used for food, 25 to 28 percent for animal feed, 2 to 3 percent in industry, and 14 percent was waste (Balogun 2002).

However, the use of cassava for different purposes varies considerably by region. In Africa and Asia only 6 percent of cassava utilization is accounted for by feed, while in Latin America and the Caribbean the animal feed share rises to some 47 percent, reflecting high feed usage in Paraguay and Brazil. In parts of Latin America it is also used commercially for the production of animal feed and starch-based products.

In some areas of the world, cassava is ground and used directly as an animal feed.

In other places cassava meal, a residue from the extraction of starch from cassava roots, is included in cattle and pig rations. Starch and pomace (cassava meal) are used extensively for pigs in Southeast Asia where they are regarded as valuable feeds. Cassava is used in concentrations of up to 10 percent in poultry rations to cut the richness of other feed grains. Brazil uses nearly half of its cassava in swine, poultry, and fish farming production systems. According to the FAO (2002), one of the attractive properties of cassava is that it can be added to animal feeds at concentrations of 15 percent and protect the feed from insect losses.

Cassava starch is also derived from the tubers. Cassava is the fourth largest source for starch production after corn, wheat, and potato. The tubers can be processed as a source of commercial starch for use in the food, textile, pharmaceutical, brewing, cosmetic, and paper industries. As a basic foodstuff, the starch may be converted by acid and enzyme hydrolysis to dextrins and glucose syrups. The starch can be used as a thickener in cooking and can be extracted and dried to produce tapioca. Tubers are also used for the production of flour, which in turn is used mainly for making porridge or bread. It is also possible to mix wheat and cassava flour. This blend has excellent baking qualities.

Cassava starch is also used in the production of monosodium glutamate, an important flavor-enhancing agent in some Asian cooking and many processed foods. Various processing methods such as grating, sun drying, and fermenting are used to reduce the cyanide content. Most of the cassava starch industries are located in Asia with the exception of tapioca, which is produced in Brazil.

Cassava meals, flours, and starches are used to produce a wide variety of secondary products including biscuits, chips, and noodles. These are currently being made from cassava flour in Nigeria, Madagascar, Tanzania, and Uganda.

The whole cassava plant (including the root, stems, and leaves) can be chopped and stored in simple pit silos for dry-season feeding at the village level. This tends to be done more often in Asia and Africa than in Latin America.

Finally, cassava also has nonfood and industrial uses. Cassava is used to produce alcohol for human consumption as well as for automobiles in Brazil. There is also experimentation with the starches from different varieties of cassava to make polymers with many different uses. Polymers from cassava starches are being used as filters in a wide range of industrial processes including the refining of sugar.

BOX 20.1. PRODUCTS DERIVED FROM DIFFERENT FORMS OF CASSAVA

The different forms of cassava can be processed into many different products. Some of these products include:

- *From fresh and dried roots*: Food, flour, animal feed, alcohol
- *From normal starch*: Glue, plywood, paper, textiles, monosodium glutamate
- *From modified starch*: Sweeteners, prepared foods, medicines, biodegradable products

---

SUBSTITUTES

Most grains such as corn, wheat, rice, sorghum, etc. and most other roots and tubers such as potatoes act as substitutes for cassava both for human food and for animal feed. Starches from corn, white potato, and sweet potato compete with the cassava starch industry as well. Sugar and corn compete with cassava in the manufacture of alcohol. Cassava products for animal feed are facing increasing competition from grains on international markets. However, the price of cassava makes it relatively competitive with most substitutes. The issue is the caloric value and palatability for human and animal feed uses.

MARKET CHAIN

Well-developed market access infrastructure is crucial for cassava marketing. Cassava roots are bulky and, with about 70 percent moisture content, are both unnecessarily heavy and very perishable. Rapid deterioration can begin as soon as 24 hours after harvest. Fresh cassava roots are traditionally marketed without postharvest treatment or protection and therefore have to reach consumers within a short time before deterioration becomes a problem. It is therefore imperative that effective marketing channels be available to avoid postharvest losses. Alternatively, cassava tubers can be dried at the point of production. For the most part they are ground into flour and used dried or rehydrated as necessary at the point of use.

As cassava is mostly produced in rural areas of developing countries, market infrastructure has not been very well developed. Marketing problems are becoming exacerbated as increasing urbanization is placing both distance and time between producers and consumers. Being produced mostly in developing regions with unstable local economies, cassava is often not available in the market with the same regularity and predictability as, for instance, corn. Furthermore, cassava is found in many grades and qualities that are often not comparable and even highly variable internally. Its price, especially in recent years, has fluctuated considerably.

Cassava marketing varies tremendously around the world. In some parts of the world, the cassava market has not evolved much. Typically, farmers transport their farm produce to the market on their heads, on bicycles, or in trucks that happen to pass by. Cassava can be purchased in the ground with traders supplying their own labor to harvest the crop when required. However, the quantities handled by cassava traders are usually low. By contrast, Thailand, the major exporter of cassava in the world, has a very well developed market chain that has enabled the industry to experience a pattern of growth similar to that of other agricultural commodities, especially grains.

In Ghana cassava marketing chains have evolved to cope with the perishability of the root. Rapid marketing is ensured through a wide range of operators, including producers, itinerant traders, transporters, intermediaries, market traders, and market chiefs. Operators such as farmers, traders, and consumers are often connected through complex systems of information flow, credit, and transportation.

Also, there are some hopeful signs for niche marketing. Cassava flour produced in Cruzeiro do Sul, in the extreme westernmost part of the state of Acre in the Brazilian Amazon, is sold as far away as the city of Belém at the mouth of the Amazon River for a premium price. In fact, the river towns all along the Amazon carry at minimum twenty and sometimes fifty different varieties of cassava flour distinguished by their texture (e.g., fine to course or even or uneven), color (e.g., white, gray, yellow or gold), additives (e.g., coconut, peach palm flour, etc.), or origin. Prices range accordingly and can vary by more than 300 percent.

MARKET TRENDS

The combination of high marketing costs for cassava and market interventions such as subsidized cereal prices often leads to high relative cassava prices. This can reduce urban demand because of the significant cross-price elasticities between cassava and major grains. This is when production or marketing subsidies lower the price of grains which tends to dampen the price of cassava.

FAO projections to 2005 point toward sustained growth in cassava production. This is likely to be dominated by a rise in productivity but little change in area planted. This will depend, however, on whether the dissemination of the new technologies among farmers gains momentum and these technologies become more popular. World cassava production is projected to maintain an annual growth rate of 2 percent. The rate of expansion is anticipated to be in the range of 3 percent in Africa, while it is likely to be a more modest 1 percent in Latin America and the Caribbean and in Asia. The utilization of cassava, on the other hand, is projected to remain small relative to output and to involve only a few countries. Total utilization is projected to grow at 2.2 percent annually, with food consumption representing about 59 percent of the total.

About 15 percent of global production in 1993 was exported to Europe and Japan, mainly for industrial purposes and to mix with vegetable meals as animal feed. These uses are expected to continue to grow gradually (UNCTAD 1994). Changes in European agricultural subsidies and import policies could greatly affect this crop.

Cassava has a high potential as a basic food crop during periods of civil and economic stress or when markets are otherwise disrupted. It requires very few inputs, and it has a long storage life in the ground and out if properly processed. It will, in all likelihood, long be a basic food crop in areas where it is currently known, particularly for the rural and urban poor. However, as people's incomes increase, they tend to eat far less cassava. Therefore, cassava markets will expand only if it is used increasingly in animal rations or if new industrial applications for the product are discovered.

ENVIRONMENTAL IMPACTS OF PRODUCTION

Though cassava, particularly the sweeter varieties, is subject to pests, few producers who grow it can afford chemical pesticides. Partly for this reason, the main environmental problems from its production are habitat conversion and soil erosion.

Habitat Conversion and Soil Erosion

In most parts of the world the environmental impacts of cassava production are related to who grows it and where it is grown. Cassava tends to be a poor peoples' food that is most often grown by poor people. As a consequence it is most often grown on lower-value, more marginal lands. In short, these are often lands that people have claimed because others do not want them. However, such lands often have a high biodiversity value.

Cassava's requirements are few and as a consequence it is frequently cultivated where few other cultivated crops could survive much less yield food for the producer or for sale. The cassava plant does not produce enough vegetation to cover the soil well. Even if other crops are interplanted, the early crops tend to be harvested within a few months or the first year at the latest. For both these reasons, the production of cassava can result in considerable soil erosion during the entire life of the plant. Because little else grows on such soils, the erosion often continues well after the cassava is harvested.

BETTER MANAGEMENT PRACTICES

For most producers, cassava is an ancillary crop, for example, it is not often the main crop being produced but rather a sequence crop that is grown in association with others. Consequently, it allows producers the possibility of gaining a little more production of either food or marketable crop from an area that would otherwise be allowed to return to fallow. As a result, the most effective conservation strategies are those that are aimed at the primary agricultural crops that are causing the environmental problems that need to be addressed. Even so, there are at least two strategies that can be pursued with regard to cassava.

Reduce Habitat Conversion

The most effective way to reduce the habitat conversion associated with both cassava production in marginal areas and with slash and burn cultivation (where agricultural plots are cleared from forests or secondary growth and planted to food crops and then abandoned for a period of time to allow the soil to rejuvenate) is to increase the fallow cycle time for poor farmers and to plant leguminous trees or other plants that will more actively rejuvenate the soils. When land is relatively abundant compared to overall population in many parts of the world, the fallow cycle for shifting cultivation is often ten to fifteen years or even more. In many parts of the world where more people now depend on less land for agriculture, the fallow cycle may be as little as three to seven years. Such rapid reuse of often relatively marginal areas does not give them sufficient time to rebuild fertility. Over time, shorter fallow cycles deteriorate soil productivity.

Cassava can be harvested for up to three years after planting; it is a crop that extends the productive life of agricultural land into the fallow cycle. Cassava, like fruit trees and other perennials, provides a crop on agricultural lands long after annual crops have been harvested. Such plants not only extend the productive life of otherwise abandoned agricultural plots during their fallow cycle, they attract insects and animals which are important sources of protein for the rural poor (Clay 1988). For farmers that plant cassava the strategy should be to increase soil productivity during fallow periods through planting strategies that increase biomass and nutrient production during both production periods as well as fallows.

Reduce Soil Erosion

Intercropping of cassava has great potential to decrease soil loss substantially. For example, intercropping tree crops with cassava has the benefit of reducing runoff and soil loss. However, aggressive, fast-growing tree crops such as eucalyptus effectively utilize available nutrients and moisture at the expense of companion crops. Chemical assays of plant parts indicate that cassava utilizes more soil nutrients when planted alone, and considerably less when grown with fast-growing trees. As a consequence other tree crops might be examined, particularly leguminous trees (such as leucaena) that hold the soil as well as fix nitrogen, provide fodder and fuelwood, and perhaps even attract game. If cassava were intercropped with such tree crops, the fertility of the soil could be maintained without deterioration. Other studies have shown that runoff and soil loss were effectively reduced when cassava was grown on staggered soil mounds along with eucalyptus or leucaena, due to better canopy coverage of the soil surface.

Standard soil conservation techniques can also help. Contour ridges alone or in combination with live barriers and no-till farming have provided effective erosion control under experimental conditions in Latin America and Asia (Panfilo Tabora, personal communication). Mulching or other effective ground covers could also reduce erosion and conserve water in the soil.

OUTLOOK

In the 1960s and 1970s it was assumed that the benefits of green revolution technology would "trickle down" to the rural poor. People would increasingly be able to "eat up" the food chain, substituting more nutritious crops for less nutritious ones. To some extent this happened. But the benefits of green revolution crops did not spread evenly around the world or even within a single country. It became clear that many poor were so disengaged from the market economy that they could not benefit from crops that required purchased inputs. Instead of eating up the food chain, many of the poorest people in both rural and urban areas began to eat down it.

Cassava and other tubers have become the most important food staples for poor and economically marginalized people around the world. There are several reasons for this that are not likely to go away any time soon. Cassava is very productive and consequently very cheap to buy. It grows almost anywhere. This is increasingly important for subsistence farmers and shifting cultivators who find themselves not only farming smaller plots but more marginal ones as well. Cassava is also important because it produces more calories on less land (ten times as much as corn) without inputs than almost any other crop. This frees land for other crops (particularly horticulture and tree crops) that can be sold or eaten or both. Finally, cassava can be left in the ground until it is needed without spoiling. The main drawback with cassava, however, is its poor overall nutritional value when it is the only or even the main food in the diet.

As commodity prices continue to decline, as small farmers are further marginalized, and as there is more rural labor than is necessary for increasingly mechanized agriculture, cassava will be the most important crop between the rural poor and starvation. However, if prices decline cassava is likely to become more attractive for other industrial uses such as alcohol for internal combustion engines, feed supplements for animals in intensive feedlots, and starch-based polymers with an increasing range of applications. While it is doubtful that any of these uses will markedly increase the overall demand for cassava, given the downturn in the global economy it is unlikely that cassava consumption will decline any time soon.

REFERENCES

Balogun, S. A., J. B. Bodin, N. Bikangi, I. Rafiqul, and I. Jarlebring. 2002. *Cassava—the Ultimate Future Crop*. Available at http://www.nutrition.uu.se/studentprojects/group97/cassava/cassava.htm. Accessed 2002.

Clay, J. W. 1988. *Indigenous Peoples and Tropical Forests—Models of Land Use and Management from Latin America*. Cambridge, MA: Cultural Survival.

Cock, J. H. 1985. *Cassava: New Potential for a Neglected Crop*. Boulder, CO: Westview Press.

Collinson, C., K. Wanda, A. Muganga, and R. S. B. Ferris. 2000. *Cassava Marketing in Uganda: Constraints and Opportunities for Growth and Development. (Part I: Supply Chain Evaluation)*. Draft. Chatham, UK: International Institute of Tropical Agriculture, Natural Resources Institute.

Dostie, B., J. Randriamamonjy, and L. Rabensasolo. 1999. *Cassava Production and Marketing Chains: the Forgotten Shock Absorber for the Vulnerable*. Working Paper No. 100. November. Ithaca, NY: Cornell Food and Nutrition Policy Program. Available at http://www.he.cornell.edu/cfnpp/.

FAO (Food and Agriculture Organization of the United Nations). 2002. *FAOSTAT Statistics Database*. Rome: UN Food and Agriculture Organization. Available at http://apps.fao.org.

FAO/GIEWS (Food and Agriculture Organization of the United Nations and Global Information Early Warning System). 2001. *Manihot Esculenta*. Animal Feed Resources Information System. Available at http://www.fao.org.

———. 2000. Championing the Cause of Cassava. News Highlights. Available at http://www.fao.org.

FAO/IFAD. 2000. *The World Cassava Economy—Facts, Trends, and Outlook*. A joint report by International Fund for Agricultural Development and the Food and Agriculture Organization of the United Nations. Rome: UN Food and Agriculture Organization.

Ghosh, S. P., B. Mohan Kumar, S. Kabeerathumma, and G. M. Nair. 1989. Productivity, Soil Fertility, and Soil Erosion under Cassava-Based Agroforestry Systems. *Agroforestry Systems* 8: 67–82.

Grace, M. R. 1977. Cassava Processing. FAO Plant Production and Protection Series No. 3. Rome: UN Food and Agriculture Organization.

Moreno, R. A. 1992. Recent Developments in Cassava Agronomy. In *Roots, Tubers, Plantains, and Bananas in Animal Feeding.* Proceedings of the FAO Expert Consultation held in CIAT, Cali, Colombia. January 21–25, 1991. D. Machin and S. Nyvold (eds.). Rome: UN Food and Agriculture Organization. Available at http://www.fao.org.

Ross, J. Editor. 2002. *Building Partnerships for Food Security.* Rome: UN Food and Agriculture Organization. Available at http://www.rdfs.net/linked-docs/booklet/bookl_all_en.pdf.

UNCTAD (United Nations Commission on Trade and Development). 1994. *Handbook of International Trade and Development Statistics, 1993.* Geneva, Switzerland: UNCTAD.

Wenham, J. E. 1995. *Postharvest Deterioration of Cassava—A Biotechnology Perspective.* FAO Plant Production and Protection Paper 130. Rome: UN Food and Agriculture Organization.

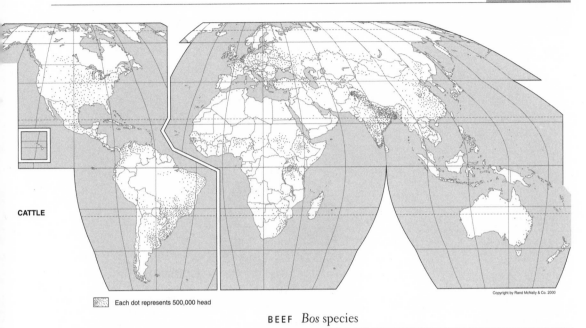

CATTLE

Each dot represents 500,000 head

BEEF *Bos* species

PRODUCTION		INTERNATIONAL TRADE	
Area under Cultivation	3,459.8 million ha	Share of World Production	23% (meat)
Herd	1,346.4 million (animals)	Exports	
Meat	56.5 million MT	Meat	13.1 million MT
Slaughter	277.8 million (animals)	Hides	2.0 million MT
Milk	488.2 million MT	Average Price	
Average Productivity	21 kg/ha (meat)	Meat	$2,322 per MT
	203 kg/animal (carcass)	Hides	$1,675 per MT
	21% (slaughter/herd)	Value	
Producer Price	$1,309 per MT (meat)	Meat	$30,434 million
Producer Value at Slaughter	$73,958 million (meat)	Hides	$3,325 million

PRINCIPAL PRODUCING COUNTRIES/BLOCS
(by weight)

United States, European Union, Brazil, China, Argentina, Australia, Russia, India

PRINCIPAL EXPORTING COUNTRIES/BLOCS

Boneless — Australia, United States, New Zealand, Ireland, Canada

Bone In — Germany, France, United States, Netherlands, Ukraine

PRINCIPAL IMPORTING COUNTRIES/BLOCS

Boneless — United States, Japan, Mexico, Canada, Egypt, South Korea, Germany

Bone In — Italy, Russia, France, South Korea, United States, Greece

MAJOR ENVIRONMENTAL IMPACTS

Habitat conversion
Overgrazing
Feedlot pollution
Production of feed grains

(continued on page 462)

(continued from page 461)

POTENTIAL TO IMPROVE

Good
Organic guidelines exist
Natural beef certification exists
Grass-fed beef can reduce pressure for grain-fed beef
Better practices known for some areas

Source: FAO 2002. All data for 2000.

Cattle Herd Size (Million animals)

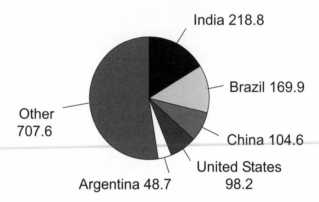

India 218.8

Brazil 169.9

Other 707.6

China 104.6

United States 98.2

Argentina 48.7

OVERVIEW

Cave paintings in France and Spain dating to 30,000 years ago show aurochs—wild cattle—and human hunters. Today's cattle are descendents of the giant aurochs, which were over two meters at the shoulder and had lyre-shaped horns.

Cattle were domesticated about 10,000 years ago in Mesopotamia, where they were both worshiped and sacrificed in religious ceremonies. World cattle breeds fall into two groups derived from two species. Cattle of the species *Bos taurus* were first yoked in Mesopotamia and used for traction to pull sleighs and wagons. *Bos indicus* was domesticated about the same time in India. The Indian cattle are better adapted to the tropics. Their genetic origins are separated from European and African cattle by more than half a million years, implying that genetic differences preexisted domestication. Today there are nearly a thousand cattle breeds and varieties around the world, with some 270 recognized as the most important.

People have prayed to cattle as gods and have sacrificed them to gods. In fact, people prayed to cattle gods for more of history than to any other type of god. Over the millennia cattle have provided food, clothing, fuel, and shelter. They have been beasts of burden and have plowed fields.

In India some 200 million cows are allowed to roam freely because Hindus consider them the mothers of life. By the mid-1990s the government had created old-age homes for half a million cattle. Killing a cow in India is a crime punishable by life in prison, but that is better than the death penalty in effect until recently in Kashmir (Rifkin 1992).

Throughout history cattle have represented wealth. It is likely that cattle were one of the earliest forms of currency. In fact, through the middle of the twentieth century cattle were still used as money in parts of Africa. The word "cattle" has the same root as "chattel" and "capital." In Spain the word for cattle also means property, and in Latin the word for money comes from the word for cattle. In Sanskrit the term for battle translates to the "desire for cattle," and a successful warlord was often referred to as a "lord of cattle" (Rifkin 1992).

The Phoenicians spread the worship of cattle throughout their colonies. They held cattle in such high regard that the first letter of their alphabet, A, was the image of a bull's head. The Minoan civilization of Crete also worshiped a bull god as did the Egyptians, Sumerians, Greeks, and even the Hebrews, who were praying to a golden calf when Moses descended from Mount Sinai.

The name for Italy, *Italia*, means "land of the cattle." In the days of the Roman Empire, the Mithraic myth of the ritual slaying of the bull attracted many Roman soldiers as followers. According to this myth, Mithra received divine guidance to kill the bull god. After he finally succeeded, a number of miracles occurred. The bull's body gave rise to all the plants and herbs that people use. The spinal cord was transformed into wheat, the staff of life, and the blood turned into the grapevine and wine. Evil, resentful of all this bounty for man, attacked the bull's testicles. In the process all of the animals on earth were created. Finally, according to the myth, the soul of the bull returned to the heavens, where it became the guardian of the herds (Rifkin 1992).

The cult of the bull god was so strong in the West that early Christians felt the need to demonize it, transforming the Mithraic bull into the new symbol of darkness. In effect, it became the devil incarnate (Rifkin 1992). In 447 A.D., the Council of Toledo (as cited in Rifkin 1992) wrote the first official description of the devil—"a large black monstrous apparition with horns on his head, cloven hoofs—or one cloven hoof—ass's ears, hair, claws, fiery eyes, terrible teeth, an immense phallus, and a sulphurous smell."

The struggle between the followers of cattle gods and the followers of other gods is mirrored in the struggle between herders and farmers, and even between East and West. About 4400 B.C. pastoralist horsemen from the Eurasian steppes first attacked and conquered Neolithic farmers in southern and eastern Europe. After these positions were consolidated, central Europe was attacked from 3400 to 3200 B.C., and a third penetration into western Europe and Scandinavia occurred between 3000 and 2800 B.C. Beginning in the first century A.D., nomadic invasions from the east pushed through much of Europe as well as India, Persia, and the far east of China.

Pastoralists were successful in these campaigns because of their speed and mobility against the people of the large grain-producing areas of the Middle East and North Africa and the small village-based agriculturalists of Europe (Rifkin 1992).

In the West, pastoralists and agriculturalists have also been at odds historically over how land would be used. This conflict evolved along class lines. Elites tended to own cattle and eat meat while peasants depended on grains. As population increased the competition grew more severe, leading to degradation of both pasture and agricultural lands. Only after plagues greatly reduced populations and labor became scarce did cattle increase to the point where meat was cheap and peasants could afford to purchase it. For the most part, however, land pressure had reached extreme proportions given the technologies of the day by the time of European colonization. The colonies not only provided an outlet for population but also a breeding ground for cattle, which, after slaughter, were salted and dried for export. After the development of the refrigerated steamer in 1878, beef could be shipped frozen. As many as 5,500 frozen carcasses of the 13 million cattle in the Pampas of Argentina were shipped to Europe in a single boat.

Today, livestock production remains an integral part of most food production systems, and its relative importance will increase if the projected increased demand for foods from animals in developing countries is to be met (Bradford 2001). What is interesting about cattle in particular is that they can transform materials that are inedible for humans into nutritious food. Thus from pasture, hay, crop residues, by-products from food and fiber processing, and waste products humans get meat, milk, and cheese in addition to leather and fertilizers such as bone meal.

The ability of cattle to serve as important nutrient recyclers clearly adds to the quantity and quality of human food supply. Feeding cattle crop production wastes greatly reduces the cost of waste disposal, and it also reduces the environmental costs (Bradford 2001). When manure is returned to the soil, it increases the content of organic matter in the soil and reduces the need to use chemical fertilizer. This alleviates pollution and improves soil fertility.

The impact of cattle on the environment is enormous. World Wildlife Fund (Clay 2000) found that globally, cattle affect more ecoregions of significant biodiversity than any other single agricultural commodity. More land is used for pasture than for any other single use. Some pasture is created from the direct conversion of natural habitat, as is the case throughout much of the Amazon. Other pasture is created only after land has been degraded by other agricultural activities.

In some instances, existing natural grasslands and open woodlands are used to pasture cattle. Where cattle have taken the place of other large ungulates, their overall impact on the ecosystem is relatively small. Only in this case is the elimination of cattle likely to yield biodiversity and ecosystem functions that are anywhere near their "original" state. Globally, in most instances, however, cattle are simply the final desperate attempt to wring a little more from an environment that has already been severely degraded. In general, research shows that the longer converted natural habitat is used for any agricultural or pastoral activities, the less likely it is ever to regenerate as anything like its original plant and animal communities.

Pasturing is not the only reason that cattle production has such a large impact on the world environment. About one-third of the global cereal grain supply and protein supplements are used to feed cattle. This reduces potential food supplies. On average, the ratio of the weight of grain fed to the weight of food produced is between six and three to one for meat, two to one for eggs, and less than one to one for milk (Bradford 2001). This latter ratio means that dairy cattle, for example, produce more than one unit of human food energy or protein for each unit of human-edible energy or protein in their feed. This can happen because the diet of dairy cows includes sufficient quantities of water and products inedible to humans (e.g., grasses, silage) that the animals convert to products that are palatable for people. In meat production, however, the amount of materials that could be utilized by people but that are fed instead to cattle (e.g., grains) is three times the weight of the final product.

The effect of animal production on human food supply depends on the species of animal, the product, the relative amounts of human-edible and human-inedible materials used as feed, the level of husbandry and management, and numerous other factors (Bradford 2001). According to Bradford, the best estimate is that feeding human-edible material to food-producing animals results in a slight reduction of total human food supply. This reduction perhaps is not as much as claimed by some who fail to recognize the quantity of inedible products that are used over the full life cycle for animals slaughtered after grain-intensive feeding operations in feedlots.

PRODUCING COUNTRIES

According to the FAO (2002), the largest producer of cattle in the world is India, which has 218.8 million animals. Brazil is second (169.9 million), followed by China (104.6 million), the United States (98.2 million), and Argentina (48.7 million). These five countries accounted for 47.5 percent of all live cattle in 2000. These countries also accounted for 47.2 percent of all animals slaughtered and 49.4 percent of all beef produced. In the United States about 10 percent of the total cattle herd is for dairy. While all of these animals eventually end up being sold for meat, dairy operations and their environmental impacts are not the subject of this chapter.

Put another way, the total world cattle population is estimated at about 1.34 billion animals. About 33 percent are in Asia, 22 percent in South America, 15 percent in Africa, 13 percent in North and Central America, and 12 percent in Europe.

Total annual, global beef production was estimated at 56.5 million metric tons (carcass weight equivalent) in 2000 from 277.8 million animals slaughtered (FAO 2002). The major producers are the United States (37.6 million animals slaughtered), China (36.1 million animals), Brazil (31.1 million animals), the European Union (26.9 million animals), the countries of the former USSR (26.5 million animals), Argentina (12.3 million animals), Australia (8.6 million animals), and Canada (3.8 million animals). The largest net exporters of bone-in and boneless beef are Australia, the European Union, New Zealand, Brazil, and Argentina. According to UNCTAD data (1999), while the United States is one of the main exporters of beef in the world, it is, on balance, a net importer.

CONSUMING COUNTRIES

From the beginning of the 1970s to the middle of the 1990s meat consumption in developing countries increased by 70 million metric tons, almost triple the increase in developed countries (Delgado et al. 2001). The market value of the increases in beef and milk consumption totaled some $155 billion (in 1990 U.S. dollars), more than twice the market value of increased cereal consumption during the green revolution (Delgado et al. 2001).

The increases in consumption can be linked to population growth, urbanization, lower real commodity prices, and income growth. In East and Southeast Asia—where income grew at 4 to 8 percent per year from 1980 to 1998, population grew at 2 to 3 percent per year, and urbanization grew at 4 to 6 percent per year—meat consumption increased at between 4 and 8 percent per year. Such trends are expected to continue well into the millennium. If sustained over time these trends will create what has been labeled a livestock revolution, i.e. a dramatic increase in productivity *and* consumption (Delgado et al. 2001).

What is significant, too, is that the demand for livestock products is not limited to beef. In fact, the consumption of milk and milk products is much larger (in kilograms/person/year) and growing more quickly than the demand for meat. The consumption trends associated with the livestock revolution have implications for agriculture. Clearly, more land will be devoted to livestock use and to feed production. Much attention has focused on feed grains, but in India the production of sorghum for silage for dairy animals is also quite high and increasing.

While the increased consumption of milk is mostly cow's milk, the increased consumption of meat is mostly of meats other than beef. Per capita beef consumption is increasing in developing countries but declining in developed ones (Delgado et al. 2001). Overall, however, total beef consumption is increasing due to increased population.

Net meat imports into developing countries from developed countries are also projected to expand by a factor of ten by 2020. Beef prices are expected to fall by 4 to 7 percent, whereas milk prices are projected to fall by 12 percent during the same period (Delgado et al. 2001).

The main beef-importing countries are the United States, Japan, Russia, Mexico, Egypt, and South Korea. While large importers of beef, the European Union and Canada are both net exporters.

PRODUCTION SYSTEMS

Beef production is becoming more intensive. Since 1970 the United States, for example, has fewer head of cattle and produces fewer calves each year but still produces more beef. The United States is the most efficient of the larger producers. With only 7.4 percent of the global herd, it slaughtered 13.3 percent of all animals and produced 21.6 percent of the beef in the world in 1999. However, by 1995 the average producer had to be about three times more productive to earn the same net income as in 1975 (Nations 1997), and it was estimated that less than 25 percent of beef producers made a profit ten years in a row.

Between 1970 and 2000 cattle numbers declined by 12.6 percent in the United States, from 112,369,008 to 98,198,000 head. During the period from 1990–98, the total number of beef operations in the United States declined from 932,000 to 855,000, and those that remained became more intensive. The average herd size increased by 11 percent from thirty-six to forty. Texas, California, Kansas, and Arizona have 32 percent of all feedlots, but they contain 75 percent of all lots with 16,000 or more animals (Conner et al. 1999).

In Argentina, too, the total herd size is declining, down 14 percent in the past ten years. This is in part due to falling prices (prices declined by one-third in 1998) and ranchers dumping animals on the market rather than to more efficient production. Meat packers and processors in that country are also in trouble. Consumption of beef has fallen by 34 percent in the country since 1980. Exports take up some of the slack (they rose 17 percent in 1998), but still they represent only 10 percent of total production. Traditionally, Argentines fed their beef only on grass on the pampas, arguing that it was leaner, tastier, and had less cholesterol. Quality issues aside, however, cattle come off the grass at only half the weight of North American animals, and the price per pound is only half as high. Many ranchers in the pampas are getting out of ranching. Since the pampas are fertile, this likely means a shift to agricultural crops that will cause more severe environmental problems than livestock (Giménez-Zapiola 1997).

To date, the expansion of beef production in developing countries has resulted primarily from a rapid increase in the number of animals rather than increased carcass weight (Delgado et al. 2001). This has resulted in a large number of animals in or near urban areas as well as in areas where land is "free" (e.g., the African Sahel). The number of animals in developing countries has increased for two reasons. First, intensification as well as the expansion of pasture areas, particularly in Latin America and Africa, have allowed herd size to increase. Second, the number of animals that are present throughout farming areas has increased as a way to recycle wastes and to make use of marginal grass areas. For the most part, these cattle represent income diversification strategies. In general, however, more intensive stocking of animals on the same land areas without other changes in management tends to degrade overall productivity and increase environmental impacts.

In 1997–98 developing countries accounted for only 36 percent of cereals used in animal feed, but they were projected to account for 47 percent by 2020 (Delgado et al. 2001). By contrast, cereal grain use in developed countries is expected to fall by 2020. Global use will continue to increase through 2020. On a per capita basis, cereal feed use will be about 362 kilograms in developed countries in 2020 but only 71 kilograms in developing countries.

Intensive livestock feeding operations have mostly been established in countries where capital is cheap relative to land (e.g., the Netherlands and the United States). Intensive operations increase nutrient loading and other environmental problems, but the actual costs of the problems are not borne by the producer (and passed to the consumer); rather, they are passed on to society at large (Delgado et al. 2001). Subsidized lending and subsidized grain production have aggravated problems that arise from feedlot operations.

A fundamental factor affecting investment patterns in the U.S. beef industry is the linkage between cattle feeding and the feed grain sector. In the United States corn comprised more than 83 percent of feed grain fed in the last five years, with the remainder accounted for by sorghum, feed wheat, barley, and oats. Oilseed meal (e.g., soybean meal, among others) is also used as a feed ingredient (Flora 1999). About two-thirds of beef cattle are fed on grains for 120–140 days, so proximity to and prices for high-quality feed ingredients are key drivers of investment. As a consequence, feedlot operations are increasingly located near the sources of feed; likewise, slaughterhouses also tend to be located closer to the feedlots. It is more efficient (and it results in higher-quality meat products) to move animals relatively short distances for slaughter, and it is cheaper to operate slaughterhouses in less-populated areas than in areas that are more urban. This is due not only to the cost of land and labor but also to the costs of waste disposal and the conflicts over traffic, smells, and other factors that happen when such operations are located in more urban settings.

The trend in the more efficient feedlot operations is for buyers to buy calves by the truckload. Such buyers prefer truckloads of animals that are uniform in terms of size and genetics because for the most part, uniform animals will move through the system (e.g., growing, feeding, and marketing) more efficiently. This makes feedlot operations easier to run and more efficient, so truckloads of uniform animals are worth a higher price. In fact, a mixed truckload, if purchased at all, will often be valued at the price of the lowest-priced animals in it rather than the average, which penalizes producers who cannot create truckload lots of uniform animals. This, too, tends to encourage larger-scale production.

In 1999 in the United States there were a total of 50,000 feedlots. The largest 400 accounted for more than 65 percent of the nation's marketed fed cattle. Of these, about ninety, each with a capacity of more than 32,000 head, marketed 35 percent of fed cattle (Flora 1999).

PROCESSING

In the developed world, typical beef calves are raised in cow/calf operations (i.e. enterprises that produce the calves that are subsequently fattened and slaughtered) on individual farms or ranches, sold to feedlots where they are fattened, and then sold to abattoirs that slaughter cattle and process the meat. Leather tanneries and other industries process the hides and other by-products. In developing countries, by contrast, the animals are grown on farms and ranches until they reach the appropriate age or size to be sold directly to the slaughterhouse. Increasingly, though, animals are sold to pasture owners who specialize in finishing the animals, for example, fattening them with grass or small amounts of grain for slaughter.

Processing beef begins when cattle reach the slaughterhouse. In developed countries most slaughterhouses are semimechanized, and this makes the operations very quick and efficient. In many parts of the world, however, slaughterhouses are, in a word, gruesome. They are torturous for the animals, inefficient, and unhealthy. Such slaughterhouses are often major sources of localized, untreated pollution. Regardless of where the slaughter occurs, much processing still consists of the labor-intensive cutting up of carcasses. Meat is classified based on where on the carcass the cut came from. Entrails and other body parts are also separated and sold by type. Blood, hide, bones, etc. are sold separately and, in developed countries, are equal to the profit margin of most slaughter operations. In other words, if slaughterhouses didn't sell by-products, they would make no profit.

The meat-packing sector, to an even greater degree than the rest of the food industry, is becoming increasingly concentrated in the hands of a few companies. In 1980 in the United States, for example, four companies accounted for 41 percent of the nation's slaughtered cattle. In 2000 the top four companies (Tyson/IBP Inc., ConAgra Beef Companies, Cargill/Excel Corporation, and U.S. Premium Beef) reportedly slaughtered 81 percent, and the single largest company slaughtered 35 percent of all feedlot cattle (USDA 2000, as cited in Hendrickson and Heffernan 2002; see also Table 1.5 in Chapter 1).

In 1997 the fourteen largest U.S. slaughtering companies, each with an annual capacity of over 1 million head, processed 94 percent of the 192.3 million head slaughtered at federally inspected facilities. Only meat from federally inspected plants can be sold in interstate or international commerce. From 1991–94, the number of firms slaughtering cattle fell 25 percent to 239, and twenty plants handled 58 percent of commercial slaughter.

The impact of U.S. companies extends beyond the country's borders. By 1999 two U.S.-owned plants handled more than 50 percent of all cattle slaughtered in Canada and 83 percent of all those slaughtered in Alberta (Flora 1999). While these trends were accentuated as a result of conditions established under the North American Free Trade Agreement (NAFTA), they also illustrate increased vertical integration globally that has its own economic rationale.

Slaughterhouses are equipped with cold storage to preserve the meat. A considerable amount of water is used to wash the incidental blood away. As the meat is cut up, blood is then washed away. This water becomes effluent. Because of the blood, the effluent is loaded with organic matter. When the intestines are processed and cleaned another type of waste is generated. In some parts of the world, this waste is separated and destined for a "septic" tank where it generates methane gas, or it is applied to fields as fertilizer. Many slaughterhouses lack these types of operations, in which case rumen matter from intestines becomes part of the effluent stream. In the worst operations, this material is simply flushed into freshwater bodies, where it causes nutrient loading and consumes most of the oxygen in the water as it degrades.

In many developing countries, slaughter occurs very near the point of sale, and animals are cut up as needed or desired by the customer. In many instances the price of beef varies only depending on whether bones are included or not. Even when carcasses are shipped to other areas, they tend to be shipped as whole, half, or quarter carcasses and then are cut up at the point of sale.

In the past, processing plants were slaughterhouses. For the most part, animals were killed, skinned, and hung, usually as half or quarter carcasses. The carcasses were then shipped on to retail chains and others who cut the meat prior to sale to consumers. This is still the situation today in much of the developing world. However, in the developed world slaughterhouses have become automated meat factories (Flora 1999), which produce and pack specific cuts of beef in boxed lots. The cuts are then vacuum-packed and shipped in boxes to hotels, restaurants, institutions, and retail groceries.

In more developed countries meat tends to be processed into standard "cuts" at the slaughterhouse as a way to reduce the bulk of shipping carcasses as well as the labor costs at the point of sale. "Boxed beef," in which cuts are standardized, is sold directly from packing plants to retail operations. This allows slaughterhouses to be moved out of areas more directly linked to markets and into areas that are closer to where the animals are raised (Flora 1999). Iowa Beef Processors (IBP) began this process in 1960 and opened the first large-scale dedicated plant for boxed production in 1967.

The change in processing has had a number of implications. First, the cost of unionized labor in more established companies can be lowered by moving into rural areas. Second, less processing is undertaken in supermarkets, and thus supermarkets need to employ fewer, if any, butchers at their meat counters. Third, meat prices can be lowered because packers can shorten the supply chain and buy more directly from producers, eliminating middlemen and reducing transportation costs. In short, it has meant that processing was moved closer to the animals and, in the case of beef, closer to the sources of grain used in feedlot operations. This approach to processing has tended to encourage a clustering of intensive feedlot operations within easy transportation range of the plant to take advantage of the economics of the system.

This type of processing has eliminated more than 100 kilograms of fat, bones, and trimmings that were part of the carcass, but were of little value to the customer. Boxed meat has improved quality, provided quicker and easier merchandising, improved shelf life, saved energy costs, and reduced transportation and labor costs. Buyers can purchase only the cuts they want and only the amounts they can use. The value of meat prepared this way exceeds the costs of preparing it.

The slaughter of cattle results in a number of products and by-products, including retail cuts with or without bones (42 percent of live weight), organs (4 percent), edible fats (11 percent), blood (4 percent), inedible raw materials (17 percent), hide and hair (8 percent), and waste (4 percent) (Conner et al. 1999). In the United States slaughterhouses are required to treat their waste effluent just as intensively as any other industry. Most solids are captured because they have economic value when turned into by-products. Even blood and other sludge have a market for use as fertilizer or land application (Verheijer et al. 1996, as cited in Conner, et al. 1999).

The main substitutes for beef are other meats (e.g., pork, chicken, or seafood), dairy products such as cheese, and vegetable-based products such as pasta and starches, which, in combination with vegetables such as beans, provide the same quality proteins that are found in meats. Globally, beef production is a close second to pork production, with 35 percent share of total meat protein. In terms of trade, beef exports are a distant second to poultry (26 percent versus 40 percent).

Much has been written about the decline in demand for red meat, particularly beef, in comparison to poultry. The consumption of poultry throughout the world has increased very rapidly during the past forty years while the consumption of beef, in developed countries at least, has remained fairly constant. Beef production relative to pork and poultry is declining for three reasons. First, the cost of production is higher than for poultry or pork. Second, cattle carcasses are large and not as easily divisible as smaller animals. Third, there is the perception that beef is not as healthy as poultry or pork. In fact, grass-fed beef has about the same amount of cholesterol as chicken.

It is quite possible that the biggest declines in beef consumption in developed countries are due to the substitution of pasta and other foods. In developing countries as incomes rise the consumption of meats (including beef) increases while the consumption of rice, wheat, cassava, and other vegetable staples declines as a portion of total caloric intake.

Wild game, buffalo, and water buffalo meat are the closest red meat substitutes for beef. Consumption of meat from game is increasing in developed countries, but only as a niche market. Unsustainable bush meat production is common in many areas of Africa and to a lesser extent in Latin America. The meat is destined for local markets, though, and the markets for bush meat decline when beef becomes more widely available. Buffalo from North America is destined for niche markets. Most water buffalo is produced on small farms in Asia, except where it is produced in large-scale operations in Brazil and Venezuela. Very little water buffalo meat is exported.

MARKET CHAIN

There has been considerable concentration of holdings within the cattle industry from ranches and cow/calf operations through feedlot systems all the way to the slaughterhouses. While much of the focus has been on the U.S. industry, this is common throughout the developed world and probably in the developing world as well, though there is far less documentation. In addition, the vertical integration of the beef industry simply reflects similar changes throughout the global food industry.

Case studies from Canada and Australia illustrate the consolidation and integration that are occurring in the beef industry. In 1987 ConAgra purchased the operations of the dominant beef processor in the northern Great Plains, Monfort. Shortly after, Cargill moved across the border into Alberta to set up a large beef slaughter system. At that time, Canadian Packers (CP) was Canada's largest manufacturer of livestock and poultry feeds, the largest cattle slaughterer, the only national poultry processor, and Ontario's largest hog slaughterer. Due to the new competition, CP began to experience difficulties and was sold to Hillsdown Holdings, Europe's largest fresh meat processor and manufacturer of value-added egg and poultry products and largest canner of fruits and vegetables. Hillsdown already owned Maple Leaf Mills, Canada's second largest flour miller. Hillsdown Holdings then announced that Canadian Packers was getting out of the fresh beef markets. Mitsubishi has indicated it is interested in slaughter operations in Canada, but a large part of Canadian cattle still moves through ConAgra's feedlots, slaughter facilities, or both (Heffernan 1994).

At the same time as it was consolidating its operations in Canada, ConAgra purchased a half interest in Elders of Australia. Elders was the dominant beef slaughter operation in Australia and the largest exporter of beef and lamb in the world. Soon after ConAgra's move into Australia via Elders, Mitsubishi began to invest in the beef slaughter industry in Australia, and Cargill has purchased beef slaughter facilities in the country as well. Cargill currently has beef operations in Brazil, Honduras, and Mexico among others. ConAgra has trading offices in twenty-three countries, and Cargill currently operates in more than sixty. Three multinationals are moving into position to control the world beef industry (Heffernan 1994).

In the United States's meat-retailing sector, the number of grocery stores began a dramatic decline in the 1930s due primarily to the demise of small grocery stores, which were unable to compete with the larger grocery retailers such as chain supermarkets. By 1965 supermarkets were the dominant form of grocery business, accounting for 70 percent of total grocery sales. The United States's grocery store industry of the 1990s was characterized by large supermarkets representing less than 25 percent of the grocery stores but accounting for more than 75 percent of grocery sales.

It is generally thought that beef producers, like most other farmers, are price takers not price makers. They resign themselves to accept whatever prices the market has to offer. Ranch operations can be streamlined and increased in scale, and pastures improved, but beef producers who market beef conventionally still fall prey to low prices brought on by fluctuating cattle cycles, which affect herd sizes at regular intervals. Producers adopt many ways of marketing to sell beef. Direct marketing of the beef to the customers, cooperative marketing through co-ops, and niche markets that sell specialty meats such as organic, natural, or pasture-fed beef are some of the common marketing methods adopted by beef producers as a way to add value to their production or to avoid selling into conventional markets.

MARKET TRENDS

Between 1960 and 2000 the average consumer price of beef declined from $3.56 per kilogram to $1.92 per kilogram, or some 46 percent. Real beef prices have fallen by a third in the 1990s alone.

Demand for beef has declined in many countries in recent years, but global demand has risen due to growing demand in developing countries. Supply in Europe has declined 4 percent since 1997; likewise, Argentinean production has been declining for a decade. Global exports have started to decline as well. In China, on the other hand, production increased 8 percent in 1998 alone. Currently, growing demand from developing countries is balanced by shrinking demand in developed ones. However, as urbanization, overall population, and disposable income continue to increase, global demand for meat is predicted to rise by as much as 50 percent by 2020 (Delgado et al. 2001). Most of this growth will take place in developing countries, where beef imports are predicted to increase by 1 million metric tons by 2020.

Livestock production has been one of the main factors in stabilizing world cereal supply (Delgado et al. 2001). If the price of cereals goes down, the number of animals fed increases. If the price of cereals goes up, the number of animals fed decreases. While this would tend to reduce the amount of meat on the market and raise prices accordingly, this effect is mitigated by the culling of herds, which supplements meat supply and keeps prices down. If this relationship holds, meat will continue to be supplied without dramatic price increases. The issues then are not whether sufficient animal products and cereals will be available, but rather what impact increased production and consumption of both will have on the environment, human health, and the poor (Delgado et al. 2001).

The projected increase in animal protein consumption in developing countries by 2020 will require large net imports of cereals by developing countries, of about the same magnitude as the annual U.S. corn crop (i.e. 200 million metric tons). About half of these net imports will be corn and cereals other than rice and wheat. Net meat imports into developing countries from developed countries are also projected to expand (by a factor of ten), but from a smaller base (Delgado et al. 2001). This would amount to about 5 million metric tons per year by 2020.

These trends play out differently in different countries. Global trends, coupled with a 30 percent devaluation of the currency in Russia (the second biggest beef importer), resulted in a 25 percent decline in imports to Russia from the world's major exporters. It is doubtful that the Russian situation will turn around any time soon.

Local markets have also followed predictable historical trends, but it is not clear if these will continue with globalization. For example, in the United States beef cattle numbers normally run on a ten-year cycle. As cattle numbers peak, prices of slaughter cattle and feeder calves decline. Reduced prices for feeder calves tend to cause more cows to be sold for meat. This increases the total meat supply and dampens overall beef prices. Reduced prices for feeder calves also tend to be linked to increased grain prices. When grain prices are high, feedlot operators (who tend to bid on calves at a rate that will give them a finished calf at about the same price as it had been historically) will purchase fewer calves, reducing demand. For the past three years, global corn production has been down and reserves are dwindling. This will tend to increase grain prices, reduce the price of calves, and force the sale of cows as ranchers cull their herds to prepare for the next upswing of the market.

In the United States the future of the cattle and beef industry will likely depend on such questions as:

1. How can the beef industry compete more effectively with other meat industries (especially chicken and pork and, increasingly, seafood) in consumer-driven markets that started in developed countries but are now spreading to less-developed countries as well?
2. How can it remain competitive and expand the sales of beef in international markets?
3. How can it produce a product that meets the concerns of health-conscious consumers while maintaining product quality and consistency?
4. How can it develop industry-wide technological and structural changes that reduce the cost of production?
5. How can it work more effectively with regulatory agencies to assure food safety and animal disease control, and provide for the long-term integrity of the environment?

ENVIRONMENTAL IMPACTS OF PRODUCTION

Beef production has several distinct and significant impacts on the environment. The impacts vary somewhat from one country to another and depend on the specific part of the beef production process being considered. Perhaps most important, unlike many other agricultural commodities, cattle have significant impacts on a wide range of ecosystems because they can be produced under such a variety of conditions and are literally capable of walking themselves to market.

Globally, the largest environmental impact of agriculture in general is the use of land for pasture. More pasture is used for cattle than all other domesticated animals and crops combined. In addition, cattle eat an increasing proportion of grain produced from agriculture, are one of the most significant contributors to water pollution, and are a major source of greenhouse gas emissions. Finally, processing cattle into meat, meat by-products, and leather is a major source of pollution in many countries.

Habitat Conversion or Modification

The most significant direct impacts of beef production on habitat are the conversion of forest habitat to pasture, the alteration of the composition of native plant communities in grasslands, and the wholesale removal of native vegetation (e.g., forests, scrublands, and grasslands) as habitat is converted to seeded or planted pasture. Currently two-thirds of the world's agricultural land is used for maintaining livestock. One-third of the world's land is suffering desertification due in large part to deforestation, overgrazing, and poor agricultural practices. An area of the world's rainforest larger than New York State is estimated to be destroyed each year to create grazing land. This not only alters the composition or existence of native plant communities but also the species of wildlife that existed in those plant communities.

Plant communities are altered over much of the world, often as a result of direct intervention such as plowing native grassland vegetation and establishing either single-species or mixed-species pastures of introduced species. The species composition of natural grasslands is transformed by continuously overstocking native rangeland with livestock, enrichment planting (e.g., sowing seeds of introduced species in native grassland), and eliminating intentional burning.

Pasture can be created from temperate or tropical forests or savannas. This often involves converting native habitat and introducing grass and forage species that provide more food for cattle. In natural grasslands the biggest impact is the alteration of the native plant communities and the associated impacts on wildlife and other biodiversity. In addition, cattle are increasingly fed hay and grains to supply food during the dry or winter seasons, or to fatten them before slaughter. While forests have been cleared to make way for livestock throughout the world, the most significant impacts recently have been in the Amazon, where massive clearing of tropical forests has had a tremendous impact on biodiversity, ecosystem functions, and even local climate.

Maintaining desired pasture composition in created pastures often requires tillage, chemical fertilizers, and pesticides. Continuous grazing causes plants to produce more leaf biomass and less root biomass. This reduces their ability to survive during periods of stress (e.g., extended cold, hot, or dry spells). Watershed protection also suffers as plant cover and leaf litter diminish, leaving the soil exposed and erodible. In areas where pastures are not maintained, woody plants tend to dominate over time, not only affecting ecological balance but also reducing the carrying capacity for cattle.

Another source of pasture is degraded agricultural land. In many areas, once land can no longer produce agricultural crops, it is used for livestock. Such land is already degraded. However, converting it to pasture degrades it even further, virtually ensuring that it will not return to anything near its natural state.

Cattle production can cause habitat conversion indirectly as well. In some instances cattle are a "push" factor, displacing the rural poor into fragile areas. In Central America, for example, the conversion of labor-intensive, cash crop-producing areas to cattle production caused many landless poor to move into and clear tropical forest areas for subsistence production. In a variation on this pattern, the rural poor and landless in the Amazon often clear land, grow a crop or two and then plant the land to pasture to sell to ranchers.

Cattle Feedlots

In the United States and, increasingly, in other parts of the world, cattle feeding operations present perhaps the greatest potential environmental threat of the beef industry. The reason feedlot production is of such concern is that it is one of the fastest-growing beef production practices in the world. The direct impacts of cattle feeding include contributions to air pollution through methane, odors, and dust, and to pollution of surface and ground water through nutrient loading from improper handling of manure. In addition, other environmental impacts such as the use of antibiotics can be intensified given the large concentrations of animals in a confined space.

About 1.4 billion metric tons of solid manure are produced by U.S. farm animals each year—130 times the amount produced by the human population. Put another way, U.S. animal feedlots produce 100,000 metric tons of manure per minute. This figure includes pigs and chickens as well as cattle, but even so cattle are the single largest source. In Texas 7.5 million head of cattle in feedlots consume more than 7 million metric tons of feed containing more than 150,000 metric tons of nitrogen and 25,000 metric tons of phosphorus. It would take 8,000 hectares of corn silage (or a similar crop) to absorb the manure from a feedlot with 50,000 head of cattle (Conner et al. 1999). If the manure cannot be added as a soil amendment, it has to be treated and disposed of another way to avoid contaminating land or water.

Disposal of organic wastes without proper treatment leads to the pollution of water resources. Eutrophication (nutrient enrichment) of water systems can cause large-scale algal blooms that kill fish and other aquatic life. Such serious situations have been encountered in northwest Europe and off the East Coast of the United States. In addition, ammonia released from manure and slurry can, through its interaction with sulphuric and nitric acids, contribute to acid rain.

Production of Feed Grains

One of the major impacts of the beef industry occurs indirectly, through the production of grain used to feed cattle. As discussed elsewhere in this book, the production of feed grains generates significant habitat conversion, soil degradation, water pollution, and other environmental impacts.

Competition for food resources (i.e., raising grain for cattle feed versus human food) is a serious concern about beef production. Globally, one-third of the world's cereal harvest (wheat, corn, rice) is fed to farm animals. While the use of feed is not broken out by type of animal, it is clear that a significant portion is used to feed cattle. In the United States some 95 percent of soybean meal is used as livestock feed. In addition, the U.S. beef industry utilized about 11 percent of the U.S. corn supply in 1992.

The switch from grass to feed grain finishing results in a more consistent product even when starting out with inferior animals or genetics. Consequently, the beef herds in the United States, for example, have shifted markedly away from genetically superior meat producers such as Angus and Hereford, which dominated United States markets in the 1950s, until today, when they represent less than half all beef cattle. More and more, beef cattle are hardier species, but their meat is of inferior quality. Given that most beef is used for hamburger this is not a serious problem. These hardy animals can tolerate more heat, less water, and a wider range of less nutritious vegetation. This change in genetics of the beef cattle herd has resulted in the expansion of pasture-based beef production into harsher and more marginal, biodiverse, and ecologically fragile areas.

Data from Sweden illustrate some of the tradeoffs between the different beef production systems (e.g., grass-fed, grain-fed feedlot-fattened, or a combination of the two) and their overall impacts. Calves produced through intensive feedlot feeding systems can be slaughtered in twelve to thirteen months at a weight of 450–475 kilograms live weight. Fed protein-rich concentrates, such animals gain more than 1 kilogram per day and can be produced with less total feed (25 megacalories per kilogram slaughtered weight). By contrast, grass-fed beef live longer and eat more roughage (grass, hay, and silage); these animals reach 525–550 kilograms (live weight) at the time of slaughter, but this requires about 18 months. Because they take longer to reach slaughter size, their overall feed consumption is somewhat higher (35 megacalories per kilogram slaughtered weight) than that of feedlot-produced beef (Tengnas and Nilsson 2002). The feed for grass-fed animals is cheaper, more locally produced roughage. However, grass-fed animals also require more land area for their production, even taking into account the land used for cereal production with more intensely fed beef. From an environmental point of view, the overall impacts depend on how the range is managed and how fragile and biodiverse it is to start with versus how cereal production is managed and how the waste issue in feedlots is addressed.

Use of Antibiotics and Growth Hormones

Antibiotics and growth hormones are increasingly used in feedlots. Antibiotics are used in feed as well as in injections, vitamins, vaccinations, and parasite controls. Antibiotics are generally administered for ninety days or more in feedlots in the United States. Animals arriving to feedlots are given antibiotics in their water for eight days or so. In the United States, less than 20 percent of all animals in feedlots were given antibiotic injections, but about 60 percent received vitamin injections. Most cattle in U.S. feedlots are given growth hormones to increase their weight gain.

In short, there is significant use of antibiotics, vaccinations, growth hormones, and vitamins in the beef industry without sufficient understanding of their overall impacts. It is well known, however, that the prophylactic use of antibiotics can lead to bacterial resistance in the animals and in the environment, and that this resistance can even be passed on to bacteria that infect humans. Similarly, the effects of growth hormones in the production of meat may be passed on to people who consume the meat (Program on Breast Cancer and Environmental Risk Factors in New York State, 2000). Unfortunately, virtually no research has been undertaken on the impact of these inputs on the wider environment, either in the vicinity of feedlots or in areas where waste from feedlots or slaughterhouses is disposed.

Water Use and Quality

Dr. Jim Oltjen of the University of California at Davis and Dr. Jon Beckett, formerly of UC-Davis, found that, including direct consumption, irrigation of pastures, and crops and carcass processing, it can take as much as 3,682 liters of water to produce 1 kilogram (2.2 pounds) of boneless beef in the United States. Given impending water shortages in many parts of the world, the price of water is likely to increase. This will either result in more expensive meat or, more likely, encourage more efficient use of water.

In addition to total water use, there is increasing concern about water pollution, especially the harmful effects on surface water and groundwater quality of pesticides used to maintain or improve pasture areas or to increase feed grain production. In addition to contaminating waterways, groundwater, and even marine environments, those who use pesticides and live in rural areas tend to contaminate not only the water supplies of their own livestock operations and those of their neighbors, but also their own water supplies. Many people living on farms in the United States cannot safely drink their own well water.

Soil Loss and Degradation

Livestock farming is one of the main activities responsible for soil erosion around the world. In 1994, for example, soil loss in Brazil's Alto Taquari watershed was estimated at 70.39 metric tons per hectare per year, which is a high erosion rate. The degree of erosion increases proportionally to the increase in deforestation in the basin. From 1977 to 1991 a 50 percent increase in habitat conversion was recorded. This has led to extensive degradation of the flooded valley-bottom vegetation. Pasture establishment results in the exposure of soil to the elements for several months, often during the rainy season. So while pasture itself may not result in as high soil erosion rates as annual agricultural crop production, the initial conversion to pasture can lead to extreme erosion with loss of topsoil and organic matter that could take decades or centuries to replace.

Overgrazing damages soil structure and causes erosion. In many parts of the world, larger herds of cattle are being kept on ever smaller amounts of land, for longer periods of time. Overgrazing is a particular problem on slopes, where soils are more easily eroded and some grasses are crushed by the animals' hooves. This is the case in many parts of the world where hillsides covered with cattle show the contoured signs of erosion and soil displacement. Overgrazing also thins and eventually removes ground cover so that the impact of wind and rain erosion increases.

Another cause of soil degradation and erosion from cattle is their repeated trampling over the same areas. The result is compaction or "soil pugging" due to the impact of cattle hooves. Soil compaction can destroy soil structure and results in resistance to root penetration, reduced water infiltration, and reduced aeration. All of these impacts harm beneficial soil microorganisms. Compaction is considered to be inevitable with cattle production. However, the severity varies with the soil type, and is worst on wet soil that has a high clay content. Severe compaction provides a site for surface runoff that can result in serious erosion and even the creation of deep trenches, a process called gullying.

An additional environmental problem resulting from soil erosion is the intense degradation of surface waters. For example, some 80 percent of the cleared areas of the Brazilian Amazon and the cerrado (the savanna and forest-covered tableland that lies between the coastal forest zone and the Amazon) has been converted to pasture. Creation of these pastures has resulted in the increased siltation of streams and rivers.

Greenhouse Gas Emissions

Beef production has a considerable effect on global warming due to the emission of greenhouse gases such as methane, nitrous oxide, and carbon dioxide. Methane is released from the cow's rumen and manure. Nitrous oxide is released from the soil by the microbial decomposition of manure and artificial fertilizers. Carbon dioxide is released by direct energy consumption through mechanized feed crop production and the herding and movement of animals (the average beef calf sees more of the U.S. than the average cattle farmer).

Globally, ruminant livestock produce about 80 million metric tons of methane annually, accounting for about 22 percent of global methane emissions from human-related activities. Cattle in the United States emit about 6 million metric tons of methane per year into the atmosphere. The cow/calf sector is the largest emitter of methane within the U.S. beef industry. It accounts for 54 percent of the total methane emissions from cattle, while the feedlots and stocker calves account for 21 percent, and dairy accounts for 25 percent (Ruminant Livestock Efficiency Program, 2002).

While cattle release huge quantities of methane into the environment, it is not clear that they produce more methane than similar wild animal populations did 200 years ago. Globally, however, as the number of cattle increases, it could well exceed historical levels of methane emitted by wild animals.

Impacts of Slaughter and Tanning Industries

The expansion of the global cattle industry has been paralleled by the vigorous growth of the beef slaughter and leather industries. The waste from both slaughterhouses and tanneries is rich in organic matter and hence poses serious public health concerns if discharged into the environment without appropriate treatment.

In the United States more than 20,000 cattle hides are tanned per day. Some 23.5 percent of these are processed with vegetable tannins. The remainder is tanned with chromium, a pollutant categorized as a heavy metal. Though tanneries in the United States are also required to treat their effluent before it is discharged (Conner et al. 1999), tannery effluents in many parts of the world are high in chromium and biochemical oxygen demand (BOD) levels. Chromium contamination of the water sources of the surrounding areas harms both humans and wildlife.

BETTER MANAGEMENT PRACTICES

There are a number of ways to reduce the environmental impacts of beef production. As with most operations, perhaps the key to reducing subsequent impacts is to site and construct operations well. Once built, however, there are still a number of management practices that can reduce environmental damage. These include maintaining vegetative cover, avoiding overgrazing, protecting riparian areas, reducing waste and disposing of waste in the least harmful ways, reducing the use of chemicals and antibiotics, reducing wastewater and improving water effluent quality, and reducing soil compaction.

There are several specific ways to address many of these issues. However, some of the important, more general approaches include aligning production needs with natural processes, improving the feed conversion of animals from any feed source, producing and marketing cattle with more meat and less fat, and integrating beef production with other activities to increase overall carrying capacity and productivity.

Site and Construct Operations Well

Where producers locate their operations is often the single largest factor that contributes to subsequent environmental impacts. Most nonpoint-source pollution problems occur in the vicinity of watering and supplemental feeding, and along fences or resting areas where cattle tend to congregate. Such concentrations can reduce vegetative cover and can compact the soil so that erosion is more likely and water percolation is diminished (Florida Cattlemen's Association 1999). There are several ways to manage the placement of such activities so as to reduce their impacts. For example, placing supplemental feeding and mineral stations a reasonable distance (30 meters) away from stormwater drains, streams, drainage canals, ponds, lakes, wetlands, wells, and sinkholes can prevent such problems. The development of alternative water sources can also attract animals away from streams, drainage canals, and lakes (Florida Cattlemen's Association 1999).

Leaving or planting small, scattered clusters of trees in upland areas of pastures can provide shade and keep cattle away from water sources as a way to keep cool. In general, feeding stations, portable water troughs, and shade structures should be moved periodically to prevent waste accumulation, loss of cover, and compaction of soil.

In some cases it is impossible to locate facilities outside of sensitive areas. In those cases, other techniques should be employed to help keep sediment, nutrients, and organic matter out of surface waters. Biological filters (biofilters) of marshes, ponds, or other natural or constructed wetlands can assimilate many nutrients and sediments. In some cases it will be necessary to reestablish natural flow patterns, plug drainage canals, or divert water to recreate the natural hydrology of an area to take advantage of bioremediation options.

Locations for any temporary holding areas should also be carefully planned, as they have the potential to concentrate large amounts of pollution. Cow pens and other temporary holding areas should be located more than 60 meters (200 feet) away from waterways and water sources to prevent runoff and contamination. For existing holding areas that cannot be moved and that are located near water bodies, filter strips, sediment traps, grass planting in seasonal waterways, retention and detention ponds, and planting or berms can minimize the transport of pollutants to water bodies.

Cattle are not the only causes of soil erosion or water quality problems in beef production systems, however. Human activities such as land clearing; culvert installation; road, ditch, and canal construction and maintenance; pasture renovation; and cultivation of forage crops can all expose soil and contribute to nutrient loading. Planting cover crops immediately after removing vegetation for infrastructure development should be standard practice. Strips of grass should be maintained along drains and ditches. The number of vehicle and animal crossings of streams and canals should be minimized. To discourage erosion, vegetation should not be cut too short near waterways and clippings should be kept from waterways.

Cow/calf operations are generally low-intensity forms of agricultural production with relatively low levels of pollutants discharged off the farms. Cow/calf operations may contribute to elevated levels of phosphorus, nitrogen, sediment, bacteria, and biological oxygen demand (BOD) in surface waters, though at much lower levels than feedlot operations. Manure from cow/calf operations can also contribute to water quality problems both from runoff and direct contamination (Florida Cattlemen's Association 1999).

The potential for discharges from cow/calf operations to cause water quality violations varies greatly, depending on soil type, slope, drainage features, stocking rate, nutrient management, pest management, or activities in wetlands. In general, areas where cattle tend to congregate or have access to water bodies have the greatest potential for pollution (Florida Cattlemen's Association 1999). Proper siting of these operations is the best way to maintain water quality. By contrast, low-density grazing on native range has the lowest pollution potential. There are better practices that minimize water quality concerns, but it will also probably be necessary to work with a number of ranches in a watershed to address cumulative impacts rather than working one ranch at a time.

Cattle can cause significant compaction of soils. One way to reduce this problem is to use mobile water, feeding, or mineral supplement locations. Rotating pasture use is also a way to avoid prolonged impacts. Some ranchers use moveable fences or herders to keep herds from compacting soils in key areas. Finally, some heavier, clay soils are more subject to compaction; if pastures are located on such soils, every effort should be made to move cattle onto lighter soil when heavy rain is likely.

Avoid Overgrazing

There are several ways to control grazing so as to mitigate environmental impacts. Controlled grazing or management-intensive grazing (MIG, also known as rotational grazing) can be adopted to check unlimited access of animals to pastures and also to manage the grazing land effectively. Sustainable pasture management practices, which include a balance of matching forage and livestock resources, resource management, proper breed selection, and looking for alternative feeds, can all help to reduce the deleterious effects of overgrazing.

Properly managed grazing can have some benefits. Cattle manure fertilizes pastures. In addition, grazing can encourage regrowth and prevent the spread of noxious weeds. In South Africa ranchers have found that native grasses germinate best in corridors where cattle have trampled the most. Ranchers have found that cattle hooves break up ground that left alone would be too hard for seeds to penetrate and find a place to germinate. Ranchers using this system have been able to double the carrying capacity of their pastures. Also they have a higher percentage of perennial grasses (which produce more biomass) as ground cover than land ranched conventionally (Spark 1994).

Properly managed grazing maintains healthy vegetation, which helps to filter pollutants from runoff, reduce runoff velocity, and control soil erosion. Management practices that help to maintain vegetative cover involve distributing cattle so that they do not overgraze portions of pasture and allowing for recovery of the vegetation following a grazing period. Using prescribed or rotational grazing systems can minimize the impact of grazing. Adjusting the stocking rate seasonally, particularly in sensitive watershed areas, can also reduce the impacts.

Protect Riparian Areas

Cattle ranchers need to protect the natural vegetation near streams from prolonged cattle grazing, as this vegetation keeps stream banks from eroding and prevents nutrients from entering and polluting streams. This can be done through fencing, creating alternative watering locations, building bridges over streams, and in general more closely monitoring pasture management. Sustainable resource management of riparian zones pays off in long-term environmental gains, reduced expenditures to repair stream-related infrastructure, and overall economic gains.

Improve Assimilation of Feeds

Cattle that produce less waste because they have an enhanced capacity to assimilate feed should be encouraged as part of an overall conservation strategy. There are at least two ways to address this issue. One way to is to identify microorganisms that break down feed more completely into amino acids and other nutrients that are more easily digested and utilized by cattle. Another way is to use breeding programs to create animals that have an improved capacity to assimilate feed, or plants that are more easily digested and assimilated by the animals. With better assimilation of feed, less land would be needed for the production of grains and for pastures.

Growth hormones that are injected in animals directly or put into feeds to stimulate rapid growth cause the animals to convert feed fats to proteins and therefore grow faster. They do not, however, reduce total feed used. Another, perhaps more effective, strategy would be to use microorganisms to ferment the feed prior to feeding so that it can be utilized better by animals. This could potentially result in the same weight gain without the use of growth hormones. The fact that many markets are rejecting beef produced with hormones could be an added incentive to adopt better

sources on a single operation. However, integrated farming can take place at the landscape level as well. This would allow wastes and resources to be used more efficiently while also allowing farmers to specialize to achieve sufficient scale in areas where they have a comparative advantage over their neighbors so that they can compete in larger markets. Unless single operations are going to buy out all their neighbors, then scaling up integrated farming to the landscape level will require much better information management as well as waste and product flow systems.

Improve Pasture Management and Rotations

There are indications that improved pastures and pasture management can reduce the amount of open pastureland devoted to cattle. There is the possibility of increasing the current global carrying capacity of 1.5 head of cattle per hectare to as much as ten head per hectare with improved pasture management and better feeding regimes. The strategy of semistabled, semigrazing projects in small farms has indicated that carrying capacities of ten cattle per hectare can be achieved. Integrated systems of grazing and feeding and the use of wastewater for pasture fertilization improve the palatability and feed quality of pasture crops. With more efficient carrying capacities, former pasture areas can be liberated for other crops or for other uses entirely, including habitat restoration.

In addition, more and more farms are finding that different animals (e.g., sheep, goats, pigs, chickens, and rabbits) will eat different parts of the pasture, so sequential rotation patterns not only improve pasture over time, they allow farmers to better utilize the full economic returns from pastures.

Protect or Improve Water Quality

Improved control of input use and efficiency can minimize off-site discharge of pollutants and therefore improve water quality. Pollutants come from manure, organic matter, fertilizers, sludge application, pesticides, chemicals, and fuels. If these materials are properly stored, applied, and disposed of, there is less chance that they will become part of runoff. The development of nutrient management plans can reduce the nutrient loading in runoff. Nutrient tests that allow producers to determine the most appropriate timing and rates for application of fertilizers can reduce use of these inputs, which can reduce expenditures for inputs in addition to reducing the nutrient content of runoff. Fertilizers and pesticides should not be applied near water bodies and drainage ditches or prior to forecasted heavy rainfall (Florida Cattlemen's Association 1999).

An important strategy to reduce impacts from pesticide use on pastures is for producers to be able to compare overall pesticide toxicity. Information is not generally available to producers that would allow them to select pesticides that are less toxic and less likely to have negative impacts on water quality. In addition, information about which pesticides are better suited to solving which problems with which associated risks would allow producers to make more informed decisions about how to reduce the overall impact of pesticide use. Some of the most important factors to consider when selecting pesticides include the soil properties of the site in question, the mobility and persistence of pesticides, and the toxicity of pesticides to humans, wildlife, and aquatic species. The selection of the proper pesticides will decrease chances of adversely affecting surface and ground-water quality. For example, certain combinations of soil and pesticide properties (along with weather conditions) can pose a significant potential hazard to water quality (Hornsby et al. 1998).

OUTLOOK

There is every indication that in the long term the global production of beef will increase. This will be driven primarily by increases in consumption in developing countries. There are also indications that consumption in developed countries will continue to decline, while production there will remain relatively stable due to an increased volume of exports to developing countries. Production in developed countries will be achieved with ever smaller herd sizes due to improved overall management of herds and inputs.

There are two outstanding issues that could have considerable effects on beef production globally and ultimately on its environmental impacts. The first is changes in production and marketing subsidies in developed countries, either direct subsidies for beef or indirect subsidies for feed grain or pasture inputs. The elimination of such subsidies could encourage a rapid increase in the production of beef for export in developing countries. The second major issue that would affect global beef consumption is a dramatic change in the global economy. A depression would reduce overall consumption and trade of beef; economic growth in developing countries, particularly in China, would stimulate demand for beef.

If current trends continue, beef consumption will gradually continue to increase globally, with most increases coming in developing countries where consumers will prefer the cheapest products possible. Such demand will be met with inefficient production from inferior animals. And while there may be some instances of increases in herd efficiency, by and large production increases will be based on increased herd size, at least in the short term. Such increases in total numbers of animals will have increased impacts on the global environment, particularly in more marginal areas of developing countries where cattle farming will be considered a productive economic activity regardless of its environmental impacts.

REFERENCES

Bradford, E. 2001. Inputs and outputs from livestock production. Paper presented at the AAAS Symposium 6010. San Francisco, CA.

Clay, J. 2000. Agricultural production in priority ecoregions. Unpublished document. World Wildlife Fund–U.S. Washington, DC.

Conner, J. R., R. A. Dietrich, and G. W. Williams. 1999. *The U.S. Cattle and Beef Industry and the Environment.* A report to the World Wildlife Fund. Washington, D.C. May. Report CI-1-99. College Station, TX: Texas Agricultural Market Research Center.

Delgado, C., M. Rosegrant, and S. Meijer. 2001. *Livestock to 2020—The Revolution Continues.* Paper presented at the International Agricultural Trade Research Consortium (IATRC). Auckland, New Zealand. January 18–19.

Fang, L. 2000. *Environmental Effects of the Beef Industry.* Agricultural and Natural Resource Economics Discussion Paper 4/00. St. Lucia, Australia: University of Queensland. Available at http://www.nrsm.uq.edu.au/agecon/pub/pub/discussion/2000/ANREDP400.pdf.

FAO (Food and Agriculture Organization of the United Nations). 2002. *FAOSTAT Statistics Database.* Rome: UN Food and Agriculture Organization. Available at http://apps.fao.org.

Flora, C. B. 1999. *The Industrial Ecology of Beef Production in the U.S.* A report to the World Wildlife Fund–U.S. Ames, IA: Iowa State University.

Florida Cattlemen's Association. 1999. *Water Quality Best Management Practices for Cow/Calf Operations in Florida.* Kissimmee, FL: Florida Cattlemen's Association. June.

Giménez-Zapiola, M. Views on USA Graziers from an Argentine Colleague. *Grass Farmer.* 54(12) December. Jackson, MS: Mississippi Valley Publishing Corporation.

Heffernan, W. D. 1994. Agricultural Profits: Who Gets Them Now, And Who Will in the Future? Paper presented at the Fourth Annual Conference on Sustainable Agriculture. Iowa State University, Ames, IA. August 4.

Hendrickson, M. and W. Heffernan. 2002. *Concentration of Agricultural Markets Table.* February. Department of Rural Sociology. Colombia, MO: University of Missouri. Available at http://www.foodcircles.missouri.edu/consol.htm.

Hornsby, A. G., T. M. Buttler, D. L. Colvin, F. A. Johnson, R. A. Dunn, and T. A. Kucharek. 1998. *Managing Pesticides for Pasture Production and Water Quality Protection.* Gainesville, FL: University of Florida and the Institute of Food and Agricultural Sciences (UF/IFAS). Available at http://edis.ifas.ufl.edu/SS032.

National Cattlemen's Association. 1997. *The Beef Handbook—Environment.* Available at http://www.beef.org/beef_handbook.

Nations, A. 1997. Allan's Observations. *Grass Farmer.* 54(2) February. Jackson, MS: Mississippi Valley Publishing Corporation.

Oltjen, J. W., and J. L. Beckett. 1993. Estimation of the Water Requirement for Beef Production in the United States. *Journal of Animal Science* 71:818–826.

Program on Breast Cancer and Environmental Risk Factors in New York State. 2000. Consumer concerns about hormones in food. Ithaca, NY: Cornell University. http://www.envirocancer.cornell.edu.

Queensland Department of Primary Industries (QDPI). 2001. *Queensland Dairy Farming Environmental Code of Practice.* Queensland, Australia: QDPI.

Rifkin J. 1992. *Beyond Beef—The Rise and Fall of the Cattle Culture.* New York: Penguin Books USA Inc.

Ruminant Livestock Efficiency Program. 2002. Ruminant livestock and the global environment. Washington, DC: U.S. Environmental Protection Agency. http://www.epa.gov/ghginfo/topics/topic2.htm.

Runge, C. F. 1994. *The Livestock Sector and the Environment: Basic Issues and Implications for Trade.* Rome: UN Food and Agriculture Organization. July 7.

Simmonds, G. No date. Matching Cattle Nutrient Requirements to a Ranch's Forage Resource, or Why We Calve Late. Deseret Land and Livestock.

Spark, D. 1994. Using Cattle to Repair the Land. *Financial Times.* London. November 4. Page 40.

Tengnäs, B. and B. R. Nilsson. 2002. *Soya Bean: Where Does It Come From and What Are Its Uses?* A report prepared for WWF Sweden.

Turner, J. 1999. *Factory Farming and the Environment*. A report for Compassion in World Farming. Hampshire, England: Compassion in World Farming Ltd. 53 pages.

UNCTAD (United Nations Conference on Trade and Development). 1999. *World Commodity Survey, 1999–2000*. Geneva, Switzerland: UNCTAD.

USDA (U.S. Department of Agriculture). 2000. *Assessment of the Cattle and Hog Industries*. Washington, D.C.: USDA.

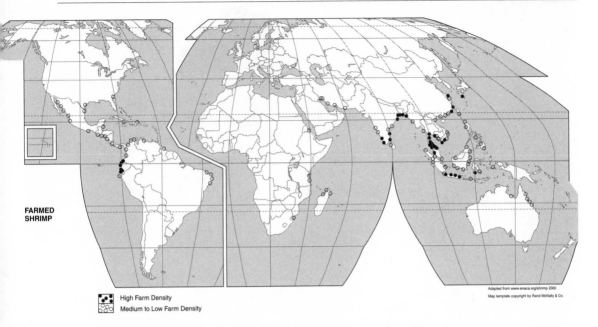

FARMED
SHRIMP

High Farm Density
Medium to Low Farm Density

Adapted from www.enaca.org/shrimp 2000
Map template copyright by Rand McNally & Co.

SHRIMP *Penaeus monodon* and *P. vannamei*

PRODUCTION		INTERNATIONAL TRADE	
Area under Cultivation	1.8 million ha	Share of World Production	91%
Global Production	1.1 million MT	Share of Exports (with wild caught)	45%
Average Productivity	611 kg/ha	Exports	1 million MT
		Average Price	$6,334 per MT
		Value	$6,334 million

PRINCIPAL PRODUCING COUNTRIES (by weight)

Thailand, China, Indonesia, Vietnam, Bangladesh, India, Ecuador

PRINCIPAL EXPORTING COUNTRIES

Thailand, China, Indonesia, Vietnam, Bangladesh, India, Ecuador

PRINCIPAL IMPORTING COUNTRIES/BLOCS

United States, European Union, Japan

MAJOR ENVIRONMENTAL IMPACTS

Destruction of coastal wetlands and mangroves
Introduction of species into new areas
Water pollution in coastal areas
Depletion of fish stocks used for feed

POTENTIAL TO IMPROVE

Good
Prices are high enough to cover BMP costs
Better practices are being identified
Declining prices encourage greater efficiency

Source: FAO 2002. All data for 2000.

Total Production (MMT)

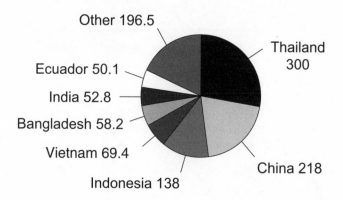

Other 196.5

Thailand 300

Ecuador 50.1

India 52.8

Bangladesh 58.2

Vietnam 69.4

Indonesia 138

China 218

Source: World Bank et al. 2002.

OVERVIEW

Aquaculture is the fastest-growing form of food production in the world. Carp and mollusks, produced mostly in China and to a lesser extent in India, dominate aquaculture production by volume and area of land used. But shrimp dominate aquaculture production by value. Consequently, shrimp aquaculture is one of the fastest-growing forms of aquaculture. From 1982 to 1992 shrimp production increased ninefold (FAO 2002). In spite of dramatic increases in production, the price of shrimp did not decline over the same period. Demand has more than kept pace with increased supply. In fact, one of the major factors contributing to unsustainable production practices, at least until recently, is that farmers have simply made too much money doing shrimp aquaculture the wrong way for anyone to be able to convince them to change. There is evidence, however, that this situation is changing.

The shrimp aquaculture industry can cause considerable damage to fragile coastal wetlands. The types of habitat that have been affected include mangroves, salt flats, mudflats, estuaries, tidal basins, and coastal marshes. While few of these areas appear to be hot spots of biodiversity, they are nonetheless essential hunting, nesting, breeding, and migratory homes to many fish, invertebrates, migratory birds, and other species. Furthermore, these coastal areas are extremely important for regulating the ecological interactions between land and water. They buffer the impact of storms and high tides on land, and they trap sediments and other organic matter on land, preventing them from choking aquatic life near the shore or deeper in the oceans. Shrimp aquaculture has often changed fundamentally the hydrology that is the basis of these ecosystem functions.

The most reliable estimates of the amount of land used for shrimp aquaculture indicate that the overall area in production at any given time is about 1.8 million hectares, and that at any one time another 300,000 to 400,000 hectares of ponds are either idle or temporarily abandoned (World Bank et al. 2002). The countries with the most land devoted to shrimp aquaculture, in descending order, are Vietnam, Indonesia, China, Bangladesh, Ecuador, Thailand, the Philippines, Mexico, Honduras, Brazil, and Colombia (Rosenberry 2000, 2003).

The top producers by weight are Thailand, China, Indonesia, Vietnam, Bangladesh, India, and Ecuador. Other countries that produce large amounts of shrimp from aquaculture include the Philippines, Mexico, Brazil, Malaysia, Colombia, Honduras, Venezuela, Taiwan, Sri Lanka, and Nicaragua. There are several tropical countries where shrimp aquaculture is being practiced on a small scale but has the potential to expand. The most rapid expansion of shrimp aquaculture, however, is occurring in countries that are already significant producers, for example, China, Indonesia, Mexico, and Brazil.

CONSUMING COUNTRIES

At this time, 75 percent of the shrimp produced in the world comes from trawling; the remainder comes from aquaculture. However, because so much of the trawled product is consumed locally, shrimp aquaculture accounts for almost half of all internationally traded shrimp. Shrimp is the most valuable traded fish product in the world today. By value, it accounts for nearly half of all seafood imports to the United States (Johnson and Associates 2002).

The main consuming countries of shrimp produced from aquaculture are the United States, the countries of the European Union, and Japan. These constitute the final markets for more than 90 percent of shrimp produced from aquaculture. Only in Japan has per capita shrimp consumption been declining for the past decade.

Shrimp is a highly differentiated product. There are more than seventy classifications of shrimp in the United States alone based on size and degree of processing. The European Union and Japan have similar classifications. Even so, most shrimp is classified only by size or processing. There is rarely any reference to species, country of origin, or whether the shrimp was produced in a pond or caught in the ocean.

PRODUCTION SYSTEMS

Globally, the number of shrimp producers has been estimated at some 400,000 (World Bank et al. 2002). These range from individual small-scale producers, to cooperatively held operations, to operations owned by corporations.

There are three main shrimp aquaculture production systems, characterized by the intensity of their resource use—extensive, semi-intensive, and intensive. In general these different intensities of production are classified according to the density of shrimp stocked in the ponds as well as the nature and type of feed used, the rate of water exchange, and whether aeration is used to increase the oxygen levels of the water.

Extensive Production

So-called extensive systems of aquaculture raise fewer than five shrimp for each cubic meter of pond water. Extensive producers tend to be characterized by their reliance on cheap land and labor, naturally occurring seed stock and feeds, and the tidal exchange of water. Individuals or families set up their operations with few inputs and little technical know-how. They construct impoundments or large ponds, of up to 100 hectares, in coastal areas where land is inexpensive. The most primitive forms of containment for extensive aquaculture consist of constructed "plugs" or dams in natural watercourses or channels that create pools or ponds.

Extensive ponds are stocked with post-larvae (PL, or juvenile shrimp) either caught in nearby estuaries or brought into ponds on the incoming tides and trapped. In these systems, shrimp eat the feed that grows in the pond. Shrimp are harvested by draining the pond and catching the shrimp as they pass through the break in the dam. Diseases are rare due to the low density at which animals are stocked in the ponds.

Because these production systems are built in tidal areas and because they are extensive relative to their yields, they contribute to considerable habitat conversion relative to the income they generate in coastal areas. Yet, because no feed is added to the system and there is little water exchange, very few effluents are put into the environment. Extensive farmers do not use chemicals or medicines of any kind. Production is less than 1,000 kilograms per hectare per year in extensive operations. Often production reaches only a few hundred kilograms per hectare.

Semi-Intensive Production

Semi-intensive shrimp aquaculture involves stocking densities beyond those that the natural environment can sustain without additional inputs. Most of these operations stock at densities of 2.5 to 20 animals per cubic meter. Semi-intensive systems depend on a regular supply of larvae, and more control over the water. Ponds with regular shapes and depth are constructed with levees or dikes and are much easier to harvest.

Semi-intensive production systems require more capital for construction, labor, maintenance, larvae, feed, and energy for pumping water. As the shrimp grow, supplemental feeding is required. Because the shrimp are stocked at higher densities, the risks from disease are higher than in extensive systems.

There are different environmental risks associated with this production system. Habitat conversion, for example, is less than with extensive systems. Other potential environmental impacts include production of nutrient-rich effluents, the use of chemicals and medicines, and increased water use. Production in semi-extensive systems can average from 1,000 to 2,000 kilograms per hectare per year (Clay 1996; World Bank et al. 2002).

Intensive Production

Intensive operations are stocked with densities that exceed 20 animals per cubic meter and sometimes reach as much as 150 animals per cubic meter. Intensive ponds are usually much smaller (0.01 to 5 hectares) and require far more inputs to maintain a healthy environment for the shrimp.

In intensive operations shrimp must be fed the entire time they are in the ponds. Because so many nutrients are being brought into the system, the water has to be aerated in order to keep the oxygen levels high enough for the shrimp to survive. In the most intensive operations production can exceed 22,000 kilograms per hectare per crop, with 2.4 crops per year (World Bank et al. 2002).

In intensive production systems there is little room for management error; the systems can crash in a matter of hours. Diseases have been a major problem with intensive shrimp aquaculture. Production averages from 3,000 kilograms per hectare per year to more than 50,000 kilograms per hectare per year depending on stocking densities, the length of time to harvest, survival rates, and the number of crops per year.

For the semi-intensive and intensive systems of production, pond construction costs range from $10,000 to $50,000 or more per hectare. Even in extensive production where family labor is often used to build ponds, the investment is considerable. For this reason there is a financial incentive for the operations to last as long as possible. This is perhaps the most important incentive for shrimp farmers to adopt better practices regardless of the scale or intensity of their production.

The single most intensive shrimp operation today is run by a company in Belize, but on average, Thailand has the most intensive shrimp aquaculture countrywide. Other Asian countries often have a mix of extensive aquaculture practiced by undercapitalized producers and intensive aquaculture practiced by larger, better-financed operations. Semi-intensive production dominates Latin America. There are, of course, exceptions to these generalizations.

Virtually no domestication has taken place for any of the species produced in aquaculture systems. About 2 percent or less (the proportion is declining) of all producers depend on wild-caught post-larvae to stock their ponds (World Bank et al. 2002). This is most common in Bangladesh, India, and Ecuador where hatcheries are not required by law. Globally, some 98 percent or more of all post-larvae used by the industry are produced in hatcheries. The vast majority of hatcheries still depend on the capture of wild brood stock. Breeding programs are just beginning. For the first time, brood stock are being hatched and raised in captivity. This provides an opportunity for selective breeding for specific characteristics that will improve the overall performance of the different shrimp species when produced by aquacultural methods (World Bank et al. 2002).

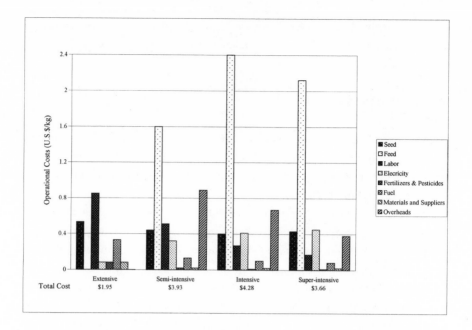

Figure 22.1. Comparison of Operational Costs of Four Types of Shrimp Aquaculture Productive Systems.

Figure 22.1 compares the operational costs of four different intensity shrimp aquaculture production systems—extensive, semi-intensive, intensive, and super-intensive. While the extensive system has the lowest costs (labor and seed being the two largest costs), it is also the least productive. For the other three systems, the most expensive single input is feed, which represents from 40 to 60 percent of all costs. For these systems, profits will depend on the ability of producers to reduce or use more efficiently the various inputs.

PROCESSING

Most of the larger shrimp producers (e.g., those with 300 hectares or more of semi-intensive operations, or their equivalent in intensive production) have their own on-farm processing facilities. Smaller producers sell their production to nearby processing plants. Globally there are probably thousands of processing companies, but the market is dominated by no more than 100. The value of the production from processing is about $7 to $8 billion depending on the amount of value added to the product (World Bank et al. 2002).

The most important issue that arises from shrimp processing is the creation of waste. More than 40 percent of the weight of shrimp, including the heads, is waste. Removing the heads creates a disposal problem. This material can be ground up and used for fish meal or animal feed. The shells can also be used for other processed items. Unfortunately, in many processing plants most of the heads and other waste material are dumped into nearby water bodies. This can spread disease and cause nutrient loading and biological oxygen demand (BOD) problems.

Increasingly, shrimp are shipped frozen with heads on and are then processed in other countries under Hazard Analysis and Critical Control Point (HACCP) approved conditions. It has been proven, however, that frozen shrimp can also carry diseases. This is of particular concern when frozen shrimp act as vectors to bring diseases into new regions (e.g., from Asia into the Americas or vice versa). While no one has documented the impacts of such disease transmission, there is some concern that white spot (a common disease in Asia) was brought to the Americas, where it has caused billions of dollars of damage in the past two years (World Bank et al. 2002).

When shrimp are first harvested, they are soaked in metabisulfite, an antioxidant, to increase their shelf life and reduce the activity of oxygen on the shrimp's body or shells. This chemical is used because it helps to prevent spots from developing, and any spots reduce the value of shrimp. After soaking, the chemical is dumped and readily finds its way into natural water systems. Its impact has not been studied. However, since the chemical is used to retard bacterial growth, it is presumed to have some impact in natural ecosystems.

SUBSTITUTES

The main substitute for shrimp produced from aquaculture is shrimp that are caught in the wild. At this time, the largest shrimp cannot be produced in aquaculture systems; they can only be captured from the wild. Smaller sizes do tend to be substituted for each other, but aquaculture shrimp are generally cheaper to produce and therefore cause tighter profit margins for the wild-caught shrimp rather than the other way around.

Other seafood items are perhaps also functional substitutes for shrimp. Few wild-caught fish, however, are available year-round, as consistently cheap, as sought out by consumers, or as available in many different sizes. All of these factors make shrimp ideal for direct purchase by consumers or for purchase by larger institutions and restaurants. In effect, shrimp has become the most commonly consumed fresh or frozen seafood in the United States, Europe, and Japan because it can be favorably substituted for so many other seafoods.

MARKET CHAIN

Figure 22.2 shows the market chain for Ecuadorian shrimp of one size as they move from the producer to the consumer. The size and species has some impact on the value in the market chain for shrimp, but this chart gives a good indication of how and where value is added as shrimp moves through the system.

What the chart does not indicate is the number of players involved at each stage of the operation. There are some 400,000 shrimp aquaculture producers. While there have been a few attempts by corporations to build central facilities and work with many small producers, this is not the norm. Most producers are independent. In countries where financial crises have occurred (e.g., Indonesia, Thailand), input suppliers have extended their goods as a form of credit and producers are then

Headed Shrimp Market Chain, 1996
Ecuador Whites (21–30)

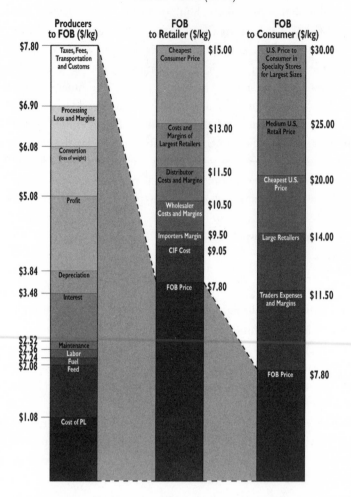

Figure 22.2.   The Components of Shrimp Price at Each Stage of
the Market Chain from Producer to Consumer.

required to sell to certain buyers and/or processors. In general, however, direct involvement in production is seen as too risky by most of the feed, seed, or processing companies. They want access to the product, but they do not want to produce it themselves.

There are probably 1 to 1.5 million people employed directly by the industry. Most research shows that they tend to be paid double, or more than double, the going rate for labor in their areas. Consequently, the turnover rate for labor is relatively low. Another million people depend on the industry for a major portion of their livelihood through the capture of brood stock and wild shrimp post-larvae, the latter of which are used to stock 20 percent of all the shrimp aquaculture operations in the world. Employment for these people is seasonal and earnings are much lower. This work tends to be dominated by women and children and is most common in Bangladesh, India, and Ecuador, where there is no law requiring shrimp hatcheries in order to protect local fisheries during post-larvae collection (World Bank et al. 2002).

In general, the overall employment multiplier effect (i.e. how many jobs are created in the economy as a whole for each job in the shrimp industry itself) of the industry is probably five or six, particularly if employment in input and processing industries is included. This would mean a rough estimate of some 5 to 9 million jobs. Since the industry is only twenty-five years old, and much less in some countries, good data does not yet exist on this topic.

The number of people involved declines through the processing and export/ import activities and then begins to increase again through distributors, retail outlets, and restaurants before reaching the end consumer. For example, there are probably a few thousand shrimp processors in the world. Many of these also process wild-caught shrimp. There are hundreds of exporters and probably hundreds of importers. Some are much larger than others, however. Ocean Garden Products, for example, is estimated to import 25 percent of all shrimp into the United States (Johnson and Associates 2002). Often one or two companies will dominate imports and distribution, even though there may be a number of other smaller companies as well. There are more distributors than there are importers, and more retailers than wholesalers. Globally there are probably thousands of distributors, tens of thousands of wholesalers, and millions of retailers who sell shrimp produced by aquaculture. There are far more retailers and restaurants, for example, than there are shrimp producers.

Relatively little shrimp is purchased and prepared directly by the consumer in the main consuming countries. In general, the main buyers of shrimp in the United States, Europe, and Japan are institutions (e.g., restaurants, educational institutions, government facilities, the military, etc.). They purchase some 70 percent or more of all shrimp sold in these markets. In the United States, Darden (owner of the Red Lobster and Olive Garden restaurant chains) purchases half of all shrimp used by the top ten restaurant chains. SysCo, a large food distribution company, reportedly buys half of all the shrimp distributed in the United States. These institutional buyers and distributors are important because only one or two individual purchasers for each make the decisions about which shrimp will be purchased for millions of consumers who subsequently eat in restaurants. Diners are not given the choice of which kind of shrimp produced in which way they would like in their meal. Strategically, then, this is the most efficient place to address consumer issues (Clay 1996; Johnson and Associates 2002; David and Lucile Packard Foundation 2001).

MARKET TRENDS

The main market trends in shrimp production are increased production, increased demand, and increased interest in product quality. To date, there has been a constantly increasing supply of shrimp from aquaculture and wild-caught sources, and there has been a fairly continuous decline in prices. While there have been occasional instances of short-term increases in prices due to disease outbreaks, within a year or two the prices continue their downward trend (Rosenberry 2003).

However, as shrimp producers adopt better practices, production will both expand and become more stable. This will cause a continuation of overall price declines and will force shrimp producers, like other commodity producers, to intensify and increase their scale to work on smaller margins. An important countertrend that could tend to lift prices would be if the overall price of shrimp became too low and the fleet for harvesting wild shrimp were reduced.

Since September 11, 2001, shrimp prices have declined. This does not appear to be a short-term fluctuation but rather the beginning of a trend. Price declines initially led to speculation (e.g., purchasing and holding the product, waiting for the price to rise), which has caused prices to decline even further and stockpiles to increase. This will affect markets at least through 2004, but most analysts do not believe that the markets will ever recover fully.

Another trend of note is the increasing concerns regarding chemical residues found in shrimp. Shrimp from China and Vietnam have been confiscated and destroyed in the European Union due to the discovery of chloramphenical, an antibiotic. As a consequence, shrimp shipments are under increasing scrutiny in the European Union as well as Japan and the United States and for a wider range of contaminants (including PCBs, dioxins, and the antimicrobial drug furazolidone).

ENVIRONMENTAL IMPACTS OF PRODUCTION

As with any production system, there are hundreds if not thousands of environmental impacts from shrimp aquaculture. However, there are six to ten key environmental impacts from shrimp aquaculture worldwide, with only three to five being significant on any given farm. The extent of the impacts depends on the intensity and scale of production, the laws of the specific country, and when and where the ponds were built. The most important impacts are discussed below.

Siting and Coastal Habitats

There is good evidence that as many as 90 percent of all of the environmental problems from shrimp aquaculture arise from where the ponds are built (Boyd and Clay 1998). The impacts include the loss of habitat, interference with ecosystem hydrology, and loss of coastal barriers that prevent storm damage. Siting can also accentuate the eventual impacts of on-farm management practices. If an operation is not built in the right place it will be hard to get to, and roads and other necessary infrastructure may cause problems unnecessarily. Poor siting can also lead to poor water quality or the inability to evacuate effluents. Siting is also the single largest source of conflicts with local communities and other resource users.

For those who can afford to buy the best-suited pieces of land for shrimp aquaculture, the overall siting of operations is not generally an issue, at least from an environmental point of view. In addition, more is known today about the best sites for shrimp aquaculture. New ponds established after a search for better sites are not generally a problem. However, for those small-scale producers who are trying to grow shrimp on the piece of land they already own, this can be a very important issue.

Without simply closing poorly sited ponds or even entire operations in some instances, it is very difficult to address their problems. Some problem sites can be remedied if the ponds are relatively small and if the water can be pumped to another containment area or discharged into a canal where it can be treated.

New ponds that are built after the best sites have already been developed can also pose environmental threats. By themselves, many such operations would probably be able to operate within acceptable environmental parameters. However, the cumulative impacts of having many operations built in the same area often exceed the local carrying capacity. This is most often a problem when producers take chances on developing more marginal areas because they are the only unoccupied sites available in an area where everyone is apparently getting rich from producing shrimp.

Shrimp farming alone appears to be responsible for some 5 to 10 percent of the global loss of mangrove habitat (Boyd and Clay 1998). Yet in some countries it has caused as much as 20 percent of the damage to mangrove areas, and in some watersheds shrimp farming accounts for virtually all mangrove destruction. In part this situation resulted from the advice of experts who initially thought that because shrimp spend part of their lives there, mangrove habitats were the best sites for shrimp farms. Over time, it became clear that the acid soils of mangrove habitats were not suitable for shrimp ponds. But the damage to these ecosystems had already been done.

Another factor complicating this issue was that many development agencies wanted to help the poor. Building shrimp ponds in tidal areas saved farmers the expense of water pumps and long-term pumping costs. Today, while it is generally known that mangrove areas are not the best sites for shrimp ponds, farmers will still build there because land is cheap or free, there are no conflicts with other agriculturalists, and profits from even a couple of years make the efforts attractive financially.

Shrimp aquaculture has also been responsible for the conversion of other fragile coastal habitats in the tropics. Little attention has been given to the loss of mud and salt flats, coastal estuaries, and wetlands to shrimp aquaculture. It is now known that while such areas are not the permanent homes for much biodiversity, many are seasonally quite important in the life cycles of some species. Furthermore, they are important from an ecosystem point of view.

A few hundred thousand hectares of coastal areas may have been converted to shrimp farms that failed and were then abandoned. These areas include a wide range of former habitats. Unfortunately, most failed shrimp operations are not required to recreate the hydrology of the coastal areas that they degraded. In many cases, simply opening the dikes so that the water could flow would reestablish the areas again.

Capture of Larvae and Brood Stock from the Wild

Most shrimp farms depend on the capture of brood stock (adult females and males that are captured and spawn in captivity) from the wild, while only about 2 percent of all post-larvae used to stock shrimp ponds are captured in the wild. Little is known about the impact of these activities. However, there is some evidence that for every captured post-larva, twenty to forty other living organisms are killed (Boyd and Clay 1998; World Bank et al. 2002). Post-larvae are captured near shore and are taken to land to sort. By-catch is discarded on the land, where it dies. For an industry that depends on more than a trillion post-larvae annually, this could be a significant issue. However, little is known about recruitment of the different species in the surf areas where post-larvae are harvested. Because brood stock must be taken live from the ocean (and are caught at night), much more care is taken and by-catch is not seen as a serious issue.

Introduction of Nonnative Species

Little is known about the overall impact of the introduction of shrimp species from aquaculture. For some time, the species *Penaeus vannamei* from the west coast of Latin America has been farmed along the Caribbean and Atlantic coasts from South Carolina to Brazil and more recently in Asia. Likewise, shrimp of the species *P. monodon* from Asia have been transported throughout Asia and brought to Latin America. *P. monodon* shrimp from Africa have been taken to Asia and the Pacific, and there has been a flow of this same species from Southeast Asia to South Asia and vice versa. Escapes, however, have not yet shown up in wild shrimp catch.

The introduction of shrimp from different regions, even of the same species, introduces new DNA and characteristics that have not evolved in situ. These interactions are probably insignificant within ponds, but when shrimp escape during water exchange or harvest they could cause genetic pollution that could alter the inbred characteristics, and perhaps the viability, of wild populations.

The introduction of disease pathogens from other areas is equally important. Diseases previously found only in Taiwan and China have now spread throughout Asia and even into Latin America, where they have caused billions of dollars in damage each year. The impact of disease pathogens on wild stocks is not documented, but anecdotal information suggests that it may be serious. For example, in 1992–93 when diseases reduced shrimp aquaculture production in China by 60 to 70 percent, the production of wild-caught shrimp in that country also declined by 90 percent. It is not clear whether the disease was transmitted from the wild to the ponds or vice versa, but there does seem to be some direct relationship.

Pathogens can be introduced through the transportation of infected larvae or brood stock that are released without proper quarantine and handling. In addition, diseases have been found to be viable in processed frozen product that is shipped to another region for further processing.

Pollution from Effluents

Pollution from effluents comes from many sources in shrimp production. Perhaps the most important is the feed. Semi-intensive and intensive production systems require about 2 kilograms of feed to produce 1 kilogram of shrimp. This feed averages the same space, is another possibility for reducing overall effluent load in the water. This can either be done in the shrimp ponds themselves (which is difficult in commercial operations attempting to maximize shrimp production) or undertaken sequentially with the effluent so that different species are part of the treatment process. At this time simultaneous shrimp polyculture systems produce shrimp and fish such as tilapia (an omnivore) or milkfish (an herbivore). Sequential polyculture systems tend to focus on bivalves, seaweed, etc. In some areas, sequential polyculture is similar to crop rotation, as different species (e.g., milkfish, tilapia, red claw, or bloodworms) are produced as alternate crops in the same ponds. This also functions as a form of fallowing. The same water could be recycled into these ponds. Polyculture is practiced in several countries, but it has become most common in countries that have disease problems. In general, polyculture is still not financially viable for most large-scale, commercial shrimp producers. However, as farmers begin to understand better the costs of disease and pollution, or as they are required by government to improve their performance or pay fines, this situation may change.

Encourage Domestication of Shrimp

The collection of wild post-larvae should be prohibited. This will eliminate by-catch issues and is an important step toward closing the production system. Brood stock will still be captured in the wild, but most scientists agree that their capture has little impact on the marine environment. There is no by-catch from this activity, and the numbers of individuals involved are so small as to have an insignificant impact on populations and reproduction even in a local area.

Domestication is the best way forward. This will entail raising the brood stock in hatcheries. Selective breeding programs can help develop animals that convert feed better, require less protein in general and fish meal in particular, perform better under stress, mature and gain weight more quickly, and have disease resistance. It is even possible that domestication programs could be used to make shrimp dependent on specific trace elements so that they could not survive if they escaped from the ponds. This would have a positive impact on local biodiversity.

Regulate Introduction of Species

There are several protocols on the shipment, handling, and quarantine of nonnative species. Many shrimp are now not native in the areas where they are being produced, so producers, their associations, and governments should insure that escapes are eliminated. In this regard, shrimp producers should be required to use fine-meshed screens when bringing water into their ponds or releasing it back into natural water bodies. This would prevent all but the tiniest organisms from entering the aquaculture system from the outside, and it would prevent similar organisms from the pond escaping into the wild.

Reduce Use of Chemicals and Medicines

Shrimp producers should not use chemicals or medications prophylactically. Furthermore, they should not use any chemicals or medications that are banned in the countries that buy their product. The use of such items should be in response to specific problems that have been identified. When chemicals or medications are used, the water should be held for the amount of time required for that substance to break down before it is released into the environment.

Most informed producers realize that the routine use of medicines can create resistance so that the same medicine will not be effective when it is needed. It is not clear that smaller, less-educated producers understand this concept. Any problem is accentuated when illiterate producers cannot read the labels. The real problems arise during periods of new disease outbreak when most producers will do anything to protect the animals and their investment. During such periods, a wide range of medicines and home remedies are tried to see if any will help reduce the risk of total crop loss.

The best way to maintain healthy ponds is to avoid overstocking or overfeeding. Good health is directly related to reduced stress or the conditions that lead to it. Consequently, the most profitable operations are those that have found ways to promote better management by creating worker incentives that are tied directly to monitoring and maintaining the health and density of the animals stocked, and to reducing the feeding levels to what is actually eaten so that water quality can be maintained. Feeding trays help producers monitor overall feed intake as well as feeding habits. Some farmers put all their feed in feeding trays (which requires much more labor); others only use feeding trays as an indicator of what is happening in the pond.

More intensive shrimp producers tend to fertilize their ponds to stimulate growth of aquatic organisms that young shrimp eat. Over time in such operations, the natural feed in the water column is supplemented with manufactured feed. Historically, traditional agricultural fertilizers were used. There are some attempts to use "bokashi" (fermented organic matter that includes effective microorganisms) as a substitute for inorganic fertilizer. The beneficial microbes in the bokashi digest the organic matter, which in turn promotes the development of plankton and diatoms as natural feed for the shrimp (Panfilo Tabora, personal communication).

Reduce Use of Fish Meal

Most shrimp farmers are becoming more efficient in managing the feed conversion ratios (FCRs) on their farms. As a consequence, they are using less fish meal. A smaller number of producers are actually trying to reduce not only overall feed use, but the actual proportion of fish meal in the feed as well. Ideally, farmers should produce their shrimp with an amount of fish meal that represents a weight of wild fish that is equal to or less than the shrimp being produced. It may take some time to achieve this ideal throughout the industry, however.

their title to rivers, ownership was turned over to friends of the King who worked the fisheries for profit. By 1669 some £200,000 of Scottish salmon were exported annually (Netboy 1974). After the union of Scotland and England in 1707, the trade in salmon increased even more, aided by improved transportation links.

In the seventeenth century, the Dee and Don rivers in the United Kingdom produced 170 metric tons of salmon annually, with exports going to Germany, Spain, Portugal, Holland, and even as far away as Venice. The local price for salmon was 2 pence, but in London they sold for 6 shillings and 6 pence, a 36-fold increase. By the early eighteenth century, the annual rental fee for the salmon and eel fishers of the lower River Don alone amounted to £30,000 (Netboy 1974).

There have long been efforts to maintain salmon production. In the Middle Ages, river managers developed wise-use practices to protect salmon runs. Kings in England and Scotland forbade blocking migratory routes and even taking fish in what they believed (usually erroneously) to be the spawning season. In 1030 King Malcolm II of Scotland established a closed season for salmon from the end of August to Martinmas (November 11). Richard the Lionhearted made a statute that all rivers must have a free-flowing gap in the middle of at least the length of a three-year-old pig (Netboy 1974).

Under Edward XIII in 1285 salmon were in decline in some areas, and the idea of closed seasons was introduced in England. By 1376, only authorized nets were allowed to take salmon. Under Elizabeth I, only fish longer than 41 centimeters (16 inches) were legal catch. Even so, by the early 1800s excessive netting in estuaries, damming of waterways to provide power to mills, pollution from the industrial revolution, and raw sewage dumped by pipes from growing cities all took a toll on the salmon population. Off seasons, weekly closings, mesh and net size regulations, and even water bailiffs could not prevent the demise of the fishery from these larger environmental impacts. The last salmon was caught on the Thames River in England in 1833.

Even in more remote Scotland, pollution was a problem. By 1850 there were eleven distilleries on the Spey River that consumed 2,270 barrels of malt in addition to grain and other organic material to make whiskey; the waste from this whiskey production was subsequently released into the river. By 1900 twenty-seven distilleries consumed 50,000 barrels of malt a week as well as all the other organic ingredients used in whiskey. As early as 1861 the situation for salmon was so bad that Charles Dickens wrote "Salmon in Danger" in his weekly magazine *All the Year Round* (Netboy 1974).

Salmon felt the impact of environmental degradation in North America as well. By 1900 Atlantic salmon populations in the United States ran afoul of industrial and sewage pollution and damming, as had affected England and Scotland, as well as agricultural expansion and soil erosion. At that time, salmon had become extinct in both the Salmon River and Lake Ontario, and the Hudson River no longer had a viable commercial salmon fishery.

In only a short time, salmon have gone from abundance to depleted stocks to abundance again. What has made this possible is aquaculture and the ability of humans to farm the seas. In only a generation, salmon has become semidomesticated through intensive breeding programs. While farming took some 6,000 years of learning to get where it is today, salmon aquaculture has been created in only thirty! And a food once affordable only by kings is now available to most families at home or in restaurants every night of the week.

Making salmon a commodity has not come without costs. There have been steep learning curves, and many believe that salmon production is not sustainable as it is now practiced. Even so, vast improvements in production techniques have been made. More importantly, what has been learned from salmon aquaculture (and the problems it raises) is relevant to most other overfished or depleted fish stocks whose status is now similar to salmon when it first started being produced through aquaculture.

Today nearly one-third of fish eaten in the world is produced by aquaculture. During the 1990s aquaculture production increased by 150 percent (Johnson and Associates 2001). The performance of salmon aquaculture was similar. In 1980 farmed salmon made up a negligible percentage of world salmon supply, but by 2000 more than 1 million metric tons of total production and more than 50 percent of global salmon supply was farmed (Johnson and Associates 2001). (See Figure 23.1.) Atlantic salmon is the fastest-growing high-value farmed species, with annual production exceeding 1 million metric tons (Packard Foundation 2001).

PRODUCING COUNTRIES

In 1999 farmed salmon production surpassed wild salmon catch for the first time ever. In 2000 farmed salmon production exceeded 453 million kilograms (1 billion

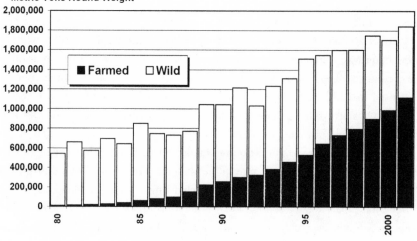

Figure 23.1
Source: Johnson and Associates 2001.

pounds), and by 2002 farmed salmon was expected to account for more than two-thirds of total salmon production. In other words, two farmed salmon would be produced for every one caught in the wild.

Aquaculture production of Atlantic salmon has long been dominated by Europe, where it was invented. Total farmed salmon production reached 668,601 metric tons in 2001 (Aquamedia 2002). Norway was by far the largest producer at 460,000 metric tons, followed by Scotland (140,000 metric tons), the Faroe Islands (40,000 metric tons), Ireland (23,000 metric tons), and Iceland (5,600 metric tons). Other important global producers included Chile with 245,000 metric tons and Canada with 75,000 metric tons (Johnson and Associates 2001). In Chile production has grown at some 9 percent per year for a decade or more, while prices have dropped a total of 30 percent. In 2002 Chile surpassed Norway as the world's leading producer of farmed salmon.

CONSUMING COUNTRIES

In 2001 sales of farmed salmon amounted to more than $2 billion (Packard Foundation 2001). The European Union, Japan, and the United States are the three biggest markets for the product. In 1998, for example, the United States imported more salmon than it exported for the first time in history. In the United States, per capita salmon consumption increased 285 percent from 1987 to 2000, rising from 0.2 kilograms per capita to 0.73 kilograms per capita. Salmon ranks behind shrimp and tuna as the third most commonly consumed seafood in the United States (Anderson 2001). Although per capita consumption of all seafood remains relatively stable in developed countries, total consumption is increasing as a function of population growth (Packard Foundation 2001).

PRODUCTION SYSTEMS

The biggest challenge for salmon aquaculture is that producers have had to find ways to produce a wild species in captivity. This includes not only growing fingerlings, but also producing them in hatcheries from captive brood stock. It also means finding the right feed formulations to insure growth, flavor, and color and still be financially viable; identifying ways to treat diseases as they arise from confining animals; and doing all this while reducing or mitigating impacts on the environment and other species. Salmon aquaculture falls into two categories: hatchery production and grow-out operations. Wild salmon are also discussed to provide context.

Wild Salmon

Wild Atlantic salmon spend most of their adult life in the ocean but return to fresh water to reproduce. In fact, 99 percent return to the river where they were spawned and reared. Each female produces about 800 eggs per pound of body weight so an average salmon of 9 to 10 pounds produces some 7,000 to 8,000 eggs. Unlike

Pacific salmon that die after spawning, Atlantic salmon can spawn as many as four or five times and live eleven years.

Once hatched, the majority of wild Atlantic salmon stay in fresh water for two years (a smaller number stay three or more years in fresh water) before migrating to the estuaries and open oceans. Salmon must undergo profound physical changes to adapt from a fresh to a salt water habitat, at which stage they are known as smolts. Size, rather than age, appears to be key to this transition. How long they stay in fresh water depends entirely on how fast they grow, and how fast they grow depends on how far north they are spawned and how abundant food is. In some cases, the small smolt will remain in freshwater streams for as much as seven to eight years until they grow to the size at which their bodies change to live in salt water.

Hatcheries

Salmon aquaculture production mimics, but compresses, the life cycle of wild salmon. As Fred Whoriskey (2000) describes it, salmon production starts in freshwater hatcheries. The development of fertilized eggs is typically accelerated by the use of heated water so that the fish hatch in February. Salmon are carnivores. As young fry in the wild, they eat plankton, insects, and eventually sand lice, herring, capelin, shrimp, and other fish. In captivity salmon must be fed a balanced diet starting as soon as they absorb their yolk sacs. The fish are reared at high densities in tanks. Larger tanks are used as the fish grow. Liquid oxygen is often injected into the water until the young fish reach smolt size (60 to 125 grams or larger). As juveniles, salmon are vaccinated against a variety of diseases. Each fish is injected individually, by hand, and vaccine formulations often carry antigens for four or more major diseases. Though it usually takes two years for salmon to reach the size at which they adapt to salt water, that length of time is not feasible for hatchery-produced fish. Given the costs of producing salmon in aquaculture and the current market price, the goal of breeding salmon is that the offspring can make this transition at one year or even less.

In general, smaller producers buy their product when it is time to stock their net cages. Large companies tend to maintain adult brood stock and sell eggs or recently hatched animals. The sales from hatcheries are more profitable than producing mature salmon. Moreover, well-run hatchery operations offset the cost of stock through profits from the sale of excess production.

Breeding programs have been publicly supported to create salmon with genetic characteristics that make them perform better within standard aquaculture operations. The artificial selection of salmon has begun to change their genetic characteristics. Genetic work has been undertaken in Norway, the United Kingdom, and Canada.

Different countries have instituted different laws regulating imports of eggs, sperm, and live fish. Some countries allow the import of eggs, sperm, and stock, usually from Europe. The United States used to import eggs, sperm, and stock; howev-

er, since most such imports were banned ten years ago, these items are now produced domestically.

To date, there have been no controlled scientific trials, in North America at least,
to compare the performance in culture of domesticated North American salmon
lines versus wild salmon, hybrids, or European lines. This issue is important because
European lines have tended to dominate the aquaculture industry. However, because
of the Gulf Stream, European growing conditions are far milder during the winter
than in many other salmon-producing regions. For this reason, researchers have been
experimenting with genetically modifying salmon. To allow salmon to grow in the
winter in less favorable climates than Europe, DNA from the Arctic char has been put
into salmon as a kind of "antifreeze." To date, no country has allowed transgenic
salmon to be produced commercially. Some industry players are interested because
transgenic salmon would not only grow faster and have a shorter time to market, but
also they could be grown on farms in far less hospitable areas near the poles.

Grow-out Production Cages

When the smolts have made the transition to salt water, they can be stocked into
containment areas in the ocean. Farmed salmon are most commonly grown in cages
or pens in sheltered coastal areas such as bays or sea lochs. The cages are designed to
hold salmon but are open to the marine environment. These tend to be large, float-
ing mesh cages. While a wide variety of brands and sizes exist, the trend historically
was for cages to get bigger. Now just the opposite is true; cages are getting smaller so
that operations can become more efficient. Mesh size generally starts at 1.9 centimeters (0.75 inch) stretched net and is changed periodically as the fish grow (to
2.54-centimeter or 1-inch and then 3.81-centimeter or 1.5-inch mesh) to improve
water circulation. The water circulation improves oxygen levels and washes away feces, uneaten feed, and other waste. At this time, waste disposal costs farmers nothing.

Large steel cages with mesh nylon nets are usually laid out in double rows. A typical salmon farm in British Columbia has between eight and twenty cages. Cages are
usually 30 meters square by 20 meters deep.

Smolts are used to stock the cages. The cages are stocked at high densities. The
number of fish that are stocked in a net cage varies depending on the age of the fish
and the size of the net cage. In many operations 180,000 to 250,000 animals per
cage are stocked initially. Formerly, harvests from single net cages were often on the
order of 160,000 8-kilogram (9-pound) animals. Now the overall size of net cages has
peaked and is decreasing.

A site includes all the net cages that are supported with a common structure and
feeding system. Multiple sites are often owned by a single farm or company. Some
countries even limit the total biomass per site. Norway, for example, does not allow
more than 300 metric tons of fish at any site. In Canada, however, sites have nearly
800 metric tons of fish (Ellis and Associates 1996). The size of operations at sites in
Chile can be considerably larger.

Farmed salmon are raised in high densities, which results in rapidly spreading diseases and severe problems with parasites such as sea lice. Sea lice were once controlled by pouring toxic substances into pens. They can now be suppressed through special additives in salmon food or by co-stocking fish species that eat lice. Other diseases are also common. For example, furunculosis is a bacterial disease against which salmon are vaccinated. Infectious salmon anemia (ISA) is a viral disease for which there is no treatment other than culling infected and exposed fish.

In addition to the grow-out operations, most farms also have a two-story float house that serves as a lab and storage area on the bottom floor with worker accommodations above. If farms have electricity, a generator is usually housed in the float house. Most operations have separate storage facilities for their feed. Feed deliveries take place every week or two, and feeding systems are nearly all automated at this time. Increasingly, feeding operations have photo sensors at the bottom of the net cages to determine when salmon have stopped feeding so that the feed system shuts down automatically to avoid wasting food.

Galvanized steel gangways typically provide access to all net cages so that it is easier to observe operations and to make any necessary repairs or adjustments. The walkways tend to be built on plastic barrels and located about 0.5 meters above the water level (Ellis and Associates 1996). The pens have upright supports around the edge that extend 1.5 meters and form a safety barrier. Nets also extend over the cage to prevent fish from jumping out of the cages.

All net cages are anchored to the bottom, to beaches (when near land), or to both. Anchor lines may extend more than 100 meters, depending on the depth of the water at the site. Buoys are used to mark the location of anchors and lines.

Algal blooms can affect salmon aquaculture production systems. In the wild, fish can swim away from potentially toxic algal blooms. When confined in net cages, however, that is not possible. Some production areas with more moderate summer temperatures are more susceptible to blooms. In salmon-producing areas susceptibility can increase due to the nutrient-rich environment around net cages that is created from the feces and feed waste. Producers in areas where this is a consistent problem must monitor the situation closely in order not to lose their entire crop.

A number of wild animals including seals, river otters, sea lions, herons, and kingfishers commonly attempt to take salmon from net cages. Most operations have developed various defenses to convince animals that it would be easier to eat elsewhere, and these often work. However, some individual animals will become persistent problems when they simply choose to live off of the fish in the salmon operations rather than continue their normal seasonal migrations. These animals are destroyed. Depending on the animal and the country, permits may be required before this can happen. Of course, if operations are built in or near areas that are traditionally used by specific species, then conflicts will be more intense. It is not clear what the impact of salmon aquaculture production is on other wild animal species.

The number of salmon-farming operators in some countries has increased, but the average size of operation has increased far more dramatically. If anything the trend is toward a smaller number of much larger producers. Scotland exemplifies this trend. In 1994 only 19 percent of farms produced more than 1,000 metric tons of salmon. By 1999 the figure had risen to 59 percent. During the same period the number of farming operations declined by 29 percent. Even more consolidation is likely given that by 1996, 106 of the firms operating had a combined production of only 4 percent of the total. Another trend represented by Scotland is that 47 percent of output of Scottish salmon was produced by foreign-owned companies (Berry and Davison 2001). In 2000 William Crowe, general secretary of the Scottish Salmon Growers, remarked, "The fundamental economics of this industry mean that one can envisage that there would [eventually] be five or six large global companies" (Berry and Davison 2001).

There are other production trends of note as well. In Scotland, as elsewhere, productivity has increased from 39.8 metric tons per worker in 1993 to 97.2 metric tons in 1999 (Berry and Davison 2001). Production has also become more intensive. However, the size of individual net cages has not been increasing. If anything, after some initial attempts to increase their size the average size has decreased so that each unit can be managed more efficiently.

Production Costs

The operating costs for salmon farming, while variable, generally fall into the following proportions. Feed is the largest single expense at about 34 percent of total costs. The smolts for stocking represent about 23 percent of the cost with wages, overhead, and depreciation amounting to 13 to 15 percent each. It is these latter three costs that give Chile a comparative advantage over Norway and other producers in developed countries. Efficiency of feed manufacture and use as well as hatchery operations are what has allowed Norway to remain competitive with Chile. In Norway, operating costs fell significantly between 1985 and 1999 (see Figure 23.2).

Labor is a significant expense of salmon farming. Over time producers have automated their production as a way to avoid high labor costs. In addition, they have shifted from full-time employees to part-time employees. By 2000, for example, Chile employed 15,000 full-time workers directly in the salmon industry and another 8,000 as seasonal employees (Claude and Oporto 2000). In Chile some twenty-one people were originally employed per production center (a cluster of net cages), whereas today only eleven are. The main changes have come about from the use of automated feeding systems and the use of prefabricated PVC (polyvinyl chloride) cages instead of plastic and wood ones that had been made by hand on the farms. What this means is that in Chile some 40 percent fewer people are employed by the industry than in 1998. From 1990–95 salaries decreased as a percentage of total costs

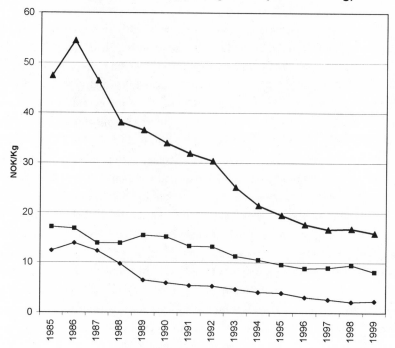

Norwegian Salmon Operating Costs (1997 NOK/Kg)

Figure 23.2
Source: Norwegian Directorate of Fisheries 1999, as cited in Anderson, no date.

of production by 8.4 percent, and taxes decreased by 3.6 percent. Profits, by contrast, increased by 11.9 percent during the same period (Claude and Oporto 2000).

The cost of raising salmon, of course, depends on the size of the operation. Over the years, costs have declined. By the late 1990s production costs in the United States were about $4.40 per kilogram, while they were only $3.28 in Norway. Since that time costs have continued to decline.

Net-cage culture is vulnerable to storms that can break anchors or mooring lines. This can result in damage to equipment or crops and even in escapes. Insurance for crop, plant, equipment, and liability ranges from 2.5 to 5 percent of fish inventory values at any time (Ellis and Associates 1996).

PROCESSING

For five days before harvest, salmon are not fed. This is done to empty the gut, reduce the fat on the animal, and firm the flesh. The fish are collected in baskets or pumped out of the net cages with fish pumps. Fish are killed by tranquilizing them with carbon dioxide or salt brine and then bled through a cut near the gill arch. In Chile only 12 percent of processors treat their blood-laden discharge water.

Farmed salmon are usually sold fresh with the heads still on; they are shipped on ice in 27-kilogram (60-pound) Styrofoam boxes with ice or gel packs. All net-cage salmon are graded based on texture, color, and factors such as oil content, according to grading standards developed by the salmon farming industry (Ellis and Associates 1996). Depending on the size and the location of the production facility, salmon may be processed nearby or at some distance from the farms. Once processed, it is common for salmon to be repackaged at larger distribution centers. Since salmon are sold fresh, they are often shipped by air from more distant production areas, and this of course increases producer/processor prices. Most of the larger salmon producers also have their own processing plants. Smaller producers are forced to sell into the processing plants of larger companies. Processors have the ability to trace salmon back to specific farms and often to specific net cages.

Farmed salmon processors have developed value-added products such as boneless and skinless fillets, salmon burgers, complete dinners, premarinated steaks and fillets, precooked portions, and breaded steaks. Some of the products can be made with trimmings or other "waste" products with far less market value than whole fish. Only salmon burgers have gained a tiny foothold (a quarter to half a million kilograms sold per year). The production of value-added products is hampered by the fact that salmon flesh is difficult to stabilize, and the cost of new frozen brand products at retail is prohibitive for most companies.

SUBSTITUTES

Any fish or even meat protein is, at some level, a substitute for aquaculture salmon. However, even with the price of salmon falling considerably, most terrestrially produced meats are cheaper. The main substitute for aquaculture salmon is wild salmon. Aquaculture salmon is nearly always cheaper. Blind taste tests, however, suggest that most consumers prefer wild salmon to aquaculture salmon. The significant advantages of aquaculture product are convenience, uniformity, and price.

The market share of seafood in the United States remains relatively stable at 7.7 percent of the protein market share (Johnson and Associates 2001). This shows that the increase in seafood consumption is due to population increase rather than a per capita change in eating habits, and that increase in salmon consumption is potentially accompanied by a decrease in consumption of other seafood, especially because of the high substitutability of other fish. Seafood is price-elastic, which means an increase in price leads to a decrease in demand.

In the early days of the industry, aquaculture salmon were harvested during the off season, when wild-caught salmon was not available in the market to depress prices. At that time, the two did not compete directly in the fresh category. This has changed in recent years as aquaculture salmon production has increased and as dedicated markets have come to rely on farmed salmon.

Wild salmon is not losing out just because of price, either. Wild-caught salmon is not keeping pace with consumer preferences. The short harvest season (eight weeks), the small airfreight shipping capacity, and the lower prices for all salmon as a result of farming have contributed to the decline of wild salmon in many markets. More than 90 percent of Alaskan salmon, the largest source of wild salmon in the world, was sold as lower-priced frozen or canned product in 2001 (Alaska Department of Revenue, as cited in Johnson and Associates 2001). Most Alaskan processors have stopped trying to increase fillet production to compete with aquaculture producers. In Alaska, labor and production costs are higher than on farms. In 2001 only 357,000 pounds of frozen fillets were produced from a harvest of 765 million pounds; in fact, 1 percent of all Alaskan fillets were exported and reprocessed in China and Thailand in 2001.

MARKET CHAIN

The farmed salmon industry has consolidated. Over half of aquaculture salmon is produced by five companies: Nutreco Holding N.V. (165,000 metric tons in 2001), Fjord Seafood ASA (102,000 metric tons), Pan Fish ASA (97,000 metric tons), Stolt Sea Farm SA (55,000 metric tons), and Statkorn Holding ASA (53,000 metric tons) (Packard Foundation 2001). All of these companies have operations in one or more of the following regions: Canada, the United Kingdom, Chile, the United States, and Norway. Production figures for these companies for 2000 are shown in Table 23.1.

Many large producer/processor companies attempt to distribute their own product, but so much salmon is consumed in so many types of markets that this is not really possible. The top seafood distributor in the United States is SYSCO with $1.3 billion in seafood sales (out of total sales of $21 billion). SYSCO has 400,000 customers. They sell eleven types of shrimp and six salmon products. Ahold-owned US Foodservice is a distant second with $760 million in seafood sales. Performance Food Group has $250 million in sales, and the next seven distributors (Morey's Seafood International, Inland Seafood, East Coast Seafood, Supreme Lobster &

TABLE 23.1. The World's Main Salmon-Producing Companies

Company (Headquarters)	Production (MT)	2000 Sales (million U.S.$)	Production Locations
Nutreco Holding N.V. (Netherlands)	141,500	$855	Canada, United Kingdom, Chile
Pan Fish ASA (Norway)	64,200	$527	Norway, Faroe Islands, British Columbia, Scotland, United States (Washington)
Stolt Sea Farm SA (Luxembourg)	47,000	$310	Canada, Chile, United Kingdom, United States (Maine)
Fjord Seafood ASA (Norway)	39,120	$258	Norway, Chile, United States
Statkorn Holding ASA (Norway)	35,000	$150	Norway
Salmones Pacifico Sur SA (Chile)	27,000	$69	Chile
George Weston Ltd. Connors Bros. (Canada)	27,700	$560	Canada, Chile, United States
Modnor Group AS (Norway)	19,300	$56	Norway
Camanchaca SA (Chile)	19,000	$70	Chile
Multiexport SA (Chile)	18,000	$85	Chile

Source: Johnson and Associates 2001.

some pressure off of comparable wild stocks in the ocean, it has a number of other detrimental impacts that, on the whole, may actually endanger wild stocks. These impacts occur at the level of individual salmon farms, but due to the open nature of oceans many can have much more far-reaching impacts as well.

Ecological Footprint

A recent report by Michael Weber (1999) suggests that for every metric ton of Atlantic salmon from aquaculture, 10.6 hectares of marine area and 3.0 hectares of terrestrial area were required to support or provide the inputs to make it possible. For example, some 99 percent of the marine requirement for production of salmon is dedicated to the production of organisms that are caught and made into salmon feed. On the terrestrial side, some two-thirds of the land required was actually to assimilate the 7 metric tons of carbon dioxide created during the production of a metric ton of salmon from aquaculture. Most of the remaining terrestrial impacts resulted from the production of crops that were converted to salmon feed.

Waste and Nutrient Loading

Salmon produced from aquaculture are efficient at converting feed to flesh. For example, 1 kilogram of salmon can be produced with as little as 0.9 to 1.1 kilograms of feed and only 0.27 to 1.1 kilograms of waste. Even so, because salmon are produced in a water column that can be up to 20 meters deep, wastes can accumulate and degrade water quality. This in turn can smother plant and animal communities living beneath the net cages (Weber 1997).

Waste from feces and uneaten food results in increased nitrogen and phosphorus released into marine environments. In 1998 Scotland produced 115,000 metric tons of Atlantic salmon. Nutrient inputs to the marine environment for that year were 6,900 metric tons of nitrogen and 1,140 metric tons of phosphorus (i.e., 1 metric ton of salmon released 60 kilograms of nitrogen and 10 kilograms of phosphorus). Ellis and Associates (1996) found that each metric ton of salmon production resulted in waste equivalent to that from nine to twenty people. So for nitrogen, the total nutrient input for 1998 was equivalent to the sewage from 3.2 million people, and for phosphorus, 9.4 million people. In 1997 Scotland's human population was 5.1 million people (MacGarvin 2000).

Nutrient pollution leads to eutrophication, which often results in increased plant growth. Even small changes in nutrients can have major impacts on phytoplankton communities. Increased phytoplankton populations reduce light availability below the surface, and as a result, threaten seaweed and eelgrass communities. Elevated nutrient concentrations, along with climatic conditions, can contribute to blooms of plankton and toxic algae (MacGarvin 2000).

Blooms can have devastating effects on farmed fish. Some plankton species have sharp spicules (needlelike pointed structures) that can damage gill tissue, making fish more susceptible to disease. Depending on cage depth, salmon raised in net cages may not be able to evade the surface plankton (Ellis and Associates 1996). The frequency of mortalities due to algal blooms around salmon farms is increasing. When these mortalities occur, salmon farmers suffer huge financial losses but can also make compensation claims. In certain areas, the evidence suggests that salmon farm pollution is the main, or at least a contributing, cause of toxic algal blooms (Staniford 2002).

Algal toxins can also be transmitted via plankton-feeding fish up the food web to other marine species including birds and marine mammals (MacGarvin 2000). In Scotland, fecal waste from fish farms has been linked to toxic algal blooms and out-breaks of the algal toxins that cause diseases in humans, most notably amnesiac shellfish disease (ASD). Both diarrhetic shellfish poisoning (DSP) and paralytic shellfish poisoning (PSP) are also of concern in the region. Such blooms have se-verely depressed the shellfish farming industry in Scotland.

Fish farm sediments are deposited on the ocean floor, disrupting and altering the community of macrofauna that live there. Benthic communities play important roles in sediment nutrient cycling. The structure of the community can change as species with low tolerance to pollution, or species that are no longer suited to the or-ganically enriched environment, die or move to other areas. The rapid deposition of waste can overwhelm organisms that promote aerobic decomposition on the ocean floor. Anaerobic decomposition by a different community is then favored, causing a drastic shift in the ability of the original benthic community to survive in the area (Ellis and Associates 1996).

Incorporating seaweed and/or shellfish into the salmon farming system can help to solve the waste problem, since these organisms filter and utilize waste products. An integrated system of, for example, salmon and seaweed or salmon and shellfish could reduce nutrients significantly. There is some question, however, as to whether such systems could reduce significantly the overall impact of having so much organ-ic matter concentrated in one place. In addition, such a system does not help solve the problem of toxic chemicals entering the marine environment (Staniford 2002).

Increased Pressure on Wild Fisheries

Salmon aquaculture is often touted as a precursor of aquaculture production systems that could relieve pressure on other wild fisheries. However, because salmon are car-nivorous, they require a diet high in fish meal and fish oil. Fish oil use is now domi-nated by aquaculture, which takes 60 percent of total production (FAO 2000). Salmon aquaculture is by far the largest user. Analysts have suggested that aquacul-ture will use more than 90 percent of fish oil by 2010.

At this time, some 20 to 25 percent of annual global seafood supply is converted to fish meal and fish oil (Packard Foundation 2001). Though a relatively new indus-try in 1994, the carnivorous aquaculture farms used approximately 15 percent of the global fish meal output (Ellis and Associates 1996). By 1997 aquaculture used 33 percent of fish meal supplies (Jacobs et al. 2002). Changes in feed formulation have focused on oil to provide energy for fish to swim and meal to result in weight gain.

Salmon feed can contain up to 40 percent fish oil. This has increased from 8 percent in 1979 (Staniford 2002). Net-cage rearing of 1 kilogram of salmon is estimated to use anywhere from 4 to 5.5 kilograms of wild fish. While this is probably a far better ratio than salmon in the wild (where the ratio could easily be 8 to 1, 10 to 1, or even higher), responsible businesses must strive to use finite resources more efficiently. Finally, the tradeoffs for this issue are further compounded if the fish used to make fish meal could be consumed directly by people rather than converted to more high-value products for wealthier consumers.

Feed accounts for 30 to 50 percent of a salmon producer's annual expenses. Because of the high quantities of fish meal and fish oil used, farmers look for low prices for these products, putting pressure on the South American fish-meal industry. Japan, Chile, Peru, and the Commonwealth of Independent States (former Soviet constituent states) account for approximately two-thirds of all fish meal production. Three species of fish (anchoveta, sardine, and jack mackerel) constitute 85 percent of South American fish meal production, and they are susceptible to large fluctuations in population due to El Niño. The anchoveta population collapsed in the 1970s, 1980s, and 1990s. When this happens, it puts more pressure on other species to make up the difference for industries that are dependent on feeds based on fish meal.

Interactions Between Wild and Farmed Fish

Caged salmon escape virtually everywhere that salmon are farmed. The introduction of a species to an area inevitably has unforeseen consequences. Salmon that escape from aquaculture operations can cause a wide range of impacts including competition for food and spawning habitat with both wild salmon and other species. Escapes can interbreed and cause genetic pollution that reduces the hardiness of wild salmon. Also, they can spread diseases that either did not previously exist in the area or were not previously a problem for wild populations.

There are large numbers of escapes. In Norway as many as 1.3 million salmon escape each year, and a full third of the salmon spawning in coastal rivers are of escaped origin. In 1997, 300,000 salmon escaped in Puget Sound in a single instance when net cages were ripped open accidentally (Weber 1997). In 2000 an estimated 500,000 fish escaped in Scotland (Berry and Davison 2001). The year before, there were sixteen reported escape incidents involving 440,000 farmed fish. Often the escapes involve much smaller numbers (only 10,000 or so might escape as a result of a single accident), but the cumulative impact on an ecosystem over the course of a year can be quite large.

There is considerable evidence that in some areas the escapes are becoming significant populations in their own right. The number of escapes in Scotland has increased more than threefold since 1998, but less than 60,000 wild salmon were caught in 1999. On the West Coast of Scotland, an estimated 22 percent of the "wild" catch is, in fact, escaped farmed salmon (Staniford 2001). In some of Norway's rivers, there are as many as four escaped farmed salmon for every wild one (Ellis and Associates 1996).

With an estimated half a million escapes in 2000 off the Scottish coast, farmed fish and wild fish may be interbreeding. As farmed fish are selectively bred for characteristics favorable for aquaculture, breeding between the two populations could alter the genetic makeup of wild fish and decrease their fitness to survive in the wild environment.

The significance of escapes can be demonstrated by the example of New Brunswick, Canada, where the first salmon farms were built in 1979. Within four years, 5 percent of the salmon in the nearby Magaguadavic River were escaped salmon from the farming operation. By 1995, 90 percent of the salmon in the river were escapes (Weber 1997). When escapes are an insignificant portion of the population in the wild, they probably pose a rather limited risk. However, when they dominate the numbers in the wild, they can very quickly become one of the major reasons for the demise of wild populations.

Farmers have every incentive to eliminate escapes because escapes represent significant costs for buying and feeding animals. However, in some instances the releases are not accidental. Some 4 million salmon are estimated to have escaped in Chile since the industry started (Claude and Oporto 2000). In 2002 when salmon prices declined, Chilean producers actually released hundreds of thousands of salmon rather than pay to harvest them.

There is another important issue, however, when discussing the issue of escapes. This is the impact of deliberate releases from hatcheries that are intended to increase or even create salmon runs in specific river systems. For more than a century, salmon species have been released throughout the world into a wide range of river systems that did not include salmon previously. In the eastern United States, Alaskan salmon were released more than a hundred years ago in an attempt to reintroduce salmon in rivers where Atlantic salmon were extinct. In the case of Chile, salmon were released into the wild in 1905, 1914, 1946, and 1952 in an attempt to colonize river systems thought to be suitable but with no comparable fish populations. None of the Chilean releases were successful (Claude and Oporto 2000). The implications of this for wild species are not clear.

Sea lice and other diseases spread by farmed salmon can have a devastating effect on wild salmon and other fish. Researchers found that 86 percent of wild migrating juvenile salmon in two Norwegian fjords died as a direct result of sea lice infestations that they contracted while migrating past salmon farms (Pearson and Black 2001, as cited in Berry and Davison 2001). This contributes to the continuing decline of wild salmon, which in turn upsets the ecological balance in marine and freshwater systems. It also reduces revenue from commercial harvesting and sport fishing.

The application of biological engineering to salmon has resulted in the creation and patenting of transgenic, or genetically altered salmon. There has been pressure on the U.S. Food and Drug Administration to consider a petition to farm and sell the salmon within the United States. The farming of such fish further increases the likelihood of salmon aquaculture affecting wild salmon populations as well as other organisms within the environment (Kay 2002).

Contamination with Toxic Compounds

The farming of fish high up the food chain can tend to concentrate contaminants (Staniford 2002). The artificial food chain built by feeding oil-rich and animal-derived diets to salmon has resulted in elevated levels of such contaminants as dioxins and polychlorinated biphenyls (PCBs) in farmed salmon compared to their wild counterparts. The term *dioxins* refers to over 200 different polychlorinated dibenzo-*para*-dioxins and dibenzofurans, seventeen of which are considered toxic. Dioxins are produced as unwanted by-products, while PCBs are manufactured for use in transformers and insulators (CFIA 2002). Chlorinated hydrocarbon compounds can accumulate in the fatty tissues of fish, so fish oil has relatively high levels of these compounds (especially if derived from fish from contaminated areas). Any of these toxins can pose serious risks to human health.

PCBs and many organochlorine pesticides (which have been found in aquaculture salmon) have been banned in most of the world, but they still affect humans through their diet. European farmed salmon can be a significant source of these toxins in the diet (Jacobs et al. 2002). The European Union's Scientific Committee on Food found that fish can represent up to 63 percent of the average daily exposure to dioxins. The Food Standards Agency of the United Kingdom recommends that people consume only one portion of oily fish per week (Staniford 2002).

A recent study of PCB concentration in salmon showed that some farmed salmon had relatively high concentrations of the compound. However, wild salmon captured from polluted water had even higher levels of PCBs. Variation in farmed-salmon PCB levels is attributed to the variation in the level of contamination in fish meal. Fish meal from Peru had PCB concentrations ten to twenty times lower than those from Denmark and the Faroe Islands (Jacobs et al. 2002). Farmed salmon in Scotland were shown to have relatively high concentrations of dioxins and PCBs, presumably due to the sources of the fish meal and oil used for feed. Concentrations of the compounds in salmon were higher than those of other species such as cod, because salmon have a higher fat content than other species. Thus, salmon retain more toxins per pound of fish than do fish with lower fat levels since the compounds accumulate in the fatty tissues of the fish (Jacobs et al. 2002). In addition, farmed salmon have four to five times more fat content than wild salmon (Staniford 2002).

In addition to these contaminants, toxic heavy metals can also accumulate in the fatty tissues of fish. These metals can be concentrated further through the rendering of fish meal and fish oil and further still in the animals that eat feed made from them. Mercury is a good example; once consumed by humans, it is readily absorbed into the gastrointestinal tract. Symptoms associated with the consumption of low levels of heavy metals may not appear until later in life (Quig 2002).

Many studies have examined the concentrations of toxins in fish, fish meal, and fish oil. Results vary considerably. One study in Canada showed that fish meal and fish oil do not contain high levels of dioxins, PCBs, DDT, or mercury (CFIA 2002), while the authors of a study in Scotland recommend that measures be taken to lower these levels because they are too high. An analysis of dioxin toxicity of thirteen categories of food (such as beef, chicken, ocean fish, freshwater fish, butter, eggs, etc.) found that the freshwater fish (in which the study included many farmed species and salmon) had the highest dioxin toxicity. In fact, freshwater fish toxicity was 50 percent higher than butter, which had the second highest toxicity. All of the other products had less than half the toxicity of butter (Schecter et al. 2001).

Use of Antifoulants

Salmon net-cage operations generally have steel cage superstructures with knotless nylon nets suspended within. While the net cages can vary considerably by area, they tend to be some 20 meters deep. One of the main problems that the net cages pose is the potential for fouling. Shellfish and marine algae grow on the nets and can make them extremely heavy. This makes the lifting and cleaning of the nets very difficult, and it shortens the lifetime of the investment (Ellis and Associates 1996).

To avoid fouling and to prolong the life of the cage, growers often use antifouling paints. Such paints, by definition, are highly toxic given that that is how they prevent organisms from growing on painted structures. The most commonly used antifoulants (organotin or copper-based compounds) are toxic to bivalves and could be harmful to fish species as well (Cripps and Kumar 2003). Titanium, copper, and tributylin (TBT) have been used in marine paints, and are known to be harmful to shellfish. However, some of these paints are also known to accumulate in the tissues of fatty fish such as salmon and are therefore inappropriate for use around fish intended for human consumption. Tributylin has been shown to be highly toxic to marine life, causing reproductive failure and growth abnormalities in molluscs. In addition, paints containing oxytetracycline should be prohibited from salmon aquaculture operations because they are known to result in increased antibiotic resistance (Ellis and Associates 1996).

Use of Chemical Inputs

In addition to the antifoulants discussed earlier, chemical inputs in salmon farming include antibiotics and insecticides such as organophosphates and synthetic pyrethroids. Therapeutic chemicals may be applied as a bath treatment or administered in feed, but in both cases the chemicals eventually make their way outside the salmon cage into the wider marine environment. The effects of chemicals on the greater marine environment are not well known. The ecological impacts resulting from the use of antibiotics in salmon farming have not been studied. It is conceivable that antibiotics could accumulate in the tissue of wild fish and invertebrates, while also leading to resistance in target pathogens and other microbial species. Scotland is known for being the strictest country when giving out permits to salmon farmers. Their typical discharge consent, however, allows the use of over fifty different chemicals. The number of drugs permitted for use by the Veterinary Medicines Directorate has increased from three to forty from 1989 to 2002 (Staniford 2002). In short, "the global advance of intensive salmon farming has meant that farmed fish have become agents of pollution rather than biological indicators of pollution" (Staniford 2002).

Several different drugs and chemicals are used to combat diseases and parasites in the production of salmon. Over time the industry has learned how to produce more salmon using fewer drugs and chemicals. However, the learning curve has tended to be repeated in each new area of culture. For example, from 1985–87, antibiotic use in salmon farms in Norway increased from 17 to 48 metric tons per year, more than the combined use of all antibiotics for humans and terrestrial animals in the country (Weber 1997). In 1999 in the United Kingdom, 4 metric tons of antibiotics were used in salmon farming compared to 11 metric tons in cattle rearing and less than 1 metric ton with sheep (Berry and Davison 1999). As vaccines have been developed and as management systems have been improved, these levels have declined drastically.

In Chile, however, the reduction in the use of antibiotics has been slower, even though most of the major investors are Norwegian. In 1990 the salmon industry used 13 metric tons of antibiotics, by 1995 usage had increased to 65 metric tons, and by 1998 it was 100 metric tons. In 1993 Chile used seventy-five times more antibiotics per kilogram of salmon produced than Norway (Claude and Oporto 2000).

In the early years, most antibiotics were put in the manufactured feed, and as late as 1999 medicated feed was still common in Chile (Claude and Oporto 2000). At least three-quarters of antibiotics in feed are lost to the environment, whether the feed is eaten or not (Weber 1997). Little is known about the impact of these drugs on ecosystems in general or on individual species in particular.

The prophylactic use of drugs can lead to growth of drug-resistant strains of pathogens in both wild and cultivated fish populations. The abuse of antibiotics through prophylactic use can also build up pathogenic resistance in humans. In 1991, 50 percent of the bacteria responsible for the fish disease furunculosis were resistant to two compounds used to treat the disease. Scientists disagree about the extent to which resistance has developed, but they agree that resistance will increase as antibiotic use increases—and that this resistance can be passed on to human pathogens. In addition, there are a limited number of compounds that are effective on aquatic pathogens, which means there will be even graver consequences if resistance develops.

The chemicals are not always even appropriate. For example, the chemicals used to treat sea lice have largely been developed for terrestrial use, and little research has been done on their use in the marine environment. In Scotland salmon producers used a chemical delousing agent called dichlorvos to reduce infection of salmon by sea lice. Later research suggested that this chemical killed oysters, mussels, and other shellfish and crustaceans within 75 meters of the salmon cages (Weber 1997).

Mort Disposal

The disposal of salmon that die before harvest (morts) has both environmental and health implications. Approximately 20 percent of salmon die during grow-out, some of them from diseases that could potentially be spread from the improper disposal of morts. A variety of disposal methods is used; the principal ones are landfilling, composting, and ensilage (a liquification of the morts that is then used in animal feed or fertilizer). Some companies dump morts into the ocean, where the chemicals ingested by the salmon before death, as well as any diseases that may be present, are released into the environment.

Impact on Predators

Salmon in net cages attract predators such as seals, river otters, sharks, kingfishers, eagles, cormorants, and great blue herons. The effect on these animals of consuming salmon that have antibiotics and other chemicals in them is not known. Nor is it clear how greatly they have been impacted by various methods salmon farmers employ to keep them away. One of the methods is simply to kill them. Seals, for example, can be shot by salmon farmers in British Columbia, though the farmers must obtain permits to do so. It is estimated that at least 500 are shot by salmon farmers each year in British Columbia, where harbor seals are estimated to cost the industry $10 million a year (Weber 1997). In Scotland the industry estimates that 350 seals are shot each year, while environmentalists put the figure at 5,000 (Weber 1997).

From the 1980s to the mid-1990s, some 5,000 to 6,000 sea lions were killed in Chile by salmon farmers. In addition, an unknown number of dolphins and even an occasional minke whale were killed (Claude and Oporto 2000). According to one study (Brunetti et al. 1998, as cited in Claude and Oporto 2000), sea lions cost the Chilean salmon industry about $21 million in damage annually (in direct costs as well as the cost of security, etc.). This amount was some 3 percent of sales.

Farmers also use predator nets, or nets above and around the salmon cage, to prevent predators from getting too close to the cage. Netting used to exclude marine mammals and birds can entangle and drown animals. Some producers leave the dead animals there as a way to scare away others. Acoustic devices that emit a high-pitched sound can be used to scare away seals and sea lions. In some instances these devices have been so successful that they have also caused the withdrawal of resident populations of harbor porpoise and whales. The extent of the impact of any of these methods on bird and mammal populations is unclear, but potentially they could have a great impact, especially in areas where salmon farms are highly concentrated.

Poorly Run Hatcheries

Well-run hatcheries should not have environmental impacts. Unfortunately, not all hatcheries are well run. In many parts of the world, hatcheries are allowed to dispose of waste without treatment. This damages the environment not only by causing nutrient overloading, but also by introducing diseases into the marine or freshwater environment that can affect both wild salmon and salmon farming operations.

In Chile hatcheries were established in large freshwater lakes rather than in closed systems as in most other parts of the world. In Southern Chile, where most salmon are produced, five of eight lakes are polluted, and the salmon aquaculture industry appears to be the main cause (Claude and Oporto 2000).

BETTER MANAGEMENT PRACTICES

There are BMPs for salmon aquaculture at both the site and the landscape level. Clearly, making sure that net cages are put in the least damaging places and that they are operated in ways to reduce their impacts are both important strategies to reduce the overall environmental impact of the industry.

Other factors are also important. For example, salmon aquaculture has received rather fewer public resources than might be expected given the phenomenal growth of the industry. This suggests that there may be some room to negotiate with governments to help fund some of the transition costs to more sustainable production. To date, technologies have been developed and deployed around the world faster than the understanding of their consequences or unintended impacts on nontarget organisms or ecosystem functions (Whoriskey 2000).

Improve Siting of Operations

The salmon aquaculture industry is centered in areas where many wild salmon populations are in crisis. While the industry may not have been the primary cause for the decline of wild salmon populations, the first step to their effective recovery will have to be to eliminate, or at least reduce substantially, the impacts of the aquaculture industry (Whoriskey 2000). One way to do this at both the farm and the landscape level is to integrate risk analysis into the review process for siting hatcheries and farms.

Use Closed Production Systems

Some have suggested that in the final analysis, completely closed systems for the containment of contaminated wastes is the only sustainable solution for salmon production (Staniford 2002). Enclosed, land-based salmon farming can reduce or eliminate many of the problems specific to net-cage production systems. Salmon farmed in net cages escape into the wild. This impact, and the genetic and disease issues that it raises, would be eliminated with on-shore closed systems. Similarly, wastes that are discharged into the ocean in the net-cage system would be captured as the water leaving the land-based tank is filtered. These nutrient-rich wastes could potentially be recycled for agricultural use. The industry, and ultimately the consumer, rather than the environment or the "public" more broadly, would pay for the cost of waste disposal.

AgriMarine Industries Inc. in Canada recently made the first sale of Pacific salmon raised in a land-based, closed containment system (Smyth 2002). The company raises salmon in concrete tanks, in which seawater is pumped in and oxygenated and outgoing water is filtered. This system produces healthy salmon but is far more expensive to operate than the standard cage production system. While Agri-Marine's salmon was sold at a higher price and marketed as "eco-friendly," it is not clear that such a system is economically viable over the long term (Smyth 2002).

The most complete study, to date, on the viability of land-based salmon aquaculture was undertaken in the Bay of Fundy (ADI Ltd. et al. 1998). The large tides in the region were seen as an asset because they could move water into reservoirs from which it could flow by gravity into salmon tanks with no pumping costs. Pumping water is a very large expenditure for land-based aquaculture systems.

The study assumed that production would follow standard industry practices (e.g., stocking densities, feeding and growth rates, etc.). It was also assumed that the factors that would most affect such operations were the price of salmon, the rate of return, the up-front capital costs (e.g., investing in dams and seawalls to hold and move the water), the growth of the fish, and the cost of money. The only scenarios modeled that showed a positive cash flow in five years were those that grew transgenic salmon. These salmon grow faster and far bigger than the animals used today. Even with transgenic animals the scale of operations would have to be increased considerably to make the operations profitable. While such systems may not work for salmon unless the price increases (which is unlikely), it may work for other, higher-valued species. In fact, this system might well have worked for salmon early on when prices were much higher than they are today. What this means is that with the current level of environmental subsidies for salmon aquaculture in many parts of the world, it may be impossible to go back to more sustainable production systems.

Another important issue is that the proposed closed system has some unique implications. It must be located in areas with severe tides, and these areas must, in turn, be located near rather flat terrestrial areas where land-based farms can be established. More importantly, the land-based systems require production units and large tracts of land that would be rarely available in coastal areas anywhere in the world without considerable conflict with existing residents.

Norway is reported to have considered land-based systems, but the country eventually abandoned the idea based on the belief that sufficient land was not available. As a consequence it was assumed that producers using closed systems would be forced to stock at higher densities. Such densities, it was felt, would lead to very real risks of disease outbreaks (Whoriskey 2000).

Another closed system that may offer more hope is the use of closed containment systems in the open water. These systems amount to little more than large plastic bags in the water column. Water is pumped into and out of the bag to provide oxygen for the fish. The shape is maintained by the force created by a small hydraulic head pumped into the bag. Such bags offer a number of environmental benefits. Seals and other predator attacks are reduced because animals no longer see the fish through the opaque bags. Waste can be collected and removed from the bottom of the bag rather than released into the water. Finally, fish raised in bags have fewer sea lice problems than those raised in open net cages (Whoriskey 2000). Closed-bag systems have been experimented with in both eastern and western Canada. To date, this system of production appears to be expensive to install and operate. This is a deadly combination given the overall decline in salmon prices.

Reduce Escapes

Escapes can have a huge impact on wild salmon populations, particularly where those populations have been depressed. For example, if a river has a salmon run of 5,000 animals a year, then a 5 percent level of escapes within that run (say twenty-five animals) would not be a large impact. If, however, a river only has twenty-five animals in its annual run, then the twenty-five escapes would have a much larger impact. To date, there is no science available to define what impact levels would be "acceptable" for what reasons.

Similarly, moving the industry away from the mouths of rivers with major runs of salmon or other species would help to reduce contact between wild and caged fish and, consequently, the spread of disease from either. Norway has adopted a much more thoughtful approach to the siting of operations as well as to the size of operations allowed in any one site since the early days of the industry when it developed with less planning and fewer controls. Even so, at this time the majority of streams and rivers in Norway no longer have salmon runs. While aquaculture was not the only or even the primary cause of the demise of these runs, it did have an effect on at least some of them.

One way to address the issue of escapes is through a code of conduct. A code could address such issues as improved cage engineering, better operating regimes, education of workers on their roles regarding this issue, improved monitoring, enforced and prompt reporting, contingency planning, and more effective recovery programs. Specific targets could be set and monitored. Whoriskey (2000) has suggested an overall escape reduction target of 10 percent per year for five years. Once the goal is set, let the industry find the best ways to meet it. Proper siting could reduce the chance that escapes would happen at all, much less enter rivers with salmon runs.

Another way to reduce the impact of escapes from salmon aquaculture is to stock only sterile fish. Sterile fish programs are not foolproof; there is no way to guarantee that the organisms are always sterile. However, as long as escapes persist, and perhaps even after they cease, sterile fish should be stocked to help to insure that escapes do not cause genetic pollution of wild salmon runs. Such programs will at least reduce the overall risk of interbreeding from escapes.

Encourage Organic Net-Cage Production

No chemicals are used in organic net-cage production, including no medication for the animals or chemical treatment of the cages to prevent "fouling." An organic net-cage operation in Canada had 30 percent losses of salmon during the grow-out phase of production compared to the industry average of 20 percent. Despite these high mortalities, the operation also had lower-than-normal production costs since it was not buying chemicals. More importantly, the product fetched a higher market price. Product on the farm is harvested each week so as not to saturate markets; some 100 metric tons are sold each year. A non-organic producer operating on the same site could produce some 700 metric tons and still not have the same net profit. While such an operation does not eliminate the dangers and impacts of escaped fish or the fish meal issue, the low-density production and lack of chemicals cause the system to have considerably lower environmental impacts than the standard net-cage production system. Total production by volume, however, is only about 15 percent of the standard system operating on the same area (Ellis and Associates 1996).

Encourage Fallowing in Net-Cage Production Systems

Fallowing can reduce, but not eliminate, the overall impact of net-cage production systems. Fallowing does not mean leaving net cages unused, but rather moving them from one area of recent production to another area. This practice spreads the impacts of production over a wider area and gives the ecosystem time to flush and disburse the wastes that accumulate below the net cage. In general, it is not advisable to produce fish in the same location over long periods of time, as there is an increasing chance of disease.

This practice is equivalent to agricultural fallow systems. An area is not used for production for a number of years (up to five for salmon aquaculture) in order to let nature recover from the effects of production (Ellis and Associates 1996). Provided there is sufficient area for moving net cages, fallowing does not have to reduce overall production, although there would be downtime while moving and setting up net cages in new locations.

Reduce Use of Fish Oil and Fish Meal

Considerable work has been done to achieve truly phenomenal results in improving the feed conversion ratios for salmon production. The industry norm at this time is nearly one-to-one: one kilogram of feed produces one kilogram of product. Work still needs to be done, however, to change the formulation of the feed to reduce the total quantities of wild fish needed to supply the oil and meal. Today, it takes four or five kilograms of wild fish to make one kilogram of farmed salmon. In order to reduce overall environmental impacts and use resources more efficiently, this proportion needs to be changed. Given that salmon are carnivorous, it is not clear how much progress can be made.

Replacing part of the fish oil component of fish feed with vegetable-based oils would be a good start and could have a number of benefits. It could decrease the toxins from fish oil that farmed salmon currently consume. Ultimately, this means that humans would consume fewer of these harmful toxins as well. While the accumulation of residues from vegetable-based oils is possible, it is much less of a problem than from fish oil (Jacobs et al. 2002).

The use of fish meal and fish oil in salmon diets has also been linked to eutrophication and pollution problems. A vegetable-based diet results in lower levels of pollution, even though there is still considerable organic matter. Because salmon raised on a vegetable-based diet have a different flavor, a lot of work will need to be done to maintain the flavor profile consumers have come to expect (Staniford 2002).

Alitec, a leading Chilean feed producer, says that it will begin to reformulate its feed so that it will contain significant quantities of vegetable oil by 2004, thus reducing the amount of fish oil used. Though some salmon farmers are skeptical, the company believes that the reformulation will have benefits, one of which will be a lower-priced feed.

Implement Measures to Reduce Diseases

Disease is one of the main threats to salmon aquaculture operations. The development and widespread adoption of a code of conduct could help producers both prevent diseases and contain them if they occur. Producers need their own systems for quarantining animals before introduction if countries do not have their own rules or if such rules are inadequate or are not enforced. The point here, however, is not simply to obey the law. Diseases can wipe out operations, so there is too much at stake to hide behind laws. Producers must develop their own programs that exceed those of most countries because producers stand to lose if things go wrong. Once procedures are established, workers need to understand their role in containment and disease transference issues, whether they work in hatcheries or net-cage operations. Vaccination programs should be mandatory, as should the quarantine of sick animals. Fish that are untreatable should be killed and properly disposed of so there is no chance that they will infect other fish, either within the aquaculture production system or in the wild.

Diseases in salmon operations are also a threat to wild fish populations. Consequently, diseases should be addressed quickly and effectively both to maintain the economic viability of the producer and to avoid the potential impact of disease outbreaks on wild populations (Whoriskey 2000). Diseases should be monitored systematically on all farms as well as within the proximity of farms to better identify and understand the role that farms play in maintaining or extending disease vectors. If disease issues cannot be addressed through management, medication, and vaccines, then they may have to be addressed through a total reduction of net cages in any given area.

Operate Systems for Continuous Improvement

A process of continuous environmental improvement, similar to the management systems that are endorsed by the ISO certification and standards processes, would help to make salmon aquaculture less of a concern (Whoriskey 2000). Such systems, however, require written procedures, measurement of impacts, and ongoing monitoring. Thus systematic, timely, and effective monitoring is required not only for each net cage or even each farming operation, but also for larger ecoregions where cumulative impacts of the entire industry may be significant.

OUTLOOK

In the space of three decades, salmon aquaculture has found ways to take a seasonal, high-value wild species and produce a year-round product at half the price. The growth of the salmon aquaculture industry has been remarkable. Currently, it threatens the viability of the wild salmon fishery in most parts of the world, especially Alaska. However, it also offers insights into the opportunities and problems with the intensive aquaculture production of other high-value, carnivorous wild fish species.

Since the founding of the salmon aquaculture industry, its environmental impacts have been tremendous. There have also been tremendous efforts and accomplishments in reducing those impacts. Norway, more than all the other producing countries combined, has taken the lead in these efforts. In 2002, for the first time ever, Norwegian production was eclipsed by that of Chile. While Chile has benefited historically from Norwegian investments and expertise, it is not clear that at this time either Chile's government or its salmon aquaculture industry has the same financial and human resources to invest in continuing the efforts to make the industry more sustainable. If anything, Chile has been lax in monitoring issues such as siting and carrying capacity. That country has also been willing to let producers cut corners such as overlooking the improper disposal of wastes and excessive use of antibiotics, both of which cause unacceptable impacts on both fresh and salt water ecosystems. Cutting these corners (in effect, subsidizing production through damage to the environment) has allowed Chile to be the lowest-cost producer of salmon. Cutting corners has also put the industry at greater risk in a place where neither the government nor the industry is prepared to address, much less anticipate, future crises as they arise. This is an explosive situation for any industry, but it is especially so in aquaculture where disaster can strike quickly and thoroughly.

For its part, Norway is still a very large producer of salmon, but the lessons being learned in Norway today are less relevant to the direction the industry has taken in Chile, for example, larger-scale, more intensive, growth-led production. Furthermore, in Norway the industry and government are both diversifying their interests. As they experiment with other high-value fish species in aquaculture, hopefully they will be able to avoid many of the problems associated with salmon farming today. Ideally, however, because of the value of many of the new species, both Norway and its producers will have the resources, as well as the inclination, to identify better ways to produce such fish in aquacultural systems, ways that can also be applied to salmon aquaculture. Getting salmon aquaculture right will be the litmus test for whether humans will be able to take pressure off wild, carnivorous finfish fisheries while reducing environmental and social impacts to acceptable levels.

REFERENCES

ADI Ltd., Canadian Aquaculture Systems Inc., and NATEC Environmental Services, Inc. 1998. *Land-Based Aquaculture in Intertidal Areas Feasibility Study.* Final report. Prepared for the Huntsman Marine Science Centre, St. Andrews, New Brunswick, Canada.

Anderson, J. 2001. *Fishing, Farming, and Fish Farming: Competitiveness and Regulation.* Presented at How to Farm the Seas II: The Science, Economics, and Politics of Aquaculture on the West Coast workshop organized by the Atlantic Institute for Market Studies and Canadian Aquaculture Institute, Vancouver, February 15–17, 2001.

Anderson, J. No date. *Fisheries, Aquaculture and the Future.* Powerpoint presentation, University of Rhode Island.

Aquamedia. 2002. *National Aquaculture Production.* Website of the Federation of European Aquaculture Producers. National Statistics. Available at http://www.aquamedia.org/production/countries/default_en.asp.

Berry, C. and A. Davison. 2001. *Bitter Harvest: A Call for Reform in Scottish Aquaculture.* Perthshire, Scotland: WWF Scotland.

CFIA (Canadian Food Inspection Agency). 2002. *Summary Report of Contaminant Results in Fish Feed, Fish Meal, and Fish Oil.* Available at http://www.inspection.gc.ca/english/anima/feebet/dioxe.shtml.

Claude, M. and J. Oporto, editors. 2000. *La Ineficiencia de la Salmonicultura en Chile.* Santiago, Chile: Terram Publications.

Cripps, S. and M. Kumar. 2003. *Environmental and Other Impacts of Aquaculture.* In Southgate, P. C. and J. S. Lucas (eds.) *Aquaculture: Fish and Shellfish Farming.* Oxford: Fishing News Books.

Ellis, D. and Associates. 1996. *Net Loss: The Salmon Netcage Industry in British Columbia.* The David Suzuki Foundation. Vancouver, BC.

FAO (Food and Agriculture Organization of the United Nations). 2000. *The State of World Fisheries and Aquaculture.* Rome: UN Food and Agriculture Organization.

Jacobs, M. N., A. Covaci, and P. Schepens. 2002. Investigation of Selected Persistent Organic Pollutants in Farmed Atlantic Salmon, Salmon Aquaculture Feed, and Fish Oil Components of the Feed. *Environmental Science & Technology.* 36(13): 2797–2805.

Jacobs, M. N., J. Ferriaro, and C. Byrne. 2002. Investigation of Polychlorinated Dibenzo-*p*-dioxins, Dibenzo-*p*-furans, and Selected Coplanar Biphenyls in Scottish Farmed Atlantic Salmon. *Chemosphere.* Vol. 47: 183–191.

Johnson, H. M., and Associates. 2000. *Annual Report on the United States Seafood Industry.* Eighth Edition. Jacksonville, OR: H. M. Johnson & Associates.

——. 2001. *Annual Report on the United States Seafood Industry.* Ninth Edition. Jacksonville, OR: H. M. Johnson & Associates.

——. 2002. *Annual Report on the United States Seafood Industry.* Tenth Edition. Jacksonville, OR: H. M. Johnson & Associates.

Kay, J. 2002. Frankenfish Spawn Controversy: Debate Over Genetically Altered Salmon. *San Francisco Chronicle.* 29 April.

MacGarvin, M. 2000. Scotland's Secret? Aquaculture, Nutrient Pollution Eutrophication, and Toxic Blooms. *Modus Vivendi.* WWF Scotland. Available at http://www.wwf.no/english/aquaculture/scotlands_secret.pdf.

Netboy, A. 1974. *The Salmon—Their Fight for Survival.* Boston: Houghton Mifflin Company.

Packard Foundation (The David and Lucile Packard Foundation). 2001. *Mapping Global Fisheries and Seafood Sectors.*: Los Altos, CA: The Packard Foundation.

Quig, D. 2002. *Cysteine Metabolism and Heavy Metal Toxicity.* Available at http://www.thorne.come/altmedrev/fulltext/tox3-4.html.

Schecter, A., P. Cramer, K. Boggess, J. Stanley, O. Papke, J. Olson, A. Silver, and M. Schmitz. 2001. Intake of Dioxins and Related Compounds from Food in the U.S. Population. *Journal of Toxicology and Environmental Health* Part A, 63.

SeaWeb. 2001. *Consumer Attitudes*. Case Presented at the Salmon Aquaculture Strategy Assessment Workshop. Washington, D.C. March.

Smyth, M. 2002. Eco-Friendly Fish Thriving in Unique Land-Based Farm. *The Province*. June 16. Available at http://www.creativeresistance.ca/awareness/2002-june16-land-based-salmon-farm.htm.

Staniford, D. 2001. *Intensive Sea Cage Fish Farming: The One That Got Away*. Friends of the Earth, Scotland. Presented at Coastal Management for Sustainability—Review and Future Trends workshop. University of London, January 24–25, 2001.

———. 2002. *A Big Fish in a Small Pond: The Global Environmental and Public Health Threat of Sea Cage Fish Farming*. Paper presented at "Sustainability of the Salmon Industry in Chile and the World" workshop organized by the Terram Foundation and Universidad de los Lagos in Puerto Monte, Chile. June 5–6.

Weber, M. 1999. Ecological Footprint and Energy Inputs Associated with the Intensive Culture of Salmon in British Colombia. Draft. June 27.

———. 1997. *Farming Salmon: A Briefing Book*. San Francisco, CA: The Consultative Group on Biological Diversity.

Whoriskey, F. 2000. The North American East Coast Salmon Aquaculture Industry: The Challenges for Wild Salmon. Atlantic Salmon Federation. Draft report. February.

CONCLUSION

Proceeding along its current trajectory, agricultural production will eventually expand onto and degrade most of the habitable areas of the planet. As a consequence, most biodiversity and ecosystem services will be lost. In the worst-case scenario, life as we know it will cease to exist. Many will argue that this will never be allowed to happen, that people always have seen the errors of their ways. The question, then, is what will it take not just to slow the current trajectories but to change them altogether?

This book has suggested a number of ways to make global agriculture more sustainable. Eleven general areas in which policy could stimulate more widespread use of the practices and techniques recommended in this book are summarized below. The list is not exhaustive. Nor will the impacts of global agriculture be corrected or even blunted by pursuing in isolation activities in one or two of the areas that are highlighted here. Fortunately there are synergies among many of the approaches suggested. The goal is not to tell people what to think about how to reduce the impacts of agriculture, but rather to expose them to new ways of thinking that can be adapted to their own realities and spheres of influence. This is the new agriculture.

1. IMPLEMENT LAND USE ZONING AND REGULATIONS TO MINIMIZE DAMAGE

The greatest environmental impacts of agriculture by far (50 to 90 percent of impacts, depending on what is measured) occur because of where operations are sited rather than how they are managed. Governments and producers both have an interest in siting operations in areas where they have the best chance of success. Failed investments are not good for anyone. Considerable information is available about the conditions under which crops can be produced sustainably. While using these criteria for zoning and land use planning will not prevent all failures (e.g., some factors may change in the future—management effectiveness, prices, input costs, etc.), they can prevent mistakes based on the best available information at the time. For example, with no major commodity prices increasing in real terms since 1960, it is possible to predict with increasing accuracy where they can be produced profitably and sustainably.

With proper zoning and land use planning most biodiversity, biological corridors, and ecosystem functions can be protected. Research still needs to be undertaken to demonstrate the value of such zoning (e.g., reduction of poor or failed investments, or reduction in the ratio of costs to value of production) to current producers and planners alike. Such research could also document existing examples of the income that can be derived from the management and sale of new products such as biodiversity conservation, watershed protection, or carbon sequestration. Finally, there are a number of costs incurred by society when critical habitat that is unsuitable for farming is converted to agricultural uses. These include increased expenditures for road maintenance and dredging, reduced fisheries production, higher costs of fresh water, and loss of tourism revenues, among others.

2. RETIRE MARGINAL LANDS

Data from research on numerous, very diverse crops suggests that when producers stop farming the 5 to 15 percent of their land that is most marginal for agriculture, they usually end up producing more and being more profitable. Equally important, when farmers stop farming marginal areas (even as little as 5 percent of the total), they can reduce their environmental impacts by as much as 50 percent or more.

So why don't more farmers do this? Most are not familiar with the concept. Many producers, if they have the equipment or ability to farm a greater area, always see it as an advantage to expand production. In some cases, farmers have borrowed money for land or invested in machinery and believe that any return on these investments is worthwhile. Perhaps the most important reason, however, is that most farmers do not keep the kinds of records that would allow them to evaluate accurately where their operations are profitable or not. Few farmers keep production data at all; fewer still have disaggregated data.

For producers (and society as a whole) to take advantage of these possibilities, it is important to generate disaggregated data for different commodities to determine where producers lose money with their current practices. The more the financial and environmental realities can be documented for different producers and crops, the more credible the approach will be with others.

Such data can also be used by government officials to shape policies and programs. Such information, for example, can be used to sharpen land use planning and zoning programs and can be the basis for identifying and retiring the least productive, most polluting areas.

3. REHABILITATE DEGRADED LANDS

Much of the research on rehabilitation of degraded lands has focused on how much it costs to rehabilitate biodiversity or ecosystem functions. These are important issues. However, this approach assumes that degraded land cannot be rehabilitated for agriculture. There is increasing evidence that this assumption is false. Not only do technology and management practices exist that allow land to be rehabilitated, the market encourages it as well. The increased price of land in most parts of the world encourages the rehabilitation and reclamation of degraded areas.

Rehabilitation of land for agricultural production is an area where producers have tremendous potential to save money and reduce their environmental damage. Bringing degraded lands back into production is cost-effective even when only part of a property can be rehabilitated. Producers have found that it is cheaper to rehabilitate degraded land than to clear natural habitat. In addition to saving the costs of clearing land, they can buy such land cheaply, build its capacity, and then have both the asset with increased value as well as the crops that they can produce on it. The approach reduces the environmental damage of agricultural expansion in many parts of the world by "retiring" degraded land that, without intervention, would only continue to be degraded and affect downstream ecosystems, or would require producers to move to and clear new areas of natural habitat.

Research that documents the parameters of degradation and demonstrates which land can be rehabilitated is extremely important at this time. What soils can be rehabilitated, with what slope, rainfall, wind, average temperatures, etc.? What specific practices best rehabilitate land for different crops? What is the range of costs for degraded land, as well as the overall cost of rehabilitating degraded land? What are the ranges of variables that affect those overall costs? And what can farmers do to generate cash flow to cover these costs in the short term? What government incentives can be used to encourage rehabilitation? This information is key in convincing producers to undertake such programs, to make them bankable, and to make government officials take note of how they might encourage such practices through policies and regulations.

4. FARM WITH NATURE

Historically, most farmers tolerated or even encouraged biodiversity. They planted crops side by side (i.e., polyculture), and accepted biodiversity within their fields and on their farms. This was true of producers of annual and perennial crops, as well as mixed systems that incorporated both. In many parts of the world, farming systems evolved and were adapted from extended fallows that were little more than enriched forest plantings. These systems not only produced annual and perennial crops but also attracted animals which were utilized by producers as well.

Most producers around the world still farm with biodiversity in many of these same ways. They plant multiple crops in the same fields and utilize plants and animals that are tolerated and even encouraged within their fields. However, the production and sale of an ever increasing amount of product from farms has tended to erode the tolerance for biodiversity within farming systems. Farmers who are dedicated commercial producers of commodities for more distant markets tend to fight biodiversity the most within their operations. These producers, often encouraged by subsidies, tend to plant single commodities year after year. These systems are not sustainable.

Farming with biodiversity starts not with planting or accommodating a wide variety of plants and animals, but with the soil itself. The value of maintaining soils and soil fertility is now well understood even by those producers who grow only single crops at a time. To mimic at least some of the positive attributes of farming with biodiversity, many commodity producers plant sequences of crops in the same year and use additional crop rotation strategies over two to five year cycles. Similarly, they plant legumes and other crops to maintain or increase soil fertility and to provide ground cover. These practices are common throughout the world. Finally, many integrated pest management programs are based on finding the right balance with nature and biodiversity rather than trying to dominate it. Such practices are increasingly common on farms that are trying to mimic nature yet still remain competitive in the marketplace.

There are several indications that in the future, the trend in agriculture will be to find the right balance between maintaining biodiversity and soil fertility on the one hand and being competitive in global markets on the other. The practices will be driven by a number of factors: consumer desires for fewer pesticides; downstream water users' desire for cleaner water; cultural values of increased biodiversity in farm landscapes; and, not least, overall efficiency and reduced costs. Currently, most of the trends toward monoculture cropping are driven by subsidies. While there are countless examples of producers who are pursuing a more thoughtful path regarding farming with nature, the question of production and export subsidies will have to be addressed before most producers will be able to adapt or find new ways to farm with nature.

5. ELIMINATE SUBSIDIES AND MARKET BARRIERS

Subsidies restricting market access and similar programs are tools that governments can use to encourage producer behavior to provide societal benefits. Unfortunately, many such programs have become entitlements—producers are paid to continue to farm rather than to produce something that is good for society. In fact, many such programs actually harm both people and the environment. There is little doubt, for example, that subsidies and market barriers have maintained or even increased global inequity.

What have received less attention are the environmental consequences of such programs. In developed countries, producers are subsidized to farm areas that would not otherwise be profitable. And in developing countries some producers, who receive few if any subsidies, try to compete by being more efficient, but most cut corners—which causes environmental degradation. Subsidies and restricted market access in developed countries are the most important barriers to the adoption of better management practices.

The elimination of production, export, input, credit, and infrastructure subsidies as well as market barriers is essential if global agriculture is to become more sustainable. Research that documents the impact of such policies on the global environment and the productive base for current and future agricultural producers will be essential for introducing such issues into the current debate on subsidies and market barriers.

Governments have legitimate interests in protecting the productive resource bases of their countries and in making sure that their citizens are not held hostage to the uncertainty of food production in other parts of the world. However, both of these issues can be addressed by using the money that is currently paid for subsidies and market barriers to pay farmers to provide environmental services that are beneficial to all members of society, both for this generation as well as for future ones.

6. DEVELOP PAYMENTS FOR ENVIRONMENTAL SERVICES

Some producers already realize that farming marginal areas is not a viable production strategy. Most, however, either have not yet come to this realization or are faced with the reality that most if not all of the land that they farm already is marginal. The creation of payments for environmental services such as maintaining water quality and quantity, protecting biodiversity, maintaining watersheds, and sequestering carbon (in either soil or plant biomass) could stimulate producers to rethink their current production strategies.

Documenting the range of existing environmental service payment systems could help producers, and those interested in public policies, better understand the role of such payments in maintaining farmer income while reducing environmental impacts. Research could also evaluate the scope and effectiveness of public expenditures to correct environmental impacts resulting from destructive farming practices or farming marginal areas.

Given the political clout of producers and the legitimate need of society to ensure food and fiber supplies, subsidies and market barriers are not likely to be eliminated in the short term. Many of the payments for current agricultural support programs such as subsidies and market protection can be usefully shifted to environmental service payments. Delinking such payments from the production of specific commodities and the prices for those commodities would tend to reduce the overall effect of subsidies globally while still supporting producers. Such shifts in payments would be welcomed in most parts of the world provided they do not distort trade. In the short term, any shift of payments would probably reduce the current distortions caused by subsidies and market barriers. Over time, however, producers and governments will want the market-distorting impacts of such programs to be reduced further still.

7. PROMOTE BETTER MANAGEMENT PRACTICES (BMPs)

As competition in the global economy increases, the producers that survive are going to be those that are the most efficient. They will be defined by their ability to invent, identify, or adapt practices that reduce input use as well as waste and pollution. Such producers will be more profitable, or at the very least will remain competitive, in the face of globally declining prices.

Efficiency will not be limited to the largest or the smallest producers, or to the wealthiest or the poorest. The producers that remain competitive will be those who learn from other, often more innovative, producers. Those who remain competitive will also not merely focus on how to produce a single commodity better. They will focus on their overall production system, and they will evaluate periodically what crops they can produce to best utilize their physical, financial, and market advantages. These crops, like the practices used to produce them, will change.

Because of their importance to the overall sustainability of agriculture, the adoption of better management practices cannot be left to the market alone. Most producers in the world will not make the transition without support. Government subsidies can, in the short term, provide incentives for the adoption of BMPs. Government regulatory and permitting systems can also encourage the identification and adoption of these practices. Most producers learned to farm from their parents, who in turn were taught by their parents and so on. Such lessons are important, but in a world of global markets, limited resources, and increased demand, those producers who survive will take the best of the traditional approaches and graft on new lessons, approaches, and technology from others.

8. PROMOTE SOCIAL- AND EQUITY-BASED BMPs

Those who fear globalization have legitimate concerns about the impact of "free" trade on the rural poor. However, the proposed solutions to address rural poverty—to improve the viability of small farmers, or simply protect them—miss the mark and, furthermore, cease when funding ends. In many rural areas, the truly poor are not landowners. Few have the skills, capital, or market access to take advantage of well-intentioned programs.

A new approach is needed to reduce rural poverty. Fortunately, research suggests that social- and equity-based better agricultural practices are not only important for reducing the impacts of producers around the world but for increasing profits as well. Such programs, while increasingly common, vary incredibly and include such approaches as worker incentive programs, bonuses, equity positions, employee stock option plans (ESOPs), and benefits. Such programs result in increased productivity and reduced costs as well as increased product quality, reduced input use, and maintenance of the resource base.

Some of the greatest gains in agricultural production are likely to come not from technology, but rather from rewarding people who think. In addition to more traditional worker incentive and bonus programs, line workers are increasingly empowered to make management suggestions about how to improve production and production efficiency. Management rewards valuable ideas from line workers, not just increased productivity. This can increase overall profits and has great potential for reducing environmental damage from day-to-day production decisions.

Many agricultural employers have even found that it is cost-effective to extend their benefit packages to nearby communities. Education programs, for example, not only help companies reduce the often costly mistakes that arise from illiteracy, but also create more qualified worker pools. Such programs also help communities and future generations develop skills that they would not otherwise have. Similarly, community health programs not only reduce worker sick days or the time they spend with sick family members but also increase their productivity as well as that of their family.

9. BASE REGULATORY STRUCTURES AND PERMITTING SYSTEMS ON BMPs

Agriculture is the most polluting activity in most countries. Governments realize this. Because it is often very difficult to identify the particular source of pollution (e.g., nonpoint-source air or water pollution), many governments require producers to adopt good or best management practices with a goal of achieving minimally acceptable performance levels.

The types of better practices that have been identified in this volume could serve as the basis of government BMP-based regulations, permitting, or licensing programs. Better practices should give government insights about what they can achieve through regulations in the future. After all, today's better practices will be tomorrow's norm. The role of government should be to identify the main environmental costs of agriculture. Policies can then be either prescriptive (e.g., prescribe practices to be adopted that are known to reduce those impacts) or results-based, in which performance levels are measured and producers achieve those performance levels in whatever way they see fit. In general, more innovative solutions will come from the latter than from the former.

Unfortunately, when governments set levels of performance that are required from producers, the results sought are rarely close to those that could be achieved through the adoption of better management practices that are already known and understood at the time. In general, such standards are designed to achieve the minimal performance levels required to meet such laws or regulations. As such, they do not encourage continued improvement, rather just minimal performance levels that comply with the law. Furthermore, such approaches are inevitably out of date with current realities.

10. BASE INVESTMENT, INSURANCE, AND PURCHASE SCREENS ON BMPs

Increasingly, there is interest on the part of investors, insurers, and major purchasers to look to BMP-based screens to reduce their risks from exposure to the environmental and social impacts of commercial agricultural and aquacultural production. Because overall management quality is the key factor in producing consistent profits, investors already evaluate management as a condition of investment in commercial operations. The adoption of BMP-based screens is, in fact, little more than a more precise way of evaluating the specific management practices of a business as they relate to critical impacts. The intent is not to reduce impacts per se, but rather to reduce liability, costs, and wastes as well as to increase profits and returns on investment. In the end, however, the two are often one and the same.

For insurers, the identification and adoption of BMP-based screens can reduce risks by determining whether producers have adopted practices that reduce overall liability. The liability could be related to impacts that contribute to crop failure, personal liability, or injury for workers; that reduce the life or productivity of soil, permanent crops, or machinery; that relate to downstream/downwind liability resulting from erosion, agrochemical runoff, or smoke from burning; or that arise from chemical residues that affect consumer safety.

With increasing concerns about food quality and safety, a number of food manufacturers and retailers are developing BMP-based screens to guide their purchases and reduce their liability resulting, for example, from pesticide residues on food products. Another important factor for manufacturers and retailers is to be able to trace problems back to their source. This also reduces liability. These concerns have resulted in the development of producer contracts that require producers, as a cost of doing business, to adopt certain practices that limit or restrict agrochemical use. Previously, producer contracts required the prophylactic use of chemicals. Today's contracts are just the opposite—some chemicals are banned altogether, and the use of other chemicals is reduced and often limited to the treatment of specific issues as they arise.

Such self-developed programs, however well-intentioned and comprehensive, have limited credibility with consumers. Thus, the first move of such companies should be to unite with similar companies to increase the market share of products using better management practices. This at least will enhance their programs' credibility in the eyes of producers, if not consumers. Over time, however, credibility with consumers will be based on third-party certification and independent, measurable standards.

While each set of actors has different reasons for pursuing BMP-based screens, their actions can be mutually reinforcing. Such synergies allow market-based approaches to be adopted very rapidly. Successful efforts to develop complementary BMP-based screens for investors, insurers, and purchasers will send signals to producers from every part of the market chain. To the extent that BMPs pay for themselves, result in market premiums, or improve market share, they will shape producer practices.

Voluntary BMP-based certification programs can support the enforcement of regulations and permits. Most, for example, require producers to obey the law. In such instances, by insisting on certified products consumers would ultimately ensure that the costs of compliance with all regulations and permits would be covered by players in the market chain rather than local governments, which may or may not be able to enforce them.

11. IMPROVE CERTIFICATION AND ECO-LABELS

There has been a tremendous growth in the number of certification and eco-label programs developed over the past twenty to thirty years. This has come about partly from producers and intermediates in the market chain, as they look for ways to differentiate products in the marketplace based on how they are produced. Consumers are also concerned about the quality of the products they are consuming, and to a lesser extent, the overall production processes or the social and environmental impacts of producing them.

The question, then, is whether eco-labeled or certified products actually deliver on their promises. Most certification programs cannot back up their claims. They certify production processes, not products. Most of the standards by which results of the programs are measured are subjective. At this time, no certification program has entirely measurable standards. While most address environmental issues, few address social ones. Many programs are guaranteed by third-party certifiers, but the programs themselves were developed by a small number of interested parties through processes that were decidedly not transparent. No program yet focuses on the cumulative impacts of production at the larger landscape or ecosystem level. To date, at least, all are focused on the individual farm or fields. Finally, very few programs are financially self-sufficient. They are, in fact, highly subsidized. For these reasons and more, most certification programs will disappear.

The side-by-side comparison of certification programs will allow interested parties to evaluate the relative comprehensiveness of the programs. Such comparisons have already been made or are in the works for bananas, coffee, wood pulp, and shrimp. Given the wide number of programs that have been developed, it is only a question of time before other comparisons are made. Such comparisons make very transparent which claims are actually based on measurements.

So, what will the surviving certification programs look like, and will they have a positive impact on agriculture? Certification programs that are credible to consumers will be objective and will have crop-specific standards that are based on measurable standards. They will be developed through a wide consultative process, with considerable transparency and room for public comment and discussion. The next generation of certification programs will focus on the known major impacts from the production of specific crops and will require that those impacts be reduced as a condition of certification. As such they will not be exhaustive, but rather will address the eight to twelve key social and environmental impacts that account for the vast bulk of subsequent impacts. They will also have to assess carrying capacity issues at the landscape level and not just focus on the fields or farms of individual producers. Certification will be driven by major actors in the market chain and, thus, will not be aimed at niche markets. Furthermore, consumers buy products, not production processes. Consequently, successful certification systems will have to stand behind their programs and the products delivered through them.

Governments, retailers, and manufacturers are all being asked by consumers to become more involved in certification. At the very least they are being asked to explain the differences between certification programs, expose fraud, and identify those that are credible. In addition, governments, buyers, retailers, and insurers in many parts of the world are developing BMP-based screens on their own to reduce liability or to achieve societal goals. In effect, if certification programs do not exist they will have to be created. Most such programs at the present are second-party certified. Credible, third-party certification programs offer a tremendous advantage for such players. In the near future, each of these groups will be actively involved in enforcing, adapting, creating, and/or implementing agricultural certification programs. There is a tremendous opportunity for producers, nongovernmental organizations (NGOs), and community groups to work with producers and retailers to create credible programs that have the potential to capture significant market share.

THE WAY FORWARD

The agriculture of the future will not be the same as the agriculture of today. However, it will not be entirely different either. Success in making agriculture more sustainable will be based on taking the best of the past and melding it with the best of the present and the future. This requires producers who have been exposed to new ideas and approaches and who have the confidence, and the incentives, to be innovative. However, it also will require government officials, investors, buyers, researchers, and others who can also recognize and encourage innovation.

So what can be done to encourage and promote such innovation? It is important to increase the number of people who think in the ways that are highlighted in this book. Producers must have access to information about innovations that might be relevant to their own management decisions. Finally, students must learn to be both entrepreneurial and respectful of tried-and-true production methods at the same time. These students may become producers in their own right or may work for other producers. They may study producers, disseminate lessons to them, invest in them, buy from them, or regulate them. In the end, sustainable agricultural production is about thinking and doing. It is not just about new seed varieties and inputs. Societies are spending all their money on the latter, when it is the human skills more broadly that will ultimately make agriculture deliver societies' needs—food, fiber, and livable environments.

Jason Clay is currently Vice President, Center for Conservation Innovation, World Wildlife Fund. Over the course of his career he has worked on a family farm, taught at Harvard University, worked in the U.S. Department of Agriculture, and spent more than twenty years working with human rights and environmental nongovernmental organizations (NGOs). Clay spent more than a decade developing research methods to document and predict human rights abuses, genocide and ethnocide, social conflict, and human-made famines. In the 1980s he was one of the inventors of green marketing and established a trading company within Cultural Survival in which he developed markets for rainforest products with nearly 200 companies in the United States and Europe (including such products as Rainforest Crunch with Ben & Jerry's). In 1996 he began to research ways to reduce the impacts of shrimp aquaculture. In 1999 he created the Shrimp Aquaculture and the Environment Consortium (with the World Wildlife Fund, World Bank, Food and Agricultural Organization, and National Aquaculture Centres for Asia and the Pacific), which he codirected, to identify and analyze better management practices that address the environmental and social impacts of shrimp aquaculture. Clay received his B.A. in anthropology at Harvard University, studied economics and geography at the London School of Economics, and anthropology and international agriculture at Cornell University, where he received his Ph.D. in 1979. Clay was founder and editor (1980–1992) of the award-winning *Cultural Survival Quarterly*, the largest-circulation anthropology and human rights publication in the world. He has previously written *Generating Income and Conserving Resources, Politics and the Ethiopian Famine—1984–1985* (with Bonnie Holcomb and Peter Niggli), *Indigenous Peoples and Tropical Rainforests,* and *Greening the Amazon—Communities and Corporations in Search of Sustainable Business Practices* (with Anthony Anderson).